T0184911

Israel Gohberg
Peter Lancaster
Leiba Rodman

Indefinite Linear Algebra and Applications

Birkhäuser
Basel · Boston · Berlin

Authors:

Israel Gohberg
School of Mathematical Sciences
Raymond and Beverly Sackler
Faculty of Exact Sciences
Tel Aviv University
Ramat Aviv 69978, Israel
e-mail: gohberg@math.tau.ac.il

Peter Lancaster
Department of Mathematics and Statistics
University of Calgary
Calgary, Alberta T2N 1N4, Canada
e-mail: lancaste@ucalgary.ca

Leiba Rodman
Department of Mathematics
College of William and Mary
P.O. Box 8795
Williamsburg, VA 23187-8795, USA
e-mail: lxrodm@math.wm.edu

2000 Mathematics Subject Classification 32-01

A CIP catalogue record for this book is available from the
Library of Congress, Washington D.C., USA

Bibliographic information published by Die Deutsche Bibliothek
Die Deutsche Bibliothek lists this publication in the Deutsche Nationalbibliografie;
detailed bibliographic data is available in the Internet at <http://dnb.ddb.de>.

ISBN 3-7643-7349-0 Birkhäuser Verlag, Basel – Boston – Berlin

© 2005 Birkhäuser Verlag, P.O. Box 133, CH-4010 Basel, Switzerland
Part of Springer Science+Business Media
Cover design: Micha Lotrovsky, CH-4106 Therwil, Switzerland
Printed on acid-free paper produced from chlorine-free pulp. TCF ∞
Printed in Germany
ISBN-10: 3-7643-7349-0
ISBN-13: 978-3-7643-7349-8

e-ISBN: 3-7643-7350-4

9 8 7 6 5 4 3 2 1

www.birkhauser.ch

We fondly dedicate this book to family members:

Israel Gohberg: To his wife, children, and grandchildren.

Peter Lancaster: To his wife, Diane.

Leiba Rodman: To Ella, Daniel, Ruth, Benjamin, Naomi.

Preface

The following topics of mathematical analysis have been developed in the last fifty years: the theory of linear canonical differential equations with periodic Hamiltonians, the theory of matrix polynomials with selfadjoint coefficients, linear differential and difference equations of higher order with selfadjoint constant coefficients, and algebraic Riccati equations. All of these theories, and others, are based on relatively recent results of linear algebra in spaces with an indefinite inner product, i.e., linear algebra in which the usual positive definite inner product is replaced by an indefinite one. More concisely, we call this subject *indefinite linear algebra*.

This book has the structure of a graduate text in which chapters of advanced linear algebra form the core. The development of our topics follows the lines of a usual linear algebra course. However, chapters giving comprehensive treatments of differential and difference equations, matrix polynomials and Riccati equations are interwoven as the necessary techniques are developed.

The main source of material is our earlier monograph in this field: *Matrices and Indefinite Scalar Products*, [40]. The present book differs in objectives and material. Some chapters have been excluded, others have been added, and exercises have been added to all chapters. An appendix is also included. This may serve as a summary and refresher on standard results as well as a source for some less familiar material from linear algebra with a definite inner product. The theory developed here has become an essential part of linear algebra. This, together with the many significant areas of application, and the accessible style, make this book useful for engineers, scientists and mathematicians alike.

Acknowledgements

The authors gratefully acknowledge support from several projects and organizations: Israel Gohberg acknowledges the generous support of the Silver Family Foundation and the School of Mathematical Sciences of Tel-Aviv University. Peter Lancaster acknowledges continuing support from the Natural Sciences and Engineering Research Council of Canada. Support from N. J. Higham of the University of Manchester for a research fellowship tenable during the preparation of this work is also gratefully acknowledged. Leiba Rodman acknowledges partial support by NSF grant DMS-9988579, and by the Summer Research Grant and Faculty Research Assignment provided by the College of William and Mary.

Contents

Chapter 1

Introduction and Outline

This book is written for graduate students, engineers, scientists and mathematicians. It starts with the theory of subspaces and orthogonalization and then goes on to the theory of matrices, perturbation and stability theory. All of this material is developed in the context of linear spaces with an indefinite inner product. The book also includes applications of the theory to the study of matrix polynomials with selfadjoint constant coefficients, to differential and difference equations (of first and higher order) with constant coefficients, and to algebraic Riccati equations.

The present book is written as a graduate textbook, taking advantage of our earlier monograph *Matrices and Indefinite Scalar Products* [40] as the main source of material. Materials not to have been included are chapters on the theory of canonical selfadjoint differential equations with periodic coefficients, and on the theory of rational matrix functions with applications. Material on the analysis of time-invariant differential and difference equations with selfadjoint coefficients has been retained. In the interests of developing a clearer and more comprehensive theory, chapters on orthogonal polynomials, normal matrices, and definite subspaces have been introduced, as well as sets of exercises for every chapter. We hope that these changes will also make our subject more accessible.

The material of this book has an interesting history. The perturbation and stability results for unitary matrices in a space with indefinite inner product, and applications to the theory of zones of stability for canonical differential equations with periodic coefficients were obtained by M. G. Krein [61]. The next development in this direction was made by I. M. Gelfand, V. B. Lidskii, and M. G. Neigaus [27], [77]. Further contributions were made by V. M. Starzhinskii and V. A. Yakubovich [108], W. A. Coppel and A. Howe [15] as well as N. Levinson [75]. The present authors have made contributions to the theory of linear differential and difference matrix equations of higher order with selfadjoint coefficients and to the theory of algebraic Riccati equations.

All of these theories are based on the same material of advanced linear algebra: namely, the theory of matrices acting on spaces with an indefinite inner product. This theory includes canonical forms and their invariants for H-selfadjoint, H-unitary and H-normal matrices, invariant subspaces of different kinds, and different aspects of perturbation theory. This material makes the core of the book and makes up a systematic *Indefinite Linear Algebra*, i.e., a linear algebra in which the linear spaces involved are equipped with an *indefinite* inner product. Immediate applications are made to demonstrate the importance of the theory. These applications are to the solution of time-invariant differential and difference equations with certain symmetries in their coefficients, the solution of algebraic Riccati equations, and to the analysis of matrix polynomials with selfadjoint coefficients.

The material included has been carefully selected to represent the area, to be self-contained and accessible, to follow the lines of a standard linear algebra course, and to emphasize the differences between the definite and indefinite linear algebras. Necessary background material is provided at the end of the text in the form of an appendix.

Naturally, this book is not of encyclopaedic character and is not a research monograph. Many subjects belonging to the field are not included. Readers interested in a broader range of material may wish to consult our first book [40], the book by V. M. Starzhinskii and V. A. Yakubovich [108] for applications, and, of course, the original papers. For the first chapters of a standard linear algebra course we also recommend F. R. Gantmacher [26], I. M. Glazman and Yu. I. Lyubich [29], and A. I. Mal'cev [80].

1.1   Description of the Contents

The first chapter contains the introduction, notation and conventions. In the second chapter the basic geometric ideas concerning spaces with an indefinite inner product are developed; the main topics being orthogonalization and classification of subspaces. Orthogonalization and orthogonal polynomials are studied in the third chapter. The fourth chapter is concerned with the classification of linear transformations in indefinite inner product spaces. Here, H-selfadjoint, H-unitary, and H-normal linear transformations are introduced together with the notion of unitary similarity. The fifth chapter is dedicated to canonical forms and invariants of H-selfadjoint and H-unitary matrices. The sign characteristic, the canonical forms of linear pencils with selfadjoint coefficients, and invariant maximal non-negative subspaces are introduced and examined. The theory of real selfadjoint matrices and real unitary matrices is presented in Chapter six. The seventh chapter is dedicated to the functional calculus in spaces with an indefinite inner product. The canonical forms and sign characteristic of functions of matrices are studied, and special attention is paid to the logarithmic and exponential functions.

The eighth chapter is "H-normal matrices". The structure of normal matrices in spaces with indefinite inner product is very complicated and even "wild" in a certain sense. A detailed analysis is presented. Following this, the ninth chapter is dedicated to perturbation and stability theory for H-selfadjoint and H-unitary matrices. This theory takes on a specific character and form in our context, and is quite different from the well-known general perturbation theory. This topic is important in applications to the study of stable boundedness of solutions of differential and difference equations. Applications for differential equations of first order appear in Chapter eleven. "Matrix Polynomials" is the subject of the twelfth chapter. It contains an introduction to the general theory of matrix polynomials with selfadjoint or symmetric coefficients. The latter theory is based on the results of the previous chapters. Applications of this theory to time-invariant differential and difference equations of higher order are presented in the thirteenth chapter. This includes a description of the connected components of differential or difference equations with stably bounded solutions. The last chapter contains the theory of algebraic Riccati equations. The appendix serves as a refresher for some parts of linear algebra and matrix theory which are used in the main body of the book, as well as a convenient location for some less-familiar technical results. The book concludes with the bibliography and index.

1.2 Notation and Conventions

Throughout the book, the following notation is used.

Fonts and Sets

- The sans serif font is used for the standard sets C (the complex numbers), R (the real numbers), T (the unit circle).

- $\Re z$ and $\Im z$ denote the real and imaginary parts of the complex number z: $z = \Re z + i\Im z$.

- $\bar{z}$ is the complex conjugate of a complex number z.

- $\arg z$ is the argument of a nonzero complex number z; $0 \leq \arg z < 2\pi$.

- Matrices are denoted by capital letters $A, B, \ldots$.

- The calligraphic font is used for vector spaces and subspaces: $\mathcal{H}, \mathcal{G}, \mathcal{M}$ etc.

- $\mathcal{M} = \mathcal{M}_0 \dot{+} \cdots \dot{+} \mathcal{M}_k$ indicates that the subspace $\mathcal{M}$ is a direct sum of its subspaces $\mathcal{M}_1, \ldots, \mathcal{M}_k$.

- := the left hand side is defined by the equality. =: the right hand side is defined by the equality.

- Set definition: $\{\, A \mid B \,\}$ or $\{\, A \ : \ B \,\}$ is the set of all elements of the form A subject to conditions (equalities, containments, etc.) B.

- $\subseteq$, $\supseteq$ set-theoretic inclusions.

Matrices and Linear Transformations

- The terminology "invertible matrix" and "nonsingular matrix" will be used interchangeably.

- We often identify a matrix with the linear transformation generated by the matrix with respect to the standard orthonormal basis.

- The spectrum of a matrix (=the set of eigenvalues, including nonreal eigenvalues of real matrices) A will be denoted $\sigma(A)$.

- Range A is the range of a matrix or linear transformation A (the set of vectors of the form Ax).

- Ker A is the kernel (null-space) of a matrix or a linear transformation A.

- diag $(X_1, \ldots, X_r)$ or $X_1 \oplus X_2 \oplus \cdots \oplus X_r$ denotes the block diagonal matrix with blocks $X_1, \ldots, X_r$ on the main diagonal (in the indicated order).

- The restriction of a matrix A (understood as a linear transformation) to its invariant subspace $\mathcal{V}$ is denoted by $A \mid_{\mathcal{V}}$.

- The transpose of a matrix A is denoted by A^T, and A^* denotes the conjugate transpose of A, which coincides with the adjoint of the linear transformation induced by A with respect to the standard orthonormal basis.

- $\overline{A}$ is the matrix whose entries are the complex conjugates of those of matrix A.

- The $p \times p$ identity matrix is written I_p or I.

- Sip matrix of size n (see Example 2.1.1):

$$
S_n := \begin{bmatrix}
0 & 0 & \cdots & 0 & 1 \\
0 & 0 & \vdots & 1 & 0 \\
\vdots & \vdots & \ddots & \vdots & \vdots \\
0 & 1 & \cdots & 0 & 0 \\
1 & 0 & \cdots & 0 & 0
\end{bmatrix}.
$$

- $\leq$, $\geq$ between hermitian matrices denotes the Loewner order: $A \leq B$ or $B \geq A$ means that the difference $B - A$ is positive semidefinite.

- Similarly, $A < B$ or $B > A$ means that the difference $B - A$ is positive definite.

- $i_+(H), i_-(H)$ is the number of positive (resp. negative) eigenvalues (counted with multiplicities) of a hermitian matrix H.

- $i_0(H) = \dim \mathrm{Ker}(H)$ is the number of zero eigenvalues (counted with multiplicities) of a hermitian matrix H.

- Inertia of a hermitian matrix H: $(i_+(H), i_-(H), i_0(H))$.

- Signature of a hermitian matrix H:

$$\mathrm{sig}\, H = i_+(H) - i_-(H).$$

- $\mathcal{R}_\lambda(A)$ is the root subspace of a matrix or linear transformation A corresponding to the eigenvalue λ:

$$\mathcal{R}_\lambda(A) = \mathrm{Ker}(A - \lambda I)^n,$$

where n is the size of A.

- $\mathcal{R}_{\mathsf{R},\lambda}(A)$, or $\mathcal{R}_{\mathsf{R},\mu\pm i\nu}(A)$, is the real root subspace of a real matrix A corresponding to its real eigenvalue λ, or to a pair of nonreal complex conjugate eigenvalues $\mu \pm i\nu$.

Vectors

- Span $\{x_1, \ldots, x_k\}$ is the subspace spanned by the vectors $x_1, \ldots, x_k$.

- For typographic convenience we sometimes represent column vectors $x = \begin{bmatrix} x_1 \\ x_2 \\ \vdots \\ x_n \end{bmatrix} \in \mathsf{C}^n$ in the form $x = \langle x_1, x_2, \ldots, x_n \rangle$. A row vector x with components $x_1, \ldots, x_n$ is denoted by $[x_1 \ x_2 \ \cdots \ x_n]$.

- $e_k = \langle 0, 0, \ldots, 0, 1, 0, \ldots, 0 \rangle \in \mathsf{C}^n$ is the kth standard unit vector (with 1 in the kth position). The dimension n is to be understood from the context.

- The standard inner product in C^n is denoted by $(.,.)$:

$$(x, y) = \sum_{j=1}^n x^{(j)} \overline{y^{(j)}}, \quad x = \langle x^{(1)}, \ldots, x^{(n)} \rangle, \quad y = \langle y^{(1)}, \ldots, y^{(n)} \rangle \in \mathsf{C}^n.$$

Norms

The following norms will be used throughout:

- Euclidean vector norm:

$$\|x\| = \sqrt{(x, x)}, \quad x \in \mathsf{C}^n.$$

- Operator matrix norm:

$$\|A\| = \max\{\|Ax\| \, : \, x \in \mathbf{C}^n, \quad \|x\| = 1\}$$

for an $m \times n$ complex matrix A.

$\|A\|$ coincides with the largest singular value of A.

Miscellaneous

The sign function: $\operatorname{sgn} x = 1$ if $x > 0$, $\operatorname{sgn} x = -1$ if $x < 0$, $\operatorname{sgn} x = 0$ if $x = 0$.

Chapter 2

Indefinite Inner Products

In traditional linear algebra the concepts of length, angle, and orthogonality are defined by a definite inner product. Here, the definite inner product is replaced by an *indefinite* one and this produces substantial changes in the geometry of subspaces. Thus, the geometry of subspaces in this context is fundamental for our subject, and is the topic of this chapter.

As in the definite case, when an inner product is introduced on C^n, then certain $n \times n$ matrices (seen as linear transformations of C^n) have symmetries defined by the inner product. If the inner product is definite this leads to the usual classes of hermitian, unitary, and normal matrices. If the inner product is indefinite, then analogous classes of matrices are defined and will be investigated in subsequent chapters.

2.1 Definition

Let C^n be the n-dimensional complex Hilbert space consisting of all column vectors x with complex coordinates $x^{(j)}$, $j = 1, 2, \ldots, n$. The typical column vector x will be written in the form $x = \langle x^{(1)}, x^{(2)}, \ldots, x^{(n)} \rangle$. The standard inner product in C^n is denoted by $(.,.)$. Thus,

$$(x, y) = \sum_{j=1}^{n} x^{(j)} \overline{y^{(j)}}$$

where $x = \langle x^{(1)}, \ldots, x^{(n)} \rangle$, $y = \langle y^{(1)}, \ldots, y^{(n)} \rangle$ and the bar denotes complex conjugation.

A function $[.,.]$ from $C^n \times C^n$ to C is called an *indefinite inner product* in C^n if the following axioms are satisfied:

(i) Linearity in the first argument;

$$[\alpha x_1 + \beta x_2, y] = \alpha[x_1, y] + \beta[x_2, y]$$

for all $x_1, x_2, y \in \mathsf{C}^n$ and all complex numbers α, β;

(ii) antisymmetry;

$$[x, y] = \overline{[y, x]}$$

for all $x, y \in \mathsf{C}^n$;

(iii) nondegeneracy; if $[x, y] = 0$ for all $y \in \mathsf{C}^n$, then $x = 0$.

Thus, the function $[.,.]$ satisfies all the properties of a standard inner product with the possible exception that $[x, x]$ may be nonpositive for $n \neq 0$.

It is easily checked that for every $n \times n$ invertible hermitian matrix H the formula

$$[x, y] = (Hx, y), \quad x, y \in \mathsf{C}^n \tag{2.1.1}$$

determines an indefinite inner product on C^n. Conversely, for every indefinite inner product $[.,.]$ on C^n there exists an $n \times n$ invertible and hermitian matrix H such that (2.1.1) holds. Indeed, for each fixed $y \in \mathsf{C}^n$ the function $x \to [x, y]$ ($x \in \mathsf{C}^n$) is a linear form on C^n. It is well known that such a form can be represented as $[x, y] = (x, z)$ for some fixed $z \in \mathsf{C}^n$. Putting $z = Hy$ we obtain a linear transformation $H : \mathsf{C}^n \to \mathsf{C}^n$. Now anti-symmetry and nondegeneracy of $[.,.]$ ensure that H is hermitian and invertible. The space C^n with an inner product defined by a nonsingular hermitian matrix H will sometimes be denoted by $\mathsf{C}^n(H)$.

Note that here, and whenever it is convenient, an $n \times n$ complex matrix is identified with a linear transformation acting on C^n in the usual way.

The correspondence $[.,.] \leftrightarrow H$ established above is obviously a bijection between the set of all indefinite inner products on C^n and the set of all $n \times n$ invertible hermitian matrices. This correspondence will be widely used throughout this book. Thus, the notions of the indefinite inner product $[.,.]$ and the corresponding matrix H will be used interchangeably.

The following example of an indefinite inner product will be important.

Example 2.1.1. *Put* $[x, y] = \sum_{i=1}^{n} x_i \overline{y_{n+1-i}}$, *where* $x = \langle x_1, \ldots, x_n \rangle \in \mathsf{C}^n$, $y = \langle y_1, \ldots, y_n \rangle \in \mathsf{C}^n$. *Clearly,* $[.,.]$ *is an indefinite inner product. The corresponding* $n \times n$ *invertible hermitian matrix is*

$$\begin{bmatrix} 0 & 0 & \cdots & 0 & 1 \\ 0 & 0 & \vdots & 1 & 0 \\ \vdots & \vdots & \ddots & \vdots & \vdots \\ 0 & 1 & \cdots & 0 & 0 \\ 1 & 0 & \cdots & 0 & 0 \end{bmatrix}$$

This matrix will be called the sip matrix *of size* n *(the standard involutary permutation).* □

The discussion above could equally well be set in the context of R^n in which case other inner products (whether definite or indefinite) are associated with non-singular real symmetric matrices H, and the resulting space is denoted by $\mathsf{R}^n(H)$.

2.2 Orthogonality and Orthogonal Bases

Let $[.,.]$ be an indefinite inner product on C^n and $\mathcal{M}$ be any subset of C^n. Define the orthogonal companion of $\mathcal{M}$ in C^n by

$$\mathcal{M}^{[\perp]} = \{x \in \mathsf{C}^n \mid [x,y] = 0 \quad \text{for all} \quad y \in \mathcal{M}\}.$$

Note that the symbol $\mathcal{M}^{[\perp]}$ will be reserved for the orthogonal companion with respect to the indefinite inner product, while the symbol $\mathcal{M}^{\perp}$ will denote the orthogonal companion in the original inner product $(.,.)$ in C^n, i.e.,

$$\mathcal{M}^{\perp} = \{x \in \mathsf{C}^n \mid (x,y) = 0 \quad \text{for all} \quad y \in \mathcal{M}\}.$$

Clearly, $\mathcal{M}^{[\perp]}$ is a subspace in C^n, and we will be particularly interested in the case when $\mathcal{M}$ is itself a subspace of C^n. In the latter case, it is not generally true (as experience with the euclidean inner product might suggest) that $\mathcal{M}^{[\perp]}$ is a direct complement for $\mathcal{M}$. The next example illustrates this point.

Example 2.2.1. *Let* $[x,y] = (Hx,y)$, $x,y \in \mathsf{C}^n$, *where* H *is the sip matrix of size* n. *Let* $\mathcal{M}$ *be spanned by the first unit vector,* e_1, *in* C^n(*i.e.,* $e_1 = \langle 1,0,\ldots,0\rangle$). *It is easily seen that* $\mathcal{M}^{[\perp]}$ *is spanned by* $e_1, e_2, \ldots, e_{n-1}$ *and is not a direct complement to* $\mathcal{M}$ *in* C^n. $\qquad\square$

In contrast, it is true that, for any subspace $\mathcal{M}$,

$$\dim \mathcal{M} + \dim \mathcal{M}^{[\perp]} = n. \tag{2.2.2}$$

To see this observe first that

$$\mathcal{M}^{[\perp]} = H^{-1}(\mathcal{M}^{\perp}). \tag{2.2.3}$$

For, if $x \in \mathcal{M}^{\perp}$ and $y \in \mathcal{M}$ we have

$$[H^{-1}x,y] = (HH^{-1}x,y) = (x,y) = 0 \tag{2.2.4}$$

so that $H^{-1}(\mathcal{M}^{\perp}) \subseteq \mathcal{M}^{[\perp]}$. Conversely, if $x \in \mathcal{M}^{[\perp]}$ and $z = Hx$ then, for any $y \in \mathcal{M}$,

$$0 = [x,y] = [H^{-1}z,y] = (z,y).$$

Thus, $z \in \mathcal{M}^{\perp}$ and $x = H^{-1}z$ so that $\mathcal{M}^{[\perp]} \subseteq H^{-1}(\mathcal{M}^{\perp})$ and (2.2.3) is established. Then (2.2.2) follows immediately.

It follows from equation (2.2.2) that, for any subspace $\mathcal{M} \subseteq \mathsf{C}^n$,

$$(\mathcal{M}^{[\perp]})^{[\perp]} = \mathcal{M}. \tag{2.2.5}$$

Indeed, the inclusion $(\mathcal{M}^{[\perp]})^{[\perp]} \supseteq \mathcal{M}$ is evident from the definition of $\mathcal{M}^{[\perp]}$. But (2.2.2) implies that these two subspaces have the same dimension, and so (2.2.5) follows.

A subspace $\mathcal{M}$ is said to be *nondegenerate* (with respect to the indefinite inner product $[.,.]$) if $x \in \mathcal{M}$ and $[x,y] = 0$ for all $y \in \mathcal{M}$ imply that $x = 0$. Otherwise $\mathcal{M}$ is *degenerate*. For example, the defining property (iii) for the indefinite inner product $[.,.]$ ensures that C^n itself is always nondegenerate. In Example 2.2.1 the subspace $\mathcal{M}$ is degenerate because $\langle 1, 0, \dots, 0 \rangle \in \mathcal{M}$ and $[\langle 1, 0, \dots, 0 \rangle, y] = 0$ for all $y \in \mathcal{M}$ (if $n \geq 2$).

The nondegenerate subspaces can be characterized in another way:

Proposition 2.2.2. $\mathcal{M}^{[\perp]}$ *is a direct complement to* $\mathcal{M}$ *in* C^n *if and only if* $\mathcal{M}$ *is nondegenerate.*

Proof. By definition, the subspace $\mathcal{M}$ is nondegenerate if and only if $\mathcal{M} \cap \mathcal{M}^{[\perp]} = \{0\}$. In view of (2.2.2) this means that $\mathcal{M}^{[\perp]}$ is a direct complement to $\mathcal{M}$. $\quad\square$

In particular, the orthogonal companion of a nondegenerate subspace is again nondegenerate.

Let $P : \mathsf{C}^n \to \mathcal{M}$ be the orthogonal projection onto subspace $\mathcal{M}$ in the sense of $(.,.)$ and consider the hermitian linear transformation $PH \mid_{\mathcal{M}} : \mathcal{M} \to \mathcal{M}$. Nondegenerate subspaces can be characterized in another way using this transformation, namely: subspace $\mathcal{M}$ is nondegenerate if and only if $PH \mid_{\mathcal{M}} : \mathcal{M} \to \mathcal{M}$ is an invertible linear transformation.

If $\mathcal{M}$ is any nondegenerate nonzero subspace, Proposition 2.2.2 can be used to construct a basis in $\mathcal{M}$ which is *orthonormal* with respect to the indefinite inner product $[.,.]$, i.e., a basis $x_1, \dots, x_k$ satisfying

$$[x_i, x_j] = \left\{ \begin{array}{ll} \pm 1 & \text{for} \quad i = j \\ 0 & \text{for} \quad i \neq j \end{array} \right. .$$

To start the construction observe that there exists a vector $x \in \mathcal{M}$ such that $[x,x] \neq 0$. Indeed, if this were not true, then $[x,x] = 0$ for all $x \in \mathcal{M}$. Then the easily verified identity

$$[x,y] = \frac{1}{4} \left\{ [x+y, x+y] + i[x+iy, x+iy] - [x-y, x-y] - i[x-iy, x-iy] \right\} \tag{2.2.6}$$

shows that $[x,y] = 0$ for all $x, y \in \mathcal{M}$; a contradiction.

So it is possible to choose $x \in \mathcal{M}$ with $[x,x] \neq 0$ and write $x_1 = x / \sqrt{|[x,x]|}$ so that $[x_1, x_1] = \pm 1$. By Proposition 2.2.2 (applied in $\mathcal{M}$), the orthogonal companion $(\mathrm{Span}\,\{x_1\})^{[\perp]}$ of $\mathrm{Span}\,\{x_1\}$ in $\mathcal{M}$ is a direct complement of $\mathrm{Span}\,\{x_1\}$ in $\mathcal{M}$ and is also nondegenerate. Now take a vector $x_2 \in (\mathrm{Span}\,\{x_1\})^{[\perp]}$ such that $[x_2, x_2] = 1$, and so on, until $\mathcal{M}$ is exhausted.

In the next proposition we describe an important property of bases for a nondegenerate subspace which are orthonormal in the above sense. By $\mathrm{sig}\,Q$, where

$Q : \mathcal{M} \to \mathcal{M}$ is an invertible hermitian linear transformation from $\mathcal{M}$ to $\mathcal{M}$, we denote the *signature* of Q, i.e., the difference between the number of positive eigenvalues of Q and the number of negative eigenvalues of Q (in both cases counting with multiplicities).

Proposition 2.2.3. *Let* $[.,.] = (H.,.)$ *be an indefinite inner product with corresponding hermitian invertible matrix* H, *and let* $x_1, \ldots, x_k$ *be an orthonormal (with respect to* $[.,.]$) *basis in a nondegenerate subspace* $\mathcal{M} \subseteq \mathbb{C}^n$. *Then the sum* $\sum_{i=1}^{k}[x_i, x_i]$ *coincides with the signature of the hermitian linear transformation* $PH \mid_{\mathcal{M}} : \mathcal{M} \to \mathcal{M}$, *where* P *is the orthogonal projection (with respect to* $(.,.)$) *of* $\mathbb{C}^n$ *onto* $\mathcal{M}$.

Proof. Since $\mathcal{M}$ is nondegenerate, the transformation $PH \mid_{\mathcal{M}}$ is invertible. Now for $y = \sum_{i=1}^{k} \alpha_i x_i$ we have

$$(PH \mid_{\mathcal{M}} y, y) = \sum_{i=1}^{k} |\alpha_i|^2 [x_i, x_i].$$

Thus, the quadratic form defined on $\mathcal{M}$ by $(PH \mid_{\mathcal{M}} x, x)$ reduces to a sum of squares in the basis $x_1, \ldots, x_k$. Since $[x_i, x_i] = \pm 1$ it follows that $\sum_{i=1}^{k}[x_i, x_i]$ is just the signature of $PH \mid_{\mathcal{M}}$. □

Note, in particular, that the sum $\sum_{i=1}^{k}[x_i, x_i]$ does not depend on the choice of the orthonormal basis.

2.3 Classification of Subspaces

Let $[.,.]$ be an indefinite inner product on $\mathbb{C}^n$. A subspace $\mathcal{M}$ of $\mathbb{C}^n$ is called *positive* (with respect to $[.,.]$) if $[x, x] > 0$ for all nonzero x in $\mathcal{M}$, and *nonnegative* if $[x, x] \geq 0$ for all x in $\mathcal{M}$. Clearly, every positive subspace is also nonnegative but the converse is not necessarily true (see Example 2.3.1 below). Observe that a positive subspace is nondegenerate. If the invertible hermitian matrix H is such that $[x, y] = (Hx, y)$, $x, y \in \mathbb{C}^n$, we say that a positive (resp. nonnegative) subspace is *H-positive* (resp. *H-nonnegative*).

Example 2.3.1. *Let* $[x, y] = (Hx, y)$, $x, y \in \mathbb{C}^n$, *where* H *is the sip matrix of size* $n > 1$, *and assume* n *is odd. Then the subspace spanned by the first* $\frac{1}{2}(n + 1)$ *unit vectors is nonnegative, but not positive. The subspace spanned by the unit vector with 1 in the* $\frac{1}{2}(n + 1)$-*th position is positive.* □

We are to investigate the constraints on the dimensions of positive and nonnegative subspaces. But first a general observation is necessary.

Let $[.,.]_1$ and $[.,.]_2$ be two indefinite inner products on $\mathbb{C}^n$ with corresponding invertible hermitian matrices H_1 and H_2, respectively. Suppose, in addition, that H_1 and H_2 are congruent, i.e., $H_1 = S^* H_2 S$ for some invertible matrix S. (Here, and subsequently, the adjoint S^* of S is taken with respect to $(.,.)$.) In this case, a subspace $\mathcal{M}$ is H_1-positive if and only if $S\mathcal{M}$ is H_2-positive, with a similar statement replacing "positive" by "nonnegative". The proof is direct: take $x \in \mathcal{M}$ and

$$[Sx, Sx]_2 = (H_2 Sx, Sx) = (S^* H_2 Sx, x) = (H_1 x, x) = [x, x]_1.$$

Thus, $[x, x]_1 > 0$ for all nonzero $x \in \mathcal{M}$ if and only if $[y, y]_2 > 0$ for all nonzero y in $S\mathcal{M}$.

Theorem 2.3.2. *The maximal dimension of a positive, or of a nonnegative subspace with respect to the indefinite inner product $[x, y] = (Hx, y)$ coincides with the number of positive eigenvalues of H (counting multiplicities).*

Note that the maximal possible dimensions of nonnegative and positive subspaces coincide.

Proof. We prove only the nonnegative case (the positive case is analogous). So let $\mathcal{M}$ be a nonnegative subspace, and let $p = \dim \mathcal{M}$. Then

$$\min_{(x,x)=1, \ x\in\mathcal{M}} (Hx, x) \geq 0. \tag{2.3.7}$$

Write all the eigenvalues of H in the nonincreasing order: $\lambda_1 \geq \lambda_2 \geq \cdots \geq \lambda_n$. By the max-min characterization of the eigenvalues of H, (Theorem A.1.6) we have

$$\lambda_p = \max_{\mathcal{L}} \min_{(x,x)=1, \ x\in\mathcal{L}} (Hx, x),$$

where the maximum is taken over all the subspaces $\mathcal{L} \subseteq \mathbf{C}^n$ of dimension p. Then (2.3.7) implies $\lambda_p \geq 0$ and, since H is invertible, $\lambda_p > 0$. So $p \leq k$ where k is the number of positive eigenvalues of H.

To find a nonnegative subspace of dimension k, appeal to the observation preceding the theorem. By Theorem A.1.1, there exists an invertible matrix S such that $S^* HS$ is a diagonal matrix of 1's and -1's:

$$H_0 := S^* HS = \operatorname{diag} (1, \ldots, 1, -1, \ldots, -1), \tag{2.3.8}$$

where the number of $+1$'s is k. Hence, it is sufficient to find a k-dimensional subspace which is nonnegative with respect to H_0. One such subspace (which is even positive) is spanned by the first k unit vectors in $\mathbf{C}^n$. $\square$

A subspace $\mathcal{M} \subseteq \mathbf{C}^n$ is called H-*negative* (where H is such that $[x, y] = (Hx, y)$, $x, y \in \mathbf{C}^n$), if $[x, x] < 0$ for all nonzero x in $\mathcal{M}$. Replacing this condition by the requirement that $[x, x] \leq 0$ for all $x \in \mathcal{M}$, we obtain the definition of a *nonpositive* (with respect to $[.,.]$) or H-*nonpositive* subspace. As in Theorem 2.3.2 it can be proved that the maximal possible dimension of an H-negative or of an H-nonpositive subspace is equal to the number of negative eigenvalues of H (counting multiplicities).

Note also the following inequality: Let $\mathcal{M}$ be an H-nonnegative or H-nonpositive subspace, then

$$|(Hy, z)| \leq (Hy, y)^{1/2}(Hz, z)^{1/2} \qquad (2.3.9)$$

for every $y, z \in \mathcal{M}$. The proof of (2.3.9) is completely analogous to the standard proof of Schwarz's inequality.

We pass now to the class of subspaces which are peculiar to indefinite inner product spaces and have no analogues in the spaces with a definite inner product. A subspace $\mathcal{M} \subseteq \mathsf{C}^n$ is called *neutral* (with respect to $[.,.]$), or *H-neutral* (where H is such that $[x, y] = (Hx, y)$, $x, y \in \mathsf{C}^n$) if $[x, x] = 0$ for all $x \in \mathcal{M}$. Sometimes, such subspaces are called *isotropic*. In Example 2.3.1 the subspaces spanned by the first k unit vectors, for $k = 1, \ldots, \frac{n-1}{2}$, are all neutral.

In view of the identity (2.2.6) a subspace $\mathcal{M}$ is neutral if and only if $[x, y] = 0$ for all $x, y \in \mathcal{M}$. Observe also that a neutral subspace is both nonpositive and nonnegative, and (if nonzero) is necessarily degenerate.

We have seen in Example 2.3.1 that the nonnegative subspace spanned by the first $\frac{1}{2}(n + 1)$ unit vectors is a direct sum of a neutral subspace (spanned by the first $\frac{1}{2}(n - 1)$ unit vectors) and a positive definite subspace (spanned by the $\frac{1}{2}(n + 1)$-th unit vector). This is a general property, as the following theorem shows.

Theorem 2.3.3. *An H-nonnegative (resp. H-nonpositive) subspace is a direct sum of an H-positive (resp. H-negative) subspace and an H-neutral subspace.*

Proof. Let $\mathcal{M} \subseteq \mathsf{C}^n$ be an H-nonnegative subspace, and let $\mathcal{M}_0$ be a maximal H-positive subspace in $\mathcal{M}$ (since $\dim \mathcal{M}$ is finite, such an $\mathcal{M}_0$ always exists). Since $\mathcal{M}_0$ is nondegenerate, Proposition 2.2.2 implies that $\mathcal{M}_0 \dot{+} \mathcal{M}_0^{[\perp]} = \mathsf{C}^n$, and hence

$$\mathcal{M}_0 \dot{+} \left(\mathcal{M}_0^{[\perp]} \cap \mathcal{M} \right) = \mathcal{M}.$$

It remains to show that $\mathcal{M}_0^{[\perp]} \cap \mathcal{M}$ is H-neutral. Suppose not; so there exists an $x \in \mathcal{M}_0^{[\perp]} \cap \mathcal{M}$ such that $[x, x] \neq 0$. Since $\mathcal{M}$ is H-nonnegative it follows that $[x, x] > 0$. Now for each $y \in \mathcal{M}_0$ we have $[x + y, x + y] = [x, x] + [y, y] > 0$, in view of the fact that $\mathcal{M}_0$ is H-positive. So $\text{Span}\{x, \mathcal{M}_0\}$ is also an H-positive subspace; a contradiction with the maximality of $\mathcal{M}_0$.

For an H-nonpositive subspace the proof is similar. $\qquad \square$

The decomposition of a nonnegative subspace $\mathcal{M}$ into a direct sum $\mathcal{M}_0 \dot{+} \mathcal{M}_1$, where $\mathcal{M}_0$ is positive and $\mathcal{M}_1$ is neutral, is not unique. However, $\dim \mathcal{M}_0$ is uniquely determined by $\mathcal{M}$. Indeed, let P be the orthogonal (with respect to $(.,.)$) projection on $\mathcal{M}$, then it is easily seen that $\dim \mathcal{M}_0 = \text{rank } PH \mid_{\mathcal{M}}$, where $PH \mid_{\mathcal{M}}: \mathcal{M} \to \mathcal{M}$ is a selfadjoint linear transformation.

One can easily compute the maximal possible dimension of a neutral subspace.

Theorem 2.3.4. *The maximal possible dimension of an H-neutral subspace is* $\min(k, l)$, *where k (resp. l) is the number of positive (resp. negative) eigenvalues of H, counting multiplicities.*

Proof. In view of the remark preceding Theorem 2.3.2 it may be assumed that $H = H_0$ is given by (2.3.8). The existence of a neutral subspace of dimension $\min(k, l)$ is easily seen. A basis for one such subspace can be formed from the unit vectors $e_1, e_2, \ldots$ as follows: $e_1 + e_{k+1}, e_2 + e_{k+2}, \ldots$.

Now let $\mathcal{M}$ be a neutral subspace of dimension p. Since $\mathcal{M}$ is also nonnegative it follows from Theorem 2.3.2 that $p \leq k$. But $\mathcal{M}$ is also nonpositive and so the inequality $p \leq l$ also applies. Thus, $p \leq \min(k, l)$. $\square$

2.4   Exercises

1. For which values of the parameter $w \in \mathbb{C}$ are the following hermitian matrices nonsingular? When they are nonsingular determine whether the determinant is positive or negative and find the number of negative eigenvalues.

   (a)
   $$\begin{bmatrix} 1 & \alpha & \alpha^2 & \cdots & \alpha^{n-1} & w \\ \overline{\alpha} & 1 & \alpha & \cdots & \alpha^{n-2} & \alpha^{n-1} \\ \vdots & & & & \vdots & \vdots \\ \overline{\alpha}^{n-1} & \overline{\alpha}^{n-2} & \cdots & & 1 & \alpha \\ \overline{w} & \overline{\alpha}^{n-1} & \cdots & \overline{\alpha}^2 & \overline{\alpha} & 1 \end{bmatrix}, \quad \alpha \in \mathbb{C}.$$

   (b)
   $$\begin{bmatrix} \alpha & 0 & \cdots & 0 & w \\ 0 & \alpha & \cdots & \beta & 0 \\ \vdots & \vdots & & \vdots & \vdots \\ 0 & \beta & \cdots & \alpha & 0 \\ \overline{w} & 0 & \cdots & 0 & \alpha \end{bmatrix}, \quad \alpha, \beta \in \mathbb{R}. \text{ The matrix here has even size.}$$

   (c)
   $$\begin{bmatrix} \alpha & \beta & \beta^2 & \cdots & \beta^{n-1} & w \\ \overline{\beta} & \alpha & \beta & \cdots & \beta^{n-2} & \beta^{n-1} \\ \vdots & \vdots & \vdots & \vdots & \vdots & \vdots \\ \overline{w} & \overline{\beta}^{n-1} & \overline{\beta}^{n-2} & \cdots & \overline{\beta} & \alpha \end{bmatrix} \quad \alpha \in \mathbb{R}, \ \beta \in \mathbb{C}.$$

2. Define an inner product $[.,.]$ on $\mathbb{C}^n$ by
   $$[x, y] = \sum_{j=1}^{n} x_j \overline{y_{n+1-j}}.$$

   (a) Describe all positive and nonnegative subspaces (with respect to this inner product).

(b) Describe all negative and nonpositive subspaces.

(c) Describe all neutral subspaces.

3. Let S_n be the $n \times n$ sip matrix and consider the following matrices:

$$H_1 = \begin{bmatrix} 0 & I_n \\ I_n & 0 \end{bmatrix}, \quad H_2 = \begin{bmatrix} I_n & 0 \\ 0 & S_n \end{bmatrix}, \quad H_3 = \begin{bmatrix} S_n & 0 \\ 0 & I_n \end{bmatrix}.$$

Find an H-orthogonal basis in each of the three cases.

4. Find all H-neutral subspaces in each of the three cases of Exercise 3.

5. Find all H-positive subspaces in each of the three cases of Exercise 3.

6. Let $\mathcal{M}_1$ and $\mathcal{M}_2$ be two subspaces of $\mathbf{C}^n$ for which $\mathcal{M}_1 \subseteq \mathcal{M}_2$ and $\dim \mathcal{M}_1 + \dim \mathcal{M}_2 = n$. Show that there is an indefinite inner product on $\mathbf{C}^n$ in which $\mathcal{M}_2 = \mathcal{M}_1^{[\perp]}$.

7. Define an indefinite inner product on $\mathbf{C}^{3n}$ in terms of

$$H = \begin{bmatrix} 0 & 0 & I_n \\ 0 & I_n & 0 \\ I_n & 0 & 0 \end{bmatrix}.$$

Consider the subspaces $\mathcal{M}_1 = \mathrm{Span}\,\{e_1, e_2, \ldots, e_n\}$, $\mathcal{M}_2 = \mathrm{Span}\,\{e_{n+1}, e_{n+2}, \ldots, e_{2n}\}$, and $\mathcal{M}_3 = \mathrm{Span}\,\{e_{2n+1}, e_{2n+2}, \ldots, e_{3n}\}$.

(a) Find all H-neutral, all H-nonnegative, and all H-nonpositive subspaces that contain $\mathcal{M}_1$.

(b) Similarly for $\mathcal{M}_3$.

(c) Find all H-positive and all H-nonnegative subspaces that contain $\mathcal{M}_2$.

8. Let $[x, y] = (Hx, y)$, where H is the $n \times n$ matrix

$$H = \begin{bmatrix} 0 & 1 & 0 & \ldots & 0 & 0 \\ 1 & 0 & 0 & \ldots & 0 & 0 \\ 0 & 0 & 1 & \ldots & 0 & 0 \\ & & & \ldots & & \\ 0 & 0 & 0 & \ldots & 0 & 1 \\ 0 & 0 & 0 & \ldots & 1 & 0 \end{bmatrix}.$$

Find the maximal H-positive and the maximal H-negative subspaces.

9. Solve Exercise 8 for the following matrices:

(a) $H_0 = \begin{bmatrix} 0 & 0 & \ldots & 0 & 1 \\ 0 & 0 & \ldots & 1 & 0 \\ \vdots & & \ldots & & \vdots \\ 1 & 0 & \ldots & 0 & 0 \end{bmatrix}.$

(b) $H_1 = \begin{bmatrix} a_1 & 0 & \cdots & 0 & 1 \\ 0 & a_2 & \cdots & 1 & 0 \\ \vdots & & \cdots & & \vdots \\ 1 & 0 & \cdots & 0 & a_n \end{bmatrix}$, $a_j \in \mathbb{R}$, $n = 2k$.

The parameters $a_1, \ldots, a_n$ are such that H_1 is invertible.

(c) $H_2 = i \begin{bmatrix} 0 & 0 & \cdots & 0 & -1 \\ 0 & 0 & \cdots & -1 & 0 \\ \vdots & & \cdots & & \vdots \\ 0 & 1 & \cdots & 0 & 0 \\ 1 & 0 & \cdots & 0 & 0 \end{bmatrix}$ (and the matrix has even size).

(d) $H_3 = P - (I - P)$, where P is hermitian and $P^2 = P$.

(e) $H_4 = i \begin{bmatrix} 2i & 0 & \cdots & 0 & 1 \\ 0 & -2i & \cdots & 1 & 0 \\ \vdots & & \cdots & & \vdots \\ 0 & -1 & \cdots & 2i & 0 \\ -1 & 0 & \cdots & 0 & -2i \end{bmatrix}$ (and the matrix has even size).

(f) $H_5 = \begin{bmatrix} 1 & \alpha & \alpha & \cdots & \alpha \\ \alpha & 1 & \alpha & \cdots & \alpha \\ \vdots & & & \cdots & \vdots \\ \alpha & \alpha & \alpha & \cdots & 1 \end{bmatrix}$, $\alpha \in \mathbb{R}$.

The number α is chosen so that H_5 is invertible. What are all such α?

10. Let $\mathcal{M}$ be a nonnegative subspace of $\mathbb{C}^n(H)$. Show that, except for the trivial cases when $\mathcal{M}$ is neutral or when $\mathcal{M}$ is positive, there is a continuum of distinct direct sum decompositions $\mathcal{M} = \mathcal{M}_0 \dotplus \mathcal{M}_1$, where $\mathcal{M}_0$ is positive and $\mathcal{M}_1$ is neutral.

11. Show that the neutral subspace $\mathcal{M}_1$ is the same for all direct sum decompositions $\mathcal{M} = \mathcal{M}_0 \dotplus \mathcal{M}_1$ of Exercise 10.

12. Let $\mathcal{L}_2^{(n)}$ be the space of all polynomials with complex coefficients of degree n or less and consider the function ω defined on the unit circle by

$$\omega(e^{i\theta}) = \sum_{j=-n}^{n} \omega_j e^{ij\theta}, \qquad 0 \le \theta < 2\pi,$$

where $\omega_{-n}, \omega_{-n+1}, \ldots, \omega_n \in \mathbb{C}$. Assume that ω is not identically zero and that the values of $\omega(e^{i\theta})$ are real.

Let

$$x(\lambda) = \sum_{j=0}^{n} \xi_j \lambda^j \in \mathcal{L}_2^{(n)}, \quad y(\lambda) = \sum_{j=0}^{n} \eta_j \lambda^j \in \mathcal{L}_2^{(n)},$$

and define the bilinear form $[.,.]$ on $\mathcal{L}_2^{(n)}$ by

$$[x,y] = \frac{1}{2\pi} \int_0^{2\pi} \sum_{j=0}^n \xi_j e^{ij\theta} w(e^{i\theta}) \overline{\sum_{k=0}^n \eta_k e^{ik\theta}} \, d\theta.$$

(a) Show that $[.,.]$ defines an inner product on $\mathcal{L}_2^{(n)}$. (The inner product is generally indefinite.)

(b) Represent the inner product in the form $[x,y] = (H_2\xi, \eta)$ where H_2 is an $(n+1) \times (n+1)$ hermitian matrix.

(c) Under what conditions on $w(t)$ are all principal minors of H_2 invertible?

(d) When is the inner product $[.,.]$ definite, i.e., $[x,x] > 0$ for every nonzero $x \in \mathcal{L}_2^{(n)}$?

13. For each of the following indefinite inner products on R^n find: (a) all nonnegative subspaces, (b) all positive subspaces, (c) all nonpositive subspaces, (d) all negative subspaces, (e) all neutral subspaces, (f) all maximal nonnegative subspaces, (g) all maximal nonpositive subspaces.

The indefinite inner products are:

(α) $[x,y] = \sum_{j=1}^n x_j y_{n+1-j}$; $x = \langle x_1, \ldots, x_n \rangle$, $y = \langle y_1, \ldots, y_n \rangle \in R^n$;

(β) $[x,y] = (H_1 x, y)$, $H_1 = \begin{bmatrix} 0 & I_m \\ I_m & 0 \end{bmatrix}$, $(m = \frac{n}{2})$;

(γ) $[x,y] = (H_2 x, y)$, $H_2 = \begin{bmatrix} I_m & 0 \\ 0 & S_m \end{bmatrix}$, $(m = \frac{n}{2})$;
and S_m is the $m \times m$ sip matrix;

(δ) $[x,y] = (H_3 x, y)$, $H_3 = \begin{bmatrix} S_m & 0 \\ 0 & I_m \end{bmatrix}$, $(m = \frac{n}{2})$;

(ε) $[x,y] = \sum_{j=1}^m (x_{2j-1} y_{2j} + x_{2j} y_{2j-1})$, $n = 2m$;

(ζ) $[x,y] = \sum_{j=1}^n x_j y_{n+1-j} + \sum_{j=1}^n a_j x_j y_j$, $a_1, \ldots, a_n \in R$;

(η) $[x,y] = a \left(\sum_{i,j=1}^n x_i y_j \right) + \sum_{j=1}^n x_j y_j$, $a \in R$.

14. Describe geometrically the curve in R^2 formed by the vectors $x \in R^2$ that satisfy the equation $(H_j x, x) = 1$, for each of the following real symmetric matrices H_j, $j = 1, 2, 3, 4, 5$:

(a) $H_1 = \begin{bmatrix} 3 & 2 \\ 2 & 3 \end{bmatrix}$; (b) $H_2 = \begin{bmatrix} 1 & 2 \\ 2 & 1 \end{bmatrix}$; (c) $H_3 = \begin{bmatrix} -1 & -1 \\ -1 & -1 \end{bmatrix}$;

(d) $H_4 = \begin{bmatrix} a & b \\ b & a \end{bmatrix}$; (e) $H_5 = \begin{bmatrix} a & b \\ b & c \end{bmatrix}$;

where a, b, c are real numbers.

15. Describe geometrically the surface in $\mathbb{R}^3$ formed by the vectors $x \in \mathbb{R}^3$ that satisfy the equation $(H_j x, x) = 1$, for each of the following real symmetric matrices H_j, $j = 1, 2, 3, 4, 5$:

$$
\text{(a)} \ H_1 = \begin{bmatrix} 3 & 2 & 0 \\ 2 & 3 & 0 \\ 0 & 0 & 1 \end{bmatrix} ; \quad \text{(b)} \ H_2 = \begin{bmatrix} 1 & 2 & 0 \\ 2 & 1 & 0 \\ 0 & 0 & 1 \end{bmatrix} ;
$$

$$
\text{(c)} \ H_3 = \begin{bmatrix} 1 & 2 & 0 \\ 2 & 1 & 0 \\ 0 & 0 & -1 \end{bmatrix} ; \quad \text{(d)} \ H_4 = \begin{bmatrix} -1 & -1 & 0 \\ -1 & -1 & 0 \\ 0 & 0 & \pm 1 \end{bmatrix} ;
$$

$$
\text{(e)} \ H_5 = \begin{bmatrix} a & b & 0 \\ b & a & 0 \\ 0 & 0 & \pm 1 \end{bmatrix} ;
$$

where a, b are real numbers.

16. Describe geometrically the surface in $\mathbb{R}^n$ formed by the vectors $x \in \mathbb{R}^n$ that satisfy the equation $(Hx, x) = 1$, for each of the following situations:

(a) $H > 0$; (b) $H = S_n$, the sip matrix; (c) $H = \text{diag}\,(\lambda_1, \lambda_2, \ldots, \lambda_n)$,

where $\lambda_1 \leq \lambda_2 \leq \cdots \leq \lambda_n$ are real numbers.

2.5 Notes

The material of this chapter is well-known, even in the infinite dimensional setting (see [5], [6], [11], [57]).

Chapter 3

Orthogonalization and Orthogonal Polynomials

Let $x_1, x_2, \ldots, x_m$ be a linearly independent set in a linear space with a definite inner product. In classical linear algebra and analysis a fundamental role is played by the construction of a mutually orthogonal set of vectors $y_1, y_2, \ldots, y_m$ for which each subset $y_1, \ldots, y_k$ $(k \leq m)$ spans the same subspace as $x_1, \ldots, x_m$. The well-known Gram–Schmidt process is of this kind. This is also the central idea in the analysis of systems of orthogonal polynomial functions. This chapter is devoted to the development of this line of thought in the context of the linear space C^n with an *indefinite* inner product.

Motivated by applications, attention will be confined to sets of vectors $y_1, \ldots, y_m$ for which $[y_j, y_j] \neq 0$ for each j; sometimes described as the case of "nonzero squares". In this case the orthogonalization will be said to be *regular*. It turns out that the results look very like those obtained in the classical case based on a definite inner product.

The chapter consists of four sections. The first contains general results concerning orthogonal (regular) systems. The second contains discussion of a fundamental theorem of Szegő for the case of a definite inner product, as well as a more general theorem of M. G. Krein for the indefinite case. The last two sections contain a proof of the Krein theorem using the "one-step case" developed by Ellis and Gohberg as well as their and Lay's analysis of fundamental determinants arising in an extension of this process.

3.1 Regular Orthogonalizations

Let C^n be a vector space with an indefinite inner product $[\cdot, \cdot]$. A vector $y \in C^n$ is called *nonneutral* if $[y, y] \neq 0$. Note first of all that any set of nonneutral vectors $y_1, y_2, \ldots, y_m$ which is orthogonal in the sense of the indefinite inner product $[\cdot, \cdot]$

is necessarily linear independent. To see this, suppose that $\sum_{j=1}^{m} g_j y_j = 0$, and hence, for $k = 1, 2, \ldots, m$,

$$\sum_{j=1}^{m} g_j \left[y_j, y_k \right] = g_k \left[y_k, y_k \right] = 0.$$

Then it follows that $g_k = 0$.

We now introduce the key definition. Let $v_1, v_2, \ldots, v_m$ be a system (i.e., an ordered set) of vectors of $\mathbf{C}^n$. A system of vectors $y_1, y_2, \ldots, y_m$ which are mutually orthogonal with respect to $[\cdot, \cdot]$ is said to be a *regular orthogonalization* of $v_1, v_2, \ldots, v_m$ if:

$$[y_j, y_j] \neq 0, \quad j = 1, 2, \ldots, m, \tag{3.1.1}$$

and

$$\text{Span } \{y_1, y_2, \ldots, y_k\} = \text{Span } \{v_1, v_2, \ldots, v_k\}, \quad \text{for} \quad k = 1, 2, \ldots, m. \tag{3.1.2}$$

As the following example shows, not every sequence of vectors $v_1, v_2, \ldots, v_m \in \mathbf{C}^n$ admits a regular orthogonalization.

Example 3.1.1. *Consider*

$$[x, y] = (S_{2m} x, y), \quad x, y \in \mathbf{C}^{2m},$$

where S_{2m} is the sip matrix of size $2m$. Let $e_1, e_2, \ldots, e_m$ be the system of m first standard unit vectors in $\mathbf{C}^{2m}$. It is easily seen that this system is orthogonal in $[\cdot, \cdot]$ and all vectors are neutral

$$[e_j, e_k] = 0 \quad j, k = 1, 2, \ldots, m. \qquad \square$$

For any system of vectors $v_1, v_2, \ldots, v_m \in \mathbf{C}^n$, the Gram matrix is defined to be the hermitian matrix

$$G(v_1, v_2, \ldots, v_m) = \left([v_j, v_k] \right)_{j,k=1}^{m}.$$

Clearly, the sip matrix of size $m \times m$ is the Gram matrix in Example 3.1.1.

Theorem 3.1.2. *The system of vectors $v_1, v_2, \ldots, v_m$ from $\mathbf{C}^n$ admits a regular orthogonalization if and only if*

$$\det G(v_1, v_2, \ldots, v_k) \neq 0, \quad for \quad k = 1, 2, \ldots, m.$$

If these conditions hold, then one such orthogonalization $y_1, y_2, \ldots, y_m$ is given by

$$y_r = \sum_{j=1}^{r} \gamma_{rj}^{(r)} v_j, \quad r = 1, 2, \ldots, m,$$

where

$$\left(\gamma_{jk}^{(r)} \right)_{j,k=1}^{r} = (G(v_1, v_2, \ldots, v_r))^{-1}.$$

Moreover

$$\gamma_{rr}^{(r)} = [y_r, y_r] \neq 0, \quad r = 1, 2, \ldots, m.$$

Proof. Suppose there exists a regular orthogonalization $y_1, y_2, \ldots, y_m$ of $v_1, \ldots, v_m$. Since

$$\text{Span}\{y_1, y_2, \ldots, y_k\} = \text{Span}\{v_1, v_2, \ldots, v_k\}, \qquad 1 \leq k \leq m,$$

there exist numbers

$$\alpha_{k,j} \in \mathbb{C}, \qquad 1 \leq j \leq k, \quad 1 \leq k \leq m,$$

such that

$$y_k = \sum_{j=1}^{k} \alpha_{k,j} v_j, \qquad 1 \leq k \leq m. \tag{3.1.3}$$

Furthermore, for $2 \leq k \leq m$, the vector y_k is orthogonal (in the sense of $[\cdot, \cdot]$) to $y_1, y_2, \ldots, y_{k-1}$ and hence also to $v_1, \ldots, v_{k-1}$. From (3.1.3) it follows that

$$[y_k, y_k] = \sum_{j=0}^{k} \alpha_{k,j} [v_j, y_k] = \alpha_{k,k}[v_k, y_k]. \qquad 1 \leq k \leq m.$$

For $k = 1, 2, \ldots, m$ we define the matrices Ω_k and A_k by

$$\Omega_k = ([y_j, v_i])_{j,i=1}^{k} \quad \text{and} \quad A_k = (\alpha_{i,j})_{i,j=1}^{k},$$

where $\alpha_{i,j} = 0$ for $1 \leq i < j \leq m$. In view of (3.1.3) we obtain

$$[y_j, v_i] = [\sum_{\ell=1}^{j} \alpha_{j,\ell} v_\ell, v_i] = \sum_{\ell=1}^{j} \alpha_{j,\ell}[v_\ell, v_i] = \sum_{\ell=1}^{k} \alpha_{j,\ell}[v_\ell, v_i],$$

and hence

$$\Omega_k = A_k G(v_1, v_2, \ldots, v_k), \qquad k = 1, 2, \ldots, m.$$

Therefore,

$$\det G(v_1, v_2, \ldots, v_k) \neq 0.$$

Let $G_r = G(v_1, v_2, \ldots, v_r)$ and

$$\Gamma_r = G_r^{-1} = \left(\gamma_{jk}^{(r)}\right)_{j,k=1}^{r}.$$

The first step is to prove that $\gamma_{rr}^{(r)} \neq 0$. It is obvious that $\gamma_{11}^{(1)} \neq 0$. For $r = 2, \ldots, m$ write G_r in the form

$$G_r = \begin{bmatrix} G_{r-1} & Z^* \\ Z & [v_r, v_r] \end{bmatrix}$$

where $Z = ([v_r, v_1] \ldots [v_r, v_{r-1}])$. It is clear that

$$G_r \begin{bmatrix} \gamma_{1r}^{(r)} \\ \gamma_{2r}^{(r)} \\ \vdots \\ \gamma_{rr}^{(r)} \end{bmatrix} = \begin{bmatrix} 0 \\ 0 \\ \vdots \\ 1 \end{bmatrix}.$$

Hence

$$G_{r-1} \begin{bmatrix} \gamma_{1r}^{(r)} \\ \vdots \\ \gamma_{r-1,r}^{(r)} \end{bmatrix} + Z^* \gamma_{rr}^{(r)} = 0$$

and

$$Z \begin{bmatrix} \gamma_{1r}^{(r)} \\ \vdots \\ \gamma_{r-1,r}^{(r)} \end{bmatrix} + [v_r, v_r]\, \gamma_{rr}^{(r)} = 1.$$

The latter two equations imply

$$\left([v_r, v_r] - Z G_{r-1}^{-1} Z^*\right) \gamma_{rr}^{(r)} = 1,$$

and it follows that $\gamma_{rr}^{(r)} \neq 0$.

In the second step it is proved that the system

$$y_r := \sum_{j=1}^{r} \gamma_{rj}^{(r)} v_j \quad (r = 1, 2, \ldots, m)$$

is a regular orthogonalization of $v_1, v_2, \ldots, v_m$. From the definition of Γ_r, $r = 1, 2, \ldots, m$ it follows that

$$\sum_{j=1}^{r} \gamma_{rj}^{(r)} [v_j, v_p] = \delta_{rp} \quad (p = 1, 2, \ldots, r),$$

where δ_{rp} is the Kronecker symbol: $\delta_{rp} = 0$ if $r \neq p$, and $\delta_{rp} = 1$ if $r = p$. Therefore

$$[y_r, v_p] = \left[\sum_{j=1}^{r} \gamma_{rj}^{(r)} v_j, v_p \right] = \sum_{j=1}^{r} \gamma_{rj}^{(r)} [v_j, v_p] = \delta_{rp}.$$

Using this equality we obtain for $l \geq k$,

$$[y_k, y_l] = \left[\sum_{j=1}^{k} \gamma_{kj}^{(k)} v_j, y_l \right] = \sum_{j=1}^{k} \gamma_{kj}^{(k)} [v_j, y_l] = \sum_{j=1}^{k} \gamma_{kj}^{(k)} \delta_{jl} = \delta_{kl} \gamma_{kk}^{(k)}.$$

In the third step, it remains to prove that equation (3.1.2) holds. From the definition of $y_1, y_2, \ldots, y_m$ it follows that

$$\mathrm{Span}\, \{y_1, y_2, \ldots, y_r\} \subseteq \mathrm{Span}\, \{v_1, v_2, \ldots, v_r\}, \quad \text{for} \quad r = 1, 2, \ldots, m.$$

The reverse inclusion will be proved by induction and, for the first step, it is clear that $\mathrm{Span}\, \{y_1\} = \mathrm{Span}\, \{v_1\}$. Suppose that

$$\mathrm{Span}\, \{y_1, y_2, \ldots, y_{r-1}\} = \mathrm{Span}\, \{v_1, v_2, \ldots, v_{r-1}\}.$$

Since $\gamma_{rr}^{(r)} \neq 0$,

$$v_r = \left(y_r - \sum_{j=1}^{r-1} \gamma_{rj}^{(r)} v_j \right) \left(\gamma_{rr}^{(r)} \right)^{-1},$$

and it follows from this relation that

$$\text{Span } \{v_1, v_2, \ldots, v_r\} \subseteq \text{Span } \{v_1, v_2, \ldots, v_{r-1}, y_r\} \subseteq \text{Span } \{y_1, \ldots, y_r\}.$$

The theorem is proved. $\square$

Theorem 3.1.3. *If the sequence of vectors $v_1, v_2, \ldots, v_m$ admits a regular orthogonalization and*

$$\{x_1, x_2, \ldots, x_m\}, \quad \{y_1, y_2, \ldots, y_m\} \in \mathbb{C}^n$$

are two such orthogonalizations, then for some $\alpha_j \in \mathbb{C} \setminus \{0\}$: $y_j = \alpha_j x_j$ $(j = 1, 2, \ldots, m)$.

Proof. Suppose the orthogonal systems $x_1, x_2, \ldots, x_m$ and $y_1, y_2, \ldots, y_m$ are given as in the theorem. Since

$$\text{Span } \{v_1, v_2, \ldots, v_r\} = \text{Span } \{y_1, y_2, \ldots, y_r\} = \text{Span } \{x_1, x_2, \ldots, x_r\},$$

for $r = 1, 2, \ldots, m$, there exist $\alpha_{rj}, \beta_{rj} \in \mathbb{C}$ such that

$$y_r = \sum_{j=1}^{r} \alpha_{rj} v_j, \qquad x_r = \sum_{j=1}^{r} \beta_{rj} v_j.$$

Furthermore

$$[y_r, y_j] = [x_r, x_j] = 0, \quad j = 1, 2, \ldots, r-1$$

and hence

$$[y_r, v_j] = 0, \quad j = 1, 2, \ldots, r-1. \tag{3.1.4}$$

Since x_j is a linear combination of $v_1, \ldots, v_j$ $(j = 1, 2, \ldots, r-1)$ we have

$$[y_r, x_j] = 0, \quad j = 1, 2, \ldots, r-1. \tag{3.1.5}$$

Now write

$$x_r = a_r y_r + \sum_{j=1}^{r-1} a_j x_j, \quad a_1, \ldots, a_r \in \mathbb{C}.$$

Taking inner products $[\cdot, \cdot]$ of this equality with y_r and with x_j $(j = 1, 2, \ldots, r-1)$, and using (3.1.4) and (3.1.5) we obtain

$$a_r = [x_r, y_r]/[y_r, y_r]$$

and

$$[x_r, x_j] = 0 = a_j[x_j, x_j], \quad j = 1, 2, \ldots, r-1.$$

Hence $a_j = 0$ $(j = 1, 2, \ldots, r-1)$, and $x_r = a_r y_r$. $\square$

Corollary 3.1.4. *Let* $v_1, v_2, \ldots, v_r$ *be the standard orthonormal system in* $\mathbb{C}^n$, *i.e.*,

$$v_k = e_k = \underbrace{\langle 0, 0, \ldots, 1, 0, \ldots, 0 \rangle}_{k},$$

and assume that $\det([e_p, e_q])_{p,q=1}^k \neq 0$, *for* $k = 1, 2, \ldots, r$. *Then the vectors*

$$y_1 = \begin{bmatrix} \gamma_{11}^{(1)} \\ 0 \\ \vdots \\ 0 \end{bmatrix}, \quad y_2 = \begin{bmatrix} \gamma_{12}^{(2)} \\ \gamma_{22}^{(2)} \\ 0 \\ \vdots \\ 0 \end{bmatrix}, \quad \ldots, \quad y_r = \begin{bmatrix} \gamma_{1r}^{(r)} \\ \vdots \\ \gamma_{rr}^{(r)} \\ 0 \\ \vdots \\ 0 \end{bmatrix}, \qquad (3.1.6)$$

from $\mathbb{C}^n$, *where*

$$\left(\gamma_{pq}^{(r)} \right)_{p,q=1}^r = G(e_1, e_2, \ldots, e_r)^{-1}$$

and

$$G(e_1, e_2, \ldots, e_r) = ([e_p, e_q])_{p,q=1}^r,$$

form a regular orthogonalization of $e_1, e_2, \ldots, e_r$ *in the inner product* $[\cdot, \cdot]$.

This corollary follows immediately from Theorem 3.1.2.

If the indefinite inner product is defined in terms of the matrix $H = (h_{pq})_{p,q=1}^m$ then, with vectors v_k of the corollary,

$$[v_j, v_k] = (He_j, e_k) = h_{kj},$$

and we obtain

$$\left(\gamma_{pq}^{(r)} \right)_{p,q=1}^r = \left((h_{qp})_{p,q=1}^r \right)^{-1}.$$

Finally, note that a regular orthogonalization of a system can be constructed with the help of the Gram-Schmidt orthogonalization process in exactly the same way as in the case of $\mathbb{C}^n$ with a definite inner product.

Example 3.1.5. *Let* $L_2^{(n)}$ *be the vector space of all polynomials of degree* $\leq n$:

$$v(t) = a_0 + a_1 t + \cdots + a_n t^n$$

on the unit circle $|t| = 1$ *with the inner product*

$$(v, u) = \frac{1}{2\pi} \int_0^{2\pi} v(e^{i\theta}) \overline{u(e^{i\theta})} d\theta.$$

Let

$$\omega(e^{i\theta}) = \sum_{j=-n}^{n} \omega_j e^{ij\theta} \neq 0$$

be a weight function for which $\omega(e^{i\theta})$ is real valued for $\theta \in [0, 2\pi]$, and define an indefinite inner product by

$$[v, u]_w = \frac{1}{2\pi} \int_0^{2\pi} v(e^{i\theta})\omega(e^{i\theta})\overline{u(e^{i\theta})}d\theta.$$

It is clear that the system

$$1, t, t^2, \ldots, t^n$$

is orthonormal in the inner product $(\cdot, \cdot)$. By direct computation we obtain

$$G(1, t, \ldots, t^k) = \begin{bmatrix} \omega_0 & \omega_1 & \cdots & \omega_k \\ \omega_{-1} & \omega_0 & \cdots & \omega_{k-1} \\ & & \cdots & \\ \omega_{-k} & \omega_{-k+1} & \cdots & \omega_0 \end{bmatrix}.$$

Assume

$$\det G(1, t, \ldots, t^k) \neq 0, \quad \text{for} \quad k = 0, 1, \ldots, n,$$

and let $G(1, t, \ldots, t^k)^{-1} = (\omega_{pq}^{(k)})_{p,q=0}^k$. The system of polynomials

$$\begin{aligned} s_0(t) &= 1, \\ s_k(t) &= \omega_{00}^{(k)} + \omega_{10}^{(k)}t + \cdots + \omega_{k0}^{(k)}t^k, \quad (k = 1, 2, \ldots, n) \end{aligned} \qquad (3.1.7)$$

form a regular orthogonalization of $1, t, \ldots, t^n$ in $L_2^{(n)}$ in the indefinite inner product $[\, , \,]_\omega$. $\square$

The last statement follows directly from Theorem 3.1.2.

Example 3.1.6. Let H_0 be a hermitian matrix defined by the equality

$$H_n = \begin{bmatrix} 1 & a_1 & a_2 & \cdots & a_{n-1} \\ \overline{a_1} & 1 & 0 & \cdots & 0 \\ & & & \cdots & \\ \overline{a_{n-1}} & 0 & 0 & \cdots & 1 \end{bmatrix}$$

and

$$[x, y]_{H_n} = (H_n x, y), \quad x, y \in \mathbb{C}^n.$$

Let $e_1, e_2, \ldots, e_n$ be the standard basis in $\mathbb{C}^n$. Assume that

$$1 - \sum_{j=1}^{k-1} |a_j|^2 > 0 \quad \text{and} \quad 1 - \sum_{j=1}^k |a_j|^2 < 0$$

for some k $(2 \leq k \leq n - 1)$. It is easy to see that

$$\det H_r = 1 - \sum_{j=1}^{r-1} |a_j|^2.$$

Hence it is obvious that the matrix H_r is indefinite for $r = k, k+1, \ldots, n$ and positive definite for $r = 1, 2, \ldots, k$. Moreover, all eigenvalues of H_n except one are positive and one is negative. (Cf. Theorems A.1.3 and A.1.4.) From the equality

$$H_r = \begin{bmatrix} 1 & b \\ b^* & I \end{bmatrix},$$

where $b = [a_1 \, a_2 \, \ldots \, a_{r-1}]$ and $b^* = \begin{bmatrix} \overline{a_1} \\ \overline{a_2} \\ \vdots \\ \overline{a_{r-1}} \end{bmatrix}$ there follows the factorization

$$H_r = \begin{bmatrix} 1 & 0 \\ b^* & I \end{bmatrix} \begin{bmatrix} 1 & 0 \\ 0 & I - b^*b \end{bmatrix} \begin{bmatrix} 1 & b \\ 0 & I \end{bmatrix},$$

and hence

$$H_r^{-1} = \begin{bmatrix} 1 & -b \\ 0 & I \end{bmatrix} \begin{bmatrix} 1 & 0 \\ 0 & B \end{bmatrix} \begin{bmatrix} 1 & 0 \\ -b^* & I \end{bmatrix} = \begin{bmatrix} 1 + bBb^* & -bB \\ -Bb^* & B \end{bmatrix},$$

where

$$B = (I - b^*b)^{-1}.$$

To compute B consider the equation

$$y - b^*by = x.$$

Let $z = by$, then $y = b^*z + x$, hence

$$z = \frac{b}{1 - bb^*} x$$

and

$$Bx = \left(\frac{b^*b}{1 - bb^*} + I \right) x.$$

So the last column in H_r^{-1} has the form

$$\begin{bmatrix} h_1^{(r)} \\ h_2^{(r)} \\ \vdots \\ h_r^{(r)} \end{bmatrix},$$

where

$$h_1^{(r)} = \frac{-a_{r-1}}{1 - bb^*}, \quad h_j^{(r)} = (1 - bb^*)^{-1} \overline{a_j} a_{r-1} + \delta_{jr} \quad \text{for} \quad j = 2, 3, \ldots, r.$$

The vectors

$$
g_r = \begin{bmatrix} \overline{h_1^{(r)}} \\ \overline{h_2^{(r)}} \\ \vdots \\ \overline{h_r^{(r)}} \\ 0 \\ \vdots \\ 0 \end{bmatrix} \in \mathbb{C}^n, \quad r = 1, 2, \ldots, n
$$

form a regular orthogonalization of the system $e_1, e_2, \ldots, e_n$. $\qquad\square$

3.2 The Theorems of Szegő and Krein

Let the weight function

$$
\omega(t) := \sum_{j=-n}^{n} \omega_j t^j, \quad |t| = 1
$$

be nonnegative (but not identically zero) on the unit circle, and let $s_k(t)$ ($k = 0, 1, \ldots, n$) be the orthogonalization of $1, t, \ldots, t^n$ in the definite inner product $[\cdot, \cdot]_\omega$. The next theorem is due to Szegő [96].

Theorem 3.2.1. *The zeros of the polynomials $s_k(t)$, $k = 1, 2, \ldots, n$ lie inside the unit circle.*

Proof. Let t_0 be a zero of $s_k(t)$, then $s_k(t) = (t - t_0)r(t)$, for some polynomial $r(t)$ of degree $k - 1$, and so

$$
s_k(t) + t_0 r(t) = t r(t).
$$

Since $[r, s_k]_\omega = 0$, we have

$$
\begin{aligned}
[r, r]_\omega &= [t r(t), t r(t)]_\omega = [s_k(t) + t_0 r(t), s_k(t) + t_0 r(t)]_\omega \\
&= [s_k(t), s_k(t)]_\omega + |t_0|^2 [r(t), r(t)]_\omega,
\end{aligned}
$$

and hence

$$
(1 - |t|_0^2)[r(t), r(t)]_\omega = [s_k(t), s_k(t)]_\omega.
$$

Since $[s_k(t), s_k(t)]_\omega$ and $[r(t), z(t)]_\omega$ are positive, it follows that $|t_0| < 1$. $\quad\square$

This theorem admits a generalization for the case in which the function $\omega(t)$ has changes of sign on the unit circle. In this case the corresponding inner product is indefinite. However, it will be assumed in this case that the matrices

$$
\Omega_k := \begin{bmatrix} \omega_0 & \omega_1 & \cdots & \omega_k \\ \omega_{-1} & \omega_0 & \cdots & \omega_{k-1} \\ & & \cdots & \\ \omega_{-k} & \omega_{-k+1} & \cdots & \omega_0 \end{bmatrix} \tag{3.2.8}
$$

are nonsingular, i.e.,

$$d_k := \det \Omega_k \neq 0 \quad (k = 0, 1, \ldots, n). \tag{3.2.9}$$

We now state Krein's theorem [60]:

Theorem 3.2.2. *Let $s_k(t)$, $k = 0, 1, \ldots, n$, be a regular orthogonalization of $1, t, \ldots,$ t^n in the inner product $[\, , \,]_\omega$, with $\omega(t) = \sum_{j=-n}^{n} \omega_j t^j$, and assume that (3.2.9) holds. For any $k \geq 1$ let β_k and $\gamma_k = k - \beta_k$ denote, respectively, the number of constancies of sign, and the number of changes of sign, in the sequence*

$$1, d_0, d_1, \ldots, d_{k-1}. \tag{3.2.10}$$

Then for $s_k(t)$,

$$n_+(s_k) = \beta_k \quad and \quad n_-(s_k) = \gamma_k, \quad if \quad d_k d_{k-1} > 0$$

and

$$n_+(s_k) = \gamma_k \quad and \quad n_-(s_k) = \beta_k, \quad if \quad d_k d_{k-1} < 0$$

where $n_+(s_k)$ $(n_-(s_k))$ is the number of zeros of s_k inside the open unit disc (outside the closed unit disc). In particular, the polynomials s_k have no zeros on the unit circle.

If all d_k's are positive then the matrix $(\omega_{j-p})_{j,p=0}^{n}$ is positive definite by Theorem A.1.3, and hence Szegő's theorem follows from Krein's theorem.

The proof of Krein's theorem will be presented in the next section. Note that from the Law of Inertia (see Theorem A.1.1) it follows that

$$i_+(\Omega_n) = \beta_n \quad for \quad d_n d_{n-1} > 0$$

and

$$i_-(\Omega_n) = \gamma_n \quad for \quad d_n d_{n-1} > 0,$$

where $i_+(\Omega_n)$ and $i_-(\Omega_n)$ are the numbers of positive and negative eigenvalues of Ω_n, respectively (counted with multiplicities).

Now Krein's theorem can be restated in a slightly different notation:

Theorem 3.2.3. *Let*

$$\Omega_k = \begin{bmatrix} t_0 & t_{-1} & \cdots & t_{-k} \\ t_1 & t_0 & \cdots & t_{-k+1} \\ & & \cdots & \\ t_k & t_{k-1} & \cdots & t_0 \end{bmatrix}, \quad k = 0, 1, \ldots, n$$

be hermitian matrices. Assume that

$$d_k := \det \Omega_k \neq 0, \quad k = 0, 1, \ldots, n,$$

and let

$$
\begin{bmatrix} \omega_0 \\ \omega_1 \\ \vdots \\ \omega_k \end{bmatrix} = \Omega_k^{-1} \begin{bmatrix} 0 \\ 0 \\ \vdots \\ 1 \end{bmatrix}.
$$

Then the polynomial

$$
s_k(t) = \omega_0 + \omega_1 t + \cdots + \omega_k t^k
$$

has no zeros on the unit circle and, furthermore,

$$
n_\pm(s_k) = i_\pm(\Omega_k) \quad for \quad d_k d_{k-1} \;>\; 0,
$$
$$
n_\pm(s_k) = i_\mp(\Omega_k) \quad for \quad d_k d_{k-1} \;<\; 0.
$$

3.3 One-Step Theorem

In this section we prove the following "one-step theorem" due to Ellis and Gohberg [18]. A matrix $A = [a_{jk}] \in \mathbb{C}^{n \times n}$ is said to be *Toeplitz* if $a_{jk} = a_{rs}$ for all ordered pairs of indices such that $j - k = r - s$. In other words, each diagonal of A which is parallel to the main diagonal has all entries equal.

Theorem 3.3.1. *Let $\Omega_n = (t_{j-k})_{j,k=0}^n$ be a hermitian Toeplitz matrix for which*

$$
\det \Omega_m \neq 0, \quad for \quad m = n - 2, n - 1, n.
$$

Let

$$
s_n(t) = \omega_0 + \omega_1 t + \cdots + \omega_n t^n
$$

and

$$
s_{n-1}(t) = u_0 + u_1 t + \cdots + u_{n-1} t^{n-1},
$$

where

$$
\begin{bmatrix} \omega_0 \\ \omega_1 \\ \vdots \\ \omega_n \end{bmatrix} = \Omega_n^{-1} \begin{bmatrix} 0 \\ 0 \\ \vdots \\ 0 \\ 1 \end{bmatrix} \quad and \quad \begin{bmatrix} u_0 \\ u_1 \\ \vdots \\ u_{n-1} \end{bmatrix} = \Omega_{n-1}^{-1} \begin{bmatrix} 0 \\ 0 \\ \vdots \\ 0 \\ 1 \end{bmatrix}.
$$

Then $\omega_n \neq 0$, $u_{n-1} \neq 0$. If we define

$$
g_0 = -\frac{1}{u_{n-1}} \sum_{k=0}^{n-2} t_{k+1} \overline{u_k},
$$

then $|t_n - g_0| \neq \left| u_{n-1}^{-1} \right|$ and:

a) *if $|t_n - g_0| < \left| u_{n-1}^{-1} \right|$ or equivalently $\det \Omega_n \det \Omega_{n-2} > 0$ then $n_-(s_n) = n_-(s_{n-1})$;*

b) *if* $|t_n - g_0| > |u_{n-1}^{-1}|$ *or equivalently* $\det \Omega_n \det \Omega_{n-2} < 0$ *then* $n_-(s_n) = n - n_-(s_{n-1})$.

First we will prove a preliminary result:

Theorem 3.3.2. *Let* $\Omega_n = (t_{j-k})_{j,k=0}^n$ *be a hermitian Toeplitz matrix and*

$$\det \Omega_n \neq 0, \quad \det \Omega_{n-1} \neq 0.$$

Then the polynomial

$$s_n(t) = \omega_0 + \omega_1 t + \cdots + \omega_n t^n$$

with

$$
\begin{bmatrix} \omega_0 \\ \omega_1 \\ \vdots \\ \omega_n \end{bmatrix} = \Omega_n^{-1} \begin{bmatrix} 0 \\ 0 \\ \vdots \\ 0 \\ 1 \end{bmatrix}
$$

has no zeros on the unit circle.

Proof. By Cramer's rule, $\omega_n = (\det \Omega_{n-1})/(\det \Omega_n)$ and hence $\omega_n \neq 0$. Let S_{n+1} be the $(n+1) \times (n+1)$ sip matrix. Then $S_{n+1}^2 = I$ and $S_{n+1} \Omega_n S_{n+1} = \overline{\Omega_n}$, where $\overline{\Omega_n} = (\overline{\omega}_{j-k})_{j,k=0}^n$. From these equalities it follows that

$$
\Omega_n \begin{bmatrix} \overline{\omega}_n \\ \overline{\omega}_{n-1} \\ \vdots \\ \overline{\omega}_0 \end{bmatrix} = \begin{bmatrix} 1 \\ 0 \\ \vdots \\ 0 \end{bmatrix}
$$

or

$$
\Omega_n \begin{bmatrix} 1 \\ \nu_1 \\ \vdots \\ \nu_n \end{bmatrix} = \begin{bmatrix} \rho \\ 0 \\ \vdots \\ 0 \end{bmatrix}, \tag{3.3.11}
$$

where $\nu_1 = \omega_n^{-1} \overline{\omega}_{n-1}, \ldots, \nu_n = \omega_n^{-1} \overline{\omega}_0$ and $\rho = \omega_n^{-1}$.

Let $p_n(t)$ be the polynomial

$$p_n(t) = 1 + \nu_1 t + \cdots + \nu_n t^n.$$

We will prove first that $p_n(t) \neq 0$ for $|t| = 1$.

Assume that $t_0 \in \mathbb{C}$ is a zero of $p_n(t)$ and hence

$$p_n(t) = (t - t_0)(g_1 + g_2 t + \cdots + g_n t^{n-1}).$$

Equating coefficients of like powers of t it is found that

$$b + t_0 c = d, \tag{3.3.12}$$

where

$$b = \begin{bmatrix} 1 \\ \nu_1 \\ \vdots \\ \nu_n \end{bmatrix}, \quad c = \begin{bmatrix} g_1 \\ g_2 \\ \vdots \\ g_n \\ 0 \end{bmatrix}, \quad d = \begin{bmatrix} 0 \\ g_1 \\ \vdots \\ g_n \end{bmatrix}.$$

In particular, $t_0 g_1 = -1$ and $\nu_n = g_n$.

For the Toeplitz matrix Ω_n we have

$$c^* \Omega_n c = [\overline{g_1}, \dots, \overline{g_n}, 0]\, \Omega_n \begin{bmatrix} \overline{g_1} \\ \vdots \\ \overline{g_n} \\ 0 \end{bmatrix} = [0, \overline{g_1}, \dots, \overline{g_n}]\Omega_n \begin{bmatrix} 0 \\ \overline{g_1} \\ \vdots \\ \overline{g_n} \end{bmatrix} = d^* \Omega_n d.$$

It follows from (3.3.11) that

$$b^* \Omega_n b = [1, \overline{\nu_1}, \dots, \overline{\nu_n}] \begin{bmatrix} \rho \\ 0 \\ \vdots \\ 0 \end{bmatrix} = \rho$$

and

$$\overline{t_0} c^* \Omega_n b = \overline{t_0}[\overline{g_1}, \dots, \overline{g_n}, 0] \begin{bmatrix} \rho \\ 0 \\ \vdots \\ 0 \end{bmatrix} = \overline{t_0}\overline{g_1}\rho = -\rho.$$

Using these equalities together with

$$(b^* + \overline{t_0}c^*)\Omega_n(b + t_0 c) = d^* \Omega_n d,$$

we obtain

$$\rho - \rho - \rho + |t_0|^2 c^* \Omega_n c = c^* \Omega_n c$$

which implies that $\rho = \left(|t_0|^2 - 1\right) c^* \Omega_n c$. Since $\rho \neq 0$ it follows that $|t_0| \neq 1$.

Finally, since

$$p_n(t) = 1 + \nu_1 t + \cdots + \nu_n t^n = 1 + w_n^{-1}\overline{w_{n-1}}t + \cdots + w_n^{-1}\overline{w_0}t^n,$$

we obtain

$$p_n(t) = w_n^{-1} t^n (\overline{w_n}t^{-n} + \overline{w_{n-1}}t^{-n+1} + \cdots + \overline{w_0}),$$

and, for $|t| = 1$, $p_n(t) = w_n^{-1}t^n\overline{s_n(t)}$. Hence $s_n(t) \neq 0$ for $|t| = 1$, and the theorem is proved. □

Now we prove the "one-step" version of Krein's theorem.

Proof of Theorem 3.3.1. As in the proof of Theorem 3.3.2, denote

$$\rho = \omega_n^{-1}, \quad \nu_1 = \omega_n^{-1}\overline{\omega_{n-1}}, \ldots, \nu_n = \omega_n^{-1}\overline{\omega_0}$$

and, similarly,

$$\mu = u_{n-1}^{-1}, \; x_1 = u_{n-1}^{-1}\overline{u_{n-2}}, \ldots, x_{n-1} = u_{n-1}^{-1}\overline{u_0}.$$

It is clear that

$$\mu = \frac{\det \Omega_{n-1}}{\det \Omega_{n-2}} \quad \text{and} \quad \rho = \frac{\det \Omega_n}{\det \Omega_{n-1}}.$$

The numbers μ and ρ are real and $\mu, \rho \neq 0$. As in the proof of Theorem 3.3.2, we build the polynomials

$$p_n(t) = 1 + \nu_1 t + \cdots + \nu_n t^n \quad \text{and} \quad p_{n-1}(t) = 1 + x_1 t + \cdots + x_{n-1} t^{n-1}.$$

By Theorem 3.3.2, the polynomials have no zeros on the unit circle, and the following relations hold:

$$\Omega_{n-1}\begin{bmatrix} 1 \\ x_1 \\ \vdots \\ x_{n-1} \end{bmatrix} = \begin{bmatrix} \mu \\ 0 \\ \vdots \\ 0 \end{bmatrix} \quad \text{and} \quad \Omega_n\begin{bmatrix} 1 \\ \nu_1 \\ \vdots \\ \nu_n \end{bmatrix} = \begin{bmatrix} \rho \\ 0 \\ \vdots \\ 0 \end{bmatrix}.$$

For any complex number c we now have

$$\Omega_n\left\{\begin{bmatrix} 1 \\ x_1 \\ \vdots \\ x_{n-1} \\ 0 \end{bmatrix} + c\begin{bmatrix} 0 \\ \overline{x_{n-1}} \\ \vdots \\ \overline{x_1} \\ 1 \end{bmatrix}\right\} = \begin{bmatrix} \mu \\ 0 \\ \vdots \\ 0 \\ a \end{bmatrix} + c\begin{bmatrix} \overline{a} \\ 0 \\ \vdots \\ 0 \\ \mu \end{bmatrix} = \begin{bmatrix} \mu + c\overline{a} \\ 0 \\ \vdots \\ 0 \\ a + c\mu \end{bmatrix},$$

where

$$a = t_n - g_0, \quad \text{with} \quad g_0 := -\frac{1}{u_{n-1}}\left(\sum_{j=1}^{n-1} t_j \overline{u_{j-1}}\right).$$

Choosing $c = -a\mu^{-1}$, we obtain

$$\Omega_n\begin{bmatrix} 1 \\ x_1 + c\overline{x_{n-1}} \\ \vdots \\ x_{n-1} + c\overline{x_1} \\ c \end{bmatrix} = \begin{bmatrix} \mu - |a|^2/\mu \\ 0 \\ \vdots \\ 0 \\ 0 \end{bmatrix},$$

and hence $\rho = \mu - |a|^2/\mu$ and

$$\begin{bmatrix} 1 \\ \nu_1 \\ \vdots \\ \nu_{n-1} \\ \nu_n \end{bmatrix} = \begin{bmatrix} 1 \\ x_1 \\ \vdots \\ x_{n-1} \\ 0 \end{bmatrix} + c \begin{bmatrix} 0 \\ \overline{x_{n-1}} \\ \vdots \\ \overline{x_1} \\ 1 \end{bmatrix}.$$

It follows from this that

$$p_n(t) = p_{n-1}(t) + ct\widetilde{p}_{n-1}(t),$$

where

$$\widetilde{p}_{n-1}(t) = t^{n-1}\left(1 + \overline{x_1}\frac{1}{t} + \cdots + \overline{x_{n-1}}\frac{1}{t^{n-1}}\right).$$

Since $\rho \neq 0$ we have $|a| \neq |\mu|$ and $|t_n - g_0| \neq |\mu|$. Also $|c| \neq 1$.

It is clear that

$$\left|ct\widetilde{p}_{n-1}(t)\right| = \left|ct^n\overline{p_{n-1}(t)}\right| = |c|\,|p_{n-1}(t)|, \quad |t| = 1.$$

If $|c| < 1$, then $|p_{n-1}(t)| > |ct\widetilde{p}_{n-1}(t)|$ for $|t| = 1$, by the theorem of Rouché (a classical theorem of complex analysis, see [82, Theorem 6.2.5], for example). It follows that

$$n_+(p_n) = n_+(p_{n-1}).$$

If $|c| > 1$ we deduce in a similar way that

$$n_+(p_n) = n_+(t\widetilde{p}_{n-1}) = 1 + n_+(\widetilde{p}_{n-1}).$$

Note that $|c| < 1$ if and only if $|a| < |\mu|$, which in turn happens if and only if $\rho\mu > 0$, or equivalently, if and only if $\det \Omega_n \det \Omega_{n-2} > 0$. To complete the proof it remains to remark that, from the equalities

$$p_n(t) = \rho t^n \overline{s_n(1/\overline{t})}, \quad p_{n-1}(t) = \text{const } t^{n-1}\overline{s_{n-1}(1/\overline{t})}, \quad t \neq 0,$$

it follows that

$$n_+(p_n) = n_-(s_n), \quad n_+(p_{n-1}) = n_-(s_{n-1}).$$

Also,

$$p_{n-1}(t) = t^{n-1}\overline{\widetilde{p}_{n-1}(1/\overline{t})},$$

and therefore in case $|c| > 1$ we have

$$n_-(s_n) = n_+(p_n) = 1 + n_+(\widetilde{p}_{n-1}) = 1 + n_-(p_{n-1}) = 1 + (n - 1 - n_+(p_{n-1}))$$
$$= n - n_-(s_{n-1}). \qquad \qquad \square$$

Proof of Theorem 3.2.3. By Theorem 3.3.2, the polynomial p_n has no zeros on the unit circle. We will use induction on n. For $n = 1$ it is easy to see that $\nu_1 = -t_0^{-1}t$, and

$$\rho = t_0 - t_{-1}t_0^{-1}t_1 = t_0(1 - |t_0^{-1}t_1|^2),$$

where ν_1 and ρ are as in the proof of Theorem 3.3.1. Therefore

$$s_1(t) = 1 - t_0^{-1}t_1 t$$

so s_1 has a zero if and only if $t_1 \neq 0$, and in that case the zero of s_1 is $t_0 t_1^{-1}$. Thus $n_+(s_1) = 1$ if and only if $|t_1 t_0^{-1}| > 1$, if and only if $\rho(t_0)^{-1} > 0$. Since t_0 is the (only) eigenvalue of Ω_0 the last condition becomes $i_-(\Omega_0) = 1$ if $\rho > 0$, and it becomes $i_+(\Omega_0) = 1$ if $\rho < 0$. Hence the theorem holds for $n = 1$.

Now take $n > 1$ and assume that Theorem 3.2.3 is true for $n - 1$. First consider the case

$$\det \Omega_n \det \Omega_{n-1} > 0.$$

By the induction hypothesis, $n_+(s_{n-1}) = i_+(\Omega_{n-2})$. Let

$$\xi = \langle t_1, t_2, \ldots, t_{n-1} \rangle \quad \text{and} \quad x = \langle x_1, x_2, \ldots, x_{n-1} \rangle.$$

Then

$$\Omega_{n-1} = \begin{bmatrix} t_0 & \xi^* \\ \xi & \Omega_{n-2} \end{bmatrix}.$$

Since

$$\begin{bmatrix} t_0 & \xi^* \\ \xi & \Omega_{n-2} \end{bmatrix} \begin{bmatrix} 1 \\ x \end{bmatrix} = \begin{bmatrix} \mu \\ 0 \end{bmatrix}$$

it follows that $t_0 = \xi^*\Omega_{n-2}^{-1}\xi + \mu$.

Consequently, the matrix Ω_{n-1} can be factorized in the form

$$\Omega_{n-1} = \begin{bmatrix} t_0 & \xi^* \\ \xi & \Omega_{n-2} \end{bmatrix} = \begin{bmatrix} 1 & \xi^*\Omega_{n-2}^{-1} \\ 0 & 1 \end{bmatrix} \begin{bmatrix} \mu & 0 \\ 0 & \Omega_{n-2} \end{bmatrix} \begin{bmatrix} 1 & 0 \\ \Omega_{n-2}^{-1}\xi & I \end{bmatrix}. \quad (3.3.13)$$

If $\mu > 0$ then $i_-(\Omega_{n-1}) = i_-(\Omega_{n-2})$ and since $n_+(s_n) = n_+(s_{n-1})$ it follows that

$$n_+(s_n) = i_-(\Omega_{n-1}).$$

Now, consider first the case when

$$\det \Omega_n \det \Omega_{n-2} > 0.$$

Then $\det \Omega_n$, $\det \Omega_{n-1}$, $\det \Omega_{n-2}$ have the same sign (positive or negative). It is clear from (3.3.13) that in this case $\mu > 0$, and according to Theorem 3.3.1

$$n_+(s_n) = n_+(s_{n-1}) + 1.$$

Taking into the consideration that $i_-(\Omega_{n-1}) = i_-(\Omega_{n-2})$ we obtain

$$n_+(s_n) = n_+(s_{n-1}) + 1 = i_+(\Omega_{n-1}) + 1 = i_+(\Omega_n).$$

Consider the second case when

$$\det \Omega_n \, \det \Omega_{n-2} < 0.$$

Then $\det \Omega_n$ and $\det \Omega_{n-1}$ have the same sign, and $\det \Omega_{n-2}$ has the opposite sign. By Theorem 3.3.1 $n_-(s_n) = n - n_-(s_{n-1})$, hence

$$n_+(s_n) = n - n_+(s_{n-1}) - 1.$$

The inductive hypothesis implies $n_+(s_{n-1}) = i_-(\Omega_{n-1})$. Since in this case $\mu < 0$, we obtain that $i_+(\Omega_{n-2}) = i_+(\Omega_{n-1})$ and hence

$$n_+(s_n) = n - (i_-(\Omega_{n-1}) + 1) = n - 1 - i_-(\Omega_{n-1}) = i_+(\Omega_{n-1}).$$

This completes the proof for the case when $\det \Omega_n \, \det \Omega_{n-1} > 0$.

Let us now pass to the case when $\det \Omega_n \, \det \Omega_{n-1} < 0$. We assume first that $\det \Omega_n \, \det \Omega_{n-2} > 0$. Then by Theorem 3.3.1,

$$n_+(s_n) = n_+(s_{n-1}) + 1.$$

The inductive hypothesis in this case implies $n_+(s_{n-1}) = i_-(\Omega_{n-2})$, so $n_+(s_n) = i_-(\Omega_{n-2}) + 1$. Since $\mu < 0$, according to (3.3.13) we have

$$i_-(\Omega_{n-1}) = i_-(\Omega_{n-2}) + 1,$$

hence $n_+(s_n) = i_-(\Omega_{n-1})$, as it is to be proved.

Assume now

$$\det \Omega_n \, \det \Omega_{n-2} < 0.$$

By Theorem 3.3.1, $n_-(s_n) = n - n_-(s_{n-1})$. The inductive hypothesis implies $n_+(s_{n-1}) = i_-(\Omega_{n-2})$, so

$$n_+(s_n) = n_+(s_{n-1}) = i_-(\Omega_{n-2}).$$

From (3.3.13) it follows that $\mu > 0$ and $i_-(\Omega_{n-1}) = i_-(\Omega_{n-2})$; hence

$$n_+(s_n) = i_-(\Omega_{n-1}).$$

This completes the induction and the proof of the theorem. $\qquad\square$

3.4   Determinants of One-Step Completions

The two conditions

$$|t_n - g_0| < |u_{n-1}^{-1}| \quad \text{and} \quad |t_n - g_0| > |u_{n-1}^{-1}|$$

used in Theorem 3.3.1 have a geometric interpretation. Define an $(n+1) \times (n+1)$
Toeplitz matrix $\Omega(\omega)$ by setting $t_n = \omega$, $t_{-n} = \overline{\omega}$, and all other entries as in Ω_n,
It turns out that, for all ω on the circle

$$|\omega - g_0| = |u_{n-1}^{-1}|$$

(which appears in Theorem 3.3.1), the matrix $\Omega(\omega)$ is singular.

In contrast, a matrix $\Omega(\omega)$ for which

$$|\omega - g_0| < |u_{n-1}^{-1}|$$

is said to be *internal.* If $|\omega - g_0| > |u_{n-1}^{-1}|$, the matrix $\Omega(\omega)$ is said to be *external.*

The properties of extensions of this kind for general hermitian matrices (not
necessarily Toeplitz) are studied in this section.

We will start with some more notations. For a matrix $M = (m_{jk})_{j,k=1}^n$,
$M(p, \ldots, q)$ will denote the principal submatrix $(m_{jk})_{j,k=p}^q$. It will be assumed
that all entries of the hermitian matrix

$$F = (f_{jk})_{j,k=1}^n \tag{3.4.14}$$

are given except for the entries f_{1n} and $f_{n1} = \overline{f_{1n}}$. Then we study the determinant
of $F(w)$, where $F(w)$ has the same entries as F except that f_{1n} and f_{n1} are replaced
by $w \in \mathsf{C}$ and $\overline{w}$, respectively.

The following theorem is due to Ellis, Gohberg, and Lay [19], [20].

Theorem 3.4.1. *Let F be the given hermitian matrix (3.4.14) and $w \in \mathsf{C}$. Assume
that*

$$d_1 := \det F(1, \ldots, n-1) \neq 0, \;\; d_2 := F(2, \ldots, n) \neq 0, \;\; d_3 := \det F(2, \ldots, n-1) \neq 0.$$

Then

$$\det F(w) = \frac{d_1 d_2}{d_3} \left(1 - \frac{|w - w_0|^2 d_3^2}{d_1 d_2} \right),$$

where

$$w_0 = \frac{1}{w_{1n}} \sum_{j=2}^{n-1} f_{nj} w_{jn} \tag{3.4.15}$$

and

$$[F(1, \ldots, n-1)]^{-1} = (w_{jk})_{j,k=1}^{n-1}.$$

Proof. Put $\rho_{n-1} = d_1/d_3$, $\mu_{n-1} = d_2/d_3$, and

$$
\begin{bmatrix} 1 \\ \nu_2 \\ \vdots \\ \nu_{n-1} \end{bmatrix} = d_1^{-1} \begin{bmatrix} \rho_{n-1} \\ 0 \\ \vdots \\ 0 \end{bmatrix}, \qquad \begin{bmatrix} u_2 \\ \vdots \\ u_{n-1} \\ 1 \end{bmatrix} = d_2^{-1} \begin{bmatrix} 0 \\ 0 \\ \vdots \\ \mu_{n-1} \end{bmatrix}.
$$

Introduce

$$
\Delta_n = w + \sum_{j=2}^{n-1} f_{nj}\nu_j, \qquad \delta_n = \overline{w} + \sum_{j=2}^{n-1} f_{1j}u_j
$$

and

$$
\varphi_n = -\frac{\Delta_n}{\mu_{n-1}}, \qquad \psi_n = \frac{\delta_n}{\rho_{n-1}}.
$$

It is easy to see that, if we define $\rho_n := \rho_{n-1} + \varphi_n \delta_n$ and $\mu_n := \mu_{n-1} + \psi_n \Delta_n$, then

$$
F(w) \left\{ \begin{bmatrix} 1 \\ \nu_2 \\ \vdots \\ \nu_{n-1} \\ 0 \end{bmatrix} + \varphi_n \begin{bmatrix} 0 \\ u_2 \\ \vdots \\ u_{n-1} \\ 1 \end{bmatrix} \right\} = \begin{bmatrix} \rho_n \\ 0 \\ \vdots \\ 0 \\ 0 \end{bmatrix}
$$

and

$$
F(w) \left\{ \psi_n \begin{bmatrix} 1 \\ \nu_2 \\ \vdots \\ \nu_{n-1} \\ 0 \end{bmatrix} + \begin{bmatrix} 0 \\ u_2 \\ \vdots \\ u_{n-1} \\ 1 \end{bmatrix} \right\} = \begin{bmatrix} 0 \\ 0 \\ \vdots \\ 0 \\ \mu_n \end{bmatrix}.
$$

It follows from these equations that

$$
\rho_n = \frac{\det F(w)}{d_2} \quad \text{and} \quad \mu_n = \frac{\det F(w)}{d_1}.
$$

Hence $\rho_n = \rho_{n-1}(1 - \varphi_n \psi_n)$ and $\mu_n = \mu_{n-1}(1 - \varphi_n \psi_n)$.

If we define (this is the same w_0 as in formula (3.4.15))

$$
w_0 = -\sum_{j=2}^{n-1} f_{nj}\nu_j = -\overline{\sum_{j=2}^{n-1} f_{1j}\mu_j},
$$

then $\delta_n = w - w_0$, $\Delta_n = \overline{w} - \overline{w_0}$ and

$$
|w - w_0|^2 = \varphi_n \psi_n \rho_{n-1} \mu_{n-1}. \tag{3.4.16}
$$

Also

$$\varphi_n \psi_n = \frac{|w - w_0|^2 d_3^2}{d_1 d_2}.$$

Since

$$\det F(w) = \rho_n d_2 = \rho_{n-1}(1 - \varphi_n \psi_n)d_2,$$

we obtain

$$\det F(w) = \frac{d_1 d_2}{d_3}\left(1 - \frac{|w - w_0|^2 d_3^2}{d_1 d_2}\right), \tag{3.4.17}$$

as required.   □

Corollary 3.4.2.

a) *If*

$$\det F(1, \ldots, n-1)\det F(2, \ldots, n-1) < 0,$$

   then for any $w \in \mathbb{C}$, $\det F(w) \neq 0$ *and*

$$\det F(w)\det F(2, \ldots, n-1) < 0.$$

b) *If*

$$\det F(1, \ldots, n-1)\det F(2, \ldots, n-1) > 0,$$

   then $\det F(w) \neq 0$ *for all* $w \in \mathbb{C}$ *except for* w *on the circle* $|w - w_0| = \rho_0$, *where*

$$\rho_0 = \frac{(\det F(1, \ldots n-1)\det F(2, \ldots, n))^{1/2}}{\det F(2, \ldots, n-1)}$$

   and

$$w_0 = -\sum_{j=2}^{n-1} f_{nj}\nu_j, \tag{3.4.18}$$

$$\begin{bmatrix} 1 \\ \nu_2 \\ \vdots \\ \nu_{n-1} \end{bmatrix} = F(1, \ldots, n-1)^{-1}\begin{bmatrix} \rho_{n-1} \\ 0 \\ \vdots \\ 0 \end{bmatrix}. \tag{3.4.19}$$

c) *Moreover, for internal completions* w *(when* $|w - w_0| < \rho_0$*) we have*

$$\det F(w)\det F(2, \ldots, n-1) > 0$$

   and for external completions w *(when* $|w - w_0| > \rho_0$*),*

$$\det F(w)\det F(2, \ldots, n-1) < 0.$$

In order to explain Corollary 3.4.2 we will use notation and formulas from the proof of Theorem 3.4.1. From the formulas

$$\det F(w) = \rho_n d_2 = \mu_n d_1$$

it follows that $\det F(w) \neq 0$ if and only if both of the numbers ρ_n, μ_n are different from zero. Since $\rho_n = \rho_{n-1}(1 - \varphi_n \psi_n)$ and $\mu_n = \mu_{n-1}(1 - \varphi_n \psi_n)$, it follows that

$$\det F(w) \neq 0 \quad \text{if and only if} \quad \varphi_n \psi_n \neq 1.$$

Now use the formula (3.4.16). Therefore the condition $\varphi_n \psi_n \neq 1$ becomes

$$|w - w_0|^2 \neq \mu_{n-1}\rho_{n-1} = \frac{d_1 d_2}{d_3^2}.$$

The statements of the corollary now follow from the final formula (3.4.17).

Example 3.4.3. *Consider the hermitian matrix*

$$F(w) = \begin{bmatrix} 1 & 0 & \cdots & w \\ 0 & 1 & \cdots & a_1 \\ \vdots & & \cdots & \vdots \\ \overline{w} & \overline{a_1} & \cdots & 1 \end{bmatrix}.$$

In this case

$$\det F(1, \ldots, n-1) = 1, \quad \det F(2, \ldots, n) = 1 - \sum_{j=1}^{n-1} |a_j|^2$$

and $\det F(2, \ldots, n-1) = 1$, and it follows easily that

$$w_0 = 0, \quad \rho_0 = \left(1 - \left(\sum_{j=1}^{n-1} |a_j|^2\right)\right)^{1/2} \quad \text{if} \quad 1 - \sum_{j=1}^{n-1} |a_j|^2 > 0$$

and $\det F(w) < 0$ if $1 - \sum_{j=1}^{n-1} |a_j|^2 < 0$.
It is easy to check directly that

$$\det F(w) = 1 - \left(|w|^2 + |a_1|^2 + \cdots + |a_n|^2\right).$$

Hence, if $1 - |a_1|^2 - \cdots - |a_n|^2 > 0$ then $\det F(w) \neq 0$ if and only if

$$|w| \neq \left(|a_1^2| + |a_2|^2 + \cdots + |a_n|^2\right)^{1/2}.$$

Here

$$w_0 = 0 \quad \text{and} \quad \rho_0 = \left(|a_1|^2 + \cdots + |a_n|^2\right)^{1/2}.$$

If $1 - |a_1|^2 - \cdots + |a_n|^2 < 0$ then $\det F(w) < 0$ for any $w \in \mathbb{C}$. $\qquad\square$

If $\Omega = (\omega_{j-k})_{j,k=0}^n$ is a Toeplitz matrix then $\Omega(1,\dots,n-1) = \Omega(2,\dots,n)$ and some formulas from this section simplify. For instance,

$$\rho_0 = \frac{|\det \Omega(1,\dots,n-1)|}{|\det \Omega(1,\dots,n-2)|}.$$

Note also that an alternative formula for w_0 arises. Namely,

$$w_0 = [f_{n2},\dots,f_{n(n-1)}]F(2,\dots,n-1)\begin{bmatrix} f_{21} \\ \vdots \\ f_{(n-1)1} \end{bmatrix}.$$

3.5 Exercises

1. Let

$$H_1 = \begin{bmatrix} 1 & a_2 & a_3 & \cdots & a_n \\ \overline{a_2} & 1 & 0 & \cdots & 0 \\ \vdots & & & \cdots & \vdots \\ \overline{a_n} & 0 & 0 & \cdots & 1 \end{bmatrix}, \quad a_j \in \mathbb{C}.$$

   (a) Under what conditions is the matrix H_1 positive or negative definite?

   (b) If the matrix H_1 is indefinite how many negative eigenvalues does it have?

   (c) Find a regular orthogonalization of the system $e_1, e_2, \dots, e_n$ in the inner product $[x,y] = (H_1 x, y)$.

2. As in Exercise 12 of Chapter 2, let $\mathcal{L}_2^{(n)}$ be the space of all polynomials on the unit circle of degree n or less, and let the inner product $[x,y]$ be defined as in that exercise. Let H_2 be an invertible hermitian matrix such that

$$[x,y] = (H_2\xi,\eta), \quad \text{where} \quad x(\lambda) = \sum_{j=0}^n \xi_j \lambda^j \in \mathcal{L}_2^{(n)}, \quad y(\lambda) = \sum_{j=0}^n \eta_j \lambda^j \in \mathcal{L}_2^{(n)}$$

   and

$$\xi = \langle \xi_0, \xi_1, \dots, \xi_n \rangle, \qquad \eta = \langle \eta_0, \eta_1, \dots, \eta_n \rangle.$$

   (a) Under what conditions on $w(t)$ are all principal minors of H_2 invertible?

   (b) Find the regular orthogonalization of $1, t, t^2, \dots, t^n$ for the weight function

$$w(t) = w_{-n}t^n + 1 + w_n t^n, \qquad w_{-n} = \overline{w}_n \in \mathbb{C}.$$

   Assume that the conditions of part (a) hold.

3. Let

$$
H_3 = \begin{bmatrix}
1 & a & a^2 & \cdots & a^n \\
\overline{a} & 1 & a & \cdots & a^{n-1} \\
\vdots & & & \cdots & \vdots \\
\overline{a}^n & \overline{a}^{n-1} & \overline{a}^{n-2} & \cdots & 1
\end{bmatrix}, \quad a \in \mathbb{C}.
$$

(a) When is the matrix positive or negative definite?

(b) How many negative eigenvalues does the matrix H_3 have?

(c) When is the matrix H_3 invertible? Find the inverse. When are all principal minors invertible?

(d) Find a regular orthogonalization of the system $e_1, e_2, \ldots, e_{n+1}$ in the inner product $[x, y] = (H_3 x, y)$.

(e) Find a regular orthogonalization of $1, t, \ldots, t^n$ in $\mathcal{L}_2^{(n)}$, when

$$
w(t) = \Re(1 + 2\sum_{j=1}^{n} a^j e^{ij\theta}).
$$

(f) Where are the zeros of the polynomials obtained in the regular orthogonalization of (e)?

4. Answer questions (a)-(d) of Exercise 3 for the following matrices:

(a) $H_4 = (\alpha\delta_{jk} + \beta_j\overline{\beta_k})^n_{j,k=1}$, $\alpha \in \mathbb{R}$, $\beta_j \in \mathbb{C}$.
Here, δ_{jk} is the Kronecker symbol: $\delta_{jk} = 1$ if $j = k$, and $\delta_{jk} = 0$ if $j \neq k$.

(b) $H_5 = \begin{bmatrix}
\alpha & \alpha & \cdots & \alpha & \beta \\
\alpha & \alpha & \cdots & \beta & \alpha \\
\vdots & & \cdots & & \vdots \\
\beta & \alpha & \cdots & \alpha & \alpha
\end{bmatrix}$; where $\alpha, \beta \in \mathbb{R}$, with even size.

(c) $H_6 = \begin{bmatrix}
\beta & \alpha & \cdots & \alpha & \alpha \\
\alpha & \beta & \cdots & \alpha & \alpha \\
\vdots & & \cdots & & \vdots \\
\alpha & \alpha & \cdots & \alpha & \beta
\end{bmatrix}, \quad \alpha, \beta \in \mathbb{R}.$

In case (c) answer also question (e) of Exercise 3 for the weight function

$$
w(t) = \beta - \alpha + \alpha \sum_{j=-n}^{n} t^j.
$$

(d) $H_7 = \begin{bmatrix}
\alpha & i & i & \cdots & i \\
-i & \alpha & i & \cdots & i \\
\vdots & & & \cdots & \vdots \\
-i & -i & -i & \cdots & \alpha
\end{bmatrix}, \quad \alpha \in \mathbb{R}.$

In case (d) answer also question (e) of Exercise 3 for the weight function

$$w(t) = \alpha + 2\Re\left(i\sum_{j=1}^{n} t^j\right).$$

(e) $H_8 = \begin{bmatrix} \alpha & 0 & \cdots & 0 & \beta \\ 0 & \alpha & \cdots & \beta & 0 \\ \vdots & & \cdots & & \vdots \\ 0 & \beta & \cdots & \alpha & 0 \\ \beta & 0 & \cdots & 0 & \alpha \end{bmatrix}$, $\alpha, \beta \in \mathsf{R}$, with even size, $2m$.

(f) $H_9 = \begin{bmatrix} \alpha_1 & \beta & \cdots & \beta \\ \overline{\beta} & \alpha_2 & \cdots & \beta \\ \vdots & \vdots & \ddots & \vdots \\ \overline{\beta} & \overline{\beta} & \cdots & \alpha_n \end{bmatrix}$, $\alpha_1, \alpha_2, \ldots \alpha_n \in \mathsf{R}$, $\beta \in \mathsf{C}$.

(g) $H_{10} = \begin{bmatrix} \alpha & z & z^2 & \cdots & z^n \\ z^{-1} & \alpha & z & \cdots & z^{n-1} \\ \vdots & & & \cdots & \vdots \\ z^{-n} & & & \cdots & \alpha \end{bmatrix}$, where $\alpha \in \mathsf{R}$ and $z = e^{i\theta_0}$, $\theta_0 \in \mathsf{R}$.

In case (g) answer also questions (e) and (f) of Exercise 3 for

$$w(t) = \alpha + 2\Re\left(\sum_{j=1}^{n} e^{ij(\theta_0+\theta)}\right).$$

(h) $H_{11} = \begin{bmatrix} I & G \\ G^* & I \end{bmatrix}$, where $G = \text{diag}\,(g_1, \ldots, g_n)$, $g_j \in \mathsf{C}$.

5. Consider the hermitian matrices

(a) $\begin{bmatrix} 1 & a & a^2 & \cdots & a^{n-1} & w \\ \overline{a} & 1 & a & \cdots & a^{n-2} & a^{n-1} \\ \vdots & & & \cdots & & \vdots \\ \overline{w} & \overline{a}^{n-1} & & \cdots & \overline{a} & 1 \end{bmatrix}$, $a \in \mathsf{C}$,

(b) $\begin{bmatrix} \alpha & 0 & \cdots & & w \\ 0 & \alpha & \cdots & \beta & 0 \\ \vdots & & \cdots & & \vdots \\ 0 & \beta & \cdots & \alpha & 0 \\ \overline{w} & 0 & \cdots & 0 & \alpha \end{bmatrix}$, $\alpha, \beta \in \mathsf{R}$, with even size,

(c) $\begin{bmatrix} \alpha & \beta & \beta^2 & \cdots & \beta^{n-1} & w \\ \overline{\beta} & \alpha & \beta & \cdots & \beta^{n-2} & \beta^{n-1} \\ \vdots & & & \cdots & & \vdots \\ \overline{w} & \overline{\beta}^{n-1} & \overline{\beta}^{n-2} & \cdots & \beta & \alpha \end{bmatrix}$, $\alpha \in \mathsf{R}, \beta \in \mathsf{C}$.

Find the values of the parameter $w \in \mathsf{C}$ for which the determinants of these matrices are nonzero. What are the signs of these determinants? How many negative eigenvalues do the matrices have?

6. *Levinson's algorithm.* (This algorithm generates orthogonal polynomials $s_1(t), s_2(t), \ldots$ by updating the degree.) Let

$$s_k(t) = w_0^{(k)} + w_1^{(k)} t + \cdots + w_{k-1}^{(k)} t^{k-1},$$

where

$$\begin{bmatrix} w_0 & w_1 & w_2 & \cdots & w_{k-1} \\ \overline{w_1} & w_0 & w_1 & \cdots & w_{k-2} \\ \vdots & & & \cdots & \vdots \\ \overline{w_{k-1}} & \overline{w_{k-2}} & \overline{w_{k-3}} & \cdots & w_0 \end{bmatrix} \begin{bmatrix} w_0^{(k)} \\ w_1^{(k)} \\ \vdots \\ w_{k-1}^{(k)} \end{bmatrix} = \begin{bmatrix} 1 \\ 0 \\ \vdots \\ 0 \end{bmatrix}.$$

Introduce

$$\nu^{(k)} = \langle 1, \nu_1^{(k)}, \ldots, \nu_{k-1}^{(k)} \rangle, \quad \text{where} \quad \nu_j^{(k)} = (w_0^{(k)})^{-1} w_j^{(k)}$$

and

$$\rho_k = (w_0^{(k)})^{-1}.$$

Prove that

$$\nu^{(k+1)} = \begin{bmatrix} 0 \\ \nu^{(k)} \end{bmatrix} - c_k \begin{bmatrix} \widetilde{\nu}^{(k)} \\ 0 \end{bmatrix},$$

where

$$\widetilde{\nu}^{(k)} = \langle 1, \nu_{k-1}^{(k)}, \ldots, \nu_1^{(k)} \rangle$$

and where

$$c_k = \frac{w_{k-1} + w_{k-2} \nu_1^{(k)} + \cdots + w_1 \nu_{k-1}^{(k)}}{\rho_k} = \frac{\Delta_k}{\rho_k},$$

$$\rho_{k+1} = \rho_k - \frac{\Delta_k^2}{\rho_k} - \rho_k (1 - |c_k|^2),$$

and

$$\widetilde{\nu}_k = \begin{bmatrix} \widetilde{\nu}_{k-1} \\ 0 \end{bmatrix} - c_k \begin{bmatrix} 0 \\ \nu_k \end{bmatrix}.$$

7. Generalize the Levinson algorithm to the case when the Toeplitz matrix is not selfadjoint.

8. For a given k with $0 \leq k \leq n$, construct an $n \times n$ hermitian Toeplitz matrix with a given number k negative eigenvalues and $n - k$ positive eigenvalues.

9. Construct an $n \times n$ hermitian Toeplitz matrix with α negative eigenvalues, β eigenvalues equal to zero, and γ positive eigenvalues, where α, β, γ are given nonnegative integers that sum up to n.

3.6 Notes

The material of the first three sections of this chapter are selected from the first chapters of the recent monograph [18]. The main source of these materials are the papers [9], [21], [20], [60], and [71]. The source for the fourth section is the paper [19].

Chapter 4

Classes of Linear Transformations

The classical development of linear algebra in the context of definite inner product spaces includes some central questions concerning linear transformations and matrices with symmetry properties. These properties arise in natural ways in the many applications of linear algebra in physics, statistics, mechanics, and so on. The most important of these symmetries include the selfadjoint and unitary properties.

In a similar way, applications requiring analysis on spaces with *indefinite* inner products also lead to the study of linear transformations and matrices with analogous symmetries. The most common of these are, once more, the (suitably defined) properties of selfadjoint and unitary transformations and their matrix representations. This chapter develops these fundamental ideas in the setting of complex matrices acting on C^n with an indefinite inner product. Some necessary geometric notions in these spaces have been developed in Chapter 2 and will be important here.

The last three sections of the chapter provide introductions to important classes of matrices which arise frequently. They are H-contractions, dissipative matrices, and symplectic matrices.

4.1 Adjoint Matrices

Let $[.,.]$ be an indefinite inner product on C^n and let H be the associated invertible hermitian matrix as in Equation (2.1.1). Let A be an $n \times n$ complex matrix considered as a linear transformation on C^n. The H-*adjoint* of A is the unique $n \times n$ matrix, written $A^{[*]}$, which satisfies

$$[Ax, y] = [x, A^{[*]}y] \qquad (4.1.1)$$

for all $x, y \in \mathbb{C}^n$. The H-adjoint of A may also be described as the adjoint of A with respect to $[., .]$. Expressing (4.1.1) in terms of H we have

$$(HAx, y) = (Hx, A^{[*]}y) \tag{4.1.2}$$

for all $x, y \in \mathbb{C}^n$, and, hence

$$(x, A^*Hy) = (x, HA^{[*]}y)$$

where (here and elsewhere) $A^* : \mathbb{C}^n \to \mathbb{C}^n$ is the usual adjoint of A (i.e., $(x, A^*y) = (Ax, y)$ for all $x, y \in \mathbb{C}^n$). It follows that

$$A^{[*]} = H^{-1}A^*H. \tag{4.1.3}$$

In particular, this representation confirms the existence of $A^{[*]}$ and shows that it is uniquely determined by A. It also follows from (4.1.3), or from the definition of $A^{[*]}$, that $(A^{[*]})^{[*]} = A$.

There are important and well-known connections between the images and kernels of A and A^*, its usual adjoint. The next proposition describes the extension of these results to the H-adjoint of A.

Proposition 4.1.1. *Let* $A : \mathbb{C}^n \to \mathbb{C}^n$ *and let* $A^{[*]}$ *be its* H-*adjoint. Then*

$$\text{Range}A^{[*]} = (\text{Ker}A)^{[\perp]}; \quad \text{Ker}A^{[*]} = (\text{Range}A)^{[\perp]}. \tag{4.1.4}$$

Proof. Let $x \in \text{Range}A^{[*]}$ so that $x = A^{[*]}y$ for some $y \in \mathbb{C}^n$. Then for every $z \in \text{Ker}A$:

$$[x, z] = [A^{[*]}y, z] = [y, Az] = 0,$$

and it follows that

$$\text{Range}A^{[*]} \subseteq (\text{Ker}A)^{[\perp]} \tag{4.1.5}$$

However, using (4.1.3) and (2.2.2),

$$\dim(\text{Range}A^{[*]}) = \dim(\text{Range}A^*) = n - \dim(\text{Ker}A) = \dim(\text{Ker}A)^{[\perp]},$$

so that the equality must obtain in (4.1.5).

The proof of the second relation in (4.1.4) is similar. $\qquad\qquad\square$

A subspace $\mathcal{M} \subseteq \mathbb{C}^n$ is said to be invariant for an $n \times n$ matrix A (considered as a linear transformation from $\mathbb{C}^n$ to $\mathbb{C}^n$), or to be A-*invariant*, if $x \in \mathcal{M}$ implies $Ax \in \mathcal{M}$.

Proposition 4.1.2. *Let* $A : \mathbb{C}^n \to \mathbb{C}^n$ *and let* $[., .]$ *be an indefinite inner product in* $\mathbb{C}^n$. *Then a subspace* $\mathcal{M}$ *is* A-*invariant if and only if its orthogonal companion* $\mathcal{M}^{[\perp]}$ *is* $A^{[*]}$-*invariant.*

Proof. Let $\mathcal{M}$ be A-invariant and let $x \in \mathcal{M}$, $y \in \mathcal{M}^{[\perp]}$. Then

$$[A^{[*]}y, x] = [y, Ax] = 0$$

since Ax is again in $\mathcal{M}$. So $A^{[*]}y \in \mathcal{M}^{[\perp]}$ and $\mathcal{M}^{[\perp]}$ is $A^{[*]}$-invariant.

To prove the converse statement apply what is already proved to $A^{[*]}$ and $(A^{[*]})^{[*]} = A$, taking into account the fact that $(\mathcal{M}^{[\perp]})^{[\perp]} = \mathcal{M}$. $\square$

Now it is natural to describe a matrix as *H-selfadjoint* (or selfadjoint with respect to $[.,.]$) if $A = A^{[*]}$. In a similar way, a matrix U is said to be *H-unitary* if it is invertible and $U^{-1} = U^{[*]}$, and matrix N is *H-normal* if $NN^{[*]} = N^{[*]}N$. Clearly, H-selfadjoint matrices and H-unitary matrices are also H-normal.

The next result shows that if matrices used in defining indefinite inner products are congruent, then matrices of these three types are transformed in a natural way.

Proposition 4.1.3. *Let H_1, H_2 define indefinite inner products on $\mathbf{C}^n$ and $H_2 = SH_1S^*$ for some invertible $n \times n$ matrix S. Then A_1 is H_1-selfadjoint (or H_1-unitary, or H_1-normal) if and only if the matrix $A_2 := (S^*)^{-1}A_1S^*$ is H_2-selfadjoint (or H_2-unitary, or H_2-normal, respectively).*

Proof. We consider the "only if" part of the statement. Proof of the converse statement is analogous. Suppose first that A_1 is H_1-selfadjoint so that, (see (4.1.3)), $H_1A_1 = A_1^*H_1$. Then

$$\begin{aligned} H_2A_2 &= (SH_1S^*)((S^*)^{-1}A_1S^*) = SH_1A_1S^* = SA_1^*H_1S^* \\ &= (SA_1^*S^{-1})(SH_1S^*) = A_2^*H_2, \end{aligned}$$

which implies that A_2 is H_2-selfadjoint.

If A_1 is H_1-unitary then $A_1^{-1} = A_1^{[*]}$, and it follows from (4.1.3) that $H_1A_1^{-1} = A_1^*H$. So we have

$$\begin{aligned} H_2A_2^{-1} &= (SH_1S^*)((S^*)^{-1}A_1^{-1}S^*) = SH_1A_1^{-1}S^* = SA_1^*H_1S^* \\ &= (SA_1^*S^{-1})(SH_1S^*) = A_2^*H_2. \end{aligned}$$

Thus, A_2 is H_2-unitary.

In much the same way, it is easily seen that if A_1 is H_1-normal, then $A_1H_1^{-1}A_1^*H_1 = H_1^{-1}A_1^*H_1A_1$ and that this implies a similar relation between A_2 and H_2. $\square$

Proposition 4.1.3 allows us to study the properties of these selfadjoint, unitary and normal matrices in the context of a canonical indefinite inner product of the form $[.,.] = (P.,.)$ where $P^* = P$ and $P^2 = I$. Indeed, given an invertible hermitian H there is an invertible S such that $P := S^*HS$ is a diagonal matrix: $P = \text{diag}(1, 1, \ldots, 1, -1, \ldots, -1)$. However, this reduction is achieved at the expense of replacing A by $S^{-1}AS$. This suggests that S may be chosen in such a way that *both* H and A are reduced to some simplest possible forms; an idea that will be developed in the next chapter.

4.2 H-Selfadjoint Matrices: Examples and Simplest Properties

Let $[.,.] = (H.,.)$ be an indefinite inner product in $\mathbb{C}^n$. As we have seen in the last section, an $n \times n$ matrix A is said to be H-selfadjoint (or selfadjoint with respect to $[.,.]$) if $A = A^{[*]}$ or, in other words, (see (4.1.3)) if

$$A = H^{-1}A^*H. \tag{4.2.6}$$

Thus, any H-selfadjoint matrix A is similar to A^*. We shall see later that the converse is also true: if a matrix A is similar to its adjoint (i.e., $A = S^{-1}A^*S$ for some S) then A is H-selfadjoint for some H. In other words, the similarity between A and A^* can be carried out by means of an invertible *hermitian* matrix. Observe also that the set of all $n \times n$ H-selfadjoint matrices form a real linear space; i.e., if A and B are H-selfadjoint then so is $\alpha A + \beta B$ where α, β are any real numbers.

For the case when $H^2 = I$ (see concluding remarks of the preceding section) it is easily seen that A is H-selfadjoint if and only if A^* is H-selfadjoint. This leads to the following observation: if $H^2 = I$ and H is hermitian, then A is H-selfadjoint if and only if both of

$$\Re A = \frac{1}{2}(A + A^*), \quad i(\Im A) = \frac{1}{2}(A - A^*) \tag{4.2.7}$$

are H-selfadjoint.

The following examples of H-selfadjoint matrices are fundamental.

Example 4.2.1. *Let* $[x, y] = (\varepsilon S_n x, y)$, $x, y \in \mathbb{C}^n$, *where* S_n *is the* $n \times n$ *sip matrix introduced in Example 2.1.1, and* ε *is either* 1 *or* -1. *Further, let*

$$J = \begin{bmatrix} \alpha & 1 & 0 & \cdots & 0 \\ 0 & \alpha & 1 & & \vdots \\ \vdots & & \ddots & \ddots & \\ & & & \alpha & 1 \\ 0 & \cdots & & 0 & \alpha \end{bmatrix}$$

be the $n \times n$ *Jordan block with* real *eigenvalue* α. *The equality* $(\varepsilon S_n)J = J^T(\varepsilon S_n)$ *is easily checked and, since* $J^T = J^*$, *it means that* J *is* εS_n-*selfadjoint.*   □

Example 4.2.2. *Let* $[x, y] = (Qx, y)$, $x, y \in \mathbb{C}^{2n}$ *where*

$$Q = \begin{bmatrix} 0 & S_n \\ S_n & 0 \end{bmatrix},$$

and S_n *is again the* $n \times n$ *sip matrix. Let*

$$K = \begin{bmatrix} J & 0 \\ 0 & \bar{J} \end{bmatrix},$$

where J is the $n \times n$ Jordan block with nonreal eigenvalue α (so that $\overline{J}$ is the $n \times n$ Jordan block with the eigenvalue $\overline{\alpha}$). Again, one checks easily that $QK = K^*Q$, i.e., K is Q-selfadjoint.   □

The H-selfadjoint matrices from Examples 4.2.1 and 4.2.2 will appear in the next chapter as elements of a canonical form for selfadjoint matrices in indefinite inner product spaces.

We describe now some simple properties of H-selfadjoint matrices.

Proposition 4.2.3. *The spectrum $\sigma(A)$ of an H-selfadjoint matrix A is symmetric relative to the real axis, i.e., $\lambda_0 \in \sigma(A)$ implies $\overline{\lambda_0} \in \sigma(A)$. Moreover, in the Jordan normal form of A, the sizes of the Jordan blocks with eigenvalue λ_0 are equal to the sizes of Jordan blocks with eigenvalue $\overline{\lambda_0}$.*

Proof. Using (4.2.6), write

$$\lambda I - A = H^{-1}(\overline{\lambda}I - A)^*H.$$

So $\lambda I - A$ is singular if and only if $\overline{\lambda}I - A$ is singular; i.e., $\lambda_0 \in \sigma(A)$ implies $\overline{\lambda_0} \in \sigma(A)$. Further, let J be the Jordan form of A with reducing matrix $T : A = T^{-1}JT$.

Observe also that J^* is similar to $\overline{J}$, and write $J^* = K^{-1}\overline{J}K$ for some invertible K. Then, using the equation above,

$$\lambda I - A = T^{-1}(\lambda I - J)T = H^{-1}\left\{T^{-1}(\overline{\lambda}I - J)T\right\}^* H$$

so that

$$\lambda I - J = (TH^{-1}T^*)(\lambda I - J^*)((T^*)^{-1}HT^{-1}) = S(\lambda I - \overline{J})S^{-1}$$

where $S = TH^{-1}T^*K^{-1}$. Thus, J and $\overline{J}$ are similar and $\overline{J}$ can be obtained from J by permutation of some of its Jordan blocks. The proposition follows.   □

Note, in particular, that nonreal eigenvalues of an H-selfadjoint matrix can only occur in conjugate pairs and, consequently, their total number, whether counted as distinct eigenvalues or according to algebraic multiplicities, must be even.

For an $n \times n$ matrix A, the *root subspace* $\mathcal{R}_{\lambda_0}(A)$ corresponding to an eigenvalue λ_0 is defined as follows:

$$\mathcal{R}_{\lambda_0}(A) = \{x \in \mathsf{C}^n \mid (A - \lambda_0 I)^s x = 0 \quad \text{for some positive integer} \quad s\}. \quad (4.2.8)$$

It is well known that $\mathcal{R}_{\lambda_0}(A)$ is indeed a subspace and that C^n is a direct sum of the root subspaces $\mathcal{R}_{\lambda_j}(A)$, $j = 1, 2, \ldots, k$, where $\lambda_1, \ldots, \lambda_k$ are the different eigenvalues of A (see Theorem A.2.4). If A is hermitian, i.e., selfadjoint with respect to the ordinary inner product $(.,.)$, then its root subspaces corresponding to different eigenvalues are orthogonal (with respect to $(.,.)$). It turns out that, with proper modification, this result extends to the case of H-selfadjoint matrices as well.

Theorem 4.2.4. *Let A be an H-selfadjoint matrix and $\lambda, \mu \in \sigma(A)$ with $\lambda \neq \bar{\mu}$. Then*

$$\mathcal{R}_\lambda(A) \subseteq (\mathcal{R}_\mu(A))^{[\perp]},$$

i.e., the root subspaces $\mathcal{R}_\lambda(A)$ and $\mathcal{R}_\mu(A)$ are orthogonal with respect to $[.,.] = (H.,.)$.

Proof. Let $x \in \mathcal{R}_\lambda(A)$ and $y \in \mathcal{R}_\mu(A)$ so that $(A - \lambda I)^s x = 0$ and $(A - \mu I)^t y = 0$ for some s and t. We are to prove that

$$[x, y] = 0. \tag{4.2.9}$$

Proceed by induction on $s + t$. For $s = t = 1$ we have $Ax = \lambda x$, $Ay = \mu y$. Then

$$\lambda[x, y] = [Ax, y] = [x, Ay] = [x, \mu y] = \bar{\mu}[x, y] \tag{4.2.10}$$

and since $\lambda \neq \bar{\mu}$, we obtain (4.2.9).

Suppose now that (4.2.9) is proved for all $x' \in \mathcal{R}_\lambda(A)$, $y' \in \mathcal{R}_\mu(A)$ such that $(A - \lambda I)^{s'} x' = (A - \lambda I)^{t'} y' = 0$ for some s' and t' satisfying $s' + t' < s + t$. Given x and y as above, put $x' = (A - \lambda I)x$, $y' = (A - \mu I)y$. Then, by the induction assumption $[x', y] = [x, y'] = 0$, which means that $\lambda[x, y] = [Ax, y]$; $\bar{\mu}[x, y] = [x, Ay]$. Now use the relations of (4.2.10) once more to complete the proof. $\square$

In particular, taking $\lambda = \mu$ nonreal in Theorem 4.2.4 we obtain the following corollary.

Corollary 4.2.5. *Let A be an H-selfadjoint matrix and let $\lambda_0 \in \sigma(A)$ be nonreal. Then the root subspace $\mathcal{R}_{\lambda_0}(A)$ is H-neutral.*

4.3 H-Unitary Matrices: Examples and Simplest Properties

As introduced in Section 4.1, an $n \times n$ matrix A is called H-unitary (or unitary with respect to $[.,.]$) if A is invertible and $A^{-1} = A^{[*]}$. In other words, A is H-unitary if and only if $[Ax, y] = [x, A^{-1}y]$ for all $x, y \in \mathbb{C}^n$, or

$$A = H^{-1}(A^*)^{-1}H; \quad A^*HA = H. \tag{4.3.11}$$

In particular, A is similar to $(A^*)^{-1}$. The converse statement is also true (as in the case of H-selfadjoint matrices): if A is similar to $(A^*)^{-1}$, then the matrix achieving this similarity can be chosen to be hermitian. This fact will be proved later.

Note that for a fixed H, the set of all H-unitary matrices form a group, i.e., if A, B are H-unitary then so are A^{-1}, B^{-1} and AB.

If H is first reduced by congruence so that $H^2 = I$ (as described at the end of Section 4.1), then it is easily seen that A is H-unitary if and only if A^* is H-unitary.

The following examples of H-unitary matrices are related to the canonical forms of H-unitary matrices to be developed in the next chapter.

Example 4.3.1. Let $[x, y] = (\varepsilon S_n x, y)$, $x, y \in \mathbb{C}^n$, where S_n is the $n \times n$ sip matrix and $\varepsilon = \pm 1$. Suppose that $\lambda \in \mathbb{C}$ with $|\lambda| = 1$ and

$$
A = \begin{bmatrix}
\lambda & 2i\lambda & 2i^2\lambda & \cdots & & 2i^{n-1}\lambda \\
0 & \lambda & 2i\lambda & & & \vdots \\
& & \lambda & & & \\
\vdots & & & \ddots & & \\
& & & & \lambda & 2i\lambda \\
0 & & \cdots & & 0 & \lambda
\end{bmatrix}.
$$

It is easily verified that $A^*(\varepsilon S_n)A = \varepsilon S_n$ so that A is εP-unitary. □

Example 4.3.2. Let $[x, y] = (Qx, y)$ for all $x, y \in \mathbb{C}^n$ where

$$
Q = \begin{bmatrix} 0 & S_n \\ S_n & 0 \end{bmatrix}.
$$

For a nonzero $\lambda \in \mathbb{C}$ such that $|\lambda| \neq 1$ put

$$
A = \begin{bmatrix} K_1 & 0 \\ 0 & K_2 \end{bmatrix}
$$

where

$$
K_1 = \begin{bmatrix}
\lambda & k_1 & k_2 & \cdots & k_{n-1} \\
0 & \lambda & k_1 & & \vdots \\
& & \ddots & \ddots & \\
\vdots & & & \lambda & k_1 \\
0 & \cdots & & 0 & \lambda
\end{bmatrix}, \quad
K_2 = \begin{bmatrix}
\overline{\lambda^{-1}} & \kappa_1 & \kappa_2 & \cdots & \kappa_{n-1} \\
0 & \overline{\lambda^{-1}} & \kappa_1 & & \vdots \\
& & \ddots & \ddots & \\
\vdots & & & \overline{\lambda^{-1}} & \kappa_1 \\
0 & & & 0 & \overline{\lambda^{-1}}
\end{bmatrix}
$$

and

$$
k_r = \lambda q_1^{r-1}(q_1 - \overline{q_2}), \quad \kappa_r = \overline{\lambda^{-1}} q_2^{r-1}(q_2 - \overline{q_1}), \quad \text{for } r = 1, 2, \ldots, n-1,
$$

and $q_1 = \frac{i}{2}(1 + \lambda)$, $q_2 = \frac{i}{2}(1 + \overline{\lambda^{-1}})$.

A direct computation shows that $A^*QA = Q$, so A is Q-unitary. Note also that $K_2 = \overline{K_1^{-1}}$. □

More generally, if K_1 is an upper triangular Toeplitz matrix which is similar to one Jordan block with a nonzero eigenvalue, and if $K_2 = \overline{K_1^{-1}}$, then $\begin{bmatrix} K_1 & 0 \\ 0 & K_2 \end{bmatrix}$ is $\begin{bmatrix} 0 & S_n \\ S_n & 0 \end{bmatrix}$-unitary.

Using the similarity of A and $(A^*)^{-1}$ for an H-unitary matrix A, one can prove the following analogue of Proposition 4.2.3 (its proof is similar to that of Proposition 4.2.3 and therefore is omitted).

Proposition 4.3.3. *Let A be an H-unitary matrix. Then $\sigma(A)$ is symmetric relative to the unit circle, i.e., $\lambda_0 \in \sigma(A)$ implies $\overline{\lambda_0}^{-1} \in \sigma(A)$. Moreover, in the Jordan normal form of A, the sizes of Jordan blocks with eigenvalue λ_0, and the sizes of Jordan blocks with eigenvalue $\overline{\lambda_0}^{-1}$, are the same.*

There is a strong connection between H-unitary and H-selfadjoint matrices. As in the case of the usual inner product, one way to describe this connection is via Cayley transforms.

Recall that if $|\alpha| = 1$ and $w \neq \overline{w}$ then the map f defined by

$$f(z) = \alpha(z - \overline{w}) / (z - w) \tag{4.3.12}$$

maps the real line in the z-plane onto the unit circle in the ζ-plane, where $\zeta = f(z)$. The inverse transformation is

$$z = (w\zeta - \overline{w}\alpha) / (\zeta - \alpha). \tag{4.3.13}$$

If $w \notin \sigma(A)$, then the function f is defined on $\sigma(A)$ and, if A is H-selfadjoint, one anticipates that $U = f(A)$ is H-unitary. This idea is developed in:

Proposition 4.3.4. *Let A be an H-selfadjoint matrix. Let w be a nonreal complex number with $w \notin \sigma(A)$ and let α be any unimodular complex number. Then*

$$U = \alpha(A - \overline{w}I)(A - wI)^{-1} \tag{4.3.14}$$

is H-unitary and $\alpha \notin \sigma(U)$.

Conversely, if U is H-unitary, $|\alpha| = 1$ and $\alpha \notin \sigma(U)$, then for any $w \neq \overline{w}$ the matrix

$$A = (wU - \overline{w}\alpha I)(U - \alpha I)^{-1} \tag{4.3.15}$$

is H-selfadjoint and $w \notin \sigma(A)$. Furthermore, formulas (4.3.14) and (4.3.15) are inverse to one another.

The transformation (4.3.15) will be referred to as the *Cayley transform.*

Proof. If A is H-selfadjoint and $|\alpha| = 1$ it is easily seen that

$$(A^* - \overline{w}I)H(A - wI) = (\overline{\alpha}A^* - \overline{\alpha}wI)H(\alpha A - \alpha\overline{w}I).$$

Premultiplying by $(A^* - \overline{w}I)^{-1}$ and postmultiplying by $(\alpha A - \alpha\overline{w}I)^{-1}$ it is found that $HU^{-1} = U^*H$ where U is defined by (4.3.14), and this means that U is H-unitary. Furthermore, it follows from (4.3.14) that

$$(U - \alpha I)(A - wI) = \alpha(w - \overline{w})I. \tag{4.3.16}$$

Thus, the hypothesis that w is not real implies that $U - \alpha I$ is invertible and so $\alpha \notin \sigma(U)$.

The relation (4.3.16) also gives

$$
\begin{aligned}
A &= wI + \alpha(w - \overline{w})(U - \alpha I)^{-1} = [w(U - \alpha I) + \alpha(w - \overline{w})I](U - \alpha I)^{-1} \\
&= (wU - \overline{w}\alpha I)(U - \alpha I)^{-1},
\end{aligned}
$$

so that (4.3.15) and (4.3.14) are, indeed, inverse to each other.

The proof of the converse statement is left to the reader. $\qquad\square$

Suppose that U is H-unitary and, as in Proposition 4.3.4, A is the H-selfadjoint matrix given by (4.3.15). Then the root subspace of U corresponding to $\lambda_0 \in \sigma(U)$ is also the root subspace of A corresponding to its eigenvalue

$$\mu_0 = (w\lambda_0 - \overline{w}\alpha)(\lambda_0 - \alpha)^{-1}. \tag{4.3.17}$$

Thus,

$$\mathcal{R}_{\lambda_0}(U) = \mathcal{R}_{\mu_0}(A). \tag{4.3.18}$$

This fact can be verified directly by using (4.3.15) but it is also a consequence of the following more general lemma. Although this is relatively well-known, a proof is included in the interests of a self-contained presentation.

Lemma 4.3.5. *Let S, T be $n \times n$ matrices with the property that $S = f(T)$, $T = g(S)$ for some complex functions f and g which are analytic in neighborhoods of $\sigma(T)$, $\sigma(S)$ respectively. Then for every $\lambda_0 \in \sigma(S)$ we have*

$$\mathcal{R}_{\lambda_0}(S) = \mathcal{R}_{g(\lambda_0)}(T). \tag{4.3.19}$$

Proof. It is well known that $g(S) = \widetilde{g}(S)$ where $\widetilde{g}$ is a polynomial which, in particular, has the property that $g(\lambda_0) = \widetilde{g}(\lambda_0)$ for $\lambda_0 \in \sigma(S)$. Let $\widetilde{g}(\lambda) = \sum_{j=0}^{p} \alpha_j(\lambda - \lambda_0)^j$, so that

$$T - g(\lambda_0)I = \widetilde{g}(S) - \widetilde{g}(\lambda_0)I = \sum_{j=0}^{p} \alpha_j(S - \lambda_0 I)^j - \alpha_0 I = W(S - \lambda_0 I),$$

where $W = \sum_{j=1}^{p} \alpha_j(S - \lambda_0 I)^{j-1}$ is a matrix commuting with S. Now

$$(T - g(\lambda_0)I)^s = W^s(S - \lambda_0 I)^s \quad \text{for} \quad s = 0, 1, \ldots,$$

so that, using the definition (4.2.8), we have $\mathcal{R}_{\lambda_0}(S) \subseteq \mathcal{R}_{g(\lambda_0)}(T)$. However, the same inclusion applies on replacing S, λ_0 by T, $g(\lambda_0)$, respectively, yielding

$$\mathcal{R}_{g(\lambda_0)}(T) \subseteq \mathcal{R}_{f(g(\lambda_0))}(f(T)) = \mathcal{R}_{\lambda_0}(S),$$

so (4.3.19) follows.   □

Now (4.3.18) is established by an application of the lemma and, in view of Theorem 4.2.4, it implies the following property:

Corollary 4.3.6. *Root subspaces $\mathcal{R}_\lambda(U)$ and $\mathcal{R}_\mu(U)$ of an H-unitary matrix U are H-orthogonal provided $\lambda \neq \overline{\mu^{-1}}$.*

In particular (cf. Corollary 4.2.5), every root subspace of U corresponding to an eigenvalue not on the unit circle is H-neutral.

It is clear that these orthogonality properties of root subspaces for H-unitary matrices could also be proved directly using arguments analogous to the proof of Theorem 4.2.4.

A more general class of unitary matrices will be useful. Suppose that *two* indefinite inner products are defined on $\mathbf{C}^n$ with associated invertible hermitian matrices H_1 and H_2. An $n \times n$ matrix A is said to be (H_1, H_2)-*unitary* if $[Ax, Ay]_{H_2} = [x, y]_{H_1}$ for all $x, y \in \mathbf{C}^n$. It is easily seen that this is equivalent to the relation

$$A^* H_2 A = H_1, \qquad\qquad (4.3.20)$$

which could be compared with (4.3.11). Note that an (H_1, H_2)-unitary matrix is necessarily invertible. Also, the relation (4.3.20) indicates that the notion of (H_1, H_2)-unitary matrices is meaningful only when H_1 and H_2 are congruent. Thus, the assertion that A is (H_1, H_2)-unitary implies the existence of an invertible S such that $H_2 = SH_1S^*$ and (4.3.20) is equivalent to the statement that S^*A is H_1-unitary or, alternatively, AS^* is H_2-unitary.

4.4   A Second Characterization of H-Unitary Matrices

It has been remarked in Section 4.2 that a matrix A is H-selfadjoint for some H if and only if the spectrum of A is symmetric with respect to the real line. With the aid of Cayley transformations it is easy to obtain the analogous result: A matrix U is H-unitary for some H if and only if the spectrum of U is symmetric with respect to the unit circle.

Now for *any* nonsingular A it is not difficult to verify that $\sigma(A^{-1}A^*)$ is symmetric with respect to the unit circle and hence $A^{-1}A^*$ must be H-unitary for some H. It turns out that this property characterizes H-unitary matrices. In proving this result, however, we can follow a different line of argument which makes less demands on spectral theory.

Note that in the next lemma A is not necessarily hermitian.

Lemma 4.4.1. *If* $U^*AU = A$, $\det A \neq 0$, *then there is an* H, *with* $H^* = H$ *and* $\det H \neq 0$ *such that* $U^*HU = H$.

Proof. Let $H = \bar{z}A + zA^*$ for some z with $|z| = 1$. Then

$$U^*HU = U^*(\bar{z}A + zA^*)U = \bar{z}A + zA^* = H$$

and $H^* = H$. To ensure that $\det H \neq 0$ observe

$$H = \bar{z}A + zA^* = zA(z^{-1}\bar{z}I + A^{-1}A^*)$$

and so we have only to choose z (with $|z| = 1$) so that $-z^{-1}\bar{z} = -\bar{z}^2 \notin \sigma(A^{-1}A^*)$. $\square$

Theorem 4.4.2. *A matrix* U *is* H-*unitary for some* H ($H^* = H$, $\det H \neq 0$) *if and only if* $U = A^{-1}A^*$ *for some nonsingular* A.

Proof. If $U = A^{-1}A^*$ then

$$U^*AU = A(A^*)^{-1}AA^{-1}A^* = A$$

and, from the lemma, U is H-unitary for some H.

Conversely, let U be H-unitary and let $A = i\beta(I - \alpha U^*)H$, where $|\alpha| = 1$, $\alpha \notin \sigma(U)$ and $\bar{\beta}/\beta = \alpha$. Then

$$AU = i\beta(I - \alpha U^*)HU = i\beta H(U - \alpha I) = i\alpha\beta H(\bar{\alpha}U - I) = -i\alpha\beta H(I - \bar{\alpha}U)$$
$$= -i\bar{\beta}H(I - \bar{\alpha}U) = A^*,$$

so $U = A^{-1}A^*$. $\square$

4.5 Unitary Similarity

Let A_1 and A_2 be $n \times n$ matrices which are H_1-selfadjoint and H_2-selfadjoint, respectively. A notion of equivalence of such matrices appears naturally. Namely, A_1 and A_2 are said to be *unitarily similar* if $A_1 = T^{-1}A_2T$, where the matrix T is invertible and (H_1, H_2)-unitary (i.e., $[Tx, Ty]_{H_2} = [x, y]_{H_1}$ for all $x, y \in \mathbb{C}^n$ or $H_1 = T^*H_2T$). In other words, A_1 and A_2 are unitarily similar if they are similar, and the similarity matrix is unitary with respect to the indefinite inner products involved.

It will be convenient to study this equivalence in the framework of the set $\mathcal{U}$ of all pairs of $n \times n$ matrices (A, H) where A is an arbitrary complex matrix, and H defines an indefinite inner product on $\mathbb{C}^n$, i.e., $H^* = H$ and $\det H \neq 0$. The pairs $(A_1, H_1), (A_2, H_2) \in \mathcal{U}$ are said to be *unitarily similar* if, for some invertible matrix T, we have

$$A_1 = T^{-1}A_2T \quad \text{and} \quad H_1 = T^*H_2T.$$

Thus, for unitarily similar pairs, A_1 and A_2 are similar, and H_1 and H_2 are congruent. In general, however, similarity of A_1 and A_2 and congruency of H_1 and H_2 do not guarantee that (A_1, H_1) and (A_2, H_2) are unitarily similar. To ensure this, the similarity and the congruence must be determined simultaneously by the same transforming matrix T.

It is easily verified that unitary similarity defines a relation on $\mathcal{U}$ which is reflexive, symmetric, and transitive, i.e., defines an equivalence relation on $\mathcal{U}$. The corresponding equivalence classes will be called the *unitary similarity classes* of $\mathcal{U}$. The observation that each such class is arcwise connected will be useful later.

Theorem 4.5.1. *A unitary similarity class of pairs of matrices from $\mathcal{U}$ is arcwise connected.*

Proof. Let (A, H) and (B, G) be in $\mathcal{U}$, and be unitarily similar. Thus, $A = S^{-1}BS$, $H = S^*GS$, for some invertible S.

Let $S(t)$, $t \in [0, 1]$ be a continuous path of invertible matrices with $S(0) = I$, $S(1) = S$. To establish the existence of such a path, let J be a Jordan form for S and for each Jordan block $J_p = \lambda_p I + K$, $\lambda_p \neq 0$, (where K is the nilpotent matrix with ones on the super-diagonal and zeros elsewhere) define $J_p(t) = \lambda_p(t)I + tK$, where $\lambda_p(t)$ is a continuous path of nonzero complex numbers with $\lambda_p(0) = 1$, $\lambda_p(1) = \lambda_p$. Then let $J(t)$ be the block diagonal matrix made up of blocks $J_p(t)$ in just the way that J is made up from the blocks J_p. Then the construction implies $J(0) = I$, $J(1) = J_p$. Now define $S(t) = TJ(t)T^{-1}$, where T is a matrix for which $S = TJT^{-1}$, and the construction is complete.

Using the path $S(t)$ construct a path of pairs of matrices $(B(t), G(t))$ in the unitary similarity class of (G, B) by

$$B(t) = S(t)^{-1}BS(t), \quad G(t) = S(t)^*GS(t),$$

for $t \in [0, 1]$. Then $(B(0), G(0)) = (B, G)$ and $(B(1), G(1)) = (A, H)$, as required. $\square$

Note also the following property of unitary similarity (which is just another formulation of Proposition 4.1.3).

Proposition 4.5.2. *Let (A_1, H_1) and (A_2, H_2) be unitarily similar. Then A_1 is selfadjoint (unitary, or normal) with respect to the indefinite inner product defined by H_1 if and only if A_2 is selfadjoint (resp. unitary, or normal) with respect to the inner product defined by H_2.*

In the sequel we shall generally apply the notion of unitary similarity to classes of pairs (A, H) in which A is H-selfadjoint, or A is H-unitary.

4.6 Contractions

Let H be a fixed $n \times n$ invertible hermitian matrix. A matrix $A \in \mathbb{C}^{n \times n}$ is called a *strict H-contraction* if the following inequality holds for all nonzero $x \in \mathbb{C}^n$:

$$(HAx, Ax) < (Hx, x).$$

In other words, if

$$H - A^* H A > 0. \tag{4.6.21}$$

When H is positive definite, (4.6.21) can be replaced by

$$\|H^{1/2} A H^{-1/2}\| < 1. \tag{4.6.22}$$

Here, $H^{1/2}$ is the unique positive definite square root of H. It follows from (4.6.22) that, in this case,

$$\max_{\lambda \in \sigma(A)} |\lambda| < 1. \tag{4.6.23}$$

The following theorem shows that the distribution of the eigenvalues of an H-contraction relative to the unit circle is determined by the distribution of positive and negative eigenvalues of H.

Theorem 4.6.1. *Let H be invertible and selfadjoint and let A be H-contractive. Then*

$$n_+(A) = i_+(H) \quad and \quad n_-(A) = i_-(H),$$

where $n_+(A)$ and $n_-(A)$ denote the number of eigenvalues of A, counted with their algebraic multiplicities, inside and outside the unit disc, respectively. Similarly, $i_+(H)$ and $i_-(H)$ denote the number of positive and negative eigenvalues of H, respectively, counted with multiplicities.

Proof. Consider first the special case in which the spectrum, $\sigma(A)$, is in the open unit disc. Introduce the notation

$$G := H - A^* H A.$$

By elementary algebra we obtain

$$H - (A^*)^2 H A^2 = G + A^* G A$$

and, more generally,

$$H - (A^*)^m H A^m = \sum_{j=0}^{m-1} (A^*)^j G A^j,$$

or

$$H - \sum_{j=0}^{m-1} (A^*)^j G A^j = (A^*)^m H A^m, \quad m = 1, 2, \ldots.$$

Since $\sigma(A)$ is in the open unit disc, (4.6.23) holds, and it is easy to see that the sequence $[A^m]_{m=1}^{\infty}$ converges to zero. Hence

$$H = \sum_{j=0}^{\infty} (A^*)^j G A^j,$$

and clearly $H > 0$. Thus, $i_+(H) = n$, and in this case the theorem is proved.

Now consider the general case. From the conditions of the theorem it follows that A has no eigenvalues on the unit circle. Indeed, if $|\lambda| = 1$ and $Ax = \lambda x$ for some nonzero x, then

$$(Gx, x) = (Hx, x) - (HAx, Ax) = 0,$$

which contradicts the positive definiteness of G.

Let $\mathcal{L}_+$ be the span of the eigenvectors and generalized eigenvectors corresponding to the eigenvalues of A in the open unit circle. The subspace $\mathcal{L}_+$ is A-invariant. With respect to the orthogonal sum $\mathbb{C}^n = \mathcal{L}_+ \oplus (\mathcal{L}_+)^{\perp}$ the matrices A, H, and G have the block decompositions

$$A = \begin{bmatrix} A_{11} & A_{12} \\ 0 & A_{22} \end{bmatrix}, \quad H = \begin{bmatrix} H_{11} & H_{12} \\ H_{21} & H_{22} \end{bmatrix}, \quad G = \begin{bmatrix} G_{11} & G_{12} \\ G_{21} & G_{22} \end{bmatrix}.$$

Hence

$$G = H - A^* H A = \begin{bmatrix} H_{11} - A_{11}^* H_{11} A_{11} & * \\ * & * \end{bmatrix},$$

and

$$G_{11} = H_{11} - A_{11}^* H_{11} A_{11} > 0.$$

Since $\sigma(A_{11})$ is in the open unit disc, the special case already proved can be used to see that $\mathcal{L}_+$ is a positive subspace in the inner product (Hx, y). According to Theorem 2.3.2 we have

$$\dim \mathcal{L}_+ \le i_+(H).$$

In the same way, it can be shown that

$$\dim \mathcal{L}_- \le i_-(H),$$

where $\mathcal{L}_-$ is the span of all eigenvectors and generalized eigenvectors of A corresponding to all eigenvalues in the exterior of the closed unit disc. Finally, since $\dim \mathcal{L}_- + \dim \mathcal{L}_+ = n$ and $i_-(H) + i_+(H) = n$, we arrive at the desired equalities $n_\pm(A) = i_\pm(H)$. $\square$

A matrix $A \in \mathbb{C}^{n \times n}$ is called a (nonstrict) H-*contraction* if

$$(HAx, Ax) \le (Hx, x) \quad \text{for all } x \in \mathbb{C}^n,$$

in other words, if

$$H - A^* H A \ge 0. \tag{4.6.24}$$

When H is positive definite, (4.6.24) can be replaced by

$$\|H^{1/2}AH^{-1/2}\| \leq 1 \tag{4.6.25}$$

and then it follows that

$$\max_{\lambda \in \sigma(A)} |\lambda| \leq 1. \tag{4.6.26}$$

The conclusions of Theorem 4.6.1 also hold for (nonstrict) H-contractions if the following additional hypothesis is made:

$$\sum_{j=0}^{m} (A^*)^j G A^j > 0 \quad \text{for some positive integer } m.$$

The proof of this statement is essentially the same as that of Theorem 4.6.1.

There is a natural dual concept: a matrix $R \in \mathbf{C}^{n \times n}$ is called a *strict H-expansion* if

$$(HRx, Rx) > (Hx, x) \quad \text{for every } x \in \mathbf{C}^n, \ x \neq 0, \tag{4.6.27}$$

and this equivalent to the inequality

$$R^*HR - H > 0.$$

Then one might anticipate that an invertible matrix $R \in \mathbf{C}^{n \times n}$ is a strict H-expansion if and only if R^{-1} is a strict H-contraction. This is, indeed, the case, and is easily proved.

When H is positive definite, every strict H-expansion R is invertible and the spectrum $\sigma(R)$ lies outside of the closed unit disc. Theorem 4.6.1 also admits an analogue for strict H-expansions. It can be proved first for invertible strict H-expansions, R, using the property that R^{-1} is a strict H-contraction, and then appealing to Theorem 4.6.1. The invertibility hypothesis can be removed by considering invertible strict H-expansions R' sufficiently close to R.

4.7 Dissipative Matrices

As in the preceding section, we fix an invertible hermitian $n \times n$ matrix H. A matrix $B \in \mathbf{C}^{n \times n}$ is called *strictly H-dissipative* if, for all nonzero $x \in \mathbf{C}^n$, the inequality

$$\Re(HBx, x) < 0$$

holds. In other words,

$$B^*H + HB < 0. \tag{4.7.28}$$

In the special case when H is positive definite, (4.7.28) means that

$$\Re(H^{1/2}BH^{-1/2}x, x) < 0 \tag{4.7.29}$$

for all nonzero x, and it follows from this that all eigenvalues of B are in the open left halfplane.

This time, the distribution of the eigenvalues of B with respect to the imaginary axis is determined by the distribution of the positive and negative eigenvalues of H.

Theorem 4.7.1. *Let $B \in \mathbb{C}^{n \times n}$ be a strictly H-dissipative matrix. Then*

$$N_-(B) = i_+(H), \quad N_+(B) = i_-(H), \tag{4.7.30}$$

where for the matrix B we denote by $N_+(B)$ and $N_-(B)$ the number of eigenvalues of B, counted with their algebraic multiplicities, in the open left and the open right half planes, respectively.

Proof. Let us first consider the case when $\sigma(B)$ is in the open left halfplane. The equality

$$B^* H + H B =: G < 0$$

(see (4.7.28)) can be rewritten in the form

$$e^{tB^*} G e^{tB} = e^{tB^*} H B e^{tB} + e^{tB^*} B^* H e^{tB} = \frac{de^{tB^*}}{dt} H e^{tB} + e^{tB^*} H \frac{de^{tB}}{dt} \tag{4.7.31}$$

$$= \frac{d(e^{tB^*} H e^{tB})}{dt}, \quad t \in \mathbb{R}.$$

After integration with respect to t we obtain

$$(e^{tB^*} H e^{tB})|_0^\infty = \int_0^\infty (e^{tB^*} G e^{tB}) dt.$$

Since we assume that $\Re \lambda < 0$ for every $\lambda \in \sigma(B)$, it follows that $\lim_{t \longrightarrow \infty} e^{tB} = \lim_{t \longrightarrow \infty} e^{tB^*} = 0$, and hence

$$H = -\int_0^\infty (e^{tB^*} G e^{tB}) dt. \tag{4.7.32}$$

All integrals appearing in this computation converge absolutely. This can be checked by first rewriting the matrices in (4.7.31) and (4.7.32) in the Jordan basis for B. In this simple form it is easily seen that, after the corresponding multiplication, the integrals of all the entries converge absolutely.

From (4.7.32) we find that $H > 0$ and hence

$$N_-(B) = n = i_+(H).$$

Now consider the general case. It follows from the hypotheses of the theorem that B has no eigenvalues on the imaginary axis. Indeed, if $Bx = \lambda x$, $x \neq 0$, where $\Re \lambda = 0$, then

$$(Gx, x) = (HBx, x) + (B^* Hx, x) = (\lambda + \bar{\lambda})(Hx, x) = 0,$$

a contradiction with the negative definiteness of G.

Let $\mathcal{M}_-$ be the span of the eigenvectors and generalized eigenvectors corresponding to all eigenvalues of B from the open left half plane. The subspace $\mathcal{M}_-$ is B-invariant. With respect to the orthogonal decomposition $\mathbb{C}^n = \mathcal{M}_- \oplus (\mathcal{M}_-)^\perp$, the matrix B has a block triangular form

$$B = \left[\begin{array}{cc} B_{11} & B_{12} \\ 0 & B_{22} \end{array} \right].$$

Also,

$$H = \left[\begin{array}{cc} H_{11} & H_{12} \\ H_{21} & H_{22} \end{array} \right], \quad G = \left[\begin{array}{cc} G_{11} & G_{12} \\ G_{21} & G_{22} \end{array} \right].$$

Hence

$$G = HB + B^*H = \left[\begin{array}{cc} H_{11}B_{11} + B_{11}^*H_{11} & * \\ * & * \end{array} \right],$$

and

$$G_{11} = H_{11}B_{11} + B_{11}^*H_{11} < 0.$$

Since $\sigma(B)$ is in the open left half plane, the part of the theorem already proved applies, and shows that $\mathcal{M}_-$ is a positive subspace in the inner product (Hx, y). Now Theorem 2.3.2 implies that $\dim \mathcal{M}_- \leq i_+(H)$. In a similar way we obtain $\dim \mathcal{M}_+ \leq i_-(H)$, where $\mathcal{M}_+$ is the span of all eigenvectors and generalized eigenvectors of B corresponding to all eigenvalues in the right half plane. Finally, since $\dim \mathcal{M}_- + \dim \mathcal{M}_+ = n$ and $i_-(H) + i_+(H) = n$, we arrive at the desired equalities $N_\pm(B) = i_\mp(H)$. $\qquad\square$

A matrix $B \in \mathbb{C}^{n \times n}$ is called (nonstrictly) H-*dissipative* if

$$\Re(HAx, x) \leq 0 \quad \text{for all } x \in \mathbb{C}^n;$$

in other words, if

$$G := HB + B^*H \leq 0. \tag{4.7.33}$$

In the case when H is positive definite, this inequality can be replaced by

$$\Re(H^{1/2}BH^{-1/2}x, x) \leq 0, \quad x \in \mathbb{C}^n,$$

and hence all eigenvalues of B lie in the *closed* left halfplane.

The result of Theorem 4.7.1 also holds for (nonstrictly) H-dissipative matrices, provided another hypothesis is made, namely, that

$$\int_0^T e^{tB^*} G e^{tB} dt < 0$$

for some positive T. This statement follows from Theorem 4.7.1.

A dual notion is that of *accretive* matrices. Thus, a matrix $S \in \mathbb{C}^{n \times n}$ is called *strictly H-accretive* if

$$\Re(HSx, x) > 0$$

for every nonzero $x \in \mathbb{C}^n$, or in a different notation, $HS + S^*H > 0$. Obviously, S is strictly H-accretive if and only if S is strictly $-H$-dissipative, or $(-S)$ is strictly H-dissipative. For a strictly H-accretive matrix S and positive definite H, the spectrum of S is in the open right half plane. All results about strictly H-dissipative matrices can be extended in the obvious way to strictly H-accretive matrices. The same can be done for (nonstrictly) *H-accretive* matrices, i.e., matrices S for which

$$\Re(HSx, x) \geq 0, \quad \text{for every } x \in \mathbb{C}^n.$$

4.8 Symplectic Matrices

In this section we provide basic properties of real symplectic matrices. Symplectic matrices arise, in particular, in the study of Riccati equations (see Section 14.11). Throughout this section we work with the real skew-symmetric matrix

$$G_{2m} = \begin{bmatrix} 0 & I_m \\ -I_m & 0 \end{bmatrix} \in \mathbb{R}^{2m \times 2m};$$

often abbreviated to G (with the subscript $2m$ understood). The following easily verified properties of G will often be used in this section:

$$G^T = -G, \quad G^2 = -I, \quad G^{-1} = -G = G^T. \tag{4.8.34}$$

A real $2m \times 2m$ matrix S is called *symplectic* if

$$S^T G_{2m} S = G_{2m}. \tag{4.8.35}$$

Example 4.8.1. *A real 2×2 matrix S is symplectic if and only if* $\det S = 1$. *Indeed, an easy computation shows that, with* $a, b, c, d \in \mathbb{R}$,

$$S^T G S = \begin{bmatrix} a & c \\ b & d \end{bmatrix} \begin{bmatrix} 0 & 1 \\ -1 & 0 \end{bmatrix} \begin{bmatrix} a & b \\ c & d \end{bmatrix} = \begin{bmatrix} 0 & ad - bc \\ -ad + bc & 0 \end{bmatrix}.$$

Consequently, $\det S = ad - bc = 1$ *implies* $S^T G S = G$ *and, conversely,* $S^T G S = S$ *implies that* $\det S = ad - bc = 1$.

Elementary properties of symplectic matrices are collected in the following proposition:

Proposition 4.8.2. *If S is symplectic, then S is invertible, and the matrices S^{-1}, S^T, $-S$ are symplectic as well.*
If $S_1, S_2 \in \mathbb{R}^{2m \times 2m}$ are symplectic, then so is $S_1 S_2$.

Proof. These results follow easily from the definition and (4.8.34). We only show that, if S is symplectic, then so is S^T. Indeed, multiply $S^T G S = G$ on left and right by G^{-1} and by S^{-1}, respectively. This results in $G^{-1} S^T G = S^{-1}$. Now multiply on left and right by S and G^{-1}, respectively, to obtain $S G^{-1} S^T = G^{-1}$ and, finally, use the property that $G^{-1} = -G$. $\qquad\square$

In particular, Proposition 4.8.2 shows that the set of $2m \times 2m$ symplectic matrices is a multiplicative group.

To make a connection with earlier topics, observe that a symplectic matrix is iG-unitary, where iG is (complex) hermitian. Thus, the results on H-unitary matrices apply. In particular, we have the following theorem:

Theorem 4.8.3. *Let $S \in \mathsf{R}^{2m \times 2m}$ be a symplectic matrix. Then $\sigma(S)$ is symmetric relative to both the real line and the unit circle, i.e.,*

$$\lambda_0 \in \sigma(S) \quad \Longrightarrow \quad \overline{\lambda_0} \in \sigma(S), \quad \overline{\lambda_0^{-1}} \in \sigma(S), \quad \lambda_0^{-1} \in \sigma(S).$$

Moreover, in the Jordan normal form of S, the sizes of Jordan blocks with eigenvalue λ_0, and the sizes of Jordan blocks with each of the following eigenvalues: $\overline{\lambda_0}$, $\overline{\lambda_0^{-1}} \in \sigma(S)$, $\lambda_0^{-1} \in \sigma(S)$, are the same.

Proof. The result follows from Proposition 4.3.3, and the fact that S is a real matrix (in particular, the spectrum of S is symmetric relative to the real axis). $\quad\square$

It will be convenient to use Theorem 4.8.3 to classify elementary real invariant subspaces of symplectic matrices into four types. Recall that

$$\mathcal{R}_{\mathsf{R},\lambda}(A) = \mathrm{Ker}\,(A - \lambda I)^n \subseteq \mathsf{R}^n,$$

and

$$\mathcal{R}_{\mathsf{R},\mu \pm i\nu}(A) = \mathrm{Ker}\,(A^2 - 2\mu A + (\mu^2 + \nu^2)I)^n \subseteq \mathsf{R}^n,$$

denote the real root subspaces of a real $n \times n$ matrix A corresponding to a real eigenvalue λ, or to a pair of nonreal complex conjugate eigenvalues $\mu \pm i\nu$, respectively. If S is symplectic, we define

$$\mathcal{RS}_\lambda(S) := \begin{cases} \mathcal{R}_{\mathsf{R},\lambda}(S) & \text{if } \lambda = 1 \text{ or } \lambda = -1 \\[2mm] \mathcal{R}_{\mathsf{R},\mu \pm i\nu}(S) & \text{if } |\lambda| = 1 \text{ and the imaginary part of} \\ & \lambda =: \mu + i\nu \text{ is positive} \\[2mm] \mathcal{R}_{\mathsf{R},\lambda}(S) \dotplus \mathcal{R}_{\mathsf{R},\lambda^{-1}}(S) & \text{if } \lambda \in \mathsf{R},\ |\lambda| > 1 \\[2mm] \mathcal{R}_{\mathsf{R},\mu_1 \pm i\nu_1}(S) \dotplus \mathcal{R}_{\mathsf{R},\mu_2 \pm i\nu_2}(S) & \text{if } \lambda \text{ has positive imaginary part} \\ & \text{and } |\lambda| > 1. \\[2mm] & \text{(Here, } \lambda =: \mu_1 + i\nu_1 \text{ and} \\ & \lambda^{-1} =: \mu_2 + i\nu_2 \text{).} \end{cases}$$

Clearly (cf. Theorem A.2.7), there is a direct sum decomposition of real invariant subspaces for a symplectic matrix S:

$$\mathsf{R}^n = \sum \dotplus \mathcal{R}\mathcal{S}_\lambda(S),$$

where the direct sum is taken over all eigenvalues of S in the set

$$\{1\} \cup \{-1\} \cup \{z \in \mathsf{C} : |z| = 1, \quad \Im z > 0\} \cup \{z \in \mathsf{R} : |z| > 1\}.$$

With this notation, we can now state an orthogonality result analogous to Corollary 4.3.6.

Theorem 4.8.4. *If S is symplectic, and if $v \in \mathcal{R}\mathcal{S}_{\lambda_1}(S)$, $w \in \mathcal{R}\mathcal{S}_{\lambda_2}(S)$, where $\lambda_1 \neq \lambda_2$, then v and w are G-orthogonal:*

$$(Gv, w) = 0. \tag{4.8.36}$$

Given that S is iG-unitary, this result is an immediate consequence of Corollary 4.3.6.

It turns out that the eigenvalues ± 1 of symplectic matrices have special structure:

Theorem 4.8.5. *For a symplectic matrix, the algebraic multiplicity of the eigenvalue 1, as well as of the eigenvalue -1, is even.*

The following proof of this theorem is based on perturbation arguments (see [74, Appendix]). A lemma will be useful:

Lemma 4.8.6. (a) *If $S_0 \in \mathsf{R}^{2m \times 2m}$ is a symplectic matrix, and if $L \in \mathsf{R}^{2m \times 2m}$ is a symmetric matrix, then for every value of the real parameter t the matrix $S(t) := e^{tGL} S_0$ is symplectic.*

(b) *If, in addition to the hypotheses of part* (a), *the matrix L is positive definite, then there exists an $\varepsilon > 0$ such that for $0 < |t| < \varepsilon$, $t \in \mathsf{R}$, the matrix $S(t) + I$ is invertible, i.e., -1 is not an eigenvalue of $S(t)$.*

Proof. Consider the derivative of the matrix function $S(t)^T G S(t)$:

$$\frac{d(S(t)^T G S(t))}{dt} = \frac{dS(t)^T}{dt} G S(t) + S(t)^T G \frac{dS(t)}{dt}$$

$$= S_0^T (GL)^T e^{t(GL)^T} G e^{tGL} S_0 + S_0^T e^{t(GL)^T} G (GL) e^{tGL} S_0.$$

Since

$$(GL)^T e^{t(GL)^T} = e^{t(GL)^T} (GL)^T$$

and $L^T G^T G = -G^2 L$ (which follows from (4.8.34) and $L^T = L$), it is found that

$$\frac{d(S(t)^T G S(t))}{dt} = 0, \quad \text{and hence} \quad S(t)^T G S(t) = \text{constant}.$$

Evaluating $S(t)^T GS(t)$ at $t = 0$ we see that $S(t)^T GS(t) = G$ for all real t, i.e., $S(t)^T GS(t)$ is symplectic.

For part (b), we argue by contradiction. Assume that $S(t_p)+I$ is not invertible for a sequence of nonzero real values t_p, $p = 1, 2, \ldots$, such that $t_p \to 0$ as $p \to \infty$. An application of Theorem A.6.4 (see also the paragraph after that theorem) with $A(t) = S(t) + I$ yields the existence of a nonzero real analytic vector function $h(t) \in \mathsf{R}^n$ in a real neighborhood $\mathcal{U}$ of $t_0 = 0$ such that

$$S(t)h(t) = -h(t), \quad t \in \mathcal{U}.$$

By differentiating we obtain

$$\frac{dS(t)}{dt}h(t) + S(t)\frac{dh(t)}{dt} = -\frac{dh(t)}{dt},$$

or

$$GLS(t)h(t) = -h_t - S(t)h_t, \quad h_t := \frac{dh(t)}{dt}.$$

The left-hand side here is equal to $-GLh(t)$; thus

$$GLh(t) = h_t(t) + S(t)h_t.$$

Now use the fact that $S(t)$ is symplectic in the following inner product to obtain:

$$
\begin{aligned}
(GLh(t), Gh(t)) &= (h_t, Gh(t)) + (S(t)h_t, Gh(t)) \\
&= (h_t, Gh(t)) + (h_t, S(t)^T Gh(t)) \\
&= (h_t, Gh(t)) + (h_t, GS(t)^{-1}h(t)) \\
&= (h_t, Gh(t)) + (h_t, G(-h(t))) = 0. \quad (4.8.37)
\end{aligned}
$$

On the other hand,

$$(GLh(t), Gh(t)) = (G^T GLh(t), h(t)) = (Lh(t), h(t)). \quad (4.8.38)$$

But (4.8.37) and (4.8.38) contradict the positive definiteness of L. $\qquad\square$

Proof of Theorem 4.8.5. It will suffice to consider the eigenvalue -1 only, since $-S$ is symplectic if and only if S is symplectic.

Arguing by contradiction, assume that a symplectic matrix S_0 has an eigenvalue -1 of odd algebraic multiplicity k. Let $0 < \varepsilon < 1$ be such that -1 is the only eigenvalue (perhaps of high multiplicity) of S in the closed disc of radius ε centered at -1. Since eigenvalues of a matrix are continuous functions of the entries of the matrix, there exists a $\delta > 0$ such that every symplectic matrix S satisfying $\|S - S_0\| < \delta$ has exactly k eigenvalues (counted with their algebraic multiplicities) in the open disc

$$\{z \in \mathsf{C} : |z + 1| < \varepsilon\}. \quad (4.8.39)$$

Since $\varepsilon < 1$, none of the k eigenvalues of S in the disc (4.8.39) is equal to 1. Also, if δ is sufficiently small then, for every eigenvalue λ of S in the disc (4.8.39), the eigenvalues $\bar{\lambda}$, λ^{-1}, and $\overline{\lambda^{-1}}$ are also in the disc (4.8.39). But it follows from Theorem 4.8.3 that the number of eigenvalues of S in the disc (4.8.39) that are different from -1 is even. Therefore S must have -1 as an eigenvalue of odd algebraic multiplicity. But this contradicts Lemma 4.8.6. $\qquad\square$

Corollary 4.8.7. *Every symplectic matrix S has determinant equal to 1.*

Proof. Observe that $S^T G S = G$ yields $\det S = \pm 1$ and, in view of Theorem 4.8.3, the case $\det S = -1$ is equivalent to S having -1 as an eigenvalue of odd algebraic multiplicity. However, this is precluded by Theorem 4.8.5. $\qquad\square$

In numerical analysis, the class of matrices that are simultaneously orthogonal and symplectic, is of importance (see Section 14.11). This class is characterized in the next lemma. It is also a preparation for the proof of Theorem 14.11.2.

Lemma 4.8.8. *A real orthogonal matrix U of size $2m \times 2m$ is symplectic if and only if*

$$U = \begin{bmatrix} Q_1 & Q_2 \\ -Q_2 & Q_1 \end{bmatrix} \qquad (4.8.40)$$

where Q_1 and Q_2 are in $\mathsf{R}^{m \times m}$.

Proof. The proof is a simple verification. It is found that U is orthogonal with the block structure of (4.8.40) if and only if $Q_1^T Q_1 + Q_2^T Q_2 = I_m$ and $Q_1^T Q_2 = Q_2^T Q_1$. But $U^T G U = G$ is also equivalent to these two relations. $\qquad\square$

4.9 Exercises

1. Let $A, H \in \mathsf{C}^{n \times n}$ with H positive definite and assume that at least one of the equalities holds:

$$A = A^{[*]}, \quad A^{-1} = A^{[*]}, \quad A A^{[*]} = A^{[*]} A.$$

   Show that A is diagonalizable in an H-orthogonal basis.

2. If the $n \times n$ matrix A is diagonalizable in C^n show that there exists a positive definite matrix H such that, in $\mathsf{C}^n(H)$, $A A^{[*]} = A^{[*]} A$.

   (a) If, in addition, all eigenvalues of A are real then H can be chosen so that $A^{[*]} = A$.

   (b) If, in addition, all eigenvalues of A are on the unit circle then H can be chosen so that $A^{-1} = A^{[*]}$.

3. Find the eigenvalues and eigenvectors of the following matrices:

(a) $H_0 = \begin{bmatrix} 0 & 0 & \cdots & 0 & 1 \\ 0 & 0 & \cdots & 1 & 0 \\ & \vdots & \cdots & & \vdots \\ 1 & 0 & \cdots & 0 & 0 \end{bmatrix}$.

(b) $H_1 = \begin{bmatrix} a_1 & 0 & \cdots & 0 & 1 \\ 0 & a_2 & \cdots & 1 & 0 \\ & \vdots & \cdots & & \vdots \\ 1 & 0 & \cdots & 0 & a_n \end{bmatrix}$, $a_j \in \mathsf{R}$, $n = 2k$.

(c) $H_2 = i \begin{bmatrix} 0 & 0 & \cdots & 0 & -1 \\ 0 & 0 & \cdots & -1 & 0 \\ \vdots & & \cdots & & \vdots \\ 0 & 1 & \cdots & 0 & 0 \\ 1 & 0 & \cdots & 0 & 0 \end{bmatrix}$, with even size.

(d) $H_3 = P - (I - P)$, where $P^* = P$, $P^2 = P$.

(e) $H_4 = i \begin{bmatrix} 2i & 0 & \cdots & 0 & 1 \\ 0 & -2i & \cdots & 1 & 0 \\ \vdots & & \cdots & & \vdots \\ 0 & -1 & \cdots & 2i & 0 \\ -1 & 0 & \cdots & 0 & -2i \end{bmatrix}$ (the matrix here has even size).

(f) $H_5 = \begin{bmatrix} 1 & \alpha & \alpha & \cdots & \alpha \\ \alpha & 1 & \alpha & \cdots & \alpha \\ & \vdots & & \cdots & \vdots \\ \alpha & \alpha & \alpha & \cdots & 1 \end{bmatrix}$, $\alpha \in \mathsf{R}$.

4. Let $\mathsf{C}^n(H)$ be a space with the indefinite inner product defined by an invertible hermitian matrix H, and let $A = (a_{jk})_{j,k=1}^n$. Find the matrix $A^{[*]}$ in the following cases:

(a) $H = \begin{bmatrix} 0 & 0 & \cdots & 0 & 1 \\ 0 & 0 & \cdots & 1 & 0 \\ \vdots & & \cdots & & \vdots \\ 1 & 0 & \cdots & 0 & 1 \end{bmatrix}$,

(b) $H = \begin{bmatrix} 1 & 0 & \cdots & 0 & 0 \\ 0 & 1 & \cdots & 0 & 0 \\ \vdots & & \cdots & & \vdots \\ 0 & 0 & \cdots & 1 & 0 \\ 0 & 0 & \cdots & & -1 \end{bmatrix}$, $\text{sig } H = n - 1$.

(c) $H = \begin{bmatrix} 0 & 0 & \dots & 0 & i \\ 0 & 0 & \dots & 1 & 0 \\ \vdots & & \dots & & \vdots \\ 0 & 1 & \dots & 0 & 0 \\ -i & & \dots & 0 & 0 \end{bmatrix}$.

(d) $H = \begin{bmatrix} 1 & 0 & \dots & 0 & 0 \\ 0 & 1 & \dots & 0 & 0 \\ \vdots & & \dots & & \vdots \\ 0 & 0 & \dots & 0 & 1 \\ 0 & 0 & \vdots & 1 & 0 \end{bmatrix}$, $\operatorname{sig} H = n - 1$.

(e) $H = \begin{bmatrix} 0 & I_k \\ I_k & 0 \end{bmatrix}$, $n = 2k$.

(f) $H = \begin{bmatrix} S_k & 0 \\ 0 & S_{n-k} \end{bmatrix}$.

(g) $H = \begin{bmatrix} I_k & 0 \\ 0 & S_{n-k} \end{bmatrix}$.

5. In Exercise 4, under what further conditions is A H-selfadjoint?

6. When is a linear transformation A in $\mathbf{C}^n(H)$ simultaneously H-selfadjoint and H-unitary?

7. In the space $\mathbf{C}^n(H)$ the linear transformation A is simultaneously H-selfadjoint and H-unitary. Find its eigenvectors and eigenvalues.

8. Let P be an H-orthogonal projection in $\mathbf{C}^n(H)$. Prove that the subspaces $\operatorname{Ker} P$ and $\operatorname{Range} P$ are H-orthogonal and $P^{[*]} = P$.

9. Let P be as in Exercise 8. Show that the linear transformation

$$S = P - (I - P),$$

is simultaneously H-unitary and H-selfadjoint.

10. Prove that, for the linear transformation

$$Ax = \sum_{j=1}^{r} (H\varphi_j, x)\psi_j$$

in $\mathbf{C}^n(H)$, where φ_j $(j = 1, 2, \dots, n)$ and ψ_j $(j = 1, 2, \dots, n)$ are systems of vectors, the H-adjoint is given by the formula

$$A^{[*]}x = \sum_{j=1}^{n} (H\psi_j, x)\varphi_j.$$

11. When is the linear transformation A from the previous exercise H-selfadjoint?

12. Are the following equalities correct for any invertible hermitian matrix H?

$$(A + B)^{[*]} = A^{[*]} + B^{[*]}, \quad (AB)^{[*]} = B^{[*]} + A^{[*]}, \quad (\alpha A)^{[*]} = \overline{\alpha} A^{[*]},$$

$$I^{[*]} = I, \quad 0^{[*]} = 0.$$

13. Let $\{\varphi_j\}_{j=1}^n$ be an H-orthonormal basis in $C^{(n)}(H)$, that is, $(H\varphi_j, \varphi_k) = 0$ if $j \neq k$ and $(H\varphi_j, \varphi_k) = 1$ if $j = k$. When is the linear transformation

$$Ax = \sum_{j=1}^{m} \alpha_j [x, \varphi_j] \varphi_j, \quad \alpha_j \in C$$

(a) H-unitary ? (b) H-selfadjoint ? (c) H-positive (i.e., $[Ax, x] > 0$ for every nonzero $x \in C^n$) ?

Find the spectrum of A in cases (a), (b), and (c).

14. Let $H = \begin{bmatrix} 0 & 1 \\ 1 & 0 \end{bmatrix}$. Under what conditions is the linear transformation

$$A = \begin{bmatrix} 1 & \alpha \\ 0 & 1 \end{bmatrix}, \quad \alpha \in C$$

(a) H-selfadjoint ? (b) H-unitary ? (c) H-normal ? (d) H-diagonalizable (i.e., there exists an H-orthogonal basis of eigenvectors) ?

15. Is an H-selfadjoint linear transformation always diagonalizable?

16. Is an H-unitary linear transformation always diagonalizable?

17. Let A be H-selfadjoint (H-unitary). Do all eigenvectors of A form an H-orthonormal basis in $C^n(H)$?

18. Under what condition on H does the following equality hold for every matrix A:
$$\text{Range } A = (\text{Ker } A^{[*]})^{\perp}.$$

19. Prove or disprove that for every linear transformation A there exists an H-orthogonal basis with respect to which the matrix of A is lower (upper) triangular.

20. (a) Find the canonical form of an H-selfadjoint matrix A in $C^2(H)$, if

$$H = \begin{bmatrix} 0 & 1 \\ 1 & 0 \end{bmatrix}.$$

(b) Determine when the H-selfadjoint matrix A is H-positive.

21. Let

$$A = \begin{bmatrix} 0 & 1 & 0 & \cdots & 0 \\ 0 & 0 & 1 & \cdots & 0 \\ & & & \cdots & \\ -a_0 & -a_1 & & \cdots & -a_{n-1} \end{bmatrix} ; \quad a_j \in \mathrm{R}.$$

Prove that A is H-selfadjoint, where

$$H = \begin{bmatrix} a_1 & a_2 & \cdots & a_{n-1} & 1 \\ a_2 & a_3 & \cdots & 1 & 0 \\ & & \cdots & & \\ a_{n-1} & 1 & \cdots & & 0 \\ 1 & 0 & \cdots & & 0 \end{bmatrix} .$$

22. Show that the matrix H of the previous exercise is never positive definite.

23. For matrices A and H of Exercise 21, find when A is H-positive.

24. Generalize and repeat Exercise 21 for the block matrix

$$A = \begin{bmatrix} 0 & I & 0 & \cdots & 0 \\ 0 & 0 & I & \cdots & 0 \\ \vdots & & & \cdots & \vdots \\ -A_0 & -A_1 & -A_2 & \cdots & -A_{n-1} \end{bmatrix} , \qquad (4.9.41)$$

where all entries are $k \times k$ matrices and $A_j = A_j^*$.

25. Let

$$W(\lambda) = I + B^* H (I\lambda - A)^{-1} B,$$

where A is a square $n \times n$ H-selfadjoint matrix. Show that $W(\lambda) = [W(\bar{\lambda})]^*$.

26. Assume that (4.9.41) is a block matrix with $k \times k$ blocks and $A_j^* = A_{n-j}$ for $j = 0, 1, \ldots, n$, where $n = 2r$ is even and where $A_0 = A_n = I$. Verify that the matrix A is H-unitary, where

$$H = i \begin{bmatrix} & & & A_n & & 0 \\ & & & A_{n-1} & \ddots & \\ & 0 & & \vdots & \ddots & \ddots \\ & & & A_{r+1} & \cdots & A_{n-1} & A_n \\ -A_0 & -A_1 & \cdots & -A_{r-1} & & & \\ & -A_0 & \cdots & -A_{r-2} & & 0 & \\ & & \ddots & \vdots & & & \\ 0 & & & -A_0 & & & \end{bmatrix} .$$

27. Extend the result of the preceding exercise (with the same matrix H) to the case when $A_j^* = A_{n-j}$ for $j = 0, 1, \ldots, n$, n is even, and A_0 (and hence also A_n) is invertible.

28. Let $X(t)$ be an $n \times n$ matrix function satisfying the initial value problem

$$E \frac{dX}{dt} = iH(t)X, \quad X(0) = I,$$

where E and $H(t)$ are selfadjoint. Show that $X(t)$ is E-unitary for any $t \in \mathsf{R}$.

29. Provide details for the proof of Proposition 4.3.3.

30. Let A be an $n \times n$ matrix with $\sigma(A)$ inside the open unit disc.

   (a) Show that for any $n \times n$ matrix D there exists a unique solution X of the matrix equation

$$X - A^* X A = D.$$

   Show that the solution can be expressed in the form

$$X = \sum_{j=0}^{\infty} (A^*)^j D A^j.$$

   (Hint: If A is not nilpotent, prove the inequalities $\|A^j\| \leq K q^j$, $j = 1, 2, \ldots$, where the positive constant K is independent of j, and $q = \max\{|\lambda| : \lambda \in \sigma(A)\}$.)

   (b) Show that A is similar to a strict contraction: $A = S^{-1} B S$ for some invertible S and some B such that $\|B\| < 1$.

31. Prove that a matrix $A \in \mathsf{C}^{n \times n}$ is a strict H-contraction for some positive definite H if and only if the spectrum $\sigma(A)$ is in the open unit disc. Is the matrix H unique?

32. Prove that a matrix $A \in \mathsf{C}^{n \times n}$ is a strict H-contraction for some selfadjoint invertible H if and only if the spectrum $\sigma(A)$ does not intersect the unit disc. Is the matrix H unique?

33. Let B be an $n \times n$ matrix with $\sigma(B)$ inside the left open half plane.

   (a) Show that for any $n \times n$ matrix D the equation

$$BX + XB^* = D$$

   has a unique solution X, and that the solution can be expressed in the form

$$X = -\int_0^{\infty} e^{tB^*} D e^{tB} dt.$$

(Hint: Prove the inequality

$$\|e^{tB}\| \leq Ke^{\sigma t}, \quad 0 \leq t < \infty,$$

where the positive constant K depends on B only, and $\sigma = \max\{\Re\lambda :$ $\lambda \in \sigma(B)\}$.)

(b) Show that the matrix B is similar to a strictly dissipative matrix, i.e., there exists an invertible S such that $\Re(S^{-1}BSx, x) < 0$ for every nonzero $x \in \mathbf{C}^n$.

34. Prove that an $n \times n$ matrix B is strictly H-dissipative for some positive definite H if and only if $\sigma(B)$ is in the open left half plane. Is the matrix H unique?

35. Prove that an $n \times n$ matrix B is strictly H-dissipative for some invertible hermitian H if and only if $\sigma(B)$ does not intersect the imaginary axis. Is H unique?

36. Let S be a symplectic matrix, and let $\mathcal{R}_{\mathbf{R},\lambda_1}(S)$, $\mathcal{R}_{\mathbf{R},\lambda_2}(S)$, $\mathcal{R}_{\mathbf{R},\mu_1\pm i\nu_1}(S)$, $\mathcal{R}_{\mathbf{R},\mu_2\pm i\nu_2}(S)$ be the real root subspaces of S corresponding to real eigenvalues λ_1 and λ_2 and pairs of nonreal complex conjugate eigenvalues $\mu_1 \pm i\nu_1$ and $\mu_2 \pm i\nu_2$. Find when:

(a) $\mathcal{R}_{\mathbf{R},\lambda_1}(S)$ and $\mathcal{R}_{\mathbf{R},\lambda_2}(S)$ are G-orthogonal;

(b) $\mathcal{R}_{\mathbf{R},\lambda_1}(S)$ and $\mathcal{R}_{\mathbf{R},\mu_1\pm i\nu_1}(S)$ are G-orthogonal;

(c) $\mathcal{R}_{\mathbf{R},\mu_1\pm i\nu_1}(S)$ and $\mathcal{R}_{\mathbf{R},\mu_2\pm i\nu_2}(S)$ are G-orthogonal.

4.10   Notes

The material of this chapter is well-known, and the greater part of Sections 4.1, 4.2, 4.3, and 4.4 is known even in infinite dimensional setting [5], [6], [11], [57].

Theorem 4.4.2 appeared in [56].

Theorems 4.6.1 and 4.7.1 are called inertia theorems in some literature. Different versions of these theorems can be found in [14], [49], [62], [86], [97], [98], and [107]. See also [16].

More details on the structure and properties of real symplectic matrices are found in [73], [4], [74, Appendix], [30], [108], [22], [78], among many other sources.

Chapter 5

Canonical Forms

The fundamental definition of an H-selfadjoint matrix A has been introduced in the preceding chapter. Also, in Section 4.5, it has been shown that pairs (A, H) of this kind can be divided into equivalence classes under the relation of "unitarily similarity". Thus, pairs (A_1, H_1) and (A_2, H_2) of this kind are in the same equivalence class if there is a matrix T such that

$$A_1 = T^{-1} A_2 T \quad \text{and} \quad H_1 = T^* H_2 T,$$

or, what is the same, there is a simultaneous congruence $H_1 = T^* H_2 T$ and $H_1 A_1 = T^*(H_2 A_2)T$. The first objective of this chapter is the description of canonical pairs in these equivalence classes. As one might expect, this investigation reveals important invariants of unitary similarity and, in particular, the notion of the "sign characteristic" of an equivalence class is found to be important. Indeed, three different characterizations of this notion are to be developed here. The stability of the sign characteristic under perturbations is also investigated.

Similar questions are studied in the case of unitary similarity of pairs (A, H) in which A is H-unitary. A broader class of matrices, the "H-normal" matrices, including the H-selfadjoint and H-unitaries will be the topic of Chapter 8.

5.1 Description of a Canonical Form

Using the notation of Section 4.5, consider a unitary similarity class of pairs of matrices $(A, H) \in \mathcal{U}$ in which A is H-selfadjoint. We seek a canonical form in such a class, i.e., a standard simple pair (J, P) in the class with the property that for each pair (A, H) in the class, there is a T such that

$$A = T^{-1} J T, \qquad H = T^* P T.$$

In fact, the canonical pair (J, P) will consist of a matrix J in Jordan normal form, and a hermitian, invertible P of simple structure (in particular, it will transpire that $P^2 = I$).

Since A is H-selfadjoint we know (Proposition 4.2.3) that the number and the sizes of Jordan blocks in J corresponding to an eigenvalue $\lambda_0 (\neq \overline{\lambda_0})$ and those for $\overline{\lambda_0}$ are the same. So we can assume that J is a direct sum of Jordan blocks with *real* eigenvalues and blocks of the type diag $(J_k, \overline{J_k})$, where J_k is a Jordan block with nonreal eigenvalue.

It will be convenient to introduce the following notation. By $J_k(\lambda)$ we denote the Jordan block with eigenvalue λ of size k if λ is real, and the direct sum of two Jordan blocks of size $\frac{k}{2}$, the first with eigenvalue λ and the second with eigenvalue $\overline{\lambda}$, if λ is nonreal. Often we shall write $J(\lambda)$ omitting the subscript k.

We have seen in Example 4.2.1 that $J_k(\lambda)$ is $\pm P$-selfadjoint, where P is the $k \times k$ sip matrix. The following theorem shows that the pair (A, H) is unitarily similar to a direct sum of blocks of types $(J(\lambda), \pm P)$ when λ is real and $(J(\lambda), P)$ when λ is nonreal.

Theorem 5.1.1. *Let $A, H \in \mathbb{C}^{n \times n}$ with H selfadjoint and invertible and let A be H-selfadjoint. Then there is an invertible $T \in \mathbb{C}^{n \times n}$ such that $A = T^{-1} JT$ and $H = T^* PT$ where*

$$J = J(\lambda_1) \oplus \cdots \oplus J(\lambda_\alpha) \oplus J(\lambda_{\alpha+1}) \oplus \cdots \oplus J(\lambda_\beta) \qquad (5.1.1)$$

is a Jordan normal form for A, $\lambda_1, \ldots, \lambda_\alpha$ are the real eigenvalues of A, and $\lambda_{\alpha+1}, \ldots, \lambda_\beta$ are the nonreal eigenvalues of A from the upper half-plane. Also,

$$P_{\varepsilon,J} = \varepsilon_1 P_1 \oplus \cdots \oplus \varepsilon_\alpha P_\alpha \oplus P_{\alpha+1} \oplus \cdots \oplus P_\beta, \qquad (5.1.2)$$

where $P_1, P_2, \ldots, P_\beta$ are sip matrices with the sizes of $J(\lambda_1), \ldots, J(\lambda_\beta)$ respectively, and $\varepsilon = \{\varepsilon_1, \ldots, \varepsilon_\alpha\}$ is an ordered set of signs ± 1. The set ε is uniquely determined by (A, H) up to permutation of signs corresponding to equal Jordan blocks.

Conversely, if for some set of signs ε, the pairs (A, H) and $(J, P_{\varepsilon,J})$ are unitarily similar, then A is H-selfadjoint.

Note that in Theorem 5.1.1 the number of blocks with each eigenvalue λ_k of A is equal to the geometric multiplicity of λ_k, i.e., the dimension of Ker $(\lambda_k I - A)$.

The set of signs ε appearing in this theorem will be called the *sign characteristic* of the pair (A, H), and recall that it consists of a $+1$ or -1 factor applied to each *real* Jordan block of the Jordan form J of A.

An alternative description of the sign characteristic can be made in terms of partial multiplicities of A associated with real eigenvalues: one sign ($+1$ or -1) is attached to each such partial multiplicity.

The proof of Theorem 5.1.1 will be given later in this chapter. Here, we note two immediate corollaries. The first gives a complete description of all the invertible hermitian matrices H for which a given matrix A is H-selfadjoint.

Corollary 5.1.2. *Let A be an $n \times n$ matrix which is similar to A^* and let J be a Jordan form for A arranged as in (5.1.1). Then A is H-selfadjoint if and only if H has the form $H = T^* P_{\varepsilon, J} T$ where $P_{\varepsilon, J}$ is given by (5.1.2) for some set of signs ε, and T is an invertible matrix for which $A = T^{-1} J T$.*

In particular, the following result follows immediately.

Corollary 5.1.3. *If an $n \times n$ matrix A is similar to A^*, then there exists an invertible hermitian H such that $A^* = H^{-1} A H$. Such a matrix is given by $H = T^* P_{\varepsilon, J} T$ where J is a Jordan normal form for A (arranged as in (5.1.1)), T is any invertible matrix for which $A = T^{-1} J T$, ε is an arbitrarily chosen set of signs, and $P_{\varepsilon, J}$ is given by (5.1.2).*

The following example is a simple illustration of Corollary 5.1.2.

Example 5.1.4. *Let*

$$J = \mathrm{diag} \left(\begin{bmatrix} 0 & 1 \\ 0 & 0 \end{bmatrix}, \begin{bmatrix} 1 & 1 \\ 0 & 1 \end{bmatrix} \right).$$

Then

$$P_{\varepsilon, J} = \mathrm{diag} \left(\begin{bmatrix} 0 & \varepsilon_1 \\ \varepsilon_1 & 0 \end{bmatrix}, \begin{bmatrix} 0 & \varepsilon_2 \\ \varepsilon_2 & 0 \end{bmatrix} \right) \quad for \quad \varepsilon = (\varepsilon_1, \varepsilon_2), \quad \varepsilon_i = \pm 1.$$

According to Theorem 5.1.1, the set $\Omega = \Omega_J$ of all invertible hermitian matrices H such that J is H-selfadjoint, splits into 4 disjoint sets $\Omega_1, \Omega_2, \Omega_3, \Omega_4$ corresponding to the sets of signs $(+1, +1)$, $(+1, -1)$, $(-1, +1)$ and $(-1, -1)$ respectively: $\Omega = \Omega_1 \cup \Omega_2 \cup \Omega_3 \cup \Omega_4$. An easy computation shows that each set Ω_i consists of all matrices H of the form

$$H = \mathrm{diag} \left(\begin{bmatrix} 0 & a_1 \varepsilon_1 \\ a_1 \varepsilon_1 & b_1 \end{bmatrix}, \begin{bmatrix} 0 & a_2 \varepsilon_2 \\ a_2 \varepsilon_2 & b_2 \end{bmatrix} \right),$$

where a_1, a_2, are positive and b_1, b_2 are real parameters; $\varepsilon_1, \varepsilon_2$ are ± 1 depending on the set Ω_i. Note also that each set Ω_i $(i = 1, 2, 3, 4)$ is (arcwise) connected. $\square$

5.2 First Application of the Canonical Form

In this section we consider some conclusions that can be drawn readily from Theorem 5.1.1. The first concerns the structure of $\mathrm{Ker}(\lambda_0 I - A)$ when $\lambda_0 \in \sigma(A)$ and A is H-selfadjoint. Note that for any eigenvalue λ_0 the eigenspace $\mathrm{Ker}(\lambda_0 I - A)$ can be written as a direct sum

$$\mathrm{Ker}(\lambda_0 I - A) = \mathcal{L}_1 \dotplus \mathcal{L}_2, \tag{5.2.3}$$

where $\mathcal{L}_1$ and $\mathcal{L}_2$ are generated by all eigenvectors of λ_0 associated with partial multiplicities equal to 1, and partial multiplicities larger than 1, respectively. These subspaces could be defined in other ways. For example,

$$\mathcal{L}_1 = \mathrm{Span}\,\{x_1,\ldots,x_k\}, \quad \mathcal{L}_2 = \mathrm{Ker}(\lambda_0 I - A) \cap \mathrm{Range}\,(\lambda_0 I - A),$$

where $\{x_1,\ldots,x_k\}$ is a maximal set of vectors in $\mathrm{Ker}(\lambda_0 I - A)$ which are linearly independent modulo $\mathrm{Range}(\lambda_0 I - A)$. Clearly, $\dim \mathcal{L}_1$ ($\dim \mathcal{L}_2$) is just the number of partial multiplicities equal to 1 (partial multiplicities larger than 1) associated with λ_0.

Our first observation is that, if A *is H-selfadjoint and $\lambda_0 \in \sigma(A)$ is real then, in the decomposition* (5.2.3), $\mathcal{L}_1$ *is H-nondegenerate and $\mathcal{L}_2$ is H-neutral.* To see this, suppose that λ_0 has p partial multiplicities equal to 1, and q partial multiplicities larger than 1. Then the corresponding submatrices of J and $P_{\varepsilon,J}$ from (5.1.1) and (5.1.2) have the form

$$J^{(0)} = \mathrm{diag}\,(\lambda_0,\ldots,\lambda_0, J_1, J_2, \ldots, J_q)$$

where λ_0 appears p times and J_j is a Jordan block of size $m_j \geq 2$ for $j = 1, 2, \ldots, q$,

$$P_{\varepsilon,J} = \mathrm{diag}\,(\varepsilon_1,\ldots,\varepsilon_p, \varepsilon_{p+1}P_{p+1},\ldots,\varepsilon_q P_q).$$

For these matrices, the eigenspace associated with partial multiplicities equal to 1, say $\mathcal{L}_1^{(0)}$ is spanned by unit coordinate vectors $e_1,\ldots,e_p$ and, in the indefinite inner product determined by $P_{\varepsilon,J}^{(0)}$, we have $[e_j,e_k] = \varepsilon_j \delta_{jk}$ for $j, k = 1, 2, \ldots, p$. It follows readily that $\mathcal{L}_1^{(0)}$ is $P_{\varepsilon,J}^{(0)}$-nondegenerate, and hence that $\mathcal{L}_1$ is H-nondegenerate.

The eigenspace of $J^{(0)}$ associated with the partial multiplicities that are larger than 1, say $\mathcal{L}_2^{(0)}$ is spanned by unit vectors e_j with $j = p + 1, p + m_1 + 1, \ldots, p + \sum_{r=1}^{q-1} m_r + 1$, and for indices j, k taking these values we clearly have $[e_j, e_k] = 0$ in the $P_{\varepsilon,J}^{(0)}$ indefinite inner product. It follows that $\mathcal{L}_2$ is H-neutral.

If $\lambda_0 \in \sigma(A)$ and $\overline{\lambda_0} \neq \lambda_0$, then the whole eigenspace $\mathcal{L}_1 \dotplus \mathcal{L}_2$ is H-neutral. This can be deducted from the presence of zeros in certain strategic positions of $P_{\varepsilon,J}$, but this statement can be strengthened to the observation that the whole root subspace $\mathcal{R}_{\lambda_0}(A) = \mathrm{Ker}(\lambda_0 I - A)^n$, where n is the size of A, is H-neutral. This can be "seen" from the canonical pair $J, P_{\varepsilon,J}$, but has also been proved in Corollary 4.2.5.

Another useful observation can be made in the case that $\lambda_0 \in \sigma(A)$ and $\overline{\lambda_0} \neq \lambda_0$; namely, that *the direct sum of root subspaces* $\mathcal{R}_{\lambda_0}(A) \dotplus \mathcal{R}_{\overline{\lambda_0}}(A)$ *is nondegenerate with respect to H.* This is easily verified for a canonical pair $(J(\lambda_0), P)$ and hence, in full generality.

Our next deduction from the canonical forms of Theorem 5.1.1 concerns the possibility of counting the negative (or the positive) eigenvalues of H, once the sign characteristic is known. To be precise about this, let the real Jordan blocks

$J(\lambda_1), J(\lambda_2), \ldots, J(\lambda_\alpha)$ of (5.1.1) have sizes $m_1, \ldots, m_\alpha$ and note (from (5.1.2)) that the associated signs in the sign characteristic of (A, H) are $\varepsilon_1, \ldots, \varepsilon_\alpha$. If $i_-(H)$ is the number of *negative* eigenvalues of H (and hence of $P_{\varepsilon,J}$), counted with multiplicities, it follows first from (5.1.2) that

$$\operatorname{sig} P_{\varepsilon,J} = \frac{1}{2} \sum_{i=1}^{\alpha} [1 - (-1)^{m_j}]\varepsilon_i$$

and, hence, that

$$i_-(H) = \frac{1}{2}n - \frac{1}{4} \sum_{i=1}^{\alpha} [1 - (-1)^{m_j}]\varepsilon_i. \tag{5.2.4}$$

This equation immediately implies that

$$i_-(H) \geq \frac{1}{2}n - \frac{1}{2} \sum_{i=1}^{\alpha} m_i$$

with equality if and only if the m_i are all equal to one and $\varepsilon_1 = \cdots = \varepsilon_\alpha = 1$. Now the lower bound for $i_-(H)$ is just half the number of nonreal eigenvalues of A. Thus we obtain the following corollary:

Corollary 5.2.1. *Let A be H-selfadjoint. Then H has $\frac{1}{2}n$ positive eigenvalues (and so $\frac{1}{2}n$ negative eigenvalues), counted with multiplicities, if and only if the signs associated with the Jordan blocks in the Jordan form of A that correspond to real eigenvalues and have odd size (if any) are equally divided between $+1$'s and -1's.*

5.3 Proof of Theorem 5.1.1

It is easily verified that, in the statement of Theorem 5.1.1, J is $P_{\varepsilon,J}$-selfadjoint. The converse statement of the theorem is then an immediate application of Proposition 4.1.3.

Now let A be H-selfadjoint, let $\lambda_1, \ldots, \lambda_\alpha$ be all the different real eigenvalues of A and $\lambda_{\alpha+1}, \ldots, \lambda_\beta$ be all the different eigenvalues in the open upper half of the complex plane. Decompose $\mathbb{C}^n$ into a direct sum:

$$\mathbb{C}^n = \mathcal{X}_1 \dotplus \cdots \dotplus \mathcal{X}_\alpha \dotplus \mathcal{X}_{\alpha+1} \dotplus \cdots \dotplus \mathcal{X}_\beta$$

where $\mathcal{X}_1, \ldots, \mathcal{X}_\alpha$ are the root subspaces corresponding to $\lambda_1, \ldots, \lambda_\alpha$ respectively, and for $j = \alpha + 1, \ldots, \beta$, the subspace $\mathcal{X}_j$ is the sum of the root subspaces corresponding to λ_j and $\overline{\lambda}_j$. It follows immediately from Theorem 4.2.4 that for $j, k = 1, 2, \ldots, \beta$ the subspaces $\mathcal{X}_j$ and $\mathcal{X}_k$ are orthogonal with respect to H, i.e., $[x, y] = (Hx, y) = 0$ for every $x \in \mathcal{X}_j$, $y \in \mathcal{X}_k$. From (2.2.2) it follows that for $i = 1, 2, \ldots, \beta$,

$$\mathcal{X}_i^{[\perp]} = \mathcal{X}_1 \dotplus \cdots \dotplus \mathcal{X}_{i-1} \dotplus \mathcal{X}_{i+1} \dotplus \cdots \dotplus \mathcal{X}_\beta \tag{5.3.5}$$

and then Proposition 2.2.2 implies that each $\mathcal{X}_i$ is nondegenerate.

Consider a fixed $\mathcal{X}_i$ with $i \in \{\alpha + 1, \ldots, \beta\}$, so that $\mathcal{X}_i = \mathcal{X}_i' + \mathcal{X}_i''$ and X_i', $\mathcal{X}_i''$ are the root subspaces corresponding to λ_i and $\overline{\lambda}_i$, respectively. Corollary 4.2.5 asserts that both $\mathcal{X}_i'$ and $\mathcal{X}_i''$ are H-neutral.

There exists an integer m with the properties that $(A - \lambda_i I)^m \mid_{\mathcal{X}_i'} = 0$, but $(A - \lambda_i I)^{m-1} a_1 \neq 0$ for some $a_1 \in \mathcal{X}_i'$. Since $\mathcal{X}_i$ is nondegenerate and $\mathcal{X}_i'$ is neutral, there exists a $b_1 \in \mathcal{X}_i''$ such that $[(A - \lambda_i I)^{m-1} a_1, b_1] = 1$. Define sequences $a_1, \ldots, a_m \in \mathcal{X}_i'$ and $b_1, \ldots, b_m \in \mathcal{X}_i''$ by

$$a_j = (A - \lambda_i I)^{j-1} a_1, \quad b_j = (A - \overline{\lambda}_i I)^{j-1} b_1, \quad j = 1, \ldots, m.$$

Observe that

$$[a_1, b_m] = [a_1, (A - \overline{\lambda}_i I)^{m-1} b_1] = [(A - \lambda_i I)^{m-1} a_1, b_1] = 1,$$

in particular, $b_m \neq 0$. Further, for every $x \in \mathcal{X}_i'$ we have

$$[x, (A - \overline{\lambda}_i I) b_m] = [x, (A - \overline{\lambda}_i I)^m b_1] = [(A - \lambda_i I)^m x, b_1] = 0;$$

so the vector $(A - \overline{\lambda}_i I) b_m$ is H-orthogonal to $\mathcal{X}_i'$. In view of (5.3.5) we deduce that $(A - \overline{\lambda}_i I) b_m$ is H-orthogonal to $\mathbb{C}^n$, and hence

$$(A - \overline{\lambda}_i I) b_m = 0.$$

Then clearly $a_m, \ldots, a_1$ (resp. $b_m, \ldots, b_1$) is a Jordan chain of A corresponding to λ_i (resp. $\overline{\lambda}_i$), i.e., for $j = 1, 2, \ldots, m - 1$,

$$A a_j - \lambda_i a_j = a_{j+1} \quad \text{and} \quad A a_m = \lambda_i a_m,$$

and

$$A b_j - \overline{\lambda}_i b_j = b_{j+1} \quad \text{and} \quad A b_m = \overline{\lambda}_i b_m.$$

For $j + k = m + 1$ we have

$$[a_j, b_k] = [(A - \lambda_i I)^{j-1} a_1, (A - \overline{\lambda}_i I)^{k-1} b_1] = [(A - \lambda_i I)^{j+k-2} a_1, b_1] = 1 \quad (5.3.6)$$

and, similarly,

$$[a_j, b_k] = 0 \quad \text{if} \quad j + k > m + 1. \tag{5.3.7}$$

Now put

$$c_1 = a_1 + \sum_{j=2}^{m} \alpha_j a_j, \quad c_{j+1} = (A - \lambda_i I) c_j, \quad j = 1, \ldots, m - 1,$$

where $\alpha_2, \ldots, \alpha_m$ are chosen so that

$$[c_1, b_{m-1}] = [c_1, b_{m-2}] = \cdots = [c_1, b_1] = 0. \tag{5.3.8}$$

Such a choice is possible, as can be checked easily using (5.3.6) and (5.3.7). Now for $j + k \leq m$ we have

$$[c_j, b_k] = [(A - \lambda_i I)^{j-1} c_1, b_k] = [c_1, (A - \overline{\lambda_i} I)^{j-1} b_k] = [c_1, b_{k+j-1}] = 0,$$

and for $j + k \geq m + 1$ we obtain, using $(A - \lambda_i I)^m a_1 = 0$ together with (5.3.6) and (5.3.7):

$$
\begin{aligned}
[c_j, b_k] &= [(A - \lambda_i I)^{j-1} c_1, (A - \overline{\lambda_i} I)^{k-1} b_1] \\
&= [(A - \lambda_i I)^{j+k-2} c_1, b_1] = [(A - \lambda_i I)^{j+k-2} a_1, b_1] \\
&= \begin{cases} 1, & \text{for } j + k = m + 1 \\ 0, & \text{for } j + k > m + 1 \end{cases}.
\end{aligned}
$$

Let $\mathcal{N}_1 = \operatorname{Span} \{c_1, \ldots, c_m, b_1, \ldots, b_m\}$. The relations above show that $A|_{\mathcal{N}_1} = J_1 \oplus \overline{J_1}$ in the basis $c_m, \ldots, c_1, b_m, \ldots, b_1$ where J_1 is the Jordan block of size m with eigenvalue λ_i; and

$$[x, y] = y^* \begin{bmatrix} 0 & P_1 \\ P_1 & 0 \end{bmatrix} x, \quad x, y \in \mathcal{N}_1$$

in the same basis, and P_1 is the sip matrix of size m. We see from this representation that N_1 is nondegenerate. By Proposition 2.2.2, $\mathbb{C}^n = \mathcal{N}_1 \dotplus N_1^{[\perp]}$, and by Proposition 4.1.2, $\mathcal{N}_1^{[\perp]}$ is an invariant subspace for A. If $A|_{\mathcal{N}_1^{[\perp]}}$ has nonreal eigenvalues, apply the same procedure to construct a subspace $\mathcal{N}_2 \subseteq \mathcal{N}_1^{[\perp]}$ with basis $c'_{m'}, \ldots, c'_1, b'_{m'}, \ldots b'_1$ such that in this basis $A|_{\mathcal{N}_2} = J_2 \oplus \overline{J_2}$, where J_2 is the Jordan block of size m' with a nonreal eigenvalue, and

$$[x, y] = y^* \begin{bmatrix} 0 & P_2 \\ P_2 & 0 \end{bmatrix} x, \quad x, y \in \mathcal{N}_2$$

with the sip matrix P_2 of size m'. Continue this procedure until the nonreal eigenvalues of A are exhausted.

Now consider a fixed $\mathcal{X}_i$, where $i \in \{1, \ldots, \alpha\}$ so that λ_i is real. Again, let m be such that $(A - \lambda_i I)^m |_{\mathcal{X}_i} = 0$ but $(A - \lambda_i I)^{m-1} |_{\mathcal{X}_i} \neq 0$. Let Q_i be the orthogonal projection on $\mathcal{X}_i$ and define $F : \mathcal{X}_i \to \mathcal{X}_i$ by

$$F = Q_i H (A - \lambda_i I)^{m-1} |_{\mathcal{X}_i}.$$

Since λ_i is real, it is easily seen that F is a selfadjoint linear transformation. Moreover, $F \neq 0$; so there is a nonzero eigenvalue of F (necessarily real) with an eigenvector a_1. Normalize a_1 so that

$$(F a_1, a_1) = \varepsilon, \quad \varepsilon = \pm 1.$$

In other words,

$$\left[(A - \lambda_i I)^{m-1} a_1, a_1 \right] = \varepsilon. \tag{5.3.9}$$

Let $a_j = (A - \lambda_i I)^{j-1} a_1$, $j = 1, \ldots, m$. It follows from (5.3.9) that for $j + k = m + 1$

$$[a_j, a_k] = [(A - \lambda_i I)^{j-1} a_1, (A - \lambda_i I)^{k-1} a_1] = [(A - \lambda_i I)^{m-1} a_1, a_1] = \varepsilon. \quad (5.3.10)$$

Moreover, for $j + k > m + 1$ we have:

$$[a_j, a_k] = [(A - \lambda_i I)^{j+k-2} a_1, a_1] = 0 \qquad (5.3.11)$$

in view of the choice of m. Now put

$$b_1 = a_1 + \alpha_2 a_2 + \cdots + \alpha_m a_m, \quad b_j = (A - \lambda_i I)^{j-1} b_1, \quad j = 1, \ldots, m,$$

and choose α_i so that

$$[b_1, b_1] = [b_1, b_2] = \cdots = [b_1, b_{m-1}] = 0.$$

Such a choice of α_i is possible. Indeed, the equation $[b_1, b_j] = 0$ $(j = 1, \ldots, m - 1)$ implies, in view of (5.3.10) and (5.3.11),

$$\begin{aligned} 0 &= [a_1 + \alpha_2 a_2 + \cdots + \alpha_m a_m, a_j + \alpha_2 a_{j+1} + \cdots + \alpha_{m-j+1} a_m] \\ &= [a_1, a_j] + 2\varepsilon \alpha_{m-j+1} + (\text{terms in } \alpha_2, \ldots, \alpha_{m-j}). \end{aligned}$$

Evidently, these equations determine unique numbers $\alpha_2, \ldots, \alpha_m$ in succession.

Let $\mathcal{N} = \mathrm{Span}\{b_1, \ldots, b_m\}$. In the basis $b_1, \ldots, b_m$ the linear transformation $A\,|_N$ is represented by the single Jordan block with eigenvalue λ_i, and

$$[x, y] = y^* \varepsilon P_0 x, \quad x, y \in \mathcal{N},$$

where P_0 is the sip matrix of size m.

Continue the procedure on the orthogonal companion to N, and so on.

Applying this construction, we find a basis $f_1, \ldots, f_n$ in $\mathbb{C}^n$ such that A is represented by the Jordan matrix J of (5.1.1) in this basis and, with $P_{\varepsilon,J}$ as defined in (5.1.2),

$$[x, y] = y^* P_{\varepsilon,J} x, \quad x, y \in \mathbb{C}^n,$$

where x and y are represented by their coordinates in the basis $f_1, \ldots, f_n$. Let T be the $n \times n$ invertible matrix whose i^{th} column is formed by the coordinates of f_i (in the standard orthonormal basis), $i = 1, \ldots, n$. For such a T, the relation $T^{-1} A T = J$ holds because $f_1, \ldots, f_n$ is a Jordan basis for A, and equality $T^* H T = P_{\varepsilon,J}$ follows from the construction of $f_1, \ldots, f_n$. So (A, H) and $(J, P_{\varepsilon,J})$ are unitarily similar.

It remains to prove the uniqueness of the normalized sign characteristic of (A, H); a sign characteristic of (A, H) is called *normalized* if the order of the blocks in the canonical form of (A, H) is such that for every collection of identical Jordan blocks with a real eigenvalue the signs $+1$ (if any) associated with this collection appear first, before the signs -1 (if any) associated with the same collection of blocks. So suppose that (A, H) is unitarily similar to two canonical pairs, $(J, P_{\varepsilon,J})$

and $(J, P_{\delta,J})$ where ε, δ are sets of signs. It is to be proved that ε and δ are the same up to permutation of signs corresponding to Jordan blocks with the same eigenvalue and the same size.

The hypothesis implies that

$$P_{\delta,J} = S^* P_{\varepsilon,J} S, \quad J = S^{-1} J S \qquad (5.3.12)$$

for some invertible S. The second of these equations shows that attention can be restricted to the case when $\sigma(J)$ is a singleton. (Indeed, if $J = \mathrm{diag}\,(K_1, K_2)$ where $\sigma(K_1) \cap \sigma(K_2) = \phi$, then every S commuting with J has the form $\mathrm{diag}\,(S_1, S_2)$ with partitioning consistent with that of J; see, for instance, [26, Section VIII.2].) Thus, it is now assumed that $\sigma(J) = \{\lambda\}$ and $\lambda \in \mathsf{R}$.

Let J have k_i blocks $J_{i1}, \ldots, J_{ik_i}$ of size m_i for $i = 1, 2, \ldots, t$ and $m_1 > m_2 > \cdots > m_t$. Thus, we may write

$$J = \mathrm{diag}\,\left(\mathrm{diag}\,[J_{1j}]_{j=1}^{k_1}, \mathrm{diag}\,[J_{2j}]_{j=1}^{k_2}, \ldots, \mathrm{diag}\,[J_{tj}]_{j=1}^{k_t}\right),$$

$$P_{\varepsilon,J} = \mathrm{diag}\,\left(\mathrm{diag}\,[\varepsilon_{1j} S_{m_1}]_{j=1}^{k_1}, \mathrm{diag}\,[\varepsilon_{2j} S_{m_2}]_{j=1}^{k_2}, \ldots, \mathrm{diag}\,[\varepsilon_{tj} S_{m_t}]_{j=1}^{k_t}\right),$$

where S_m is the $m \times m$ sip matrix, and

$$P_{\delta,J} = \mathrm{diag}\,\left(\mathrm{diag}\,[\delta_{1j} S_{m_1}]_{j=1}^{k_1}, \mathrm{diag}\,[\delta_{2j} S_{m_2}]_{j=1}^{k_2}, \ldots, \mathrm{diag}\,[\delta_{tj} S_{m_t}]_{j=1}^{k_t}\right).$$

Define also

$$J_i = J_{i1} \oplus \cdots \oplus J_{ik_i}, \qquad i = 1, 2, \ldots, t.$$

It follows from (5.3.12) that, for any nonnegative integer k,

$$P_{\delta,J}(J - I\lambda)^k = S^* P_{\varepsilon,J}(J - I\lambda)^k S, \qquad (5.3.13)$$

and is a relation between hermitian matrices. Consequently,

$$\mathrm{sig}\,(P_{\delta,J}(J - I\lambda)^{m_1-1}) = \mathrm{sig}\,(P_{\varepsilon,J}(J - I\lambda)^{m_1-1}). \qquad (5.3.14)$$

Observe that $(J_1 - I\lambda)^{m_1-1}$ is nilpotent of rank 1 and $(J_i - I\lambda)^{m_1-1} = 0$ for $i = 2, 3, \ldots, t$. It follows that

$$P_{\varepsilon,J}(J - I\lambda)^{m_1-1} = \mathrm{diag}\,\left(\varepsilon_{1,1} S_{m_1}(J_1 - I\lambda)^{m_1-1}, \ldots, \varepsilon_{1,k_1} S_{m_1}(J_1 - I\lambda)^{m_1-1}, 0, \ldots, 0\right),$$

and therefore

$$\mathrm{sig}\,(P_{\varepsilon,J}(J - I\lambda)^{m_1-1}) = \sum_{i=1}^{k_1} \varepsilon_{1,i}.$$

Similarly

$$\mathrm{sig}\,(P_{\delta,J}(J - I\lambda)^{m_1-1}) = \sum_{i=1}^{k_1} \delta_{1,i}.$$

Noting (5.3.14) it is found that

$$\sum_{i=1}^{k_1} \varepsilon_{1,i} = \sum_{i=1}^{k_1} \delta_{1,i}. \tag{5.3.15}$$

Consequently, the subsets $\{\varepsilon_{1,1}, \ldots, \varepsilon_{1,k_1}\}$ and $\{\delta_{1,1}, \ldots, \delta_{1,k_1}\}$ of ε and δ agree to within normalization.

Now examine the hermitian matrix $P_{\varepsilon,J}(J - I\lambda)^{m_2-1}$. This is found to be block diagonal with nonzero blocks of the form

$$\varepsilon_{1,j} S_{m_1}(J_1 - I\lambda)^{m_2-1}, \quad j = 1, 2, \ldots, k_1$$

and

$$\varepsilon_{2,j} S_{m_2}(J_2 - I\lambda)^{m_2-1}, \quad j = 1, 2, \ldots, k_2.$$

It follows that the signature of $P_{\varepsilon,J}(J - I\lambda)^{m_2-1}$ is given by

$$\left[\sum_{i=1}^{k_1} \varepsilon_{1,j}\right] \left(\text{sig}\ \left(S_{m_1}(J_1 - I\lambda)^{m_2-1}\right)\right) + \sum_{j=1}^{k_2} \varepsilon_{2,j}.$$

But again, in view of (5.3.13), this must be equal to the corresponding expression formulated using δ instead of ε. Hence, using (5.3.15), it is found that

$$\sum_{j=1}^{k_2} \varepsilon_{2,j} = \sum_{j=1}^{k_2} \delta_{2,j}$$

and the subsets $\{\varepsilon_{2,1}, \ldots, \varepsilon_{2,k_2}\}$ and $\{\delta_{2,1}, \ldots, \delta_{2,k_2}\}$ of ε and δ agree within the normalization convention.

Now it is clear that the argument can be continued for t steps after which the uniqueness of the normalized sign characteristic is established. This completes the proof of Theorem 5.1.1. $\square$

5.4   Classification of Matrices by Unitary Similarity

We now exploit the notion of unitary similarity introduced in Section 4.5 and take advantage of the canonical pairs introduced in Theorem 5.1.1. Recall the remark in Section 4.5 that if A_1 and A_2 are similar and H_1 and H_2 are congruent, it is not necessarily the case that (A_1, H_1) and (A_2, H_2) are unitarily similar. We show first that, if A_j is H_j-selfadjoint for $j = 1$ and 2, then unitary similarity of the two pairs can be characterized quite easily. Let the set $\mathcal{U}$ of pairs of matrices (A, H) be as defined in Section 4.5, i.e., $A, H \in \mathbb{C}^{n \times n}$ and H is hermitian and invertible.

Theorem 5.4.1. Let (A_1, H_1) and $(A_2, H_2) \in \mathcal{U}$ and let A_j be H_j-selfadjoint for $j = 1$ and 2. Then (A_1, H_1) and (A_2, H_2) are unitarily similar if and only if A_1 and A_2 are similar and the pairs $(A_1, H_1), (A_2, H_2)$ have the same sign characteristic.

Proof. If A_1 and A_2 are similar and $(A_1, H_1), (A_2, H_2)$ have the same sign characteristic then, by Theorem 5.1.1,

$$A_j = T_j^{-1} J T_j, \quad H_j = T_j^* P_{\varepsilon, J} T_j, \quad j = 1 \text{ and } 2.$$

Hence

$$A_1 = (T_2^{-1} T_1)^{-1} A_2 (T_2^{-1} T_1), \quad H_1 = (T_2^{-1} T_1)^* H_2 (T_2^{-1} T_1),$$

i.e., (A_1, H_1) and (A_2, H_2) are unitarily similar and the matrix $T_2^{-1} T_1$ determines the unitary similarity.

Conversely, suppose that (A_1, H_1) and (A_2, H_2) are unitarily similar. Then A_1 and A_2 have a common Jordan form J, which can be arranged as in (5.1.1). Then, if (A_2, H_2) and $(J, P_{\varepsilon, J})$ are unitarily similar for the same set of signs ε, it follows that (A_1, H_1) and $(J, P_{\varepsilon, J})$ are also unitarily similar. In particular, the sign characteristics of (A_1, H_1) and (A_2, H_2) are the same.   □

Consider now the important case of unitary similarity in which $H_1 = H_2 = H$. Thus, we say that A_1 and A_2 are H-*unitarily similar* if

$$A_1 = U^{-1} A_2 U \quad \text{and} \quad U^* H U = H;$$

in other words, if $A_1 = U^{-1} A_2 U$ for some H-unitary matrix U. It is easily seen that H-unitary similarity defines an equivalence relation on the set of all square matrices (with the size of H). Furthermore, if an equivalence class contains an H-selfadjoint matrix it is easily verified that every matrix in the equivalence class is also H-selfadjoint.

Theorem 5.4.1 shows that H-selfadjoint matrices A_1 and A_2 are H-unitarily similar if and only if they are similar and the pairs (A_1, H) and (A_2, H) have the same sign characteristic. Using this observation, a canonical representative can now be constructed in each equivalence class of H-selfadjoint matrices.

First, a set of normalized Jordan matrices is constructed, all of which have spectrum symmetric with respect to the real axis. Thus, Ξ is defined to be the set of all $n \times n$ Jordan matrices J of the form

$$J_{m_1}(\gamma_1) \oplus \cdots \oplus J_{m_\alpha}(\gamma_\alpha) \oplus J_{m_{\alpha+1}}(\gamma_{\alpha+1} + i\delta_{\alpha+1}) \oplus \cdots \oplus J_{m_\beta}(\gamma_\beta + i\delta_\beta),$$

where γ_i $(1 \le i \le \beta)$ are real, $\gamma_1 \le \gamma_2 \le \cdots \le \gamma_\alpha$, and blocks with the same eigenvalue are in nondecreasing order of size. Also, $\delta_k > 0$ for $k = \alpha + 1, \ldots, \beta$ and $\gamma_{\alpha+1} \le \gamma_{\alpha+2} \le \cdots \le \gamma_\beta$ with $\delta_k \le \delta_{k+1}$ if $\gamma_k = \gamma_{k+1}$ and, finally, all such blocks with the same eigenvalue are in nondecreasing order of size. (Recall that $J_k(\lambda)$ stands for the $k \times k$ Jordan block with eigenvalue λ if λ is real, and for the direct sum of two $\frac{k}{2} \times \frac{k}{2}$ Jordan blocks with eigenvalues λ and $\overline{\lambda}$ respectively if λ is nonreal).

With these conventions it is clear that two matrices $J', J'' \in \Xi$ are similar if and only if $J' = J''$.

Now suppose that H is given and consider the subset

$$\Xi_H = \left\{ J \in \Xi \mid TJT^{-1} \text{ is } H\text{-selfadjoint for some invertible } T \right\}.$$

For each $J \in \Xi_H$ define $\pi(J)$ to be the set of matrices $P_{\varepsilon,J}$ (constructed as in (5.3.6)) with block structure consistent with that of J, and a set of signs ε such that $P_{\varepsilon,J}$ and H are congruent.

The construction is completed by associating a unique S_P with $P \in \pi(J)$, $(J \in \Xi_H)$, such that $H = S_P^* P S_P$ and defining

$$R_H = \left\{ S_P^{-1} J S_P \mid J \in \Xi_H \text{ and } P \in \pi(J) \right\}.$$

Then we have:

Theorem 5.4.2. *The set R_H forms a complete set of representatives of the equivalence classes (under H-unitary similarity) of H-selfadjoint matrices. In other words, for every H-selfadjoint matrix A there is an $A' \in R_H$ such that A and A' are H-unitarily similar and, if $A', A'' \in R_H$ and $A' \neq A''$, then A' and A'' are not unitarily similar.*

Proof. Let A be H-selfadjoint, and let $A = S^{-1}JS$, $H = S^* P_{\varepsilon,J} S$ be the canonical form of (A, H), with normalized sign characteristic, and J chosen from Ξ. Clearly, $P_{\varepsilon,J} \in \pi(J)$. Put $A' = S_{P_{\varepsilon,J}}^{-1} J S_{P_{\varepsilon,J}} \in R_H$. Then, by Theorem 5.4.1, A and A' are H-unitarily similar .

Conversely, suppose that

$$A' = S_{P'}^{-1} J' S_{P'} \in R_H; \quad A'' = S_{P''}^{-1} J'' S_{P''} \in R_H,$$

and A' and A'' are H-unitarily similar. In particular, A' and A'' are similar, and so are J' and J''. But since $J', J' \in \Xi$ we have $J' = J''$. Now use Theorem 5.4.1 to deduce that also $P' = P''$ $(\in \pi(J') = \pi(J''))$. Clearly, $A' = A''$ (because the choice of S_P is fixed; so $P' = P''$ implies $S_{P'} = S_{P''}$). $\qquad\qquad \square$

This section is to be concluded by showing that the equivalence classes of Theorem 5.4.2 are arcwise connected. But first a lemma is needed.

Lemma 5.4.3. *The set of all H-unitary matrices is arcwise connected.*

Proof. First observe that the set of all H-selfadjoint matrices is arcwise connected because it is a real linear space. The lemma is proved by combining this observation with an appropriate use of the Cayley transformation.

Let U_1, U_2 be H-unitary matrices and let a_1, a_2 be unimodular complex numbers for which $U_1 - a_1 I$ and $U_2 - a_2 I$ are invertible. For $j = 1, 2$, define the H-selfadjoint matrices

$$A_j = (U_j - a_j I)^{-1}(wU_j - wa_j I)$$

where $w \in C \setminus R$. It follows that $w \notin \sigma(A_1) \cup \sigma(A_2)$.

Let $A(t)$, $t \in [0,1]$, be a continuous path of H-selfadjoint matrices for which $A(0) = A_1$, $A(1) = A_2$, and let $w(t)$, $t \in [0,1]$, be a continuous path in $C \setminus R$ for which $w(0) = w = w(1)$ and $w(t) \notin \sigma(A(t))$ for $t \in [0,1]$. For example, $w(t)$ can be chosen to be equal to the constant w, except for neighborhoods of those points t for which $w \in \sigma(A(t))$.

Then define

$$U(t) = (A(t) - \overline{w(t)}I)(A(t) - w(t)I)^{-1}$$

to obtain a path of H-unitary matrices connecting $a_1^{-1}U_1$ and $a_2^{-1}U_2$. Now it is clear that U_1 and U_2 are also arcwise connected. $\square$

Theorem 5.4.4. *Each equivalence class (under H-unitary similarity) of H-selfadjoint matrices is arcwise connected.*

Proof. Let A_1 and A_2 be H-selfadjoint and $A_1 = U^{-1}A_2U$ where U is H-unitary. From Lemma 5.4.3, there exists a path $U(t)$, $t \in [0,1]$ of H-unitary matrices for which $U(0) = U$ and $U(1) = I$. Then the path

$$A(t) = U(t)^{-1}A_2U(t), \quad t \in [0,1]$$

connects A_1 and A_2 in the equivalence class. $\square$

5.5 Signature Matrices

Let A be H-selfadjoint and let $(J, P_{\varepsilon,J})$ be the canonical form of the pair (A, H). We have observed that J and $P_{\varepsilon,J}$ satisfy the relations

$$P_{\varepsilon,J}J = J^*P_{\varepsilon,J}, \quad P_{\varepsilon,J}^* = P_{\varepsilon,J}, \quad P_{\varepsilon,J}^2 = I. \tag{5.5.16}$$

The question arises: to what extent do these relations determine $P_{\varepsilon,J}$ (for a fixed J of the form (5.1.1))? In other words, we are interested in solutions of the simultaneous equations

$$PJ = J^*P, \quad P^* = P, \quad P^2 = I. \tag{5.5.17}$$

The next theorem shows that the set of all solutions is a set of matrices which are unitarily similar to $P_{\varepsilon,J}$; in fact, of the form $SP_{\varepsilon,J}S^*$ where S is unitary and commutes with J. Thus, every solution can be considered as a representation of $P_{\varepsilon,J}$ in some special orthonormal basis of C^n, and this fact casts a new light on the canonical matrix $P_{\varepsilon,J}$.

Theorem 5.5.1. *If matrix P satisfies the relations (5.5.17), then there is a set of signs ε (which is unique up to permutation of signs corresponding to equal Jordan blocks in J) and a unitary matrix S commuting with J such that $P = SP_{\varepsilon,J}S^{-1}$.*

Conversely, if ε is any set of signs and $P = SP_{\varepsilon,J}S^{-1}$ where S is a unitary matrix commuting with J, then P satisfies the equation (5.5.17).

Proof. An easy calculation confirms the converse statement. Indeed, if $P = SP_{\varepsilon,J}S^{-1}$ with $S^*S = I$ and $SJ = JS$, the clearly $P^* = P$ and since $P_{\varepsilon,J}^2 = I$, also $P^2 = I$. Furthermore, we have $J^* = (S^{-1}JS)^* = S^{-1}J^*S$ so that

$$PJ = SP_{\varepsilon,J}S^{-1}J = SP_{\varepsilon,J}JS^{-1} = SJ^*P_{\varepsilon,J}S^{-1} = J^*SP_{\varepsilon,J}S^{-1} = J^*P.$$

Now consider the direct statement of the theorem and the special case in which $\sigma(J) = \{\alpha\}$ where α is real. Replacing J by $J - \alpha I$ we may assume $\alpha = 0$. Let $J = \operatorname{diag}(J_1, \ldots, J_k)$ be the decomposition of J into Jordan blocks, and let

$$P = [P_{ij}]_{i,j=1}^k$$

be the corresponding decomposition of P. Note that $P^* = P$ implies $P_{ij}^* = P_{ji}$, and write

$$P_{ij} = \left[\alpha_{pq}^{(ij)}\right]_{p,q=1}^{m_i, m_j}$$

where m_i and m_j are the sizes of J_i and J_j respectively. It is easily seen that the equation $PJ = J^*P$ implies $P_{ij}J_j = J_i^*P_{ij}$ for $i, j = 1, 2, \ldots, k$, and then this relation implies that each block of PJ has lower triangular Toeplitz form. More precisely,

$$\alpha_{pq}^{(ij)} = 0 \qquad \text{if} \quad p + q \leq \max(m_i, m_j)$$
$$\alpha_{pq}^{(ij)} = \alpha_{uv}^{(ij)} \quad \text{if} \quad p + q = u + v.$$

Now assume, without loss of generality, that the sizes of the Jordan blocks are ordered so that $m_1 = \cdots = m_s > m_{s+1} \geq \cdots \geq m_k$ for some s. Let $m = \sum_{i=1}^k m_i$ and consider the $s \times m$ submatrix A of P formed by rows $1, (m_1 + 1), \ldots, (m_1 + \cdots + m_{s-1} + 1)$. Let B be the submatrix of P formed by the last $m_{s+1} + \cdots + m_k$ columns. Since $P^2 = I$ the product AB is zero.

Because of the ordering of the m_j's the last $m_{s+1} + \cdots + m_k$ columns of A are zero and so, if A_0 is the $s \times (m_1 + \cdots m_s)$ leading submatrix of A, we have

$$A_0 B_0 = AB = 0, \tag{5.5.18}$$

where the matrix B_0 is formed by the top $m_1 + \ldots + m_s$ rows of B. Then the invertibility of P is easily seen to imply that of the $s \times s$ submatrix of A_0 made up by the columns $m_1, m_1 + m_2, \ldots, m_1 + m_2 + \ldots + m_s$ of A_0. Now (5.5.18) implies that $B_0 = 0$. In other words, $\alpha_{pq}^{(ij)} = 0$ for $i = 1, 2, \ldots, s$ and $j = s + 1, \ldots, k$.

Apply the same argument to the next group of Jordan blocks of equal size, and repeat as often as possible. The result of this process shows that P must be block diagonal with each diagonal block corresponding to a group of Jordan blocks of the same size. Therefore the problem is reduced to the case in which $m_1 = m_2 = \cdots = m_k = \ell$, say.

Let C be the $k \times k(\ell - 1)$ submatrix of P formed by rows $\ell, 2\ell, \ldots, k\ell$ and all columns except columns $1, \ell + 1, \ldots, (k - 1)\ell + 1$. If $A = [\alpha_{1\ell}^{(ij)}]_{i,j=1}^k$ then $P^2 = I$

implies $AC = 0$. Again A is invertible and we deduce $C = 0$. Thus, P is reduced to the form

$$P = [\gamma_{ij} S_\ell]_{i,j=1}^k$$

where S_ℓ is the sip matrix of size ℓ, and the matrix $\Gamma = [\gamma_{ij}]_{i,j=1}^k$ is hermitian. Then $P^2 = I$ implies that $\Gamma^2 = I$ as well. Hence there exists a $k \times k$ unitary matrix T such that

$$T^{-1}\Gamma T = \mathrm{diag}\,(\varepsilon_1, \ldots, \varepsilon_k)$$

where $\varepsilon_j = \pm 1$ for each j.

Now define $S = [t_{ij} I_\ell]_{i,j=1}^k$ where $T = [t_{ij}]_{i,j=1}^k$ and I_ℓ is the unit matrix of size ℓ. Then S is unitary and

$$S^{-1}PS = \mathrm{diag}\,(\varepsilon_1 S_\ell, \ldots, \varepsilon_k S_\ell).$$

Moreover $SJ = JS$, so the theorem is proved in the case $\sigma(J) = \{\alpha\}$, $\alpha \in \mathbb{R}$.

Now consider the case

$$J = J_0 \oplus \overline{J_0}, \tag{5.5.19}$$

where J_0 is a Jordan matrix with single nonreal eigenvalue, and let

$$P = \begin{bmatrix} P_{11} & P_{12} \\ P_{21} & P_{22} \end{bmatrix}$$

be the corresponding decomposition of P. The condition $PJ = J^*P$ can be written in the form

$$\begin{bmatrix} P_{11}J_0 & P_{12}\overline{J_0} \\ P_{21}J_0 & P_{22}\overline{J_0} \end{bmatrix} = \begin{bmatrix} J_0^* P_{11} & J_0^* P_{12} \\ J_0^T P_{21} & J_0^T P_{22} \end{bmatrix}.$$

Since $\sigma(J_0) \cap \sigma(\overline{J_0}) = \phi$, it follows from Theorem A.4.1 that the equations $P_{11}J_0 = J_0^* P_{11}$ and $P_{22}\overline{J_0} = J_0^T P_{22}$ imply $P_{11} = P_{22} = 0$. Also

$$P_{12}\overline{J_0} = J_0^* P_{12}; \quad P_{21}J_0 = J_0^T P_{21}. \tag{5.5.20}$$

But $P = P^*$ implies $P_{21} = P_{12}^*$, so the two relations in (5.5.20) are equivalent.

Let $S_0 = P_{21}^* P_0$, where $P_0 = \mathrm{diag}\,(P_1, \ldots, P_k)$, and P_i are sip matrices with sizes equal to the sizes of Jordan blocks in J_0. Clearly, S_0 is unitary. We prove that

$$S_0 J_0 = J_0 S_0.$$

Indeed, (5.5.20) implies $J_0 P_{21}^{-1} = P_{21}^{-1} J_0^T$ and since $P_{21}^{-1} = P_{21}^*$, we have $P_{21}^* J_0^T = J_0 P_{21}^*$. Now, using the fact that $P_0 J_0 = J_0^T P_0$, we have

$$S_0 J_0 = P_{21}^* P_0 J_0 = P_{21}^* J_0^T P_0 = J_0 P_{21}^* P_0 = J_0 S_0.$$

Put $S = \mathrm{diag}\,(S_0, I)$, then clearly, $SJ = JS$, S is unitary, and

$$S^{-1}PS = \begin{bmatrix} S_0^{-1} & 0 \\ 0 & I \end{bmatrix} \begin{bmatrix} 0 & P_{21}^* \\ P_{21} & 0 \end{bmatrix} \begin{bmatrix} S_0 & 0 \\ 0 & I \end{bmatrix} = \begin{bmatrix} 0 & S_0^{-1} P_{21}^* \\ P_{21} S_0 & 0 \end{bmatrix}$$

$$= \begin{bmatrix} 0 & P_0 \\ P_0 & 0 \end{bmatrix}.$$

So the theorem is proved in the case when J has the form (5.5.19) with $\sigma(J_0) = \{\alpha\}$, and α is nonreal.

We turn now to the general case of the theorem. Write as in (5.1.1):

$$J = J(\lambda_1) \oplus \cdots \oplus J(\lambda_\alpha) \oplus J(\lambda_{\alpha+1}) \oplus \cdots \oplus J(\lambda_\beta),$$

and partition P accordingly: $P = [P_{ij}]_{i,j=1}^\beta$. The equality $PJ = J^*P$ implies

$$P_{ik}J(\lambda_k) = J(\lambda_i)^* P_{ik}. \tag{5.5.21}$$

Again, by Theorem A.4.1 (5.5.21) implies that $P_{ik} = 0$ whenever $\sigma(J(\lambda_k)) \cap \sigma(J(\lambda_i)^*) = \emptyset$. So the proof of the general case is reduced to the two cases already proved.   □

Theorem 5.5.1 and its proof allow us to establish some particular cases in which every P satisfying (5.5.17) will be in the form $P_{\varepsilon,J}$ *in the same basis*, i.e., when one can always choose $S = I$ in Theorem 5.5.1. This will be the case, for instance, when there is a unique (apart from multiplication of vectors of each Jordan chain of J by a unimodular complex number, which may be different for different chains) orthonormal basis in $\mathbb{C}^n$ in which J has the fixed form (5.1.1). This property is easily seen to be true for a *nonderogatory* matrix J, i.e., when there is just one Jordan block associated with each distinct eigenvalue. So we obtain the first part of the following corollary. The second part can be traced from the proof of Theorem 5.5.1.

Corollary 5.5.2. *Assume that J is given by (5.1.1) and at least one of the following conditions holds:*

1. *J is nonderogatory;*

2. *$\sigma(J)$ is real and each distinct eigenvalue has no two associated Jordan blocks of the same size.*

Then P satisfies equations (5.5.17) if and only if $P = P_{\varepsilon,J}$ for some choice of the signs ε.

In the language of unitary similarity, Theorem 5.5.1 has the consequence:

Corollary 5.5.3. *Let H be an invertible hermitian matrix and A be a matrix with Jordan form J. Then A is H-selfadjoint if and only if (A, H) and (J, P) are unitarily similar for any matrix P satisfying the equations (5.5.17).*

Suppose now that J and ε are fixed, and consider solutions P_1 and P_2 of (5.5.17). Theorem 5.4.1 implies that (J, P_1) and (J, P_2) are unitarily similar; which can be expressed by saying that P_1 and P_2 are congruent, say $P_1 = T^*P_2T$, for some invertible T commuting with J. In fact, Theorem 5.5.1 implies that the transforming matrix T is unitary so that the congruence becomes a similarity.

Theorem 5.5.4. *Solutions P_1 and P_2 of equations* (5.5.17) *are unitarily similar with a unitary transforming matrix which commutes with J if and only if P_1 and P_2 have the same sign characteristic.*

Proof. Since J is fixed in this argument we abbreviate $P_{\varepsilon,J}$ to P_ε. If P_1 and P_2 are solutions of (5.5.17) with the same sign characteristic, ε, then there are unitary matrices S_1 and S_2 commuting with J such that

$$P_1 = S_1 P_\varepsilon S_1^*, \quad P_2 = S_2 P_\varepsilon S_2^*.$$

Defining $U = S_1 S_2^*$ it is easily verified that $P_2 = U^* P_1 U$ and that U is unitary and commutes with J.

Conversely, let P_1, P_2 be solutions of (5.5.17) with sign characteristics ε_1 and ε_2, respectively, and $P_1 = U^* P_2 U$ where $U^* U = I$ and $UJ = JU$. Then Theorem 5.5.1 implies $P_2 = S^* P_{\varepsilon_2} S$ where $S^* S = I$ and $SJ = JS$. Consequently,

$$P_1 = (SU)^* P_{\varepsilon_2}(SU)$$

where $(SU)^*(SU) = I$ and $(SU)J = J(SU)$. Then (J, P_1) and (J, P_{ε_2}) are unitarily similar so, by Theorem 5.4.1, $\varepsilon_1 = \varepsilon_2$. $\square$

5.6 The Structure of H-Selfadjoint Matrices when H has a Small Number of Negative Eigenvalues

In this section we investigate the structure of H-selfadjoint matrices A in the cases when H has $0, 1$, or 2 negative eigenvalues (counting multiplicities) The results of this section are easily obtained by inspecting the canonical form $(J, P_{\varepsilon,J})$ of (A, H), and by sorting out the cases when $P_{\varepsilon,J}$ has the required number of negative eigenvalues. Note also that the number of negative eigenvalues of the $n \times n$ sip matrix is m if $n = 2m$ or $n = 2m + 1$.

The first case is obvious: if A is H-selfadjoint with positive definite H, then the spectrum of A is real, A is similar to a diagonal matrix, and all the signs in the sign characteristic are $+1$'s.

When H has exactly one negative eigenvalue and A is an H-selfadjoint matrix, then one of the following four statements holds:

(i) $\sigma(A)$ is real, A is similar to a diagonal matrix, and all but one sign in the sign characteristic are $+1$'s;

(ii) $\sigma(A)$ is real, one Jordan block in the Jordan form of A has size 2 (with arbitrary sign in the sign characteristic), and the rest of the Jordan blocks have size 1 with signs $+1$;

(iii) $\det(I\lambda - A)$ has exactly 2 nonreal zeros (counting multiplicities), A is similar to a diagonal matrix, and all signs are $+1$;

(iv) $\sigma(A)$ is real, one Jordan block in the Jordan form of A has size 3, the rest of the Jordan blocks have size 1, and all the signs are $+1$.

Case Number	No. of nonreal eigenvalues of A (counting multiplicities)	Number and signs (in the sign characteristic) of Jordan blocks of A with real eigenvalues			
		of size 1	of size 2	of size 3	of size 4
1	0	n all but 2 signs are $+1$	0	0	0
2	0	$n-4$ all signs $+1$	2 signs arbitrary	0	0
3	0	$n-6$ all signs $+1$	0	2 both signs $+1$	0
4	0	$n-5$ all signs $+1$	1 sign arbitrary	1 sign $+1$	0
5	0	$n-4$ all signs $+1$	0	0	1 sign arbitrary
6	0	$n-3$ all signs $+1$	0	1 sign -1	0
7	0	$n-3$ all signs but one are $+1$	0	1 sign $+1$	0
8	0	$n-1$ all signs but one are $+1$	1 sign arbitrary	0	0
9	2	$n-2$ all signs but one are $+1$	0	0	0
10	2	$n-4$ all signs $+1$	1 sign arbitrary	0	0
11	2	$n-5$ all signs $+1$	0	1 sign $+1$	0
12	4	$n-4$ all signs $+1$	0	0	0

It is easily seen that, for every hermitian invertible H of size at least 3 with exactly 1 negative eigenvalue, all four possibilities can be realized, i.e., for each case (i)–(iv), there exists an H-selfadjoint A for which this case applies.

The case of two negative eigenvalues of H is somewhat more complicated. In this case, for an H-selfadjoint matrix A of size at least six, one of 12 possibilities holds. For convenience, they are tabulated above. Again, all 12 possibilities can be realized for every $n \times n$ hermitian invertible H with exactly two negative eigenvalues (counting multiplicities) and $n \geq 6$.

5.7 *H*-Definite Matrices

Another interesting class of H-selfadjoint matrices are "H-definite", and are defined as follows: Let $H = H^*$ be an invertible matrix, and let $[.,.] = (Hx, y)$ be the indefinite inner product defined by H. An $n \times n$ matrix A is called H-*nonnegative* (resp. H-*positive*) if $[Ax, x] \geq 0$ for all $x \in C^n$ (resp. $[Ax, x] > 0$ for all $x \in C^n \setminus \{0\}$). Clearly, these classes of matrices contain only (not all) H-selfadjoint matrices. So Theorem 5.1.1 is applicable, and it is easily found that A is H-nonnegative (resp. H-positive) if and only if the matrix $P_{\varepsilon,J} J$ is positive semidefinite (resp. positive definite) with respect to the usual inner product. Examining the product $P_{\varepsilon,J} J$ leads to the conclusions:

Theorem 5.7.1. *A matrix $A \in C^{n \times n}$ is H-positive if and only if the following conditions hold:*

(i) *A is H-selfadjoint and invertible;*

(ii) *the spectrum $\sigma(A)$ is real;*

(iii) *A is similar to a diagonal matrix;*

(iv) *the sign (in the sign characteristic of (A, H)) attached to all Jordan blocks in the Jordan form of A corresponding to an eigenvalue λ_0, is 1 if $\lambda_0 > 0$, and -1 if $\lambda_0 < 0$.*

Proof. We may assume from the start that A is H-selfadjoint. Then, without loss of generality we let $A = J$, $H = P_{\varepsilon,J}$. Now, the conditions (i) - (iv) are easily seen to be equivalent to the positive definiteness of the hermitian matrix $P_{\varepsilon,J} J$. Finally, use the definition of H-positive matrices. □

Theorem 5.7.2. *A matrix A is H-nonnegative if and only if the following conditions hold:*

(i) *A is H-selfadjoint;*

(ii) *$\sigma(A)$ is real;*

(iii) *for nonzero eigenvalues of A, all Jordan blocks in the Jordan form of A have size 1;*

for the zero eigenvalue of A (if zero is an eigenvalue) all Jordan blocks have sizes either 1 or 2;

(iv) *the sign attached to the Jordan blocks of a nonzero eigenvalue λ_0 is 1 if $\lambda_0 > 0$, and -1 if $\lambda_0 < 0$; the sign attached to a Jordan block of size 2 (if any) of the zero eigenvalue, is $+1$.*

The proof of Theorem 5.7.2 is similar to that of Theorem 5.7.1: One verifies that, assuming $A = J$, $H = P_{\varepsilon,J}$, the conditions (i) - (iv) are equivalent to the positive semidefiniteness of the hermitian matrix $P_{\varepsilon,J}J$.

An $n \times n$ matrix A is called *H-nonpositive* (resp. *H-negative*) is $[Ax, x] \leq 0$ for all $x \in \mathbb{C}^n$ (resp. $[Ax, x] < 0$ for all $x \in \mathbb{C}^n \setminus \{0\}$). It is left to the reader to describe *H*-nonpositive and *H*-negative matrices. These descriptions are, of course, closely analogous to Theorem 5.7.1 and 5.7.2.

5.8 Second Description of the Sign Characteristic

The sign characteristic of a pair (A, H), where A is *H*-selfadjoint, was described in Section 5.1 in terms of the canonical form. In this section the sign characteristic will be defined directly in terms of the Jordan chains of the *H*-selfadjoint matrix A.

Let λ_0 be a fixed real eigenvalue of A, and let $\Psi_1 \subseteq \mathbb{C}^n$ be the subspace spanned by the eigenvectors of A corresponding to λ_0. For $x \in \Psi_1 \setminus 0$, denote by $\nu(x)$ the maximal length of a Jordan chain of A beginning with the eigenvector x. In other words, there exists a chain of $\nu(x)$ vectors $y_1 = x, y_2, \ldots, y_{\nu(x)}$ such that

$$(A - \lambda_0 I)y_j = y_{j-1} \quad \text{for} \quad j = 2, 3, \ldots, \nu(x), \quad (A - \lambda_0 I)y_1 = 0,$$

and there is no chain of $\nu(x) + 1$ vectors with analogous properties. Let Ψ_i, $i = 1, 2, \ldots, \gamma$ ($\gamma = \max\{\nu(x) \mid x \in \Psi_1 \setminus \{0\}\}$) be the subspace of Ψ_1 spanned by all $x \in \Psi_1$ with $\nu(x) \geq i$. Then

$$\mathrm{Ker}(I\lambda_0 - A) = \Psi_1 \supseteq \Psi_2 \supseteq \cdots \supseteq \Psi_\gamma.$$

The following result describes the sign characteristic of the pair (A, H) in terms of certain bilinear forms defined on the subspaces Ψ_i.

Theorem 5.8.1. *For $i = 1, \ldots, \gamma$, let*

$$f_i(x, y) = (x, Hy^{(i)}), \quad x \in \Psi_i, \quad y \in \Psi_i \setminus \{0\},$$

where $y = y^{(1)}, y^{(2)}, \ldots, y^{(i)}$ is a Jordan chain of A corresponding to real λ_0 with the eigenvector y, and let $f_i(x, 0) = 0$. Then:

(i) *$f_i(x, y)$ does not depend on the choice of $y^{(2)}, \ldots, y^{(i)}$;*

(ii) *for some selfadjoint linear transformation $G_i : \Psi_i \to \Psi_i$,*

$$f_i(x, y) = (x, G_i y), \quad x, y \in \Psi_i;$$

(iii) *for the transformation G_i of* (ii), *$\Psi_{i+1} = \operatorname{Ker} G_i$ (by definition $\Psi_{\gamma+1} = \{0\}$);*

(iv) *the number of positive (negative) eigenvalues of G_i, counting multiplicities, coincides with the number of positive (negative) signs in the sign characteristic of (A, H) corresponding to the Jordan blocks of size i associated with the eigenvalue λ_0 of A.*

Proof. Let $(J, P_{\varepsilon, J})$ be the canonical form for (A, H) as described in (5.1.1) and (5.1.2). Then (A, H) and $(J, P_{\varepsilon, J})$ are unitarily similar for some set of signs ε, i.e.,

$$H = T^* P_{\varepsilon, J} T, \quad A = T^{-1} J T$$

for some nonsingular T. Hence, for $x, y \in \Psi_i$,

$$f_i(x, y) = (Tx, P_{\varepsilon, J} T y^{(i)}).$$

Clearly, Tx, and $Ty^{(1)}, \ldots, Ty^{(i)}$ are, respectively, an eigenvector and a Jordan chain of J corresponding to λ_0. In this way the proof is reduced to the case $A = J$ and $H = P_{\varepsilon, J}$. But in this case the assertions (i)–(iv) are easily verified. $\square$

Consider once more the Jordan matrix J of the proof above. The argument shows that the basis in which the quadratic form $f_i(x, x)$ on Ψ_i is reduced to a sum of squares consists of vectors $T^{-1} x_{i1}, \ldots, T^{-1} x_{i,q_i}$, where x_{ij}, $j = 1, \ldots, q_i$ are all the coordinate unit vectors in $\mathbf{C}^n$ such that $x_{ij} = (J - I\lambda_0)^{i-1} y_{ij}$ for some y_{ij}, and $(I\lambda_0 - J) x_{ij} = 0$.

For the case that $H = P_{\varepsilon, J}$ and $A = J$ it is not hard to obtain a formula for the transformations G_i of (ii). Namely, in the standard orthonormal basis (consisting of unit coordinate vectors) in Ψ_i,

$$G_i = P_{\Psi_i} (-I\lambda_0 + J)^{i-1} P_{\varepsilon, J} \mid_{\Psi_i}, \quad i = 1, \ldots, \gamma, \tag{5.8.22}$$

where P_{Ψ_i} is the orthogonal projection on Ψ_i. Indeed, it is sufficient to prove that

$$f_i(x, y) = (x, P_{\Psi_i} (-I\lambda_0 + J)^{i-1} P_{\varepsilon, J} y)$$

for any coordinate unit vectors $x, y \in \Psi_i$. This is an easy exercise bearing in mind the special structure of $P_{\varepsilon, J}$ and J.

For future reference, we record a corollary of Theorem 5.8.1 that makes explicit the connection between the sign characteristics of A and of $-A$.

Corollary 5.8.2. *Let A be H-selfadjoint, and let λ be a real eigenvalue of A. Then the sign characteristic of A corresponding to the odd (resp., even) partial multiplicities, d, associated with λ coincides with (resp., is opposite to) the sign characteristic of the H-selfadjoint matrix $-A$ corresponding to the same partial multiplicities d, associated with the real eigenvalue $-\lambda$ of $-A$.*

Proof. The proof is easily reduced to the situation where $A = J$ is a nilpotent Jordan matrix, and $H = P_{\varepsilon,J}$. Now observe that, for each Jordan block of size $d \times d$ in A, if $y^{(1)}, y^{(2)}, \ldots, y^{(d)}$ is a Jordan chain of A, then $y^{(1)}, -y^{(2)}, \ldots, (-1)^{d-1}y^{(d)}$ is a Jordan chain of $-A$. Now apply Theorem 5.8.1.   □

As an application of Theorem 5.8.1 we have the following description of the connected components in the set of all invertible hermitian matrices H such that a given matrix A is H-selfadjoint.

Theorem 5.8.3. *Let A be an $n \times n$ matrix similar to A^*. Suppose that there are b different Jordan blocks $J_1, \ldots, J_b$ with real eigenvalues in the Jordan form J of A, and let k_i $(i = 1, \ldots, b)$ be the number of times J_i appears in J. Then the set Ω of all invertible hermitian $n \times n$ matrices H such that A is H-selfadjoint, is a disconnected union of $k = \prod_{i=1}^{b}(k_i + 1)$ (arcwise) connected components $\Omega = \bigcup_{i=1}^{k} \Omega_i$, where each Ω_i consists of all matrices H with the same sign characteristic.*

Proof. In view of Corollary 5.1.2 we have

$$\Omega = \{H \mid H = T^*P_{\varepsilon,J}T \text{ for some } \varepsilon \text{ and invertible } T \text{ such that } TA = JT\}.$$
$$(5.8.23)$$

By Theorem 5.1.1 we can suppose that the set of signs ε in (5.8.23) is normalized. It is easily seen that the number of all normalized sets of signs is just k; designate them $\varepsilon^{(1)}, \ldots, \varepsilon^{(k)}$. Let

$$\Omega_i = \{H \mid H = T^*P_{\varepsilon^{(i)},J}T \text{ for invertible } T \text{ such that } TA = JT\}.$$

We prove now that Ω_i is connected. It is sufficient to prove that the set of all invertible matrices T such that $TA = JT$ is connected. Indeed, this set can be represented as

$$\{UT_0 \mid U \text{ is invertible and } UJ = JU\},$$

where T_0 is a fixed invertible matrix such that $T_0A = JT_0$. Now from the structure of the set of matrices commuting with J (see, for instance, [26, Chapter VIII]) it is easy to deduce that the invertible matrices U commuting with J form a connected set. Hence Ω_i is connected for every $i = 1, \ldots, k$.

It remains to prove that, for $H \in \Omega_j$, a sufficiently small neighborhood of H in Ω will contain elements from Ω_j only. This can be done using Theorem 5.8.1. In the notation of Theorem 5.8.1 the bilinear forms $f_1(x, y), \ldots, f_\gamma(x, y)$ depend continuously on H; therefore the same is true for $G_1, \ldots, G_\gamma$. So $G_i = G_i(H) :$ $\Psi_i \to \Psi_i$ is a selfadjoint linear transformation which depends continuously on $H \in \Omega$ and such that $\text{Ker}\, G_i = \Psi_{i+1}$ is fixed (i.e., independent of H). It follows from Theorem A.1.2 that the number of positive eigenvalues and the number of negative eigenvalues of each G_i remain constant in a neighborhood of $H \in \Omega_j$. By Theorem 5.8.1(iv), this neighborhood belongs to Ω_j.   □

5.9 Stability of the Sign Characteristic

As another application of Theorem 5.8.1, we present a result on the stability of the sign characteristic, in the sense of its persistence under sufficiently small perturbations. For this concept to make sense, the perturbations must be restricted to those for which the matrix A retains some of its structure after perturbation.

We arrive at the following definition. For any $A \in \mathbb{C}^{n \times n}$ with a real eigenvalue λ_0, and for a positive ε, the (λ_0, ε)-*structure preserving neighborhood* of A consists of all matrices $A_1 \in \mathbb{C}^{n \times n}$ such that $\|A - A_1\| < \varepsilon$, A_1 has exactly one real eigenvalue λ_1 (perhaps of high multiplicity) in the interval $(\lambda_0 - \varepsilon, \lambda_0 + \varepsilon)$, and the partial multiplicities of A_1 at the eigenvalue λ_1 coincide with the partial multiplicities of A at λ_0. It follows from this definition that A_1 has no nonreal eigenvalues in a complex neighborhood of λ_0.

Theorem 5.9.1. *Let there be given an H-selfadjoint matrix A, and let λ_0 be a real eigenvalue of A. Then there exist $\varepsilon > 0$ such that for every pair (A_1, H_1), where A_1 belongs to the (λ_0, ε)-structure preserving neighborhood of A, H_1 is a hermitian matrix satisfying $\|H_1 - H\| < \varepsilon$, and A_1 is H_1-selfadjoint, the sign characteristic of (A, H) at λ_0 and the sign characteristic of (A_1, H_1) at the eigenvalue λ_1 of A_1 lying in the interval $(\lambda_0 - \varepsilon, \lambda_0 + \varepsilon)$ are the same.*

By the sign characteristic of (A, H) at λ_0 we mean that part of the sign characteristic of (A, H) that corresponds to the partial multiplicities associated with λ_0 and similarly for the sign characteristic of (A_1, H_1) at λ_1. Note that by taking ε sufficiently small, the invertibility of H_1 is guaranteed.

Proof. Let A_1, H_1, and λ_1 be as in the statement of Theorem 5.9.1. Let γ be the largest partial multiplicity of A_0 at λ_0, which is also the largest partial multiplicity of A_1 at λ_1. Denote by Ψ_i, $i = 1, 2, \ldots, \gamma$, the subspace spanned by all eigenvectors of A_0 corresponding to λ_0 that generate a Jordan chain of length not less than i; in other words, the Ψ_i are the subspaces defined at the beginning of the preceding section. Denote by $\Psi_i^{(1)}$ the similarly defined subspaces for A_1 and λ_1. Since it is assumed that A_1 belongs to the (λ_0, ε)-structure preserving neighborhood of A, the dimensions of $\Psi_i^{(1)}$ and Ψ_i are equal for every fixed i.

Put

$$\mathcal{E}_\alpha = \mathrm{Ker}\,(I\lambda_0 - A)^\alpha, \quad \alpha = 1, 2, \ldots, \gamma, \qquad \mathcal{E}_0 = \{0\},$$

and for $\alpha = 1, 2, \ldots, \gamma$ choose a basis $\eta_{\alpha,1}, \ldots, \eta_{\alpha,k_\alpha}$ in a direct complement of $\mathcal{E}_{\alpha-1}$ in $\mathcal{E}_\alpha$. Then for a fixed i, $1 \le i \le \gamma$, the vectors

$$\phi_{ij} := (\lambda_0 I - A)^{i-1} \eta_{ij}, \quad j = 1, 2, \ldots, k_i$$

form a basis in Ψ_i. We claim that it is possible to choose bases $\phi_{i,1}^{(1)}, \ldots, \phi_{i,k_i}^{(1)}$ in $\Psi_i^{(1)}$, for $i = 1, 2, \ldots, \gamma$, such that the norms $\|\phi_{i,j}^{(1)} - \phi_{i,j}\|$ are as small as we wish, for sufficiently small ε.

Indeed, it follows from Proposition A.5.3 that for $\alpha = 1, 2, \ldots, \gamma$ the subspace

$$\mathcal{E}_\alpha^{(1)} := \mathrm{Ker}\,(I\lambda_1 - A_1)^\alpha$$

can be made arbitrarily close to $\mathcal{E}_\alpha$, in the sense that $\theta(\mathcal{E}_\alpha^{(1)}, \mathcal{E}_\alpha)$ can be made as small as we wish, by choosing a sufficiently small ε. From Corollary A.5.6 we find that, for every $\alpha = 1, 2, \ldots, \gamma$, there exist a basis $\eta_{\alpha,1}^{(1)}, \ldots, \eta_{\alpha,k_\alpha}^{(1)}$ in a direct complement of $\mathcal{E}_{\alpha-1}^{(1)}$ in $\mathcal{E}_\alpha^{(1)}$ such that $\eta_{\alpha,j}^{(1)}$ is arbitrarily close to $\eta_{\alpha,j}$ ($j = 1, 2, \ldots, k_\alpha$; $\alpha = 1, 2, \ldots, \gamma$). Now we can put

$$\phi_{i,j}^{(1)} := (\lambda_1 I - A_1)^{i-1}\eta_{i,j}^{(1)}, \quad j = 1, 2, \ldots, k_i$$

to produce bases $\phi_{i,1}^{(1)}, \ldots, \phi_{i,k_i}^{(1)}$ in $\Psi_i^{(1)}$, for $i = 1, 2, \ldots, \gamma$, where $\phi_{i,j}^{(1)}$ is arbitrarily close to $\phi_{i,j}$ ($j = 1, 2, \ldots, k_i$; $i = 1, 2, \ldots, \gamma$).

Since we have found bases $\phi_{i,j}^{(1)}$ in $\Psi_i^{(1)}$ that are close enough to the corresponding bases $\phi_{i,j}$ in Ψ_i, the assertion of Theorem 5.9.1 is obtained without difficulty from Theorem 5.8.1. Indeed, let $f_i(x, y)$ and G_i be the bilinear form and selfadjoint linear transformation, respectively, defined in Theorem 5.8.1 for A and λ_0, and let $f_i^{(1)}(x, y)$ and $G_i^{(1)}$ be the corresponding quantities defined for A_1 and λ_1. Then, if H_1 is close enough to H, and the bases $\phi_{i,j}^{(1)}$ are close enough to the bases $\phi_{i,j}$, then the matrix representation of $f_i^{(1)}$ in the basis $\phi_{i,1}^{(1)}, \ldots, \phi_{i,k_i}^{(1)}$ will be arbitrarily close to the matrix representation of f_i in the basis $\phi_{i,1}, \ldots, \phi_{i,k_i}$. By Theorem 5.8.1(ii) the same is true for the matrix representations of the selfadjoint linear transformation $G_i^{(1)}$ and G_i. Now Theorem 5.8.1, parts (iii) and (iv), together with Theorem A.1.2(b) ensures that for ε sufficiently small, the sign characteristic of (A, H) at λ_0 coincides with the sign characteristic of (A_1, H_1) at λ_1. $\qquad\square$

5.10 Canonical Forms for Pairs of Hermitian Matrices

Let G_1 and G_2 be hermitian $n \times n$ matrices (invertible or not). Consider the following problem: reduce G_1 and G_2 simultaneously by a congruence transformation to as simple a form as possible. In other words, by transforming the pair G_1, G_2 to the pair X^*G_1X, X^*G_2X with some invertible $n \times n$ matrix X we would like to bring G_1, G_2 to the simplest possible form.

If one of the matrices G_1 or G_2 is positive (or negative) definite, it is well known that G_1 and G_2 can be reduced simultaneously to a diagonal form. More generally, such a simple reduction cannot be achieved. However, for the case when one of the matrices G_1 and G_2 is invertible, Theorem 5.1.1 leads to the following result.

Theorem 5.10.1. *Let G_1 and G_2 be hermitian $n \times n$ matrices with G_2 invertible. Then there is an invertible $n \times n$ matrix X such that X^*G_1X and X^*G_2X have* *the forms:*

$$X^*G_1X = \varepsilon_1 K_1 \oplus \cdots \oplus \varepsilon_\alpha K_\alpha \oplus \begin{bmatrix} 0 & K_{\alpha+1} \\ K_{\alpha+1} & 0 \end{bmatrix} \oplus \cdots \oplus \begin{bmatrix} 0 & K_\beta \\ K_\beta & 0 \end{bmatrix}, \quad (5.10.24)$$

where

$$K_q = \begin{bmatrix} 0 & & & & 0 & \lambda_q \\ & & & & \lambda_q & 1 \\ & \vdots & & 1 & 0 & \\ & & & & \vdots & \\ 0 & \lambda_q & 1 & & \vdots & \\ \lambda_q & 1 & 0 & \cdots & \cdots & 0 \end{bmatrix},$$

λ_q *is real for* $q = 1, \ldots, \alpha$; λ_q *is nonreal for* $q = \alpha + 1, \ldots, \beta$ *and* $\varepsilon_q = \pm 1$
$(q = 1, \ldots, \alpha)$;

$$X^*G_2X = \varepsilon_1 P_1 \oplus \cdots \oplus \varepsilon_\alpha P_\alpha \oplus \begin{bmatrix} 0 & P_{\alpha+1} \\ P_{\alpha+1} & 0 \end{bmatrix} \oplus \cdots \oplus \begin{bmatrix} 0 & P_\beta \\ P_\beta & 0 \end{bmatrix}, \quad (5.10.25)$$

where P_q *is the sip matrix whose size is equal to that of* K_q *for* $q = 1, 2, \ldots, \beta$.
The representations (5.10.24) *and* (5.10.25) *are uniquely determined by* G_1 *and*
G_2, *up to simultaneous permutation of the same blocks in each formula* (5.10.24)
and (5.10.25).

Proof. Observe that $G_2^{-1}G_1$ is G_2-selfadjoint and let $(J, P_{\varepsilon,J})$ be the canonical
form for $(G_2^{-1}G_1, G_2)$. Thus

$$G_2^{-1}G_1 = XJX^{-1}, \quad G_2 = (X^*)^{-1}P_{\varepsilon,J}X^{-1} \quad (5.10.26)$$

for some invertible matrix X. Now (5.10.26) implies that

$$G_1 = G_2XJX^{-1} = (X^*)^{-1}P_{\varepsilon,J}JX^{-1}.$$

Taking into account the form of $P_{\varepsilon,J}J$ (see formulas (5.1.1) and (5.1.2)) we obtain
the representations (5.10.24) and (5.10.25). The uniqueness of representations
(5.10.24) and (5.10.25) follows from the uniqueness of the canonical form, as stated
in Theorem 5.1.1. □

Note that the sizes of blocks K_q $(q = 1, \ldots, \alpha)$, in (5.10.24) and the corresponding numbers λ_q are just the sizes and eigenvalues of those Jordan blocks in
the Jordan form of $G_2^{-1}G_1$ which correspond to the real eigenvalues. The sizes of
blocks K_q $(q = \alpha + 1, \ldots, \beta)$ and the corresponding numbers λ_q may be chosen
as the sizes and eigenvalues of those Jordan blocks of $G_2^{-1}G_1$ which correspond to
a maximal set of eigenvalues $C \subseteq \sigma(G_2^{-1}G_1)$ which does not contain a conjugate
pair of complex numbers.

The problem of the simultaneous reduction of a pair of hermitian matrices can also be stated as reduction (under congruence transformations) of a linear pencil of matrices. Namely, given a linear pencil $\lambda G_1 + G_2$ of $n \times n$ hermitian matrices, find the simplest form $\lambda G_1' + G_2'$ of this pencil which can be obtained from $\lambda G_1 + G_2$ by a congruence transformation: $\lambda G_1' + G_2' = X^*(\lambda G_1 + G_2)X$, where X is an invertible $n \times n$ matrix. We leave it to the reader to restate Theorem 5.10.1 in terms of linear pencils of hermitian matrices.

5.11   Third Description of the Sign Characteristic

The two ways of describing the sign characteristic already introduced result in the statements of Theorems 5.1.1 and 5.8.1. The third approach is apparently quite different in character and relies on fundamental ideas of analytic perturbation theory.

If A is H-selfadjoint we have described the associated sign characteristic as that of the pair (A, H). The discussion of the preceding section suggests that it would not be unnatural to associate the sign characteristic with the pair of hermitian matrices (H, HA), or with the pencil $\lambda H - HA$. In fact, our new description is to be formulated in terms of the (λ-dependent) eigenvalues of $\lambda H - HA$.

The fundamental perturbation theorem concerning these eigenvalues is now stated:

Theorem 5.11.1. *Let $G_1, G_2 \in \mathbb{C}^{n \times n}$ be a pair of hermitian matrices. Then there is a function $U(\lambda)$ of the real variable λ with values in the unitary matrices such that, for all real λ,*

$$\lambda G_1 + G_2 = U(\lambda)\text{diag}\,(\mu_1(\lambda), \ldots, \mu_n(\lambda))U(\lambda)^*,$$

and, moreover, the functions $\mu_j(\lambda)$ and $U(\lambda)$ can be chosen to be analytic functions of the real parameter λ.

This is a particular case of Theorem A.6.7.

Note that the functions $\mu_j(\lambda)$ of Theorem 5.11.1 are not generally polynomials in λ.

We apply Theorem 5.11.1 with $G_1 = H$, $G_2 = -HA$, where A is H-selfadjoint. Thus, the eigenvalue functions $\mu_j(\lambda)$ (which are, of course, the zeros of $\det(I\mu - (\lambda H - HA)) = 0$) are analytic in λ. This is the essential prerequisite for understanding the next characterization of the sign characteristic. Note also that λ is a real eigenvalue of A if and only if $\mu_j(\lambda) = 0$ for at least one j.

Theorem 5.11.2. *Let A be an $n \times n$ H-selfadjoint matrix, let $\mu_1(\lambda), \ldots, \mu_n(\lambda)$ be the zeros of the scalar polynomial $\det(\mu I - (\lambda H - HA))$ of degree n arranged so that $\mu_j(\lambda)$, $j = 1, \ldots, n$ are real analytic functions of real λ, and let $\lambda_1, \ldots \lambda_r$ be the different real eigenvalues of A.*

For every $i = 1, 2, \ldots, r$ write

$$\mu_j(\lambda) = (\lambda - \lambda_i)^{m_{ij}} \nu_{ij}(\lambda)$$

where $\nu_{ij}(\lambda_i) \neq 0$ and is real. Then the nonzero numbers among $m_{i1}, \ldots, m_{in}$ are the sizes of Jordan blocks with eigenvalue λ_i in the Jordan form of A, and the sign of $\nu_{ij}(\lambda_i)$ (for $m_{ij} \neq 0$) is the sign attached to m_{ij} in the (possibly nonnormalized) sign characteristic of (A, H).

The proof of Theorem 5.11.2 is based on a different set of ideas, and is therefore relegated to Chapter 12. We illustrate the theorem with an example:

Example 5.11.3. *Let*

$$H = \begin{bmatrix} 0 & \varepsilon_1 & 0 & 0 \\ \varepsilon_1 & 0 & 0 & 0 \\ 0 & 0 & 0 & \varepsilon_2 \\ 0 & 0 & \varepsilon_2 & 0 \end{bmatrix}, \quad A = \begin{bmatrix} 0 & 1 & 0 & 0 \\ 0 & 0 & 0 & 0 \\ 0 & 0 & 1 & 1 \\ 0 & 0 & 0 & 1 \end{bmatrix}$$

(cf. Example 5.1.4) where $\varepsilon_1 = \pm 1$, $\varepsilon_2 = \pm 1$. Then we have $\lambda_1 = 0$, $\lambda_2 = 1$. The four eigenvalue functions are given by

$$\mu_{1,2}(\lambda) = \frac{-\varepsilon_1 \pm \sqrt{1 + 4\lambda^2}}{2}, \quad \mu_{3,4}(\lambda) = \frac{-\varepsilon_2 \pm \sqrt{1 + 4(\lambda - 1)^2}}{2}.$$

It is easily seen that the matrix $\lambda H - HA$ has just one eigenvalue $\mu_1(\lambda)$ vanishing at $\lambda = 0$ and just one eigenvalue, say $\mu_3(\lambda)$, vanishing at $\lambda = 1$. Furthermore, in sufficiently small neighborhoods of these two points

$$\mu_1(\lambda) = (\lambda^2 - \lambda^4 + \cdots)\varepsilon_1, \quad \mu_3(\lambda) = ((1 - \lambda)^2 - (1 - \lambda)^4 + \cdots)\varepsilon_2. \qquad \square$$

5.12 Invariant Maximal Nonnegative Subspaces

Invariant subspaces of an H-selfadjoint matrix A will play an important role in subsequent parts of this book; particularly in connection with factorization problems for polynomial matrix functions. More especially, we shall be concerned with A-invariant subspaces which are nonnegative (or nonpositive) with respect to H and of maximal dimension. This section is devoted to the analysis of such subspaces.

Some notations and definitions will first be set up. Let A be an $n \times n$ H-selfadjoint matrix, and let $m_1, \ldots, m_r$ be the sizes of Jordan blocks in a Jordan form for A corresponding to the real eigenvalues. Let the corresponding signs in the sign characteristic for (A, H) be $\varepsilon_1, \ldots, \varepsilon_r$, and let $i_+(H)$ be the number of positive eigenvalues of H, counting multiplicities.

A set C of nonreal eigenvalues of A is called a *c-set* if $C \cap \overline{C} = \emptyset$ and $C \cup \overline{C}$ is the set of all nonreal eigenvalues of A. If $\mathcal{M}$ is an H-nonnegative subspace, define the *index of positivity* of $\mathcal{M}$, written $p(\mathcal{M})$, to be the maximal dimension of an H-positive subspace $\mathcal{M}_0 \subseteq \mathcal{M}$. It was noted after the proof of Theorem 2.3.3 that $p(\mathcal{M})$ does not depend on the choice of $\mathcal{M}_0$.

Theorem 5.12.1. *For every c-set C there exists an $i_+(H)$-dimensional, A-invariant, H-nonnegative subspace $\mathcal{N}$ such that the nonreal part of $\sigma(A \mid_{\mathcal{N}})$ coincides with C. The subspace $\mathcal{N}$ is maximal H-nonnegative and*

$$p(\mathcal{N}) = \frac{1}{2} \sum_{i=0}^{r} [1 - (-1)^{m_i}] \delta_i, \qquad (5.12.27)$$

where $\delta_i = 1$ if $\varepsilon_i = 1$, and $\delta_i = 0$ if $\varepsilon_i = -1$.

Proof. We make use of Theorem 5.1.1 and write $A = T^{-1}JT$, $H = T^*P_{\varepsilon,J}T$ for some invertible T. Note that the columns of T^{-1} corresponding to each Jordan block in J form a Jordan chain for A. So J has exactly $q = \frac{1}{2}[n - \sum_{i=1}^{r} m_i]$ eigenvalues (counting multiplicities) in C, and exactly r Jordan blocks $J_1, \ldots, J_r$ of sizes $m_1, \ldots, m_r$ with signs $\varepsilon_1, \ldots, \varepsilon_r$ respectively corresponding to real eigenvalues. Let $\mathcal{N}$ be the A-invariant subspace spanned by the following columns of T^{-1}:

a) the columns corresponding to the Jordan blocks with eigenvalues in C (the number of such columns is q);

b) for even m_i, the first $m_i/2$ columns corresponding to the block J_i;

c) for odd m_i, the first $(m_i + 1)/2$ or $(m_i - 1)/2$ columns corresponding to J_i, according as $\varepsilon_i = +1$ or $\varepsilon_i = -1$.

Then the dimension of $\mathcal{N}$ is found to be

$$q + \sum_{m_i \text{ even}} \frac{m_i}{2} + \sum_{m_i \text{ odd}, \, \varepsilon_i=1} \frac{m_i + 1}{2} + \sum_{m_i \text{ odd}, \, \varepsilon_i=-1} \frac{m_i - 1}{2}$$

$$= \frac{n}{2} + \frac{1}{4} \sum_{i=1}^{r} [1 - (-1)^{m_i}] \varepsilon_i.$$

Comparing this with equation (5.2.4) it is seen that $\dim \mathcal{N} = i_+(H)$. Furthermore, it is easily seen from the structure of $P_{\varepsilon,J}$ that $\mathcal{N}$ is H-nonnegative.

To check that $p(\mathcal{N})$ is given by (5.12.27), let $P_{\mathcal{N}}$ be the orthogonal projection on $\mathcal{N}$. Then in the basis for $\mathcal{N}$ formed by the chosen columns of T^{-1} we have

$$P_{\mathcal{N}} H \mid_{\mathcal{N}} = \operatorname{diag}\,(\zeta_1, \zeta_2, \ldots, \zeta_k) : \mathcal{N} \to \mathcal{N},$$

where $\zeta_i = 0$ or 1 and $\zeta_i = 1$ if and only if the i-th chosen column of T^{-1} is the $\frac{1}{2}(m_j + 1)$-th generalized eigenvector in a chain of length m_j for which m_j is odd and $\zeta_m = +1$. It is clear that, with this construction, $p(\mathcal{N}) = \sum_{i=1}^{k} \zeta_i$ and the formula (5.12.27) follows.

Finally, the maximality of $\mathcal{N}$ follows from Theorem 2.3.2. $\square$

For the case of H-nonpositive invariant subspaces of A there is, of course, another statement dual to that of Theorem 5.12.1. This can be obtained by considering $-H$ in place of H in the last theorem. The following example shows that, in general, the subspace $\mathcal{N}$ is not unique (for a given c-set C).

Example 5.12.2. *Let*

$$
A = \begin{bmatrix} 0 & 1 & 0 & 0 \\ 0 & 0 & 0 & 0 \\ 0 & 0 & 0 & 1 \\ 0 & 0 & 0 & 0 \end{bmatrix}; \quad
H = \begin{bmatrix} 0 & 1 & 0 & 0 \\ 1 & 0 & 0 & 0 \\ 0 & 0 & 0 & -1 \\ 0 & 0 & -1 & 0 \end{bmatrix}.
$$

The following 2-dimensional subspaces of C^4 are A-invariant and H-nonnegative (here the c-set C is empty);

$$
\text{Span}\,\{\langle 1\,0\,0\,0\rangle, \langle 0\,0\,1\,0\rangle\}, \quad \text{Span}\,\{\langle 1\,0\,1\,0\rangle, \langle 0\,1\,0\,1\rangle\}.
$$

Note that both subspaces are not only H-nonnegative, but also H-neutral. □

The following result includes extra conditions needed to ensure the uniqueness of the subspace $\mathcal{N}$ from Theorem 5.12.1.

Theorem 5.12.3. *Let A be an H-selfadjoint $n \times n$ matrix such that the sizes $m_1, \ldots, m_r$ of the Jordan blocks of A corresponding to real eigenvalues are all even. Then for every c-set C there exists a unique H-neutral A-invariant subspace $\mathcal{N}$ such that $\dim \mathcal{N} = \frac{1}{2}n$, $\sigma(A\,|_{\mathcal{N}}) \setminus \mathsf{R} = C$, and the sizes of the Jordan blocks of $A\,|_{\mathcal{N}}$ corresponding to the real eigenvalues are $\frac{1}{2}m_1, \ldots, \frac{1}{2}m_r$.*

Proof. Observe first, using (5.2.4) for example, that when $m_1, \ldots, m_r$ are all even the number of positive eigenvalues of H is just $i_+(H) = \frac{1}{2}n$. So choosing a c-set C, Theorem 5.12.1 ensures the existence of a subspace $\mathcal{N}$ with all the properties required by the theorem.

Let $\mathcal{N}_C$ be the sum of root subspaces of $A\,|_{\mathcal{N}}$ corresponding to the nonreal eigenvalues of $A\,|_{\mathcal{N}}$ (i.e., the eigenvalues in C), and let $\mathcal{N}_r$ be the sum of root subspaces of $A\,|_{\mathcal{N}}$ corresponding to the real eigenvalues of $A\,|_{\mathcal{N}}$. Clearly

$$
\mathcal{N} = \mathcal{N}_C \dotplus \mathcal{N}_r. \tag{5.12.28}
$$

Now form the decomposition

$$
\mathsf{C}^n = \mathcal{X}_C \dotplus \mathcal{X}_{\overline{C}} \dotplus \mathcal{X}_r, \tag{5.12.29}
$$

where $\mathcal{X}_C$ (resp. $\mathcal{X}_{\overline{C}}$) is the sum of the root subspaces of A corresponding to the eigenvalues in C (resp. in $\overline{C} := \{\lambda \in \mathsf{C} \mid \overline{\lambda} \in C\}$).

It is clear from the construction that $\mathcal{N}_C = \mathcal{X}_C$ and is uniquely determined, and also $\mathcal{N} \cap \mathcal{X}_{\overline{C}} = \{0\}$. Consequently, the only possible nonuniqueness in the determination of $\mathcal{N}$ arises in forming $\mathcal{N}_r$ of (5.12.28).

Let $\lambda_1, \ldots, \lambda_t$ be the distinct real eigenvalues of A and $\mathcal{R}_{\lambda_j}(A)$ be the (uniquely determined) root subspace of A corresponding to λ_j, $j = 1, 2, \ldots, t$. Then

$$\mathcal{X}_r = \mathcal{R}_{\lambda_1}(A) \dotplus \cdots \dotplus \mathcal{R}_{\lambda_t}(A)$$

and there is a corresponding decomposition of $\mathcal{N}_r$ (see Theorem A.2.5):

$$\mathcal{N}_r = \mathcal{N}_{\lambda_1} \dotplus \cdots \dotplus \mathcal{N}_{\lambda_r},$$

with $\mathcal{N}_{\lambda_j} \subseteq \mathcal{R}_{\lambda_j}(A))$ for each j and, moreover the dimension of $\mathcal{N}_{\lambda_j}$ is just half that of $\mathcal{R}_{\lambda_j}(A)$. These inclusions mean that uniqueness of $\mathcal{N}$ now depends on that of each $\mathcal{N}_{\lambda_j}$. Thus the proof is reduced to the case when A has just one eigenvalue, say λ_0, and λ_0 is real.

Without loss of generality assume that $\lambda_0 = 0$ and that (A, H) is in the canonical form (J, P), where $P = P_{\varepsilon,J}$. Let $\mathcal{N}_1$, $\mathcal{N}_2$ be subspaces with all the properties listed in Theorem 5.12.3, and we have to prove that $\mathcal{N}_1 = \mathcal{N}_2$.

The proof is by induction on n. Consider first the case that $m_1 = \cdots = m_r = 2$. Then there exists a unique J-invariant $n/2$-dimensional subspace $\mathcal{N}$ such that the partial multiplicities of $J \mid_{\mathcal{N}}$ are $1, \ldots, 1$ (namely, $\mathcal{N}$ is spanned by all the eigenvectors of J). So in that case Theorem 5.12.3 is evident (even without the condition of P-neutrality). Now consider the general case.

Consider the subspace $\operatorname{Ker} J$. From the properties of $\mathcal{N}_i$ it is clear that $\operatorname{Ker} J \subseteq \mathcal{N}_1$ and $\operatorname{Ker} J \subseteq \mathcal{N}_2$. Let $g_1, \ldots, g_r$ be the coordinate unit vectors with a 1 in the positions $m_1, m_1 + m_2, \ldots, m_1 + \cdots + m_r$, respectively. We show that $\mathcal{N}_i \perp$ Span $\{g_1, \ldots, g_r\}$, $i = 1, 2$, where the orthogonality is understood in the sense of the usual inner product $(.,.)$. Suppose not, so that there exists an $x \in \mathcal{N}_i$ such that $(x, g_j) \neq 0$ for some j. Let f_j be the eigenvector of J corresponding to the j-th Jordan block. Clearly, $f_j \in \mathcal{M}_i$, but $(Pf_j, x) = (g_j, x) \neq 0$; a contradiction with the P-neutrality of $\mathcal{N}_i$. So

$$\mathcal{N}_i \perp \text{Span } \{g_1, \ldots, g_r\}, \quad i = 1, 2.$$

Now define the subspaces $\mathcal{N}_i' \subseteq \mathbf{C}^n$ as follows: for every $x \in \mathbf{C}^n$, let $\varphi x \in \mathbf{C}^n$ be the vector obtained from x by putting zero instead of the coordinates of x in the positions $1, m$ and $m_1 + \cdots + m_p$ and $m_1 + \cdots + m_p + 1$ for $p = 1, \ldots, r - 1$: the rest of the coordinates of x remain unchanged under φ. Now put

$$\mathcal{N}_i' = \{\varphi x \mid x \in \mathcal{N}_i\}, \quad i = 1, 2.$$

It is clear that $\mathcal{N}_i' \perp (\operatorname{Ker} J + \text{Span } \{g_1, \ldots, g_r\})$. Moreover, since $\mathcal{N}_i \supseteq \operatorname{Ker} J$ and $\mathcal{N}_i \perp \text{Span } \{g_1, \ldots, g_r\}$, we have

$$\mathcal{N}_i' \dotplus \operatorname{Ker} J = \mathcal{N}_i, \quad i = 1, 2.$$

Evidently, it suffices to prove that $\mathcal{N}_1' = \mathcal{N}_2'$. To this end, consider the Jordan matrix $\tilde{J}$ which is obtained from J by crossing out the first and last column and row in each Jordan block. So $\tilde{J}$ has blocks of sizes $(m_1 - 2), \ldots, (m_r - 2)$. Further, define the subspaces $\tilde{\mathcal{N}}_i \subseteq \mathbb{C}^{n-2r}$ as

$$\tilde{\mathcal{N}}_i = \left\{ \tilde{x} \in \mathbb{C}^{n-2r} \mid x \in \mathcal{N}'_i \right\}, \quad i = 1, 2,$$

where $\tilde{x}$ is obtained from x by crossing out its $2r$ zero coordinates in the above mentioned positions. Clearly, the subspaces $\tilde{\mathcal{N}}_i$ are $\tilde{J}$-invariant and $\tilde{P}$-neutral, where $\tilde{P}$ is obtained from P by crossing out the first and last column and row in each of $P_1, \ldots, P_r$, and similarly $\tilde{J}$ is obtained from J. So we can apply the induction hypothesis to deduce that $\tilde{\mathcal{N}}_1 = \tilde{\mathcal{N}}_2$ and therefore also $\mathcal{N}'_1 = \mathcal{N}'_2$. $\qquad\square$

Example 5.12.2 shows that, in Theorem 5.12.3, the requirement that the multiplicities of $A \mid_{\mathcal{N}}$ are $\frac{1}{2}m_1, \ldots, \frac{1}{2}m_r$ cannot be omitted in general. However, it turns out that if we require an additional condition that the signs in the sign characteristic of (A, H) which correspond to Jordan blocks with the same eigenvalue are equal, then for every c-set C there exists an unique H-neutral A-invariant $\frac{n}{2}$-dimensional subspace $\mathcal{N}$ with $(A \mid_{\mathcal{N}}) \backslash \mathbb{R} = C$. (Obviously, this additional condition is violated in Example 5.12.2.) Moreover, in this case we are able to describe all H-neutral A-invariant $\frac{n}{2}$-dimensional subspaces, as follows.

Theorem 5.12.4. *Let A be an H-selfadjoint matrix such that the sizes $m_1, \ldots, m_r$ of the Jordan blocks of A corresponding to the real eigenvalues are all even, and the signs in the sign characteristic of (A, H) corresponding to the same eigenvalue are all equal. Let $\mathcal{M}_+$ be the sum of the root subspaces of A corresponding to the eigenvalues of A in the open upper half-plane. Then for every A-invariant subspace $\mathcal{N}_+ \subseteq \mathcal{M}_+$ there exists a unique A-invariant H-neutral $\frac{n}{2}$-dimensional subspace $\mathcal{N}$ such that*

$$\mathcal{N} \cap \mathcal{M}_+ = \mathcal{N}_+. \tag{5.12.30}$$

In other words, Theorem 5.12.4 gives a one-to-one correspondence between the set of A-invariant subspaces $\mathcal{N}_+$ of $\mathcal{M}_+$ and the set of A-invariant H-neutral $\frac{n}{2}$-dimensional subspaces $\mathcal{N}$, and is given explicitly by (5.12.30).

Proof. Passing to the canonical form of (A, H), we see that it is sufficient to consider only the two following cases (cf. the proof of Theorem 5.12.3):

(i) $\sigma(A) = \{0\}$;

(ii) $\sigma(A) = \{\lambda_0, \overline{\lambda_0}\}$, λ_0 has positive imaginary part.

Consider case (i). In this case $\mathcal{M}_+ = 0$, so Theorem 5.12.4 asserts that there is a unique A-invariant H-neutral $\frac{n}{2}$-dimensional subspace $\mathcal{N}$. To prove this, assume that $(A, H) = (J, P_{\varepsilon, J})$ is in the canonical form; so

$$J = J_1 \oplus \cdots \oplus J_r,$$

where J_i is the nilpotent Jordan block of size m_i, $i = 1, \ldots, r$. Suppose, for instance, that the signs in the sign characteristic of $(J, P_{\varepsilon, J})$ are all $+1$'s. Denote by x_{ij} the $(m_1 + \cdots + m_{i-1} + j)$-th unit coordinate vector in $\mathbb{C}^n$, $j = 1, \ldots, m_i$ (recall that $n = m_1 + \cdots + m_r$); so the vectors $x_{i1}, \ldots, x_{im_i}$ form a Jordan chain of J. Let $\mathcal{N}$ be a $P_{\varepsilon,J}$-neutral J-invariant subspace with $\dim \mathcal{N} = \frac{1}{2}n$, and let

$$x = \sum_{i,j} a_{ij} x_{ij} \in \mathcal{N}, \qquad a_{ij} \in \mathbb{C}.$$

We claim that the coefficients a_{ij} with $j > \frac{1}{2}m_i$ are zeros. Suppose not; let K be the set of all indices i ($1 \le i \le r$) for which the set $\{j \mid \frac{1}{2}m_i < j \le m_i, a_{ij} \ne 0\}$ is nonvoid and the difference

$$\max\left\{j \mid \frac{1}{2}m_i < j \le m_i, \, a_{ij} \ne 0\right\} - \frac{1}{2}m_i$$

is maximal. Denote this difference by γ. Since $\mathcal{N}$ is J-invariant, the vectors

$$y_1 = J^\gamma x, \qquad y_2 = J^{\gamma-1} x$$

are again in $\mathcal{N}$. A computation shows that

$$(P_{\varepsilon,J} y_1, y_2) = \sum_{i \in K} \left| a_{i,\frac{1}{2}m_i+\gamma} \right|^2 = 0$$

because $\mathcal{N}$ is $P_{\varepsilon,J}$-neutral. So $a_{i,\frac{1}{2}m_i+\gamma} = 0$ for all $i \in K$, which contradicts the choice of K. Thus, $a_{ij} = 0$ for $j > \frac{1}{2}m_i$. Since $\dim \mathcal{N} = \frac{1}{2}n$, this leaves only one possibility for $\mathcal{N}$; namely, $\mathcal{N} = \mathrm{Span}\left\{x_{ij} \mid 1 \le j \le \frac{1}{2}m_i; i = 1, \ldots, r\right\}$.

Now consider case (ii). Again, assume $(A, H) = (J, P_{\varepsilon,J})$ is the canonical form. Rearranging blocks in J and $P_{\varepsilon,J}$ (which amounts to a unitary similarity), we can write J and $P_{\varepsilon,J}$ in the forms:

$$J = \begin{bmatrix} J_+ & 0 \\ 0 & J_+ \end{bmatrix}$$

where $J_+ = J_1 \oplus \cdots \oplus J_r$ and J_i is the Jordan block of size p_i with eigenvalue λ_0, and

$$P_{\varepsilon,J} = \begin{bmatrix} 0 & P \\ P & 0 \end{bmatrix}$$

where $P = P_1 \oplus \cdots \oplus P_r$, and P_i is the sip matrix of size p_i. Observe that $\overline{J_+} = P^{-1}J_+^* P$; so the pair $(J, P_{\varepsilon,J})$ is unitarily similar to the pair (K, Q), where $K = \begin{bmatrix} J_+ & 0 \\ 0 & J_+^* \end{bmatrix}$, $Q = \begin{bmatrix} 0 & I \\ I & 0 \end{bmatrix}$. It is sufficient to verify Theorem 5.12.4 for $A = K$, $H = Q$.

Given a J_+-invariant subspace $\mathcal{N}_+$, put

$$\mathcal{N} := \left\{ \begin{bmatrix} x \\ y \end{bmatrix} \in \mathbb{C}^n \mid x \in \mathcal{N}_+, \, y \in \mathcal{N}_+^\perp \right\}.$$

Clearly, $\mathcal{N}$ is K-invariant, $\dim \mathcal{N} = \frac{n}{2}$ and $\mathcal{N} \cap \mathcal{R}_{\lambda_0}(K) = \begin{bmatrix} \mathcal{N}_+ \\ 0 \end{bmatrix}$. Further, $\mathcal{N}$ is Q-neutral; indeed, for $x_1, x_2 \in \mathcal{N}_+$, $y_1, y_2 \in \mathcal{N}_+^\perp$ we have

$$\left(\begin{bmatrix} 0 & I \\ I & 0 \end{bmatrix} \begin{bmatrix} x_1 \\ y_1 \end{bmatrix}, \begin{bmatrix} x_2 \\ y_2 \end{bmatrix} \right) = (y_1, x_2) + (x_1, y_2) = 0.$$

Now let $\mathcal{N}'$ be a K-invariant, $\frac{n}{2}$-dimensional, Q-neutral subspace such that

$$\mathcal{N}' \cap \mathcal{R}_{\lambda_0}(K) = \begin{bmatrix} \mathcal{N}_+ \\ 0 \end{bmatrix}. \tag{5.12.31}$$

As $\mathcal{N}'$ is K-invariant,

$$\mathcal{N}' = \left(\mathcal{N}' \cap \mathcal{R}_{\lambda_0}(K) \right) \dot{+} \left(\mathcal{N}' \cap \mathcal{R}_{\overline{\lambda_0}}(K) \right).$$

In fact

$$\mathcal{N}' \cap \mathcal{R}_{\overline{\lambda_0}}(K)) \subseteq \left\{ \begin{bmatrix} 0 \\ x \end{bmatrix} \mid x \in \mathcal{N}_+^\perp \right\}. \tag{5.12.32}$$

Indeed, if $\begin{bmatrix} 0 \\ x \end{bmatrix} \in \mathcal{N}'$ with some $x \in \mathbb{C}^{n/2} \setminus \mathcal{N}_+^\perp$, then there exists $y \in \mathcal{N}_+$ such that $(x, y) \neq 0$, and then

$$\left(\begin{bmatrix} 0 & I \\ I & 0 \end{bmatrix} \begin{bmatrix} 0 \\ x \end{bmatrix}, \begin{bmatrix} y \\ 0 \end{bmatrix} \right) = (x, y) \neq 0,$$

a contradiction with Q-neutrality of $\mathcal{N}'$. Now (5.12.31) and (5.12.32) together with $\dim \mathcal{N}' = \frac{n}{2}$, imply that $\mathcal{N}' = \mathcal{N}$.

Theorem 5.12.4 is proved. $\square$

Let A, H be as in Theorem 5.12.4. Observe that for every A-invariant H-neutral $\frac{n}{2}$-dimensional subspace $\mathcal{N}$ the sizes of Jordan blocks of the restriction $A |_\mathcal{N}$ which correspond to a real eigenvalue are just half the sizes of Jordan blocks of A corresponding to the same eigenvalue (see Theorem 5.12.3). In particular, the restriction of $A |_\mathcal{N}$ to the sum of the root subspaces of $A |_\mathcal{N}$ corresponding to the real eigenvalues is independent of $\mathcal{N}$ (in the sense that any two such restrictions, for different subspaces $\mathcal{N}$, are similar).

A special case of Theorem 5.12.4 is sufficiently important to justify a separate statement:

Corollary 5.12.5. *Let A be an H-selfadjoint matrix as in Theorem 5.12.4, and suppose, in addition, that $\sigma(A) \subseteq \mathbb{R}$. Then there exists a unique A-invariant, H-neutral, $\frac{n}{2}$-dimensional subspace.*

5.13   Inverse Problems

It has been noted in Corollary 5.1.3 that if an $n \times n$ matrix A is similar to A^* then there is an H with $\det H \neq 0$ and $H^* = H$ such that A is H-selfadjoint. Now we ask more specifically: if A is similar to A^*, is there an H with a prescribed number of negative eigenvalues such that A is H-selfadjoint? In this connection, recall equation (5.2.4)

$$i_-(H) = \frac{1}{2}n - \frac{1}{4}\sum_{j=1}^{\alpha}[1 - (-1)^{m_j}]\,\varepsilon_j \qquad (5.13.33)$$

which expresses the number of negative eigenvalues of H in terms of the sizes of Jordan blocks with the real eigenvalues, $m_1, \ldots, m_\alpha$ and the associated signs $\varepsilon_1, \ldots, \varepsilon_\alpha$ of the sign characteristic. Observe that, if the sign characteristic is prescribed along with A, then $i_-(H)$ is fixed by this relation. When the sign characteristic is not prescribed the next result holds.

Theorem 5.13.1. *Let A be an $n \times n$ matrix which is similar to A^* and let $m_1, \ldots, m_\alpha$ be the sizes of all the Jordan blocks in the Jordan form of A associated with real eigenvalues. Then there exists a nonsingular hermitian matrix H with exactly N negative eigenvalues (counting multiplicities) for which A is H-selfadjoint if and only if*

$$\frac{1}{2}n - \frac{1}{4}\sum_{j=1}^{\alpha}[1 - (-1)^{m_j}] \leq N \leq \frac{1}{2}n + \frac{1}{4}\sum_{j=1}^{\alpha}[1 - (-1)^{m_j}]. \qquad (5.13.34)$$

Proof. If A is H-selfadjoint and similar to A^* as required by the theorem, then (5.13.34) follows immediately from (5.13.33).

Conversely, if (5.13.34) holds then signs $\varepsilon_{01}, \ldots, \varepsilon_{0\alpha}$ can be chosen in such a way that

$$N = \frac{1}{2}n - \frac{1}{4}\sum_{j=1}^{\alpha}[1 - (-1)^{m_j}]\,\varepsilon_{0j},$$

and a matrix H is found accordingly.   □

Note that Theorem 5.13.1 can be re-formulated on replacing N by the number of *positive* eigenvalues of H. This is because A is H-selfadjoint if and only if A is $(-H)$-selfadjoint and the number of positive eigenvalues of H is just the number of negative eigenvalues of $-H$.

Now let 2ν be the number of nonreal eigenvalues of A, counted with multiplicities. The inequalities (5.13.34) can also be expressed in terms of ν. Thus, provided $N \leq \frac{1}{2}n$, (5.13.34) is equivalent to the two conditions:

a) $\nu + \sum_{j=1}^{\alpha}\left[\frac{1}{2}m_j\right] \leq N$,

b) the number of Jordan blocks of A corresponding to real eigenvalues of odd sizes is not less than

$$N - \nu - \sum_{j=1}^{\alpha} \left[\frac{1}{2}m_j\right].$$

Here, $[x]$ denotes the integer part of x.

This reformulation is established by examination of (5.13.34) making use of the relation

$$\nu = \frac{1}{2}\left(n - \sum_{j=1}^{a} m_j\right).$$

In particular, this reformulation leads to the corollary:

Corollary 5.13.2. *Assume that the numbers $m_1, \ldots, m_\alpha$ in Theorem 5.13.1 are all even. Then there exists an $n \times n$ hermitian H with exactly N negative eigenvalues (counting multiplicities) and for which A is H-selfadjoint, if and only if $N = \frac{1}{2}n$, i.e., sig $H = 0$.*

Note that, under the conditions of this result, n is necessarily even.

5.14 Canonical Forms for H-Unitaries: First Examples

The main result obtained earlier in this chapter was the reduction of a pair of matrices (A, H), for which A is H-selfadjoint, to a canonical form $(J, P_{\varepsilon,J})$. This is described in Theorem 5.1.1. The program for the rest of this chapter is similar, but for pairs (A, H) in which A is H-unitary.

First recall the basic ideas. Let H be an $n \times n$ invertible Hermitian matrix and $[x, y] = (Hx, y)$ be the associated indefinite inner product on $\mathbf{C}^n$. An $n \times n$ matrix A is said to be H-*unitary* if A is invertible and $[A^{-1}x, y] = [x, Ay]$ for $x, y \in \mathbf{C}^n$, or, what is equivalent, $A^{-1} = A^{[*]}$, or $A^{-1} = H^{-1}A^*H$. Some simple properties of H-unitary matrices have already been noted in Section 4.3. In particular, the spectrum of such a matrix is symmetric with respect to the unit circle, and the root subspaces of A corresponding to eigenvalues λ and μ with $\lambda\bar{\mu} \neq 1$ are orthogonal in the indefinite inner product $[.,.]$.

It has been seen in Chapter 3 that, if J is an $n \times n$ Jordan block with *real* eigenvalue λ, and if P is the $n \times n$ sip matrix, then (J, P) form a primitive matrix pair for which J is P-selfadjoint. It is natural to search for a primitive P-*unitary* matrix by examining matrices of the form $f(J)$ where f is a fractional transformation of the form (4.3.12). It is certainly the case that when λ is real $\mu = f(\lambda)$ will be the only eigenvalue of $f(J)$, and it will be unimodular.

It will be convenient to write the Jordan block J in the form $J = \lambda I + D$ (so that D has the "super-diagonal" made up of "ones," all other elements zero, and $D^n = 0$). Let $|\eta| = 1$, suppose $w \neq \overline{w}$, $w \neq \lambda$, $w \neq \overline{\lambda}$ (for the time being λ may be real or nonreal), and define the Möbius transformation

$$f(z) = \frac{\eta(z - \overline{w})}{z - w}, \qquad (5.14.35)$$

and $K = f(J)$. Then

$$f(J) = \eta(\lambda I + D - \overline{w}I)(\lambda I + D - wI)^{-1}$$

$$= \eta \left[\frac{\lambda - \overline{w}}{\lambda - w} \right] \left(I - \frac{1}{\overline{w} - \lambda}D \right) \left(I - \frac{1}{w - \lambda}D \right)^{-1}.$$

Let $\mu = f(\lambda)$ and $q = (w - \lambda)^{-1}$. Then $(\overline{w} - \lambda)^{-1} = \mu^{-1}\eta q$ and

$$K = f(J) = (\mu I - \eta q D)(I - qD)^{-1}. \qquad (5.14.36)$$

Note that, since

$$K = (\mu I - \eta q D)(I + qD + \cdots + q^{n-1}D^{n-1})$$

$$= \mu I + (\mu - \eta)qD + (\mu - \eta)q^2 D^2 + \cdots + (\mu - \eta)q^{n-1}D^{n-1},$$

it follows immediately that K is an upper triangular Toeplitz matrix. Furthermore, it is easily seen that K is similar to $\mu I + D$ (preserving the Jordan chain structure of J) if and only if the coefficient of D in the above expansion is nonzero. But this coefficient is just $(\mu - \eta)q$ and since $q \neq 0$, and $\mu = f(\lambda)$, it is found that $\mu = \eta$ only if $w = \overline{w}$, and this possibility has been excluded in the definition of f. Consequently, the matrix K of (5.14.36) is similar to $\mu I + D$ and (from Proposition 4.3.4) is P-unitary.

Consider in more detail the matrices obtained by this construction from a real λ, and from a conjugate pair $\lambda, \overline{\lambda}$ (with $\overline{\lambda} \neq \lambda$).

Case 1. ($\lambda = \overline{\lambda}$, $|\mu| = 1$).

From the definitions of q and μ, it is easily seen that $\eta q = \mu \overline{q}$, and the representation (5.14.36) becomes

$$K = \mu(I - \overline{q}D)(I - qD)^{-1}. \qquad (5.14.37)$$

We are left with the possibility of choosing the parameters w (and hence q) and η to simplify (5.14.37) as far as possible. The choice $w = \lambda - i$ (implying $q = i$) is legitimate and gives

$$K = \mu(I + iD)(I - iD)^{-1} = \mu \begin{bmatrix} 1 & 2i & 2i^2 & \cdots & 2i^{n-1} \\ 0 & 1 & 2i & & \vdots \\ \vdots & & 1 & \ddots & \vdots \\ & & & 1 & 2i \\ 0 & \cdots & \cdots & 0 & 1 \end{bmatrix}, \qquad (5.14.38)$$

which is just the matrix of Example 4.3.1. Note that, in (5.14.35), η is still undetermined.

Case 2. ($\lambda \neq \bar{\lambda}$, $\mu_1 = f(\lambda)$, $\mu_2 = f(\bar{\lambda})$, $\mu_1\overline{\mu_2} = 1$, $|\mu| \neq 1$).

Let $J_\lambda = \lambda I + D$, $J_{\bar{\lambda}} = \bar{\lambda}I + D$ and, as above, form $K_1 = f(J_\lambda)$, $K_2 = f(J_{\bar{\lambda}})$. If $q_1 = (w - \lambda)^{-1}$, $q_2 = (w - \bar{\lambda})^{-1}$, then (5.14.36) gives

$$
\left.
\begin{array}{rcl}
K_1 & = & (\mu_1 I - \eta q_1 D)(I - q_1 D)^{-1} \\
K_2 & = & (\mu_2 I - \eta q_2 D)(I - q_2 D)^{-1}
\end{array}
\right\} .
\tag{5.14.39}
$$

If Q is the matrix $\begin{bmatrix} 0 & P \\ P & 0 \end{bmatrix}$, then it follows from Proposition 4.3.4 that the matrix $\begin{bmatrix} K_1 & 0 \\ 0 & K_2 \end{bmatrix}$ is Q-unitary. In this case, one verifies that

$$
\eta q_1 = \mu_1 \overline{q_2}, \quad \eta q_2 = \mu_2 \overline{q_1},
$$

so that equations (5.14.39) can be written

$$
\left.
\begin{array}{rcl}
K_1 & = & \mu_1(I - \overline{q_2}D)(I - q_1 D)^{-1} \\
K_2 & = & \mu_2(I - \overline{q_1}D)(I - q_2 D)^{-1}
\end{array}
\right\} ,
\tag{5.14.40}
$$

and it is apparent that $K_2 = \overline{K_1}^{-1}$.

Now consider "simple" choices for w, and hence q_1 and q_2. If $w = a - i$ where a is real, then

$$
q_1 = \frac{1}{w - \lambda} = \frac{i}{2}\left[\frac{1}{w - \lambda}\right](w - \overline{w}) = \frac{i}{2}\left[\frac{1}{w - \lambda}\right](w - \lambda - (\overline{w} - \lambda)) = \frac{i}{2}(1 - \overline{\eta}\mu_1),
$$

and similarly,

$$
q_2 = \frac{i}{2}(1 - \overline{\eta}\mu_2).
$$

One can simplify further by putting $\eta = -1$ to obtain

$$
q_1 = \frac{1}{2}i(1 + \mu_1), \quad q_2 = \frac{1}{2}i(1 + \mu_2).
$$

Matrices K_1 and K_2 of (5.14.40) with this choice of parameters are just those of Example 4.3.2, namely

$$
K_1 =
\begin{bmatrix}
\mu_1 & k_1 & k_2 & \cdots & k_{n-1} \\
0 & \mu_1 & k_1 & & \vdots \\
\vdots & & \ddots & & \vdots \\
\vdots & & & \mu_1 & k_1 \\
0 & \cdots\cdots & & 0 & \mu_1
\end{bmatrix}
, \quad
K_2 =
\begin{bmatrix}
\mu_2 & \kappa_1 & \kappa_2 & \cdots & \kappa_{n-1} \\
& \mu_2 & \kappa_1 & & \vdots \\
\vdots & & \ddots & & \vdots \\
& & & \mu_2 & \kappa_1 \\
0 & \cdots\cdots & & 0 & \mu_2
\end{bmatrix}
,
$$
$$
\tag{5.14.41}
$$

where

$$
k_r = \mu_1 q_1^{r-1}(q_1 - \overline{q_2}), \quad \kappa_r = \mu_2 q_2^{r-1}(q_2 - \overline{q_1}), \quad \text{for } r = 1, 2, \ldots, n-1,
$$

and, of course, $\mu_1\overline{\mu_2} = 1$.

5.15   Canonical Forms for H-Unitaries: General Case

It has been observed that, for any H-unitary matrix A, the spectrum $\sigma(A)$ is symmetric relative to the unit circle, and moreover, the partial multiplicities of an eigenvalue λ coincide with those of the eigenvalue $\overline{\lambda}^{-1}$. The blocks of a Jordan form for A can therefore be arranged as follows:

$$J = \operatorname{diag}(J_1, J_2, \ldots, J_\alpha, J_{\alpha+1}, \ldots, J_{\alpha+2\beta}), \tag{5.15.42}$$

where each J_i is a Jordan block, $J_1, \ldots, J_\alpha$ each have their associated eigenvalue on the unit circle, the eigenvalues of $J_{\alpha+1}, J_{\alpha+3}, \ldots, J_{\alpha+2\beta-1}$ are outside the unit circle and the eigenvalue of $J_{\alpha+2j}$ $(j = 1, 2, \ldots, \beta)$ is obtained from that of $J_{\alpha+2j-1}$ by inversion in the unit circle.

Construct a block diagonal matrix

$$K_J = \operatorname{diag}(K_1, K_2, \ldots, K_\alpha, K_{\alpha+1}, \ldots, K_{\alpha+2\beta}) \tag{5.15.43}$$

in the following way. If J_j of (5.15.42) has a unimodular eigenvalue μ_j, let

$$K_j = \mu_j(I + iD)(I - iD)^{-1}$$

as in equation (5.14.38). For the pair $J_{\alpha+2j-1}, J_{\alpha+2j}$ having nonunimodular eigenvalues $\mu_1, \mu_2 = \overline{\mu_1}^{-1}$, define $K_{\alpha+2j-1}, K_{\alpha+2j}$ by (5.14.41), with $q_1 = \frac{i}{2}(1 + \mu_1)$, $q_2 = \frac{i}{2}(1 + \mu_2)$.

It is clear from our study of primitive canonical forms in Section 5.14 that, with these definitions, K_J is $P_{\varepsilon,J}$-unitary. With these preparations the first important result of this section can be stated as follows.

Theorem 5.15.1. *Let A be H-unitary, and let J be the Jordan normal form of A arranged as in (5.15.42). Then (A, H) is unitarily similar to a pair $(K_J, P_{\varepsilon,J})$, where*

$$P_{\varepsilon,J} = \operatorname{diag}\left(\varepsilon_1 P_1, \ldots, \varepsilon_\alpha P_\alpha, \begin{bmatrix} 0 & P_{\alpha+1} \\ P_{\alpha+1} & 0 \end{bmatrix}, \ldots, \begin{bmatrix} 0 & P_{\alpha+\beta} \\ P_{\alpha+\beta} & 0 \end{bmatrix}\right),$$

P_j is the sip matrix with size equal to that of J_j (and K_j) for $j = 1, \ldots, \alpha$, and equal to that of $J_{\alpha+2(j-\alpha)}$ (and $K_{\alpha+2(j-\alpha)}$) for $j > \alpha$, and $\varepsilon = (\varepsilon_1, \ldots, \varepsilon_\alpha)$ is an ordered set of signs ± 1. The set of signs ε is uniquely determined by (A, H) up to permutation of signs corresponding to equal blocks K_j.

Proof. Let $\mu_1, \ldots, \mu_{\alpha'}$ be the distinct unimodular eigenvalues of A and let $\mu_{\alpha'+1}, \ldots, \mu_{\alpha'+\beta'}$ be distinct nonunimodular eigenvalues chosen one from each conjugate pair $\mu_j, \overline{\mu_j}^{-1}$. Note that the inverse transformation of (5.14.35) is $(w\zeta - \overline{w}\eta)/(\zeta - \eta)$ (as a function of ζ).

For $j = 1, 2, \ldots, \alpha'$ define functions $g_j(\zeta)$ on $\sigma(A)$ by writing $w_j = \lambda_j - i$, $\lambda_j \in \mathbb{R}$, and $\eta_j = -\mu_j$ (cf. Case 1 of Section 5.14), and

$$g_j(\zeta) = \frac{(\lambda_j - i)\zeta + (\lambda_j + i)\mu_j}{\zeta + \mu_j}, \tag{5.15.44}$$

in a (sufficiently small) neighborhood of μ_j with $g_j(\zeta) \equiv 0$ in neighborhoods of every other point of $\sigma(A)$. Assume also that $\lambda_1, \ldots, \lambda_\alpha$ are chosen so that $\mu_j \neq \mu_k$ ($j, k = 1, 2, \ldots, \alpha'$) implies $\lambda_j \neq \lambda_k$.

For a pair of nonunimodular eigenvalues μ_j and $\overline{\mu_j}^{-1}$ (cf. Case 2 of Section 5.14), let $w_j = -i$, $\eta_j = -1$, and, in a neighborhood of μ_j and $\overline{\mu_j}^{-1}$,

$$g_j(\zeta) = -i\frac{\zeta - 1}{\zeta + 1}, \qquad (5.15.45)$$

with $g_i(\zeta) \equiv 0$ on the remainder of $\sigma(A)$.

Now define a function $g(\zeta)$ on $\sigma(A)$ by $g(\zeta) = \sum_{j=1}^{p} g_j(\zeta)$. Let $T_j = g_j(A)$ and

$$T = g(A) = \sum_{j=1}^{p} T_j.$$

We show that T is H-selfadjoint.

Observe first that, since $(A^*)^{-1} = HAH^{-1}$,

$$Hg_j(A) = (Hg_j(A)H^{-1})H = g_j(HAH^{-1})H = g_j((A^*)^{-1})H,$$

so that T_j is H-selfadjoint if $g_j(A) = g_j((A^*)^{-1})$. The latter fact is readily established using the functional calculus and the fact that $\overline{g_j(\zeta)} = g_j(\overline{\zeta}^{-1})$. Consequently, T_j is H-selfadjoint for each j, and so T is H-selfadjoint.

Furthermore, the Jordan form J_T of T is obtained from the Jordan form J of A by replacing the eigenvalue μ_j in each block of J by λ_j if $|\mu_j| = 1$, and by $i(1 - \mu_j)/(1 + \mu_j)$ if $|\mu_j| \neq 1$. Thus, for a fixed set of signs ε, we may write $P_{\varepsilon, J_T} = P_{\varepsilon, J}$. Then, applying Theorem 5.1.1, it is found that (T, H) and $(J_T, P_{\varepsilon, J})$ are unitarily similar for some set of signs ε. Thus,

$$T = S^{-1}J_T S, \qquad H = S^* P_{\varepsilon, J} S$$

for some nonsingular S.

By construction, $A = f(T)$ where the function $f(z)$ is the inverse of $g(\zeta)$ and is defined in a neighborhood of a real eigenvalue λ_j of T by

$$f(z) = \frac{-\mu_j(z - \overline{w_j})}{z - w_j},$$

where $w_j = \lambda_j - i$, and μ_j is the corresponding unimodular eigenvalue of A. In a neighborhood of a pair of nonreal eigenvalues for T, f is defined by

$$f(z) = \frac{-(z - \overline{w})}{z - w},$$

with $w = -i$. Furthermore, the construction of K_J shows that $K_J = f(J_T)$. Now we have

$$A = f(T) = S^{-1}f(J_T)S = S^{-1}K_J S,$$

and the relations $A = S^{-1}K_JS$, $H = S^*P_{\varepsilon,J}S$ show that (A, H) and $(K_J, P_{\varepsilon,J})$ are unitarily similar, as required. The uniqueness of the set of signs ε will be discussed later, making use of Theorem 5.15.4. $\square$

The set of signs ε associated with the pair (A, H), where A is H-unitary is, naturally, called the *sign characteristic* of (A, H). It is clear that the sign characteristic associated with a unimodular eigenvalue μ of A is just the sign characteristic of the real eigenvalue $\lambda = g(\mu)$ of the H-selfadjoint matrix $T = g(A)$. It also follows from the discussion of Section 5.14 that the set of J-invariant subspaces coincides with the set of K_J-invariant subspaces.

The following example confirms that the notion of the sign characteristic of (A, H) when A is H-unitary is consistent with that for the sign characteristic of (A, H) when A is H-selfadjoint, as introduced in Section 5.1.

Example 5.15.2. *Let A be a matrix which is both H-unitary and H-selfadjoint. Thus,*

$$A = H^{-1}A^*H = A^{-1}.$$

In particular, $A^2 = I$ and $\sigma(A) \subseteq \{1, -1\}$. The sign characteristic of A as an H-selfadjoint matrix is $\varepsilon^{(1)}$ say, and (A, H) is unitarily similar to the canonical pair (J_1, P_1).

Now let $\varepsilon^{(2)}$ be the sign characteristic of A as an H-unitary matrix. Thus, by definition, $\varepsilon^{(2)}$ is the sign characteristic of an H-selfadjoint matrix $T = g(A)$ where g is given by

$$g(\zeta) = \begin{cases} \dfrac{(\lambda_1 - i)\zeta + (\lambda_1 + i)}{\zeta + 1} & near \quad \zeta = 1; \\[3mm] \dfrac{(\lambda_2 - i)\zeta + (\lambda_2 + i)}{\zeta + 1} & near \quad \zeta = -1, \end{cases}$$

where $\lambda_1, \lambda_2 \in \mathbb{R}$ and $\lambda_1 \neq \lambda_2$. Note that the possible eigenvalues of T are $\lambda_1 = g(1)$, and $\lambda_2 = g(-1)$. Let (T, H) be unitarily similar to a canonical pair (J_2, P_2).
Now we have

$$A = S_1^{-1}J_1S_1, \qquad H = S_1^*P_1S_1,$$

$$T = S_2^{-1}J_2S_2, \qquad H = S_2^*P_2S_2,$$

and since $T = g(A)$, it follows that $S_2^{-1}J_2S_2 = S_1^{-1}g(J_1)S_1$, i.e.,,

$$g(J_1) = S_3^{-1}J_2S_3,$$

where $S_3 = S_2S_1^{-1}$. But it is easily seen that $P_1 = S_3^*P_2S_3$, and so $(g(J_1), P_1)$ and (J_2, P_2) are unitarily similar. It follows from Theorem 5.5.1 that P_1 and P_2 have the same sign characteristic, i.e.,, $\varepsilon^{(1)} = \varepsilon^{(2)}$, up to an allowed permutation of signs. $\square$

The result of Theorem 5.15.1 can be seen as providing a canonical form $(K_J, P_{\varepsilon,J})$ for the pair (A, H), but we prefer to reserve the phrase "canonical form" for a unitarily similar pair $(J, Q_{\varepsilon,J})$, which is the subject of the next theorem.

Consider first a lemma which specifies a reduction of a typical block of K_J (as generated by (5.14.36)) to a Jordan block.

Lemma 5.15.3. *Let K be an $n \times n$ complex matrix of the form*

$$K = \mu \begin{bmatrix} 1 & \alpha & \alpha q & \alpha q^2 & \cdots & \alpha q^{n-2} \\ 0 & 1 & \alpha & \alpha q & \cdots & \alpha q^{n-3} \\ 0 & 0 & 1 & & & \vdots \\ \vdots & & & \ddots & & \\ 0 & & & \ddots & 1 & \alpha \\ 0 & \cdots\cdots & & & 0 & 1 \end{bmatrix} = \mu(I - (q - \alpha)D)(I - qD)^{-1},$$

where $\mu\alpha q \neq 0$, and let J be the $n \times n$ Jordan block with eigenvalue μ. Then the equality

$$KX = XJ$$

holds, where

$$X = \operatorname{diag}\left(1, q^{-1}, \ldots, q^{-n+1}\right) \widehat{X}_n \operatorname{diag}\left(1, \frac{q}{\mu\alpha}, \ldots, \left(\frac{q}{\mu\alpha}\right)^{n-1}\right),$$

and where

$$\widehat{X}_n = \begin{bmatrix} 1 & 0 & 0 & 0 & 0 & \cdots & 0 \\ 0 & 1 & -1 & 1 & -1 & \cdots & (-1)^n \\ 0 & 0 & 1 & -2 & 3 & \cdots & \\ 0 & 0 & 0 & 1 & -3 & \cdots & \vdots \\ 0 & 0 & 0 & 0 & 1 & \cdots & \\ \vdots & & & & & \ddots & \\ & & & & & 1 & -n+2 \\ 0 & \cdots\cdots\cdots & & & & 0 & 1 \end{bmatrix}. \tag{5.15.46}$$

Note that the columns of $\widehat{X}_n$ are made up of the signed binomial coefficients, and $\widehat{X}_n^{-1}$ can also be written down in simple explicit form. Thus, X is invertible, and the columns of X form a Jordan chain for K. The proof of the lemma is an exercise in familiar properties of the binomial coefficients and is omitted.

The implications of Lemma 5.15.3 for the reduction of blocks of matrix K_J of (5.15.43) will now be examined for the cases of eigenvalues which are unimodular or not unimodular.

Case 1. Suppose $|\mu| = 1$ and (as in the construction of K_J) set $q = i$ and $\alpha = 2i$. A pair (J, Q) is unitarily similar to $(K, \varepsilon P)$ (where $\varepsilon = +1$ or -1, and P is the $n \times n$ sip matrix) if $J = X^{-1}KX$ and $Q = \varepsilon X^* PX$. The lemma indicates the choice

$$X = \mathrm{diag}\,\left(1, i^{-1}, \ldots, i^{-n+1}\right)\,\widehat{X}_n \mathrm{diag}\,\left(1, (2\mu)^{-1}, \ldots, (2\mu)^{-n+1}\right).$$

It is easily seen that $Q = [q_{jk}]_{j,k=1}^n$ is of triangular form with elements $q_{jk} = 0$ when $j + k \le n$.

Case 2. With K_1 and K_2 $(= \overline{K_1}^{-1})$ as defined in (5.14.41), we are to find a matrix X which will determine the unitary similarity between pairs

$$\left(\begin{bmatrix} K_1 & 0 \\ 0 & K_2 \end{bmatrix}, \begin{bmatrix} 0 & P \\ P & 0 \end{bmatrix}\right) \quad \text{and} \quad \left(\begin{bmatrix} J_1 & 0 \\ 0 & J_2 \end{bmatrix}, Q\right),$$

and hence the structure of Q. Here, J_j is the $n \times n$ Jordan block with eigenvalue μ_j for $j = 1$ and 2 where $\mu_1 \overline{\mu_2} = 1$.

Using Lemma 5.15.3, construct a matrix X_1 for which $K_1 X_1 = X_1 J_1$ by setting

$$q_1 = \frac{1}{2}i(1 + \mu_1), \qquad q_2 = \frac{1}{2}i(1 + \mu_2),$$

and

$$q = q_1, \quad \alpha = q_1 - \overline{q_2} = \frac{1}{2}i(2 + \mu_1 + \overline{\mu_2}) = \frac{1}{2}i(1 + \mu_1)(1 + \overline{\mu_2}).$$

Similarly, on setting $q = q_2$ and

$$\alpha = q_2 - \overline{q_1} = \frac{1}{2}i(1 + \overline{\mu_1})(1 + \mu_2)$$

a matrix X_2 is obtained for which $K_2 X_2 = X_2 J_2$. Then

$$\begin{bmatrix} K_1 & 0 \\ 0 & K_2 \end{bmatrix}\begin{bmatrix} X_1 & 0 \\ 0 & X_2 \end{bmatrix} = \begin{bmatrix} X_1 & 0 \\ 0 & X_2 \end{bmatrix}\begin{bmatrix} J_1 & 0 \\ 0 & J_2 \end{bmatrix},$$

and Q is defined by the congruence

$$Q = \begin{bmatrix} X_1^* & 0 \\ 0 & X_2^* \end{bmatrix}\begin{bmatrix} 0 & P \\ P & 0 \end{bmatrix}\begin{bmatrix} X_1 & 0 \\ 0 & X_2 \end{bmatrix} = \begin{bmatrix} 0 & X_1^* PX_2 \\ X_2^* PX_1 & 0 \end{bmatrix}.$$

Here, the matrix $X_1^* PX_2$ (and hence Q) is found to be nonsingular and of triangular form with zero elements above the secondary diagonal. Note also that, with $\alpha = q_1 - \overline{q_2}$ we have

$$\frac{q_1}{\mu_1 \alpha} = \frac{1}{\mu_1(1 + \overline{\mu_2})} = \frac{1}{1 + \mu_1}.$$

The next step is to put the conclusions of Cases 1 and 2 together to formulate a canonical form for (A, H) where A is H-unitary. But note first that use of the lemma implies a preferred choice of Jordan chain for K. To investigate reduction by *any* Jordan chain the matrix X can be replaced by XT, where T is an arbitrary nonsingular upper triangular Toeplitz matrix. It may be imagined that there is a propitious choice for T giving rise to a matrix Q in cases 1 and 2 of particularly simple structure, but the choice $T = I$ in Lemma 5.15.3 seems, in fact, to be the least intractable.

In formulating the canonical form suppose that block J_j of J in (5.15.42) has size n_j and P_j is the sip matrix of size n_j for $j = 1, 2, \ldots, \alpha + \beta$.

Theorem 5.15.4. *Let A be H-unitary, and let J be a Jordan form for A arranged as in (5.15.42). Let $\mu_1, \ldots, \mu_\alpha$ be the eigenvalues of $J_1, \ldots, J_\alpha$, respectively, and let $\mu_{\alpha+1}, \mu_{\alpha+2}, \ldots, \mu_{\alpha+\beta}$ be the eigenvalues of $J_{\alpha+1}, J_{\alpha+3}, \ldots, J_{\alpha+2\beta-1}$, respectively. (Thus, $|\mu_j| = 1$ for $j = 1, \ldots, \alpha$ and $|\mu_j| > 1$ for $j = \alpha + 1, \ldots, \alpha + \beta$). Then (A, H) is unitarily similar to a pair $(J, Q_{\varepsilon,J})$ where*

$$Q_{\varepsilon,J} = \mathrm{diag}\left(\varepsilon_1 Q_1, \ldots, \varepsilon_\alpha Q_\alpha, \begin{bmatrix} 0 & Q_{\alpha+1} \\ Q_{\alpha+1}^* & 0 \end{bmatrix}, \ldots, \begin{bmatrix} 0 & Q_{\alpha+\beta} \\ Q_{\alpha+\beta}^* & 0 \end{bmatrix} \right),$$
$$(5.15.47)$$

and for $j = 1, 2, \ldots, \alpha$, we have $Q_j = X_{n_j}^* P_j X_{n_j}$, where

$$X_{n_j} = \mathrm{diag}\left(1, i^{-1}, i^{-2}, \ldots \right) \widehat{X}_{n_j} \mathrm{diag}\left(1, (2\mu_j)^{-1}, (2\mu_j)^{-2}, \ldots \right),$$

$\widehat{X}_{n_j}$ *is given by (5.15.46), and $\varepsilon = (\varepsilon_1, \ldots, \varepsilon_\alpha)$ is an ordered set of signs ± 1. For $j = \alpha + 1, \ldots, \alpha + \beta$, we have $Q_j = X_{1,j}^* P_j X_{2,j}$, where*

$$X_{1,j} = \mathrm{diag}\left(1, \left\{ \frac{1}{2}i(1 + \mu_j) \right\}^{-1}, \left\{ \frac{1}{2}i(1 + \mu_j) \right\}^{-2}, \ldots \right) \widehat{X}_{n_j}$$
$$\cdot \mathrm{diag}\left(1, (1 + \mu_j)^{-1}, (1 + \mu_j)^{-2}, \ldots \right),$$

$\widehat{X}_{n_j}$ *is given by (5.15.46), and $X_{2,j}$ is obtained from $X_{1,j}$ on replacing μ_j by $\overline{\mu_j}^{-1}$.*

To illustrate, observe that the canonical blocks Q_j of sizes 2, 3, and 4 for a unimodular eigenvalue μ are as follows:

$$\frac{i}{2}\begin{bmatrix} 0 & -\overline{\mu} \\ \mu & 0 \end{bmatrix}, \quad -\frac{1}{4}\begin{bmatrix} 0 & 0 & \overline{\mu}^2 \\ 0 & -1 & \frac{1}{2}\overline{\mu} \\ \mu^2 & \frac{1}{2}\mu & -\frac{1}{4} \end{bmatrix}, \quad \frac{i}{8}\begin{bmatrix} 0 & 0 & 0 & \overline{\mu}^3 \\ 0 & 0 & -\overline{\mu} & \overline{\mu}^2 \\ 0 & \mu & 0 & -\frac{1}{4}\overline{\mu} \\ -\mu^3 & -\mu^2 & \frac{1}{4}\mu & 0 \end{bmatrix}.$$

In each case, we have $J_j^* Q_j J_j = Q_j$, where J_j is the Jordan block with the size of Q_j and eigenvalue μ.

One observes that we do not have a canonical form for H-unitary matrices A where the linear transformation representing A is expressed in terms of a Jordan form and, simultaneously, H has a simple structure comparable to that of block diagonal matrices with sip diagonal blocks (as for the case of H-selfadjoint matrices). The examples, particular cases, and theorems in this and the preceding section show that, in the case of H-unitary matrices, when we achieve one of those two goals, we do not achieve the other. This is in contrast to H-selfadjoint matrices. Yet another canonical form for H-unitary matrices is given in [44].

Next, we show that the sign characteristic of (A, H), where A is H-unitary, was properly defined in the proof of Theorem 5.15.1, i.e., that the choice of signs associated with the eigenvalues of A does not depend on the choice of the parameter w in the Möbius transformation (5.14.35). It was, of course, essential for the proof of Theorem 5.15.1 that different choices be made for w at different points of the spectrum. Also, the choice of signs was independent of the parameter η in (5.14.35).

For the purpose of the next theorem, consider two Möbius transformations

$$f_j(z) = \frac{\eta_j(z - \overline{w_j})}{z - w_j}, \quad j = 1, 2,$$

and their inverses

$$g_j(\zeta) = \frac{w_j\zeta - \overline{w_j}\eta_j}{\zeta - \eta_j},$$

where $|\eta_1| = |\eta_2| = 1$, $\eta_1 \neq \eta_2$, and $w_j \neq \overline{w_j}$. Thus, each f_j maps the reals onto the unit circle, etc.

Theorem 5.15.5. *Let A be H-unitary and suppose that, with the definitions above,*

$$T_1 = g_1(A), \qquad T_2 = g_2(A).$$

Then T_1 and T_2 are H-self-adjoint, and if $(\Im w_1)(\Im w_2) > 0$, then the sign characteristic of (T_1, H) corresponding to a real eigenvalue λ_1 of T_1 coincides with the sign characteristic of (T_2, H) corresponding to the real eigenvalue $\lambda_2 = g_2(f(\lambda_1))$.

In the proof of Theorem 5.15.4 the condition $\Im w_j = -1$ was used for each Möbius transformation employed so the hypothesis of the present theorem applies there.

Proof. It has been seen in Section 5.14 that the conditions $w_j \neq \overline{w_j}$ $(j = 1, 2)$ ensure that the sizes of the Jordan blocks of A are preserved under the transformations $A \to g_1(A) = T_1$, and $A \to g_2(A) = T_2$. Consequently, it suffices to consider a matrix A which is similar to a single $n \times n$ Jordan block with unimodular eigenvalue μ.

Choose a basis $x_1, x_2, \ldots, x_n$ of (generalized) eigenvectors of $T_1 = g_1(A)$ as a basis for $\mathbb{C}^n$ and, in this basis, the representation $\widehat{T}_1$ of T_1 is an $n \times n$ Jordan block, say J, with real eigenvalue $\lambda_1 = g_1(\mu)$. Similarly, $T_2 = g_2(A)$ has just one real eigenvalue $\lambda_2 = g_2(\mu) = g_2(f_1(\lambda_1))$. Writing $h(\lambda) = g_2(f_1(\lambda))$, the representation of T_2 in the Jordan basis constructed for T_1 has the form

$$\widehat{T}_2 = h(\widehat{T}_1) = \begin{bmatrix} \gamma_0 & \gamma_1 & \gamma_2 & \cdots & \gamma_{n-1} \\ 0 & \gamma_0 & \gamma_1 & & \vdots \\ \vdots & & \gamma_0 & \ddots & \vdots \\ \vdots & & & \ddots & \gamma_1 \\ 0 & \cdots\cdots & 0 & & \gamma_0 \end{bmatrix}, \tag{5.15.48}$$

where

$$\gamma_j = \frac{h^{(j)}(\lambda_1)}{j!} \qquad \text{for } j = 0, 1, \ldots, n-1.$$

Since

$$h(\lambda) = \frac{w_2 \eta_1 (\lambda - \overline{w_1}) - \overline{w_2} \eta_2 (\lambda - w_1)}{\eta_1 (\lambda - \overline{w_1}) - \eta_2 (\lambda - w_1)}$$

takes real values for real λ, it is clear that $\widehat{T}_2$ is a real matrix.

It is necessary to examine the element $\gamma_1 = h^{(1)}(\lambda_1)$ in some detail. First, as observed in Section 5.14, $\gamma_1 \neq 0$. A computation shows that

$$\gamma_1 = h^{(1)}(\lambda_1) = \frac{\eta_1 \eta_2 (w_1 - \overline{w_1})(w_2 - \overline{w_2})}{(\eta_1(\lambda_1 - \overline{w_1}) - \eta_2(\lambda_1 - w_1))^2}.$$

Noting that

$$\frac{\eta_1 \eta_2}{(\eta_1(\lambda_1 - \overline{w_1}) - \eta_2(\lambda_1 - w_1))^2} < 0,$$

we obtain

$$\gamma_1 = \kappa(\Im w_1)(\Im w_2), \tag{5.15.49}$$

where $\kappa > 0$.

To compare the sign characteristics of (T_1, H) and (T_2, H) (they both consist of a single sign), we shall take advantage of Theorem 5.8.1 and, to this end, note that given the Jordan basis $x_1, \ldots, x_n$ for T_1 there is a Jordan basis for T_2 of the form

$$y_1 = x_1, \quad y_2 = \gamma_1^{-1} x_2,$$

$$y_3 = \gamma^{-2} x_3 + \beta_{32} x_2, \ldots, \quad y_n = \gamma^{-n+1} x_n + \beta_{n,n-1} x_{n-1} + \cdots + \beta_{n,2} x_2. \tag{5.15.50}$$

This statement is readily verified by using formula (5.15.48).

Consider the linear transformations $G_1^{(j)}, \ldots, G_n^{(j)}$ of Theorem 5.8.1 associated with the matrices $\widetilde{T}_j$, for $j = 1$ and $j = 2$. In this case all of the $G_k^{(j)}$ act on a one-dimensional space, and it is easily seen that, in fact, $G_1^{(j)} = \cdots = G_{n-1}^{(j)} = 0$ for $j = 1, 2$, and because of (5.15.50), $G_n^{(1)}$ and $G_n^{(2)}$ correspond to multiplication by constants k_1 and k_2, respectively, where $k_2 = \gamma_1^{-n+1} k_1$.

Thus, the sign characteristics agree whenever $\gamma_1 > 0$; but this is ensured by equation (5.15.49). $\qquad\qquad\qquad\qquad\qquad\qquad\qquad\qquad\qquad\qquad\qquad\qquad\quad$ $\square$

The line of argument used in the preceding proof shows that there are other sufficient conditions guaranteeing preservation of the sign characteristics of $g_1(A)$, $g_2(A)$. For example, if all the unimodular eigenvalues of A have only Jordan blocks of size one in the Jordan canonical form for A, then the difficulties of the theorem "go away" and the sign characteristics of (T_1, H) and (T_2, H) will agree. More generally, it can be seen that the sign characteristic is preserved if and only if

$$[(\Im w_1)(\Im w_2)]^{-n_j+1} > 0$$

for each partial multiplicity n_j associated with a unimodular eigenvalue of A.

Theorem 5.15.5 demonstrates that the technique used in Theorems 5.15.1 and 5.15.4 to arrive at a sign characteristic for an H-unitary A leads to a unique definition. More simply, the sign characteristic of (A, H), where A is H-unitary can be defined as follows: Let $g(\zeta) = (w\zeta - \overline{w}\eta)/(\zeta - \eta)$, where $|\eta| = 1$, $\eta \notin \sigma(A)$, and $\Im w > 0$. Defining the H-self-adjoint matrix $T = g(A)$, the sign characteristic of (A, H) at a unimodular $\mu_0 \in \sigma(A)$ is just the sign characteristic of (T, H) at the real eigenvalue $\lambda_0 = g(\mu_0)$ of T.

5.16   First Applications of the Canonical Form of H-Unitaries

The points to be made in this section are an exact parallel for those made in Section 5.2 for H-self-adjoint matrices. Comparing the conclusions of Theorem 5.1.1 for H-self-adjoint matrices, and Theorem 5.15.4 for H-unitary matrices, one sees that matrix $Q_{\varepsilon,J}$ of (5.15.47) inherits sufficient properties from $P_{\varepsilon,J}$ of (5.1.2) to admit conclusions parallel to those of Section 5.2. The vital characteristics of $Q_{\varepsilon,J}$ are that each block Q_j is congruent to P_j and is of triangular form with all elements zero above the bottomleft-topright diagonal.

The conclusions to be drawn will simply be summarized here; the arguments justifying them are essentially the same as those used in Section 5.2. First, if A is H-unitary and $\mu \in \sigma(A)$ with $|\mu| = 1$, then $\mathrm{Ker}(\mu I - A)$ can be written as a direct sum

$$\mathrm{Ker}(\mu I - A) = \mathcal{L}_1 \dotplus \mathcal{L}_2,$$

where $\mathcal{L}_1$ is H-nondegenerate and $\mathcal{L}_2$ is H-neutral. The spaces $\mathcal{L}_1$ and $\mathcal{L}_2$ are spanned by eigenvectors of μ associated with partial multiplicities equal to one, and partial multiplicities larger than one, respectively.

If, on the other hand, $\mu \in \sigma(A)$ and $|\mu| \neq 1$, then the whole root-subspace $\mathcal{R}_\mu(A)$ is H-neutral, and the sum $\mathcal{R}_\mu(A) \dotplus \mathcal{R}_{\overline{\mu}^{-1}}(A)$ is H-nondegenerate.

If H and A are $n \times n$, and H has N negative eigenvalues, it follows (as in (5.2.4)) that

$$N = \frac{1}{2}n - \frac{1}{4}\sum_{j=1}^{\alpha}[1 - (-1)^{m_j}]\varepsilon_j,$$

and from this it is readily concluded that the number of unimodular eigenvalues of A is at least $|\mathrm{sig}\, H| = |\mathrm{sig}\, Q_{\varepsilon,J}|$. Also, H has $\frac{1}{2}n$ positive eigenvalues if and only if the signs associated with Jordan blocks of odd sizes for unimodular eigenvalues of A (if any) are equally divided between $+1$'s and -1's.

5.17 Further Deductions from the Canonical Form

It has been seen that H-self-adjoint and H-unitary matrices have in common the notion of a sign characteristic, as well as common geometrical properties of root subspaces. As a result, and in addition to the parallels drawn in the preceding section, there are other deeper analogues for theorems obtained earlier in Chapter 5.1. Three of these will be considered here.

Denote by $i_+(H)$ the number of positive eigenvalues of H (counting multiplicities), and recall that for an H-nonnegative subspace $\mathcal{L}$, the index of positivity $p(\mathcal{L})$ is the maximal dimension (≥ 0) of an H-positive subspace $\mathcal{L}_0 \subseteq \mathcal{L}$.

Theorem 5.17.1. *Let A be an H-unitary matrix, and let C be a set of nonunimodular eigenvalues of A, which is maximal with respect to the property that, if $\lambda_0 \in C$, then $\overline{\lambda_0}^{-1} \notin C$. Then there exists an $i_+(H)$-dimensional A-invariant H-nonnegative subspace $\mathcal{N}$ with the three properties:*

(i) $\mathcal{N}$ is maximal H-nonnegative;

(ii) the nonunimodular part of $\sigma(A|_{\mathcal{N}})$ coincides with C;

(iii) $p(\mathcal{N}) = \frac{1}{2}\sum_{j=1}^{r}[1 - (-1)^{m_j}]\delta_j$,

where $m_1, \ldots, m_r$ are the sizes of Jordan blocks in the Jordan form for A having unimodular eigenvalues and associated signs $\varepsilon_1, \ldots, \varepsilon_r$ in the sign characteristic of (A, H), and $\delta_j = 1$ if $\varepsilon_j = 1$, $\delta_j = 0$ if $\varepsilon_j = -1$.

This theorem is the analogue of Theorem 5.12.1 and can be obtained from it by an application of the Cayley transform. Such a procedure will be demonstrated in proving the analogue of Theorem 5.4.1.

Theorem 5.17.2. *Let U_1 be H_1-unitary and U_2 be H_2-unitary. Then (U_1, H_1) and (U_2, H_2) are unitarily similar if and only if U_1 and U_2 are similar, and the pairs (U_1, H_1) and (U_2, H_2) have the same sign characteristic.*

Proof. Let $\alpha, w \in \mathbb{C}$ with $|\alpha| = 1$, $\Im w > 0$, and such that α is not an eigenvalue of U_1 or of U_2. Let

$$g(\zeta) = \frac{w\zeta - \overline{w}\alpha}{\zeta - \alpha},$$

and let $A_1 = g(U_1)$, $A_2 = g(U_2)$. Then, by Proposition 4.3.4, A_1 is H_1-self-adjoint, and A_2 is H_2-self-adjoint.

The theorem follows from Theorem 5.4.1 if it can be shown that (A_1, H_1) and (A_2, H_2) are unitarily similar if and only if the same is true of (U_1, H_1) and (U_2, H_2). So suppose that (A_1, H_1) and (A_2, H_2) are unitarily similar; i.e., there is a nonsingular T such that

$$A_1 = T^{-1}A_2T, \quad H_1 = T^*H_2T.$$

Then

$$g(U_1) = T^{-1}g(U_2)T = g(T^{-1}U_2T),$$

and applying the inverse transformation to g (which is well defined on $\sigma(A_1)$ and $\sigma(A_2)$), it is found that $U_1 = T^{-1}U_2T$, and since $H_1 = T^*H_2T$, it follows that (U_1, H_1) and (U_2, H_2) are unitarily similar. But this argument is reversible and so the theorem is proved.   □

Finally, we quote the analogue of Theorem 5.4.4 for future reference. The proof follows the same lines as that of Theorem 5.4.4.

Theorem 5.17.3. *Each equivalence class (under H-unitary similarity) of H-unitary matrices is arcwise connected.*

Recall that H-unitary similarity defines an equivalence relation on the complex square matrices with the size of H (ref. Section 5.4). It is easily verified that if an equivalence class defined by this relation contains an H-unitary matrix, then every matrix of the class is H-unitary.

5.18   Exercises

1. Let $A \in \mathbb{C}^{n \times n}$. Show that there is a positive definite $H \in \mathbb{C}^{n \times n}$ such that (A, H) is unitarily similar to (J, I), where J is a Jordan matrix in the standard basis of $\mathbb{C}^n$.

2. Show that, for any two similar matrices A and B, there exist positive definite matrices H and G such that (A, H) and (B, G) are unitarily similar.

3. Consider two matrix pairs

$$\text{(a)} \quad A_1 = \begin{bmatrix} \lambda_1 & \alpha \\ 0 & \lambda_2 \end{bmatrix}, \quad H_1 = \begin{bmatrix} 0 & 1 \\ 1 & 0 \end{bmatrix};$$

$$\text{(b)} \quad A_2 = \begin{bmatrix} \lambda_1 & 0 \\ \alpha & \lambda_2 \end{bmatrix}, \quad H_2 = \begin{bmatrix} 1 & 0 \\ 0 & -1 \end{bmatrix}.$$

When is A_j H_j-self-adjoint, for $j = 1$ and $j = 2$? If A_j is H_j-selfadjoint, what is its canonical form? Compute the similarity matrix.

4. The same problems as in Exercise 3 for the following pairs of matrices:

(a) $A_3 = \begin{bmatrix} \lambda_1 & \alpha & \gamma \\ 0 & \lambda_2 & \beta \\ 0 & 0 & \lambda_3 \end{bmatrix}$; $H_3 = \begin{bmatrix} 0 & 0 & 1 \\ 0 & 1 & 0 \\ 1 & 0 & 0 \end{bmatrix}$;

(b) $A_4 = A_3$; $H_4 = \begin{bmatrix} 1 & 0 & 0 \\ 0 & 0 & 1 \\ 0 & 1 & 0 \end{bmatrix}$;

(c) $A_5 = A_3$; $H_5 = \begin{bmatrix} 1 & 0 & 0 \\ 0 & 1 & 0 \\ 0 & 0 & -1 \end{bmatrix}$;

(d) $A_6 = \begin{bmatrix} 0 & 1 & 0 & \cdots & 0 \\ 0 & 0 & 1 & \cdots & 0 \\ & & \cdots\cdots\cdots & & \\ -a_0 & -a_1 & -a_2 & \cdots & -a_{n-1} \end{bmatrix}$;

$H_6 = \begin{bmatrix} 0 & 0 & \cdots & 0 & 1 \\ 0 & 0 & \cdots & 1 & 0 \\ & & \cdots\cdots\cdots & & \\ 1 & 0 & \cdots & 0 & 0 \end{bmatrix}$;

(e) $A_7 = A_6$; $H_7 = \begin{bmatrix} a_1 & a_2 & \cdots & a_{n-1} & 1 \\ a_2 & & & \iddots & \\ \vdots & & \iddots & & 0 \\ 1 & & & & \end{bmatrix}$;

(f) $A_8 = \begin{bmatrix} a_0 & a_1 & \cdots & a_n \\ a_n & a_0 & \cdots & a_{n-1} \\ & & \cdots\cdots\cdots & \\ a_1 & a_2 & \cdots & a_0 \end{bmatrix}$; $H_8 = \begin{bmatrix} 0 & 0 & \cdots & 0 & 1 \\ 0 & 0 & \cdots & 1 & 0 \\ & & \cdots\cdots\cdots & & \\ 1 & 0 & \cdots & 0 & 0 \end{bmatrix}$.

The rows of A_8 are cyclic permutations of the first row.

5. Find the sign characteristic for the pairs in Exercise 3, assuming that A_j is H_j-selfadjoint.

6. Find the sign characteristic for all pairs A_j, H_j ($j = 3, 4, \ldots, 8$) in Exercise 4, assuming that A_j is H_j-selfadjoint.

7. Describe the structure of all H-unitary matrices in the following cases:

   (a) H is positive definite;

   (b) H has one negative eigenvalue;

   (c) H has two negative eigenvalues (counted with multiplicities).

8. Describe the structure of all H-nonnegative (H-positive) matrices, for a given invertible hermitian matrix H.

9. Let $\mathcal{R}_\lambda(A)$ be the root subspace of a matrix A corresponding to its eigenvalue λ.

   (a) Using the canonical form for H-selfadjoints, prove that if A is H-selfadjoint, and λ_1, λ_2 are eigenvalues of A such that $\lambda_1 \neq \overline{\lambda_2}$, then the subspaces $\mathcal{R}_{\lambda_1}(A)$ and $\mathcal{R}_{\lambda_2}(A)$ are H-orthogonal. (Cf Theorem 4.2.4.)

   (b) Using the canonical form for H-unitaries, prove that if A is H-unitary, and λ_1, λ_2 are eigenvalues of A such that $\lambda_1\overline{\lambda_2} \neq 1$, then the subspaces $\mathcal{R}_{\lambda_1}(A)$ and $\mathcal{R}_{\lambda_2}(A)$ are H-orthogonal.

10. Under what additional hypotheses is the matrix A diagonalizable in each of the following cases:

$$\text{(a)} \quad A \text{ is } H\text{-selfadjoint}; \qquad \text{(b)} \quad A \text{ is } H\text{-unitary}.$$

11. Find the canonical form of a pair of matrices (A, H), where A is simultaneously H-selfadjoint and H-unitary.

12. Let A be H_0-selfadjoint for some invertible hermitian matrix H_0. Find the smallest integer α such that A is H-selfadjoint for some H with sig $H = \alpha$.

13. Solve the preceding exercise for:

$$\text{(a)} \quad H\text{-unitary } A; \qquad \text{(b)} \quad H\text{-normal } A.$$

14. Find the canonical forms for the following pairs of hermitian matrices:

(a) $H = \begin{bmatrix} 0 & \cdots & 0 & 1 \\ 0 & \cdots & 1 & 0 \\ & \cdots \cdots & & \\ 1 & 0 & 0 & 0 \end{bmatrix}; \quad G = \begin{bmatrix} 1 & 0 & \cdots & 0 \\ 0 & 1 & \cdots & 0 \\ & \cdots \cdots & & \\ 0 & 0 & \cdots & 1 \end{bmatrix};$

(b) $H = \begin{bmatrix} 0 & \cdots & 0 & 1 \\ 0 & \cdots & 1 & 0 \\ & \cdots \cdots & & \\ 1 & 0 & 0 & 0 \end{bmatrix}; \quad G = \begin{bmatrix} 0 & 1 & & \\ 1 & 0 & & 0 \\ & & 1 & \\ & & & \ddots \\ & 0 & & 1 \end{bmatrix};$

(c) $H = \begin{bmatrix} 0 & \cdots & 0 & 1 \\ 0 & \cdots & 1 & 0 \\ & \cdots\cdots & \\ 1 & \cdots & 0 & 0 \end{bmatrix}$; $\quad G = \begin{bmatrix} 0 & 0 & 1 & & \\ 0 & 1 & 0 & & \\ 1 & 0 & 0 & & \\ & & & 1 & \\ & & & & \ddots \\ & & & & & 1 \end{bmatrix}$;

(d) $H = \begin{bmatrix} 0 & I_m \\ I_m & 0 \end{bmatrix}$; $\quad G = [a_{jk}]_{j,k=1}^{2m}$.

15. Describe the structure of all pairs G_1, G_2 of hermitian matrices:

   (a) if G_1 is positive definite;

   (b) if each of the matrices G_1 and G_2 has one negative eigenvalue;

   (c) if each of the matrices G_1 and G_2 has two negative eigenvalues.

16. Let $H \in C^{n \times n}$ be a hermitian matrix with p positive eigenvalues $\lambda_1, \ldots, \lambda_p$ and $q = n - p$ negative eigenvalues $\lambda_{p+1}, \ldots, \lambda_n$, and let $\phi_1, \ldots, \phi_n$ be corresponding eigenvectors (if some of eigenvalues λ_j, $j = 1, 2, \ldots, n$, coincide, the eigenvectors are taken to be linearly independent). Then the form $(H\phi, \phi)$ is positive definite on the subspace Span $\{\phi_1, \ldots, \phi_p\}$, and is negative definite on the subspace Span $\{\phi_{p+1}, \ldots, \phi_n\}$.

17. In Exercises 3 and 4, find the A_j-invariant maximal H_j-nonnegative subspaces, for $j = 1, 2, \ldots, 8$.

5.19 Notes

The canonical form for H-selfadjoint matrices described in this chapter was known at the end of the 19th century (see [63], [103]), and it was developed by Weierstrass in the form of Theorem 5.10.1. The canonical form was rediscovered later by many authors, [55], [100], [104] is a representative sample of works in the first half of the 20th century (by no means complete). See [101], [99] for historical remarks. More generally, canonical forms for complex hermitian matrix pairs (over C and over R) are reviewed in [68]; see also the historical remarks and bibliography in [68].

The sign characteristic as an important notion in its own right was introduced in [37], where the equivalence of the three different descriptions of the sign characteristic was obtained in a more general setting. The results of Section 5.5 appear in [40] for the first time. Theorem 5.12.3 (essentially) appeared in [66]. Theorem 5.12.4 was obtained in [91].

The material of Sections 5.14, 5.15, 5.16, 5.17 appeared in [40].

A criterion for uniqueness of the subspace $\mathcal{N}$ from Theorem 5.12.1 in the general case is given in [90].

Chapter 6

Real H-Selfadjoint Matrices

We now turn attention to the real space R^n and to an indefinite inner product $[.,.]$ on R^n defined by a real symmetric invertible matrix H. Attention is focussed on real $n \times n$ matrices acting on R^n together with such an indefinite inner product, and several results obtained in Chapters 2 to 4 are to be re-examined in this context. In particular, the reader will quickly verify that all the results and observations of Chapter 2 on the geometry of indefinite inner product spaces (when properly understood) are also valid for real spaces.

6.1 Real H-Selfadjoint Matrices and Canonical Forms

Let $[.,.]$ denote an indefinite inner product defined on R^n by a real symmetric invertible matrix H of size n, i.e.,,

$$[x,y] = (Hx, y), \qquad \text{for all } x, y \in \mathsf{R}^n.$$

Recall that $(.,.)$ stands for the standard inner product in R^n:

$$(x, y) = \sum_{j=1}^{n} x^{(j)} y^{(j)}$$

where $x = \langle x^{(1)}, \ldots, x^{(n)} \rangle$, $y = \langle y^{(1)}, \ldots, y^{(n)} \rangle$ are vectors in R^n. The adjoint $A^{[*]}$ of a real $n \times n$ matrix A is defined just as in (4.1.2) and, as in (4.1.3) it is easily seen that $A^{[*]} = H^{-1}A^*H$. Since A^* is now just the transpose of A, $A^{[*]}$ is obviously real. The following facts and definitions are all formally identical with predecessors in earlier chapters:

A real matrix A is *H-selfadjoint* if $A^* = A$, i.e., if $HA = A^*H$.

A real matrix A is *H-unitary* if $A^*A = I$, i.e., if $A^*HA = H$.

A real matrix A is *H-normal* if $A^*A = AA^*$, i.e., if $(H^{-1}A^*H)A = A(H^{-1}A^*H)$.

This section is devoted to properties of the real H-selfadjoint matrices. Some simple examples follow and, following a trend set in earlier chapters, they will form the basic blocks in real canonical forms (see Theorem 6.1.5, below).

Example 6.1.1. *Let $J_k(\lambda)$ be the $k \times k$ Jordan block with real eigenvalue λ, and let S_k be the $k \times k$ sip matrix. Then $J_k(\lambda)$ is real $\pm S_k$-selfadjoint.*   □

Example 6.1.2. *If $\sigma, \tau \in \mathbb{R}$ then $\begin{bmatrix} \sigma & \tau \\ -\tau & \sigma \end{bmatrix}$ is real S_2-selfadjoint. More generally, the matrix A of even size given by*

$$
A = \begin{bmatrix}
\sigma & \tau & 1 & 0 & & \cdots & & 0 \\
-\tau & \sigma & 0 & 1 & & & & \\
0 & 0 & \sigma & \tau & \ddots & & & \vdots \\
0 & 0 & -\tau & \sigma & & \ddots & & \\
\vdots & & & & \ddots & & 1 & 0 \\
& & & & & \ddots & 0 & 1 \\
& & & & & & \sigma & \tau \\
0 & 0 & & \cdots & & & -\tau & \sigma
\end{bmatrix}
$$

is real P-selfadjoint where P is the sip matrix with the size of A.   □

For any real matrix A, let $\lambda_1, \ldots, \lambda_t$ be the distinct real eigenvalues of A, and let $\sigma_j \pm i\tau_j$, $\tau_j \neq 0$, for $j = 1, 2, \ldots, s$, be the distinct pairs of nonreal complex conjugate eigenvalues of A. Then there is a decomposition of $\mathbb{R}^n$ into a direct sum of real A-invariant subspaces $\mathcal{X}_{ri}$:

$$
\mathbb{R}^n = \mathcal{X}_{r1} \dot{+} \cdots \dot{+} \mathcal{X}_{r,t} \dot{+} \mathcal{X}_{r,t+1} \dot{+} \cdots \dot{+} \mathcal{X}_{r,t+s},
$$

where the minimal polynomial of $A \mid_{\mathcal{X}_{r,j}}$ is a positive integer power of $\lambda - \lambda_j$, for $j = 1, \ldots, t$, and the minimal polynomial of $A \mid_{\mathcal{X}_{rj}}$ is a positive integer power of $\left((\lambda - \sigma_j)^2 + \tau_j^2 \right)$, for $j = t+1, \ldots, t+s$ (see Theorem A.2.7). Note that $\dim \mathcal{X}_{rj}$ is even for $j = t+1, \ldots, t+s$. Considering A as a linear transformation acting in $\mathbb{C}^n$, we also obtain the following decomposition:

$$
\mathbb{C}^n = \widehat{\mathcal{X}}_{r1} + \cdots + \widehat{\mathcal{X}}_{r,t} + \widehat{\mathcal{X}}_{r,t+1} + \cdots + \widehat{\mathcal{X}}_{r,t+s},
$$

where the complex subspace $\widehat{\mathcal{X}}_{rj}$ is equal to $\mathcal{X}_{rj} + i\mathcal{X}_{rj}$. Then

$$
\sigma(A \mid_{\widehat{\mathcal{X}}_{rj}}) = \{\lambda_j\} \quad \text{if} \quad j = 1, \ldots, t
$$

and

$$
\sigma(A \mid_{\mathcal{X}_{rj}}) = \{\sigma_j + i\tau_j, \sigma_j - i\tau_j\} \quad \text{if} \quad j = t+1, \ldots, t+s.
$$

(See [93, Section 6.62] for more detail of this construction). Using Theorem 4.2.4, we immediately obtain:

Theorem 6.1.3. *Let A be a real H-selfadjoint matrix, where H is a real symmetric invertible $n \times n$ matrix. Let $\mathcal{M}_1$ (resp. $\mathcal{M}_2$) be a real A-invariant subspace of $\mathbb{R}^n$ such that the minimal polynomial of the restriction $A\,|_{\mathcal{M}_1}$ (resp. $A\,|_{\mathcal{M}_2}$) is a power of an irreducible (over the real field) real polynomial $p_1(\lambda)$ (resp. $p_2(\lambda)$). Then $\mathcal{M}_1$ and $\mathcal{M}_2$ are H-orthogonal provided $p_1(\lambda) \neq p_2(\lambda)$. Moreover, the maximal real A-invariant subspace $\widehat{\mathcal{M}}_1$ with the property that the minimal polynomial of $A\,|_{\widehat{\mathcal{M}}_1}$ is a power of $p_1(\lambda)$, is H-nondegenerate.*

We shall need the following real Jordan form of a real matrix A (not necessarily H-selfadjoint). There exists a real $n \times n$ matrix S such that $SAS^{-1} = J$, where J (a real Jordan form of A) is a block diagonal matrix, each block being of one of the two following forms (see Theorem A.2.6):

$$J_p(\lambda_0), \quad \lambda_0 \in \mathbb{R}, \tag{6.1.1}$$

$$J_p(\sigma \pm i\tau), \quad \sigma, \tau \in \mathbb{R}, \quad \tau \neq 0. \tag{6.1.2}$$

Note that the minimal polynomial of the Jordan block in (6.1.1) is $(\lambda - \lambda_0)^p$ where p is the size of the block, and for the matrix in (6.1.2) the minimal polynomial is $\left\{(\lambda - \sigma)^2 + \tau^2\right\}^{p/2}$ where p is again the size of the matrix.

As in Chapter 5 a matrix $P_{\varepsilon,J}$ is constructed with the same block-diagonal structure as J. Assume that

$$J = J_1 \oplus \cdots \oplus J_t \oplus J_{t+1} \oplus \cdots \oplus J_{t+s} \tag{6.1.3}$$

where $J_1, \ldots, J_t$ are of type (6.1.1) and $J_{t+1}, \ldots, J_{t+s}$ are of type (6.1.2). Then define

$$P_{\varepsilon,J} = \varepsilon_1 P_1 \oplus \cdots \oplus \varepsilon_t P_t \oplus P_{t+1} \oplus \cdots \oplus P_{t+s} \tag{6.1.4}$$

where P_j is the sip matrix with size equal to that of J_j for $j = 1, 2, \ldots, t + s$ and $\varepsilon = \{\varepsilon_1, \ldots, \varepsilon_t\}$ and $\varepsilon_j = \pm 1$ for each j. It is easily verified that J is $P_{\varepsilon,J}$-selfadjoint.

This observation together with the real Jordan form has an immediate corollary. Let $A = S^{-1}JS$ where S is real and let ε by *any* set of signs. On forming the matrix $H = S^* P_{\varepsilon,J} S$, which is also real, it is found that A is real H-selfadjoint.

Corollary 6.1.4. *Every real matrix A is H-selfadjoint for some invertible, real, symmetric matrix H, i.e., there is such an H for which $A = H^{-1}A^T H$.*

Note that Corollary 5.1.3 asserts that A is similar to its transpose A^T with a hermitian transforming matrix H. This result says that there is, in fact, a real symmetric transforming matrix H.

Let (A_1, H_1) and (A_2, H_2) be pairs of real $n \times n$ matrices, where H_1, H_2 are hermitian and invertible. We say that (A_1, H_1) and (A_2, H_2) are *real unitarily similar* (or *r-unitarily similar* for short) if $A_1 = S^{-1}A_2 S$, $H_1 = S^* H_2 S$ for some real invertible matrix S.

The following theorem is a real version of Theorem 5.1.1.

Theorem 6.1.5. *A pair (A, H) of real matrices, where $H = H^*$ is real and invertible, and A is H-selfadjoint, is r-unitarily similar to a pair $(J, P_{\varepsilon,J})$, where J is the real Jordan form of A given by (6.1.3), and $P_{\varepsilon,J}$ is given by (6.1.4). The signs ε_i are determined uniquely by (A, H) up to permutation of signs in the blocks of $P_{\varepsilon,J}$ corresponding to the Jordan blocks of J with the same real eigenvalue and the same size.*

As in the case of complex matrices, we call the set of signs $\varepsilon = (\varepsilon_1, \ldots, \varepsilon_t)$ the *sign characteristic* of (A, H).

6.2 Proof of Theorem 6.1.5

Let $\det(\lambda I - A) = p_1(\lambda)^{\alpha_1} \cdots p_m(\lambda)^{\alpha_m}$, where $p_i(\lambda)$ are real irreducible polynomials over R. Write

$$R^n = \mathcal{X}_{r1} \dot{+} \cdots \dot{+} \mathcal{X}_{rm},$$

where $\mathcal{X}_{ri}$ is a real A-invariant subspace of R^n, and the minimal polynomial of $A \,|_{\mathcal{X}_{rj}}$ is a power of $p_j(\lambda)$, $j = 1, \ldots, m$. Using Theorem 6.1.3 we can assume (as in the proof of Theorem 5.1.1) that $m = 1$, i.e., $\det(\lambda I - A) = p(\lambda)^\alpha$ for some irreducible real polynomial $p(\lambda)$. Two cases can occur:

1. $p(\lambda) = \lambda - \lambda_0$, $\lambda_0 \in R$.

2. $p(\lambda) = (\lambda - \sigma)^2 + \tau^2$, $\tau \neq 0$, $\sigma, \tau \in R$. In this case n is even.

The first case can be proved by repeating, word for word, the proof of Theorem 5.1.1. So we focus on the second case. Consider A as a linear transformation acting in C^n; then $\sigma(A) = \{\sigma + i\tau, \sigma - i\tau\}$. Let $\lambda_0 = \sigma + i\tau$ and

$$\mathcal{X}' = \{x \in C^n \mid (A - \lambda_0 I)^n x = 0\}, \quad \mathcal{X}'' = \{x \in C^n \mid (A - \overline{\lambda_0} I)^n x = 0\}$$

be complex subspaces of C^n. Let m be the largest positive integer such that $(A - \lambda_0 I)^{m-1} \,|_{\mathcal{X}'} \neq 0$, and note that m is also the largest positive integer such that $(A - \overline{\lambda_0} I)^{m-1} \,|_{\mathcal{X}''} \neq 0$.

It will be convenient to introduce the (nonlinear) map $K : C^n \to C^n$ as follows:

$$K\langle x_1, x_2, \ldots, x_n \rangle = \langle \overline{x_1}, \overline{x_2}, \ldots, \overline{x_n} \rangle, \quad x = \langle x_1, x_2, \ldots, x_n \rangle \in C^n.$$

Since A and H are real, we have $AK = KA$, $HK = KH$.

Note that $K\mathcal{X}' \subseteq \mathcal{X}''$. Indeed, if $y \in K\mathcal{X}'$ then $y = Kx$, $x \in X'$ and, since $AK = KA$,

$$(A - \overline{\lambda_0} I)^m y = (A - \overline{\lambda_0} I)^m Kx = K(A - \lambda_0 I)^m x = 0,$$

so that $y \in \mathcal{X}''$. In fact, since we also have $K\mathcal{X}'' \subseteq \mathcal{X}'$, it follows immediately that $K\mathcal{X}' = \mathcal{X}''$.

It is now to be shown that

$$\left[(A - \lambda_0 I)^{m-1} x, K x\right] \neq 0$$

for some $x \in \mathcal{X}'$ and, of course $[x, y] = (Hx, y)$. Assuming the contrary, we have for every $x, y \in \mathcal{X}'$:

$$\begin{aligned} 0 &= \left[(A - \lambda_0 I)^{m-1}(x + y), K(x + y)\right] \\ &= \left[(A - \lambda_0 I)^{m-1} x, K y\right] + \left[(A - \lambda_0 I)^{m-1} y, K x\right]. \end{aligned}$$

But using the fact that K commutes with A and H, a direct calculation shows that

$$\left[(A - \lambda_0 I)^{m-1} x, K y\right] = \left[(A - \lambda_0 I)^{m-1} y, K x\right].$$

Consequently, $\left[(A - \lambda_0 I)^{m-1} y, K x\right] = 0$. In other words, $\left[(A - \lambda_0 I)^{m-1} y, z\right] = 0$ for every $y \in \mathcal{X}'$ and $z \in \mathcal{X}''$. Taking $y \in \mathcal{X}'$ such that $(A - \lambda_0 I)^{m-1} y \neq 0$, we observe that $(A - \lambda_0 I)^{m-1} y$ is orthogonal (with respect to H) to $\mathcal{X}' + \mathcal{X}''$. But this is a contradiction since (see the proof of Theorem 5.1.1) the subspace $\mathcal{X}' + \mathcal{X}''$ is nondegenerate.

Thus, there exists an $a_1 \in \mathcal{X}'$ such that

$$\theta \stackrel{\mathrm{def}}{=} \left[(A - \lambda_0 I)^{m-1} a_1, K a_1\right] \neq 0.$$

Replacing a_1 by αa_1 ($\alpha \in \mathbb{C}$) will replace θ by $\alpha^2 \theta$. So we can (and will) assume that

$$\left[(A - \lambda_0 I)^{m-1} a_1, K a_1\right] = 2i.$$

Let $b_1 = K a_1$ and for $j = 1, 2, \ldots, m$ define

$$a_j = (A - \lambda_0 I)^{j-1} a_1, \quad b_j := (A - \overline{\lambda_0} I)^{j-1} b_1 = K a_j.$$

As in the proof of Theorem 5.1.1 it can be shown that $a_m, \ldots, a_1$ and $b_m, \ldots, b_1$ are Jordan chains of A corresponding to λ_0 and $\overline{\lambda_0}$, respectively, and

$$[a_j, b_k] = \begin{cases} 2i & \text{if} \quad j + k = m + 1 \\ 0 & \text{if} \quad j + k > m + 1. \end{cases} \tag{6.2.5}$$

For $j = 1, 2, \ldots, m$ let $g_j = \frac{1}{2}(a_j + b_j)$ and $h_j = \frac{1}{2i}(a_j - b_j)$. Then all the vectors g_j and h_j are real (i.e., $g_j = K g_j$ and $h_j = K h_j$) and they are $\mathbb{R}$-linearly independent. The real subspace $\widetilde{\mathcal{X}_r}$ spanned by $g_1, h_1, \ldots, g_m, h_m$ is A-invariant and A has the following matrix representation in the basis $g_m, h_m, \ldots, g_1, h_1$:

$$J_0 = \begin{bmatrix} \sigma & \tau & 1 & 0 & & \cdots & & 0 \\ -\tau & \sigma & 0 & 1 & & & & \\ & & & & \ddots & & & \vdots \\ & & & & & 1 & 0 & \\ \vdots & & & & & 0 & 1 & \\ & & & & & \sigma & \tau & \\ 0 & & & & & -\tau & \sigma \end{bmatrix}. \tag{6.2.6}$$

On the other hand, the relations (6.2.5) together with $[a_i, a_k] = [b_i, b_k] = 0$ for $i, k = 1, 2, \ldots, m$ (which follows from the H-neutrality of the subspaces $\mathcal{X}'$ and $\mathcal{X}''$) imply

$$[g_j, h_k] = \begin{cases} 0 & \text{if} \quad j + k > m + 1 \\ 1 & \text{if} \quad j + k = m + 1. \end{cases} \tag{6.2.7}$$

Note also that $[g_j, g_k] = \frac{1}{2}[a_j, b_k] + \frac{1}{2}[b_j, a_k] = 0$ for $j + k \geq m + 1$, and similarly $[h_j, h_k] = 0$ for $j + k \geq m + 1$. Further,

$$[g_j, h_k] = [h_j, g_k] \quad \text{for} \quad j, k = 1, \ldots, m,$$

$$[g_j, g_k] = -[h_j, h_k] \quad \text{for} \quad j, k = 1, \ldots, m,$$

and for $j_1 + k_1 = j_2 + k_2$, $1 \leq j_1, k_1, j_2, k_2 \leq m$ we have:

$$[g_{j_1}, g_{k_1}] = [g_{j_2}, g_{k_2}], \quad [g_{j_1}, h_{k_1}] = [g_{j_2}, h_{k_2}],$$

$$[h_{j_1}, g_{k_1}] = [h_{j_2}, g_{k_2}], \quad [h_{j_1}, h_{k_1}] = [h_{j_2}, h_{k_2}].$$

Then the hermitian matrix $H_0 \overset{\text{def}}{=} [f_i, f_j]_{i,j=1}^{2m}$, where $f_1 = g_m$, $f_2 = h_m, \ldots$, $f_{2m-1} = g_1$, $f_{2m} = h_1$, has the following structure:

$$H_0 = \begin{bmatrix} 0 & & \cdots & & P_1 \\ \vdots & & & \iddots & P_2 \\ & & & & \vdots \\ 0 & P_1 & & & \\ P_1 & P_2 & \cdots & & P_m \end{bmatrix},$$

with

$$P_1 = \begin{bmatrix} 0 & 1 \\ 1 & 0 \end{bmatrix}, \qquad P_j = \begin{bmatrix} x_j & y_j \\ y_j & -x_j \end{bmatrix}, \qquad \text{for} \quad j = 2, \ldots, m,$$

for some real numbers x_j and y_j. For brevity, denote by U (resp. V) the set of all 2×2 real matrices of the form $\begin{bmatrix} x & y \\ y & -x \end{bmatrix}$ (resp. $\begin{bmatrix} x & y \\ -y & x \end{bmatrix}$); thus $P_j \in U$, $j = 1, \ldots, m$. We claim that there exist matrices $Z_2, \ldots, Z_m \in V$ such that

$$\begin{bmatrix} I & & \cdots & 0 & 0 \\ Z_2^* & I & & & 0 \\ \vdots & & \ddots & & \vdots \\ Z_m^* & \cdots & Z_2^* & I \end{bmatrix} H_0 \begin{bmatrix} I & Z_2 & \cdots & Z_m \\ 0 & I & \cdots & Z_{m-1} \\ \vdots & & \ddots & Z_2 \\ 0 & \cdots & 0 & I \end{bmatrix} = \begin{bmatrix} 0 & \cdots & 0 & P_1 \\ \vdots & & \iddots & \\ & P_1 & & \vdots \\ P_1 & \cdots & & 0 \end{bmatrix}. \tag{6.2.8}$$

Indeed, (6.2.8) is equivalent to

$$\sum_{\substack{j+k+\ell = q \\ 1 \leq j,k,\ell \leq m}} Z_j^* P_k Z_\ell = 0, \quad q = 4, 5, \ldots, m + 2, \tag{6.2.9}$$

where we write $Z_1 = I$. Rewrite (6.2.9) in the form

$$P_1 Z_{q-2} + Z_{q-2}^* P_1 = Q_q, \quad q = 4, 5, \ldots, m+2 \qquad (6.2.10)$$

where $Q_q = -\sum Z_j^* P_k Z_\ell$, and the sum is taken over all triples (j, k, ℓ) such that $j + k + \ell = q$; $1 \leq j, k, \ell \leq m$; $j < q - 2$; $\ell < q - 2$. Equations (6.2.10) can be solved for $Z_2, \ldots, Z_m$ successively; indeed, (6.2.10) with $q = 4$ is

$$P_1 Z_2 + Z_2^* P_1 = -P_2,$$

and one can take $Z_2 = \frac{1}{2} \begin{bmatrix} -y_2 & x_2 \\ -x_2 & -y_2 \end{bmatrix} \in V$ to be a solution. Suppose (6.2.10) to be already solved for $q \leq q_0$ to obtain solutions $Z_2, \ldots, Z_{q_0-2} \in V$. Then, as one checks easily, $Q_{q_0+1} \in U$, and take $Z_{q_0-1} = -\frac{1}{2} \begin{bmatrix} 0 & 1 \\ 1 & 0 \end{bmatrix} Q_{q_0+1} \in V$ to solve (6.2.10) with $q = q_0 + 1$.

Having constructed $Z_2, \ldots, Z_m \in V$ with the property (6.2.8), observe that

$$S = \begin{bmatrix} I & Z_2 & \cdots & Z_m \\ 0 & I & \cdots & Z_{m-1} \\ \vdots & \vdots & \ddots & \vdots \\ 0 & \cdots & & Z_2 \\ 0 & 0 \cdots & & I \end{bmatrix}$$

commutes with J_0. Let s_{ij} be the (i,j)-th entry of S^{-1}, and denote $\tilde{g}_i = \sum_{j=1}^{2m} s_{ji} f_j$, $i = 1, \ldots, 2m$. Then the real subspace $\mathcal{L}$ spanned by $\tilde{g}_1, \ldots, \tilde{g}_{2m}$ is A-invariant and, in the basis $\tilde{g}_1, \ldots, \tilde{g}_{2m}$, the matrix representing A is just J_0, while $[\tilde{g}_i, \tilde{g}_j] = 0$ if $i + j \neq 2m + 1$; $[\tilde{g}_i, \tilde{g}_j] = 1$ if $i + j = 2m + 1$.

Apply this construction to the pair of real linear transformations

$$(A \mid_{\mathcal{L}_1}, P_{\mathcal{L}_1} H \mid_{\mathcal{L}_1}),$$

where $P_{\mathcal{L}_1}$ is the orthogonal projection on

$$\mathcal{L}_1 := \{x \in \mathbf{R}^n \mid [x, y] = 0 \text{ for all } y \in \mathcal{L}\},$$

and so on.

Finally, the uniqueness of the sign characteristic is verified as in the proof of Theorem 5.1.1. $\qquad\square$

6.3 Comparison with Results in the Complex Case

As in the case of complex matrices, the basic Theorem 6.1.5 allows us to consider various problems concerning real H-selfadjoint linear transformations. In many cases the results and their proofs are the same for the real and for the complex cases. For instance, all results and their proofs of Sections 5.2, 5.4, 5.5 are also valid in the real case. Another description of the sign characteristic of an H-selfadjoint linear transformation A, given in Theorem 5.8.1, is valid also for the real case. Theorem 5.4.1 on classification of pairs (A, H), where A is an H-selfadjoint linear transformation, up to unitary similarity, holds also in the real case (in this case the classification is up to real unitary similarity, of course). One can obtain the real version of Theorem 5.4.1 from the complex one, using the fact that for a pair of real matrices (A, H), where A is H-selfadjoint, the sign characteristics of (A, H) as a pair of real matrices and as a pair of complex matrices are the same.

Theorem 5.10.1 on the simultaneous reduction of pairs of hermitian matrices has an analogue for pairs of real symmetric matrices. Some reformulation is required but the short proof is an exact parallel of that used for Theorem 5.10.1.

Theorem 6.3.1. *Let G_1 and G_2 be real symmetric $n \times n$ matrices with G_2 invertible. Then there is an invertible $n \times n$ real matrix X such that $X^* G_i X$, $i = 1, 2$ have the following forms:*

$$X^* G_1 X = \varepsilon_1 K_1 \oplus \cdots \oplus \varepsilon_\alpha K_\alpha \oplus K_{\alpha+1} \oplus \cdots \oplus K_\beta, \qquad (6.3.11)$$

where, for $q = 1, 2, \ldots, \alpha$, the number λ_α is real and

$$K_q = \begin{bmatrix} 0 & & & \lambda_q \\ & & \cdot^{\cdot^{\cdot}} & 1 \\ & \lambda_q & \cdot^{\cdot^{\cdot}} & \\ \lambda_q & 1 & & \end{bmatrix};$$

for $q = \alpha + 1, \ldots, \beta$, the numbers σ_q and τ_q are real with $\tau_q \neq 0$ and

$$K_q = \begin{bmatrix} 0 & 0 & 0 & 0 & \cdots & & -\tau_q & \sigma_q \\ 0 & 0 & 0 & 0 & \cdots & & \sigma_q & \tau_q \\ & & & & & & 0 & 1 \\ & \vdots & & & & 1 & 0 & \\ 0 & 0 & -\tau_q & \sigma_q & & & & \vdots \\ 0 & 0 & \sigma_q & \tau_q & & & & \\ -\tau_q & \sigma_q & 0 & 1 & & 0 & 0 \\ \sigma_q & \tau_q & 1 & 0 & \cdots & 0 & 0 \end{bmatrix}$$

and $\varepsilon_1, \ldots, \varepsilon_\alpha$ are ± 1;

$$X^* G_2 X = \varepsilon_q P_1 \oplus \cdots \oplus \varepsilon_\alpha P_\alpha \oplus P_{\alpha+1} \oplus \cdots \oplus P_\beta, \qquad (6.3.12)$$

where, for $q = 1, 2, \ldots, \beta$ P_q is the sip matrix with size equal to that of K_q. The representations (6.3.11) and (6.3.12) are uniquely determined by G_1 and G_2 up to simultaneous permutation of equal blocks in (6.3.11) and (6.3.12).

Concerning the results of Section 5.12 note that a real H-selfadjoint matrix does not always have a nontrivial invariant subspace, much less an invariant subspace which is also maximal H-nonnegative, or H-positive. For example, as a transformation on $\mathbf{R}^2$, $\begin{bmatrix} 0 & 1 \\ -1 & 0 \end{bmatrix}$ has no nontrivial subspace and is selfadjoint with respect to $\begin{bmatrix} 0 & 1 \\ 1 & 0 \end{bmatrix}$. The situation is clearer if all the eigenvalues of the real matrix A are real. If such a matrix is real H-selfadjoint, then all the conclusions of Section 5.12 can be applied with the understanding that the only candidate for a c-set is the empty set.

6.4 Connected Components of Real Unitary Similarity Classes

In Section 5.1 the set of pairs of complex $n \times n$ matrices (A, H) for which A is H-selfadjoint was introduced in order to study the equivalence classes under unitary similarity. To handle the corresponding problem in the real case consider S_r, the set of all pairs of real $n \times n$ matrices (A, H) for which A is H-selfadjoint. Then pairs $(A_1, H_1), (A_2, H_2) \in S_r$ are said to be unitarily similar (or r-unitarily similar) if there is a *real* invertible T such that $A_1 = T^{-1} A_2 T$ and $H_1 = T^* H_2 T$. Also, $(A_1, H), (A_2, H) \in S_r$ are real H-unitarily similar (H is now fixed) if $A_1 = U^{-1} A_2 U$ for some real H-unitary matrix U, i.e., for which $U^* H U = H$.

In contrast to the conclusions for the complex case described in Theorems 4.5.1 and 5.4.4, it will be shown in this section that the real unitary similarity classes and the real H-unitary similarity classes are not generally connected. The basic reason for this is the fact that the group of invertible *real* $n \times n$ matrices is not connected (see Lemma 6.4.2 below).

Theorem 6.4.1. *The r-unitary similarity class of S_r containing a pair (A, H) is connected if any real Jordan form J of A has a Jordan block of odd size with real eigenvalue and, otherwise, the class consists of exactly two connected components. In the latter case, the two connected components consist of those $(B, G) \in S_r$ for which the relations $B = T^{-1} AT$, $G = T^* HT$ hold with real matrices T having positive determinant in one case, and negative determinant in the other.*

The proof of Theorem 6.4.1 is based on the following well-known fact which it will be convenient to present with full proof.

Lemma 6.4.2. *The set $GL_r(n)$ of all real invertible $n \times n$ matrices has two connected components; one contains the matrices with determinant $+1$, the other contains those with determinant -1.*

Proof. Let T be a real matrix with $\det T > 0$ and let J be a real Jordan form for T. It will first be shown that J can be connected in $GL_r(n)$ to a diagonal matrix with diagonal entries ± 1. Indeed, J may have blocks J_p of two types: first as in (6.1.1) with nonzero eigenvalue λ_p in which case we define

$$J_p(t) = \begin{bmatrix} \lambda_p(t) & 1-t & \cdots & & 0 \\ 0 & \lambda_p(t) & \ddots & & \\ & & \ddots & \ddots & \vdots \\ \vdots & & & \ddots & 1-t \\ 0 & & \cdots & 0 & \lambda_p(t) \end{bmatrix} \tag{6.4.13}$$

for any $t \in [0,1]$, where $\lambda_p(t)$ is a continuous path of nonzero real numbers such that $\lambda_p(0) = \lambda_p$, and $\lambda_p(1) = 1$ or -1 according as $\lambda_p > 0$ or $\lambda_p < 0$.

Second, a Jordan block J_p may have the form (6.1.2) when $J_p(t)$ is defined to have the same zero blocks as J_p, while the 2×2 diagonal and superdiagonal blocks are replaced by

$$\begin{bmatrix} (1-t)\sigma + t & (1-t)\tau \\ -(1-t)\tau & (1-t)\sigma + t \end{bmatrix}, \begin{bmatrix} 1-t & 0 \\ 0 & 1-t \end{bmatrix}, \tag{6.4.14}$$

respectively, for $t \in [0,1]$. Then $J_p(t)$ determines a continuous path of real invertible matrices such that $J_p(0) = J_p$ and $J_p(1)$ is an identity matrix.

Applying the above procedures to every diagonal block in J, J is connected to J_1 by a path in $GL_r(n)$. Now observe that the path in $GL_r(2)$ defined for $t \in [0,2]$ by

$$\begin{bmatrix} -(1-t) & t \\ -t & -(1-t) \end{bmatrix} \text{ when } t \in [0,1], \quad \begin{bmatrix} t-1 & 2-t \\ -(2-t) & t-1 \end{bmatrix} \text{ when } t \in [1,2],$$

connects $\begin{bmatrix} -1 & 0 \\ 0 & -1 \end{bmatrix}$ to $\begin{bmatrix} 1 & 0 \\ 0 & 1 \end{bmatrix}$. Consequently J_1, and hence J, is connected in $GL_r(n)$ with either I or $\mathrm{diag}\,(-1,1,1,\ldots,1)$. But $\det T > 0$ implies $\det J > 0$ and so the latter case is excluded. Since $T = S^{-1}JS$ for some invertible real S, we can hold S fixed and observe that the path in $GL_r(n)$ connecting J and I will also connect T and I.

Now assume $T \in GL_r(n)$ and $\det T < 0$. Then $\det T' > 0$, where $T' = T\mathrm{diag}\,(-1,1,\ldots,1)$. Using the argument above, T' is connected with I in $GL_r(n)$. Hence T' is connected with $\mathrm{diag}\,(-1,1,\ldots,1)$ in $GL_r(n)$.   □

Proof of Theorem 6.4.1. Without loss of generality we can assume that $(A,H) = (J, P_{\varepsilon,J})$ is in the (real) canonical form, so that J is a real Jordan form of A. Denote by US_+ (resp. US_-) the set of all pairs (B,G) such that $B = T^{-1}JT$, and $G = T^*P_{\varepsilon,J}T$ for some real matrix T with $\det T > 0$ (resp. $\det T < 0$). Clearly, the r-unitary similarity class containing $(J, P_{\varepsilon,J})$ is a union of US_+ and US_-. By Lemma 6.4.2 each set US_+ and US_- is connected. Moreover, $US_+ = US_-$ if and only if there is a $(B, G) \in S_r$ which can be transformed to $(J, P_{\varepsilon,J})$ by both T_+ and T_-, say, with $\det T_+ > 0$ and $\det T_- < 0$. Thus,

$$B = T_+^{-1} J T_+ = T_-^{-1} J T_-, \quad G = T_+^* P_{\varepsilon,J} T_+ = T_-^* P_{\varepsilon,J} T_-,$$

and it follows immediately that $US_+ = US_-$ if and only if there is a real T with negative determinant such that

$$J = T^{-1} J T, \quad P_{\varepsilon,J} = T^* P_{\varepsilon,J} T. \tag{6.4.15}$$

Thus, it remains to prove that there exists a real T with $\det T < 0$ such that (6.4.15) hold if and only if J has a Jordan block of odd size with a real eigenvalue. Assume J has such a block, J_0. In an obvious notation, we can write

$$J = J_1 \oplus J_0, \quad P_{\varepsilon,J} = P_1 \oplus P_0, \tag{6.4.16}$$

where J_1 is the "rest" of J, and $P_{\varepsilon,J}$ is partitioned accordingly. Put $T = I \oplus (-I)$ with partitions consistent with those of (6.4.16). Evidently, (6.4.15) holds and, since J_0 has odd size, $\det T < 0$.

Conversely, assume that J does not have a Jordan block of odd size with real eigenvalue. It will be proved that $\det T > 0$ for every real invertible T satisfying $JT = TJ$, showing thereby that (6.4.15) never holds for a T with negative determinant.

Using Theorem A.4.1, it is sufficient to consider two cases separately:

1. $\det(\lambda I - J) = (\lambda - \lambda_0)^\alpha$, $\lambda_0 \in \mathbb{R}$;

2. $\det(\lambda I - J) = [(\lambda - \sigma)^2 + \tau^2]^m$, $\sigma, \tau \in \mathbb{R}$, $\tau \neq 0$.

Consider case 1.; then J is a Jordan matrix with eigenvalue λ_0. Let $m_1 = \cdots = m_{k_1} > m_{k_1+1} = \cdots = m_{k_2} > m_{k_2+1} = \cdots = m_{k_3} > \cdots > m_{k_{r-1}+1} = \cdots = m_{k_r}$ be the sizes of the Jordan blocks of J and, by assumption, all m_i are even. Let T be a real invertible matrix commuting with J. Then T has the following form (see, for example, [26], or [70]):

$$T = (T_{ij})_{i,j=1}^{k_r},$$

where each T_{ij} is defined by an upper triangular Toeplitz matrix in the following way: Let

$$T'_{ij} = \begin{bmatrix} t_{ij1} & t_{ij2} & \cdots & t_{ij\gamma} \\ 0 & t_{ij1} & \ddots & \\ & & \ddots & \ddots & \vdots \\ \vdots & & & t_{ij1} & t_{ij2} \\ 0 & & \cdots & 0 & t_{ij1} \end{bmatrix} \tag{6.4.17}$$

be a real $\gamma \times \gamma$ Toeplitz matrix, then:

$$1. \quad \text{if } m_i < m_j, \quad \text{we have } \quad T_{ij} = [0 \quad T'_{ij}], \tag{6.4.18}$$

$$2. \quad \text{if } m_i > m_j, \quad \text{we have } \quad T_{ij} = \begin{bmatrix} T'_{ij} \\ 0 \end{bmatrix}, \tag{6.4.19}$$

$$3. \quad \text{if } m_i = m_j, \quad \text{we have } \quad T_{ij} = T'_{ij}, \tag{6.4.20}$$

and in each case, $\gamma = \min(m_i, m_j)$.

An easy determinantal computation shows that

$$\det T = \left(\det[t_{ij1}]_{i,j=1}^{k_1}\right)^{m_{k_1}} \left(\det[t_{ij1}]_{i,j=k_1+1}^{k_2}\right)^{m_{k_2}} \cdots \left(\det[t_{ij1}]_{i,j=k_{r-1}+1}^{k_r}\right)^{m_{k_r}} \tag{6.4.21}$$

which is positive because the m_i are all even and T is invertible.

Now consider case 2. Define the set

$$\Xi := \left\{ \begin{bmatrix} a & b \\ -b & a \end{bmatrix} : a, b \in \mathsf{R} \right\}.$$

It is well-known, and easily verified, that Ξ is a subalgebra of $\mathsf{R}^{2\times 2}$ which is isomorphic to the field of complex numbers via the map

$$\phi : \Xi_2 \longrightarrow \mathsf{C}, \quad \phi\left(\begin{bmatrix} a & b \\ -b & a \end{bmatrix}\right) = a + ib.$$

Let $\Xi^{m \times m} \subseteq \mathsf{R}^{2m \times 2m}$ be the algebra of $m \times m$ matrices with entries in Ξ_2. The map ϕ extends (by applying it entrywise) to the algebra isomorphism $\phi_m : \Xi_2^{m \times m} \longrightarrow \mathsf{C}^{m \times m}$. One verifies that

$$\det T \geq 0, \quad \text{for every } T \in \Xi_2^{m \times m}, \tag{6.4.22}$$

by examining

$$\phi_m^{-1}(K) = \phi_m^{-1}(S^{-1})T\phi_m^{-1}(S) = \left(\phi_m^{-1}(S)\right)^{-1} T\phi_m^{-1}(S),$$

where K is the Jordan form of $\phi_m(T)$ with the similarity matrix $S \in \mathsf{C}^{m \times m}$:

$$K = S^{-1}\phi_m(T)S.$$

Clearly, $J \in \Xi_2^{m \times m}$. It is not difficult (but tedious) to check that

$$TJ = JT, \quad T \in \mathsf{R}^{2m \times 2m} \quad \Longrightarrow \quad T \in \Xi_2^{m \times m}. \tag{6.4.23}$$

To do this, use the following easily verifiable properties of the algebra Ξ_2:

(a) If $\alpha \in \Xi_2$ has nonzero off diagonal entries, then the equation $x\alpha = \alpha x$, $x \in \mathsf{R}^{2\times 2}$, has solutions x only in Ξ_2;

(b) If $\alpha \in \Xi_2$ has nonzero off diagonal entries, and if $\beta \in \Xi_2$, then the equation $x\alpha = \alpha x + \beta$, $x \in \mathbb{R}^{2\times2}$, has no solutions unless $\beta = 0$.

Now it follows from (6.4.22) and (6.4.23) that $JT = TJ$ holds with an invertible real T only if $\det T > 0$. $\qquad\Box$

6.5 Connected Components of Real Unitary Similarity Classes (H Fixed)

Consider now the real unitary similarity classes in the real $n \times n$ matrices obtained when the real hermitian invertible matrix H is kept fixed. The analogue for Theorem 5.4.4 turns out to be:

Theorem 6.5.1. *Let H be a fixed invertible real symmetric $n \times n$ matrix, and let A be a real $n \times n$ matrix. Then the real H-unitary similarity class*

$$US_H(A) := \{U^{-1}AU \ : \ U \ \text{ is real and } \ H - unitary\}$$

which contains A has either 1, 2 or 4 (arcwise) connected components.

The proof of Theorem 6.5.1 will follow from results on the connected components of the group of real H-unitary matrices which are to be presented in Theorem 6.5.2 below. Observe that all 3 possibilities ($US_H(A)$ connected, or has 2 or 4 connected components) may occur. When A is H-selfadjoint one can often find the exact number of connected components of $US_H(A)$ in terms of the structure of the real Jordan form of A and the sign characteristic of A with respect to H.

It has been noted (see Lemma 5.4.3) that, in the case of complex matrices, the set of H-unitary matrices is connected. In the real case, the situation is completely different. Indeed, even in the scalar case, $n = 1$, the set of H-unitary matrices consists of the two points: 1 and -1. The following result describes the connected components of the set $\mathbb{U}_r(H)$ of all real H-unitary matrices.

Theorem 6.5.2. *Let $H = H^*$ be a real invertible matrix.*

(i) *If H is neither positive nor negative definite, then $\mathbb{U}_r(H)$ has exactly 4 (arcwise) connected components, whose representatives can be described as follows: Let x_1 (resp. x_2) be a real eigenvector of H corresponding to a positive (resp. negative) eigenvalue. For every choice of the signs $n_1 = \pm1$, $n_2 = \pm1$ consider the real H-unitary matrix $A(n_1, n_2)$ which maps x_1 to $n_1 x_1$, x_2 to $n_2 x_2$ and x to itself, for every x belonging to the orthogonal complement (in the standard sense) to Span $\{x_1, x_2\}$. The 4 matrices $A(1,1), A(-1,1), A(1,-1), A(-1,-1)$ belong to the different connected components in $\mathbb{U}_r(H)$.*

(ii) *If H is either positive definite or negative definite, then $\mathbb{U}_r(H)$ has exactly 2 connected components, one consisting of matrices with determinant 1, and the second consisting of matrices with determinant -1.*

Proof. We start with (ii). Considering the case of positive definite H, we assume without loss of generality that $H = I$ (indeed, write $H = S^*S$ for some real invertible $n \times n$ matrix S. Then A is H-unitary if and only if SAS^{-1} is I-unitary). So the group $\mathbb{U}_r(H)$ becomes just the group $\mathbb{O}_n$ of (real) orthogonal $n \times n$ matrices.

Since the determinant is a continuous function of a matrix, clearly the subsets $\mathbb{O}_n^+$ and $\mathbb{O}_n^-$ consisting of all $n \times n$ orthogonal matrices with determinant 1 and -1, respectively, are disconnected in $\mathbb{O}_n$. We shall prove now that $\mathbb{O}_n^+$ is connected. Pick $A \in \mathbb{O}_n^+$. There exists an orthogonal matrix S such that $K \stackrel{\text{def}}{=} S^{-1}AS$ is in the real Jordan canonical form, as in Theorem A.2.6 but greatly simplified because A is orthogonal:

$$K = [K_1, K_2, \ldots, K_r],$$

where K_i is either the scalar ± 1, or a 2×2 matrix of the form

$$\begin{bmatrix} \cos\theta & \sin\theta \\ -\sin\theta & \cos\theta \end{bmatrix}, \quad 0 \le \theta \le 2\pi. \tag{6.5.24}$$

Since $\det K = 1$, the number of -1's is even. The 2×2 matrix $\begin{bmatrix} -1 & 0 \\ 0 & -1 \end{bmatrix}$ is of type (6.5.24) (with $\theta = \pi$), so we can assume that K_i is either 1 or has the form (6.5.24). There exists a continuous path in $\mathbb{O}_n^+$ connecting K and I. Indeed, it is sufficient to connect a block (6.5.24) with $\begin{bmatrix} 1 & 0 \\ 0 & 1 \end{bmatrix}$ by a continuous path. But this is easy: take

$$\begin{bmatrix} \cos t & \sin t \\ -\sin t & \cos t \end{bmatrix}, \quad \theta \le t \le 2\pi.$$

So there exists a continuous path $K(t)$, $t \in [0,1]$ in $\mathbb{O}_n^+$ such that $K(0) = K$; $K(1) = I$. Now the continuous path $S^{-1}K(t)S$ in $\mathbb{O}_n^+$ connects A and I. Hence $\mathbb{O}_n^+$ is arcwise connected. The arcwise connectedness of $\mathbb{O}_n^-$ is proved similarly.

The proof of part (i) is more complicated. We assume without loss of generality that

$$H = \begin{bmatrix} I_p & 0 \\ 0 & -I_q \end{bmatrix}, \quad 0 < p < p + q = n. \tag{6.5.25}$$

Given real H-unitary X, consider the polar decomposition $X = PU$, where $P = (XX^T)^{1/2}$ is real positive definite and U is real unitary. Note that since X is invertible, the polar decomposition is unique. Then P and U are also H-unitary, because $HXH^{-1} = (X^T)^{-1}$ implies

$$HPUH^{-1} = (U^T P^T)^{-1} = (U^{-1}P)^{-1} = P^{-1}U,$$

or

$$PU = H^{-1}P^{-1}UH = (H^{-1}P^{-1}H)(H^{-1}UH).$$

In view of the uniqueness of the polar decomposition of X, we have $P = H^{-1}P^{-1}H$, $U = H^{-1}UH$, so P and U are indeed H-unitary. Since the polar representation depends continuously on X (because $P = (XX^T)^{1/2}$, $U = (XX^T)^{-1/2}X$), we find that the group $\mathbb{U}_r(H)$ is homeomorphic to the product

$$\{P \in \mathbb{U}_r(H) : P \text{ is positive definite}\} \times \{U \in \mathbb{U}_r(H) : U^T U = I\}. \quad (6.5.26)$$

More precisely, the one-to-one and onto map $\psi(X) = \{P, U\}$, where $X \in \mathbb{U}_r(H)$ and $X = PU$ is the polar decomposition of X, defines the homeomorphism between $\mathbb{U}_r(H)$ and the product (6.5.26) and, in fact, both ψ and ψ^{-1} are real analytic.

We examine each component of (6.5.26) separately. A real H-unitary matrix is unitary if and only if it commutes with H. In view of the form (6.5.25), it follows that a real unitary H-unitary matrix U has the form $U = \begin{bmatrix} U_1 & 0 \\ 0 & U_2 \end{bmatrix}$, where $U_1 \in \mathsf{R}^{p \times p}$ and $U_2 \in \mathsf{R}^{q \times q}$ are I_p-unitary and $-I_q$-unitary, respectively. By the case (ii) of the theorem we see that the factor $\{U \in \mathbb{U}_r(H) : U^T U = I\}$ of (6.5.26) has exactly 4 (arcwise) connected components.

The set

$$\{P \in \mathbb{U}_r(H) : P \text{ is positive definite}\} \quad (6.5.27)$$

of (6.5.26) is connected. To verify this, note that for every positive definite $P \in \mathsf{R}^{n \times n}$ there exists a unique real hermitian Q such that $P = e^Q$. If in addition P is H-unitary, then

$$e^{-H^{-1}QH} = H^{-1}e^{-Q}H = H^{-1}(e^Q)^{-1}H = H^{-1}P^{-1}H = P.$$

Hence, by the uniqueness of the real hermitian logarithm of P we have

$$Q = -H^{-1}QH. \quad (6.5.28)$$

Writing out $Q = \begin{bmatrix} Q_{11} & Q_{12} \\ Q_{12}^T & Q_{22} \end{bmatrix}$, where $Q_{11} = Q_{11}^T \in \mathsf{R}^{p \times p}$, $Q_{22} = Q_{22}^T \in \mathsf{R}^{q \times q}$, we see that (6.5.28) holds if and only if $Q_{11} = 0$ and $Q_{22} = 0$. Thus, the set (6.5.27) is parametrized by $Q_{12} \in \mathsf{R}^{p \times q}$. In particular (6.5.27) is connected. In fact, we have proved that (6.5.27) is homeomorphic to $\mathsf{R}^{p \times q}$, with the homeomorphism being real analytic in both directions.

Finally, the characterization of the four connected components given in part (i), follows immediately from the above proof. $\square$

Let $H = H^*$ be real, invertible and neither positive definite nor negative definite. Denote the four connected components of $\mathbb{U}_r(H)$ as follows, where $A(n_1, n_2)$ are taken from Theorem 6.5.2:

$$\mathbb{U}_r^{++}(H) \ni A(1,1); \quad \mathbb{U}_r^{-+}(H) \ni A(-1,1);$$

$$U_r^{+-}(H) \ni A(1,-1); \quad U_r^{--}(H) \ni A(-1,-1).$$

Observe that $I \in U_r^{++}(H)$. Also

$$\det S \;=\; 1 \quad \text{for} \quad S \in U_r^{++}(H) \cup U_r^{--}(H);$$
$$\det S \;=\; -1 \quad \text{for} \quad S \in U_r^{+-}(H) \cup U_r^{-+}(H).$$

The multiplication between the different components of $U_r(H)$ is given by the following table:

$$
\begin{array}{c|cccc}
\cdot & ++ & +- & -+ & -- \\
\hline
++ & ++ & +- & -+ & -- \\
+- & +- & ++ & -- & -+ \\
-+ & -+ & -- & ++ & +- \\
-- & -- & -+ & +- & ++
\end{array}
\tag{6.5.29}
$$

To illustrate, this table implies that $S_1 S_2 \in U_r^{-+}(H)$ for every $S_1 \in U_r^{+-}(H)$ and $S_2 \in U_r^{--}(H)$.

Proof of Theorem 6.5.1. Assume that H is neither positive definite nor negative definite. We shall use the notation $U_r^{\xi,\eta}(H)$, $\xi, \eta \in \{1-1\}$, introduced above. Four cases can occur depending on the relation of the set $U_r(H;A)$ of all real H-unitary matrices which commute with A to the sets $U_r^{\xi,\eta}(H)$:

1. $U_r(H;A)$ is contained in $U_r^{++}(H)$,

2. $U_r(H;A)$ is contained in $U_r^{++}(H) \cup U_r^{+-}(H)$ but not in $U_r^{++}(H)$,

3. $U_r(H;A)$ is contained in $U_r^{++}(H) \cup U_r^{-+}(H)$ but not in $U_r^{++}(H)$,

4. none of the cases 1, 2, or 3 holds.

Using the multiplication table (6.5.29), one can easily see that in Case 1 the set $US_H(A)$ has exactly 4 connected components. In cases 2 and 3 this set has exactly 2 connected components, and in Case 4, $US_H(A)$ is connected.

In the case when H is either positive definite or negative definite Theorem 6.5.1 is evident in view of Theorem 6.5.2(ii). □

6.6   Exercises

1. Let H be a real symmetric invertible $n \times n$ matrix. Prove that there is a continuum of H-neutral subspaces in $\mathbb{R}^n$, except for the cases when H is definite (positive or negative) or when $n = 2$.

2. When is a linear transformation on $\mathbb{R}^n(H)$ simultaneously H-selfadjoint and H-unitary?

3. Find the real canonical form of Theorem 6.1.5 and an r-unitary similarity transformation matrix for the following pairs of real matrices (A_j, H_j), where A_j is H_j-selfadjoint:

(a) $A_1 = \begin{bmatrix} \lambda_1 & \alpha \\ 0 & \lambda_2 \end{bmatrix}$, $\quad H_1 = \begin{bmatrix} 0 & 1 \\ 1 & 0 \end{bmatrix}$, $\quad \lambda_1, \lambda_2, \alpha \in \mathbb{R}$;

(b) $A_2 = \begin{bmatrix} \lambda_1 & \alpha & \gamma \\ 0 & \lambda_2 & \alpha \\ 0 & 0 & \lambda_1 \end{bmatrix}$; $\quad H_2 = \begin{bmatrix} 0 & 0 & 1 \\ 0 & 1 & 0 \\ 1 & 0 & 0 \end{bmatrix}$, $\quad \lambda_1, \lambda_2, \alpha, \gamma \in \mathbb{R}$;

(c) $A_3 = \begin{bmatrix} 0 & 1 & 0 & \cdots & 0 \\ 0 & 0 & 1 & \cdots & 0 \\ \vdots & & & & \vdots \\ 0 & 0 & \cdots & & 1 \\ -a_0 & -a_1 & -a_2 & \cdots & -a_{n-1} \end{bmatrix}$;

$H_3 = \begin{bmatrix} a_1 & a_2 & \cdots & a_{n-1} & 1 \\ a_2 & & & \iddots & \\ \vdots & & \iddots & & 0 \\ 1 & & & & \end{bmatrix}$, where $a_1, \ldots, a_{n-1} \in \mathbb{R}$.

4. Find the real canonical forms of Theorem 6.3.1 and the transforming matrix X for the following pairs of real symmetric matrices:

(a) $G_1 = \begin{bmatrix} 0 & \cdots & 0 & 1 \\ 0 & \cdots & 1 & 0 \\ \cdots\cdots\cdots \\ 1 & \cdots & 0 & 0 \end{bmatrix}$; $\quad G_2 = \begin{bmatrix} 0 & 0 & 1 & & \\ 0 & 1 & 0 & & \\ 1 & 0 & 0 & & \\ & & & 1 & \\ & & & & \ddots \\ & & & & & 1 \end{bmatrix}$;

(b) $G_1 = \text{diag}(a_1, \ldots, a_n)$, $\quad G_2 = S_n$, where $a_1, \ldots, a_n \in \mathbb{R}$ and S_n is the $n \times n$ sip matrix;

(c) $G_1 = aE_n + bI_n$, $\quad G_2 = S_n$, where E_n is the $n \times n$ matrix of all 1's; $a, b \in \mathbb{R}$;

(d) $G_1 = [a_{jk}]_{j,k=1}^{2m}$, $\quad G_2 = \begin{bmatrix} 0 & I_m \\ I_m & 0 \end{bmatrix}$, here $a_{jk} \in \mathbb{R}$ and $a_{jk} = a_{kj}$ for all indices j and k.

5. Let G_1 and G_2 be invertible real symmetric $n \times n$ matrices, and let (6.3.11) and (6.3.12) be the real canonical form of the pair (G_1, G_2). Find the real canonical form of the pair (G_2, G_1).

6. Find the number of connected components in the r-unitary similarity classes that contain one of the following pairs (A, H), where A is H-selfadjoint:

(a) $(A, H) = (A_1, H_1)$ where the pair (A_1, H_1) is that of Exercise 3(a);

(b) $(A, H) = (A_2, H_2)$ where the pair (A_2, H_2) is that of Exercise 3(b);

(c) $(A, H) = (A_3, H_3)$ where the pair (A_3, H_3) is that of Exercise 3(c).

7. Consider the pairs (A, H) of the previous exercise. When the r-unitary similarity class containing (A, H) has more than one connected component, find a representative pair in every component.

8. Find the number of connected components in the real H-unitary similarity class that contains A in the five cases:

(a) $H = \begin{bmatrix} 1 & 0 \\ 0 & -1 \end{bmatrix}$, $A = \begin{bmatrix} 0 & 1 \\ 1 & 0 \end{bmatrix}$, $n = 2$;

(b) $H = \begin{bmatrix} 1 & 0 \\ 0 & -1 \end{bmatrix}$, $A = \begin{bmatrix} a & b \\ 0 & d \end{bmatrix}$, $n = 2$;

(c) $H = \begin{bmatrix} 1 & 0 \\ 0 & -1 \end{bmatrix}$, $A = \begin{bmatrix} 0 & b \\ c & d \end{bmatrix}$, $n = 2$;

(d) $H = S_n$, the $n \times n$ sip matrix, A an upper triangular Toeplitz matrix;

(e) $(A, H) = (A_2, H_2)$ where (A_2, H_2) is taken as in Exercise 3(b).

9. Consider the pairs (A, H) of the previous exercise. When the real H-unitary similarity class containing A has more than one connected component, find a representative matrix in every component.

6.7   Notes

The canonical form for real H-selfadjoint matrices of Theorem 6.1.5 is given in [101]. Another approach to the canonical form based on analysis of rational matrix functions is developed in [24].

A canonical form for real matrices that are selfadjoint in a skew-symmetric inner product is given in [13], [105]. Another approach to this canonical form via rational matrix functions was developed in [25]. Theorems 6.4.1 and 6.5.1 appeared in [40]. More generally, canonical forms for symmetric and skew-symmetric matrix pairs (over C and over R) are developed and reviewed in [68] and [69].

Chapter 7

Functions of H-Selfadjoint Matrices

In Section 4.3 we have made use of special functions of matrices (Moebius, or Cayley, transformations) to examine relationships between H-selfadjoint and H-unitary matrices. In this chapter the objective is to present a more systematic investigation of functions of H-selfadjoint matrices. In particular, we are to investigate how the sign characteristic is transformed.

7.1 Preliminaries

For a square matrix A with spectrum $\sigma(A)$, a function $f(\lambda)$ is said to be *defined on* $\sigma(A)$ if, for each point $\lambda_j \in \sigma(A), f(\lambda_j)$ and the derivatives $f^{(1)}(\lambda_j),\ldots,$ $f^{(m_j-1)}(\lambda_j)$ exist, where m_j is the *index* of λ_j, i.e., in a Jordan normal form for A, m_j is the size of the largest Jordan block with eigenvalue λ_j. The numbers $f^{(r)}(\lambda_j)$, $\lambda_j \in \sigma(A)$, $r = 0, 1, \ldots, m_j - 1$ are known as the *values of* f *on* $\sigma(A)$. If $f(\lambda)$ is a function of the *complex* variable and is defined on $\sigma(A)$ then, of course, for each point $\lambda_j \in \sigma(A)$, either $f(\lambda_j)$ is defined but has no derivatives (as at a branch point) and $m_j = 1$, or $f(\lambda)$ is analytic in a neighborhood of λ_j.

For any function $f(\lambda)$ defined on $\sigma(A)$, the matrix $f(A)$ is defined to be equal to $g(A)$ where $g(\lambda)$ is any polynomial which assumes the same values as f on $\sigma(A)$. This is a standard fact of the theory of functions of matrices, and can be found in [70, Chapter 9], or [26], for example. It is well-known that this procedure defines $f(A)$ uniquely and for *any* scalar polynomial $p(\lambda) = \sum_j p_j \lambda^j$, $p(A) = \sum_j p_j A^j$.

For a function $f(\lambda)$ defined on $\sigma(A)$ it is also easily seen that if $A = SJS^{-1}$ and $J = J_1 \oplus \cdots \oplus J_k$ is a Jordan form for A with blocks $J_1, J_2, \ldots, J_k$, then

$$f(A) = S[f(J_1), \ldots, f(J_k)]S^{-1}. \tag{7.1.1}$$

Furthermore, for $j = 1, 2, \ldots, k$, the diagonal blocks have triangular Toeplitz structure:

$$
f(J_j) = \begin{bmatrix}
f(\lambda_j) & f^{(1)}(\lambda_j) & \frac{1}{2!}f^{(2)}(\lambda_j) & \cdots & & \frac{1}{(m_j-1)!}f^{(m_j-1)}(\lambda_j) \\
0 & f(\lambda_j) & f^{(1)}(\lambda_j) & & & \\
& & f(\lambda_j) & \ddots & & \vdots \\
\vdots & & & \ddots & & \\
& & & & f(\lambda_j) & f^{(1)}(\lambda_j) \\
0 & & \cdots & & 0 & f(\lambda_j)
\end{bmatrix},
$$
(7.1.2)

where $\{\lambda_j\} = \sigma(J_j)$ and the block J_j has size m_j.

If $f(\lambda)$ is analytic in a neighborhood of $\sigma(A)$ it is, of course, defined on $\sigma(A)$ and an integral representation of $f(A)$ is available. Thus,

$$
f(A) = \frac{1}{2\pi i} \int_\Gamma f(\lambda)(\lambda I - A)^{-1} d\lambda
$$
(7.1.3)

where Γ is a composite contour consisting of a set of circles with sufficiently small radius; one around each distinct eigenvalue of A.

It is clear that, in this chapter, knowledge of the Jordan form of $f(A)$ in terms of that for A will be required. The first proposition together with (7.1.1) and (7.1.2) clarifies this point.

Proposition 7.1.1. *Let X be the $m \times m$ Jordan block with eigenvalue λ_0 and let $f(\lambda)$ be a function with $m - 1$ derivatives at λ_0 (i.e., $f(\lambda)$ is defined on $\sigma(X)$). Let $f^{(r)}(\lambda_0)$ $(1 \le r \le m - 1)$ be the first nonvanishing derivative of $f(\lambda)$ at λ_0 and if $f^{(j)}(\lambda_0) = 0$ for $j = 1, \ldots, m-1$, put $r = m$. Then the sizes of the Jordan blocks of $f(X)$ are given by*

$$
\left[\frac{m}{r}\right] \qquad \text{repeated} \quad r\left[\frac{m}{r}\right] - m + r \quad \text{times, and}
$$

$$
\left[\frac{m}{r}\right] + 1 \qquad \text{repeated} \quad \left(m - r\left[\frac{m}{r}\right]\right) \quad \text{times,}
$$

and $[x]$ denotes the greatest integer less than or equal to x.

Proof. Without loss of generality, it can be assumed that $\lambda_0 = f(\lambda_0) = 0$. Then, using (7.1.2), it is easily seen that $\dim \operatorname{Ker} f(X) = r$ and, more generally, if $t_j = \dim \operatorname{Ker}(f(X))^j$, then

$$
t_j = \min(m, jr).
$$
(7.1.4)

Now the sizes of Jordan blocks of $f(X)$ are uniquely determined by the sequence $t_1, t_2, \ldots, t_m$. Indeed, the number of Jordan blocks of $f(X)$ with size not less than j is just $t_j - t_{j-1}$, where $j = 1, 2, \ldots, m$ and $t_0 = 0$. This observation, together with (7.1.4) leads to the conclusion of the proposition. $\square$

Using similarity transformation, the result of Proposition 7.1.1 easily extends to matrices similar to one Jordan block. Furthermore, the Jordan form of a matrix allows one to obtain the Jordan form of a function of a matrix, by first reducing to the Jordan form with a similarity transformation, and then applying Proposition 7.1.1 to each Jordan block.

Example 7.1.2. *Let*

$$X = J_5(0) \oplus J_7(0)$$

be the 12×12 matrix which is a direct sum of two nilpotent Jordan blocks of sizes 5 and 7. Let $f(\lambda) = 1 + \sin(\lambda^3)$. Then

$$f(\lambda) = 1 + \lambda^3 - \frac{\lambda^9}{3!} + \cdots.$$

In the notation of Proposition 7.1.1, $r = 3$, and $m = 5$ (resp., $m = 7$) for the first (resp., second) Jordan block. By Proposition 7.1.1, the matrix $I + \sin(X^3)$ is similar to a direct sum of Jordan blocks with eigenvalue 1, with one block of size 3, four blocks of size 2, and one block of size 1. □

7.2 Exponential and Logarithmic Functions

It has been seen in Section 4.3 that H-unitary matrices can be obtained from H-selfadjoint matrices by means of a Cayley transform, and this fact was exploited further in Chapter 5. Another method for the transformation of H-selfadjoint to H-unitary matrices depends on properties of the exponential and logarithmic functions and will be presented in this section. But it is necessary to take some care in the definition of the logarithm of an H-unitary matrix.

First define a neighborhood Ω of $\sigma(U)$, where U is H-unitary, with the following properties:

(a) Ω is symmetric with respect to the unit circle, i.e., $\lambda \in \Omega$ implies $\overline{\lambda}^{-1} \in \Omega$.

(b) If $\sigma(U) = \{\lambda_1, \ldots, \lambda_k\}$ then $\Omega = \bigcup_{r=1}^{k} \Omega_r$ where $\Omega_1, \ldots, \Omega_k$ are disjoint neighborhoods of $\lambda_1, \ldots, \lambda_k$, respectively.

(c) $0 \notin \Omega$.

For $r = 1, 2, \ldots, k$, let $\ln_r$ denote branches of the logarithmic function. Then define a function $\ln \lambda$ on Ω by assigning $\ln = \ln_r$ whenever $\lambda \in \Omega_r$, and, if $\lambda_j \overline{\lambda_k} = 1$, then $\ln_j \lambda = \ln_k \lambda$, i.e., the same branch of the logarithm is used for domains Ω_j, Ω_k containing eigenvalues which are symmetric with respect to the unit circle. With these conventions, the function $\ln \lambda$ is defined on $\sigma(U)$.

Theorem 7.2.1. (a) *If A is H-selfadjoint then e^{iA} is H-unitary.*

(b) *If U is H-unitary and the function $\ln \lambda$ is defined on a neighborhood Ω of $\sigma(U)$ as above, then $V = -i \ln U$ satisfies $e^{iV} = U$ and V is H-selfadjoint.*

Proof. (a) Obviously $e^{i\lambda}$ is defined on the spectrum of any matrix, so e^{iA} is well-defined and, as is well known,

$$e^{iA} = I + iA + \frac{1}{2!}i^2 A^2 + \cdots .$$

Consequently, since $HA^r = A^{*r}H$ for $r = 1, 2, \ldots$,

$$He^{iA} = \left(I + iA^* + \frac{1}{2!}i^2 A^{*2} + \cdots\right) H = e^{iA^*}H = \left[\left(e^{-iA}\right)^*\right]H, = \left[\left(e^{iA}\right)^*\right]^{-1} H,$$

which implies that e^{iA} is H-unitary.

(b) Let J be a Jordan form for U and $U = SJS^{-1}$. Then

$$e^{\ln U} = Sg(J)S^{-1}$$

where $g(\lambda) = e^{\ln_r \lambda}$ when $\in \Omega_r$, and in any case $g(\lambda) = \lambda$. Hence $e^{\ln U} = U$.

We also show that
$$\ln((U^*)^{-1}) = -(\ln U)^*. \tag{7.2.5}$$

Using the Jordan form for U this is seen to be the case if and only if $\ln((J_r^*)^{-1}) = -(\ln J_r)^*$ for each Jordan block J_r of J. Using (7.1.2) we obtain

$$(\ln J_r)^* = \begin{bmatrix} \overline{\ln_r \lambda} & 0 & 0 & \cdots \\ \overline{\lambda}^{-1} & \overline{\ln_r \lambda} & 0 & \cdots \\ -\overline{\lambda}^{-2} & \overline{\lambda}^{-1} & \overline{\ln_r \lambda} & \\ \ddots & \ddots & \ddots & \end{bmatrix},$$

and a little calculation shows that

$$\ln((J_r^*)^{-1}) = \begin{bmatrix} \ln_r(\overline{\lambda}^{-1}) & 0 & 0 & \cdots \\ -\overline{\lambda}^{-1} & \ln_r(\overline{\lambda}^{-1}) & 0 & \cdots \\ \overline{\lambda}^{-2} & -\overline{\lambda}^{-1} & \ln_r(\overline{\lambda}^{-1}) & \\ \ddots & \ddots & \ddots & \end{bmatrix}.$$

But, for some integer m_r,

$$\ln_r(\overline{\lambda}^{-1}) = -\ln|\lambda| + i(\arg\lambda + 2\pi m_r) = -\overline{\ln_r \lambda},$$

since $\ln_r \lambda$ and $\ln_r \overline{\lambda}^{-1}$ are obtained using the same branch of the logarithmic function. Consequently, $\ln(J_r^{*-1}) = -(\ln J_r)^*$ and (7.2.5) holds.

Finally, defining $V = -i \ln U$ we obtain $e^{iV} = U$ and also,

$$HV = -iH\ln(U) = -iH\ln(H^{-1}(U^*)^{-1}H) = -i\ln((U^*)^{-1})H,$$

so (7.2.5) yields
$$HV = i(\ln U)^* H = V^* H. \qquad \square$$

7.3 Functions of H-Selfadjoint Matrices

The main emphasis of this section and the next is on functions of H-selfadjoint and of H-unitary matrices, respectively. We start with a general result.

Theorem 7.3.1. *If matrix A is either H-selfadjoint or H-unitary and $f(\lambda)$ is defined on $\sigma(A)$, then $f(A)$ is H-normal.*

Proof. First consider the case in which A is H-selfadjoint. Since the property of H-normality is preserved under unitary similarity (Proposition 4.1.3), it can be assumed that (A, H) is in the canonical form $(J, P_{\varepsilon,J})$ of Theorem 5.1.1. Then recall that to a Jordan block of J with real eigenvalue there corresponds a block of $f(J)$ of Toeplitz form

$$
B = \begin{bmatrix} \alpha_1 & \alpha_2 & \cdots & \alpha_k \\ 0 & \alpha_1 & \ddots & \vdots \\ \vdots & & \ddots & \alpha_2 \\ 0 & \cdots & 0 & \alpha_1 \end{bmatrix} \tag{7.3.6}
$$

as indicated in (7.1.2). Also for a pair of complex conjugate eigenvalues of J there is a pair of blocks $B_1 \oplus B_2$ of $f(J)$ where each of B_1, B_2 has the form (7.3.6) of equal size.

The proof of the theorem now reduces to verification of the facts that (1) B of (7.3.6) is εP-normal (where $\varepsilon = 1$ or $\varepsilon = -1$), and (2) that $B := B_1 \oplus B_2$ is $\begin{bmatrix} 0 & P \\ P & 0 \end{bmatrix}$-normal, and P is always the sip matrix of appropriate size. In case (1), this reduces to verifying that $BB^{[*]} = B^{[*]}B$, where B has the form (7.3.6). Since $B^{[*]} = P^{-1}B^*P$ and $P^{-1}B^*P = \overline{B}$ it is to be shown that $B\overline{B} = \overline{B}B$. This is easily checked. In case (2), $B^{[*]} = \overline{B}_2 \oplus \overline{B}_1$, and it is to be shown that B_1 and $\overline{B}_2$ commute and that B_2 and $\overline{B}_1$ commute. Again, this is easily checked.

The case when A is H-unitary is easily transformed to the H-selfadjoint case already established. Let $g(\lambda)$ be the inverse of a Cayley transform (see Proposition 4.3.4), so that $g(A)$ is H-selfadjoint and let $h(\lambda) = f(g^{-1}(\lambda))$. Then $h(\lambda)$ is defined on the spectrum of $g(A)$. By the first part of the proof it follows that the matrix

$$
h(g(A)) = f(g^{-1}(g(A))) = f(A)
$$

is H-normal. $\qquad \square$

To characterize those functions $f(\lambda)$ for which $f(A)$ is H-selfadjoint when A is H-selfadjoint, we need the following definition. The function $f(\lambda)$ defined on $\sigma(A)$ is said to be *real symmetric* on $\sigma(A)$ if for each point $\lambda_j \in \sigma(A)$ we have

$$
f^{(k)}(\lambda_j) = \overline{f^{(k)}(\overline{\lambda_j})}, \quad k = 0, 1, \ldots, m_j - 1,
$$

and m_j is the index of λ_j. For example, a scalar polynomial is symmetric on the spectrum of any H-selfadjoint matrix A if it has real coefficients. Also, if $f(\lambda) = \overline{f(\overline{\lambda})}$ and is analytic throughout a neighborhood of $\sigma(A)$, then $f(\lambda)$ is real symmetric on $\sigma(A)$.

Theorem 7.3.2. *Let A be H-selfadjoint and let $f(\lambda)$ be defined on $\sigma(A)$. Then $f(A)$ is H-selfadjoint if and only if $f(\lambda)$ is real symmetric on $\sigma(A)$.*

Proof. Let $f(\lambda)$ be real symmetric on $\sigma(A)$. Suppose that $\lambda_j \in \sigma(A)$ has index m_j and let $m = \sum_{\lambda_j \in \sigma(A)} m_j$. It is clear that there is a real polynomial $p(\lambda)$ of degree $m - 1$ satisfying the m interpolating conditions

$$p^{(k)}(\lambda_j) = f^{(k)}(\lambda_j), \quad \lambda_j \in \sigma(A), \quad k = 0, 1, \ldots, m_j - 1,$$

i.e., $p(\lambda)$ assumes the same values as $f(\lambda)$ on $\sigma(A)$. Since the definitions of $f(A)$ implies $f(A) = p(A)$ and $p(A)$ is obviously H-selfadjoint, it follows that $f(A)$ is H-selfadjoint.

Conversely, assume that $f(A)$ is H-selfadjoint. Use Theorem 5.1.1 to reduce (A, H) to a canonical pair so that $f(A)$ is given by (5.1.1) and (5.1.2). Then there are just two cases to consider:

(a) A is a single Jordan block with real eigenvalue λ_0 and size k, and $\pm H$ is the $k \times k$ sip matrix.

(b) $A = J \oplus \overline{J}$ where J is a $k \times k$ Jordan block with nonreal eigenvalue λ_0, and H is the $2k \times 2k$ sip matrix.

In case (a) it follows from (7.1.1) that $f(A)$ is H-selfadjoint, i.e., $Hf(A) = f(A)^*H$ if and only if $f(\lambda_0), f^{(1)}(\lambda_0), \ldots, f^{(k-1)}(\lambda_0)$ are real numbers. This means that $f(\lambda)$ is real symmetric on $\sigma(A)$.

In case (b) it follows from (7.1.2) that $f(A)$ is H-selfadjoint if and only if $f^{(r)}(\lambda_0) = \overline{f^{(r)}(\overline{\lambda_0})}$ for $r = 0, 1, \ldots, k - 1$ and, again, this means that $f(\lambda)$ is real symmetric on $\sigma(A)$. $\square$

It is clear that the conclusion of Theorem 7.3.2 can be paraphrased to give the statement that $f(A)$ is H-selfadjoint (when A is H-selfadjoint) if and only if $f(A) = p(A)$ for some *real* polynomial $p(\lambda)$.

For the more familiar situation in which H is positive definite, A is H-selfadjoint implies that for each $\lambda_j \in \sigma(A)$, λ_j is real and $m_j = 1$. In this case $f(\lambda)$ is real symmetric on $\sigma(A)$ if and only if the numbers $f(\lambda_j)$ are real for each $\lambda_j \in \sigma(A)$.

Now we describe an important class of functions of an H-selfadjoint matrix A for which $f(A)$ is a Riesz projection (see Section A.3). Let Λ be a set of eigenvalues of A such that $\lambda \in \Lambda$ implies $\overline{\lambda} \in \Lambda$. Let $f_\Lambda(\lambda) \equiv 1$ in a neighborhood of $\sigma(A) \setminus \Lambda$. Then $f_\Lambda(\lambda)$ is analytic in a neighborhood of $\sigma(A)$ and is real symmetric on $\sigma(A)$. By Theorem 7.3.2, $f_\Lambda(A)$ is H-selfadjoint. Formula (7.1.3) shows that $f_\Lambda(A)$ is just the Riesz projection on the sum of the root subspaces of A corresponding to the eigenvalues in Λ. Now H-selfadjointness of $f_\Lambda(A)$ implies that $\mathrm{Ker} f_\Lambda(A)$ and $\mathrm{Range} f_\Lambda(A)$ are H-orthogonal. Indeed, for $x \in \mathrm{Range} f_\Lambda(A)$, $y \in \mathrm{Ker} f_\Lambda(A)$ we have

$$(Hx, y) = (Hf_\Lambda(A)x, y) = (f_\Lambda(A)^* Hx, y) = (Hx, f_\Lambda(A)y) = 0.$$

In other words, the subspace $\mathrm{Ker} f_\Lambda(A)$, which is the sum of root subspaces of A corresponding to eigenvalues outside Λ, is an H-orthogonal complement to $\mathrm{Range} f_\Lambda(\lambda)$. We have observed this fact before (in another form) in Theorem 4.2.4.

The conditions under which a function of an H-selfadjoint matrix is H-unitary are more complicated than the counterparts of Theorems 7.3.1 and 7.3.2 in which the function is H-normal, or H-selfadjoint, respectively.

Theorem 7.3.3. *Let A be H-selfadjoint and let $f(\lambda)$ be defined on $\sigma(A)$. Then $f(A)$ is H-unitary if and only if, for each $\lambda \in \sigma(A)$,*

$$\overline{f(\lambda)} f(\overline{\lambda}) = 1 \tag{7.3.7}$$

and

$$\sum_{j=0}^{k} \binom{k}{j} \overline{f^{(j)}(\lambda)} f^{(k-j)}(\overline{\lambda}) = 0, \quad k = 1, 2, \ldots, m - 1, \tag{7.3.8}$$

and $m = m(\lambda)$ is the index of λ.

Note, in particular, that (7.3.7) implies $|f(\lambda)| = 1$ when $\lambda \in \sigma(A)$ and λ is real. Furthermore, if $f(\lambda)$ is analytic in a domain Ω containing $\sigma(A)$ and Ω is symmetric with respect to the real axis, then if (7.3.7) holds throughout Ω, (7.3.8) follows automatically. To see this, observe that for $\mu, \lambda \in \Omega$ with μ close enough to λ,

$$f(\mu) = \sum_{j=0}^{\infty} \frac{(\mu - \lambda)^j}{j!} f^{(j)}(\lambda), \quad f(\overline{\mu}) = \sum_{k=0}^{\infty} \frac{(\overline{\mu} - \overline{\lambda})^k}{k!} f^{(k)}(\overline{\lambda}).$$

Then comparing coefficients of powers of $(\overline{\mu} - \overline{\lambda})$ in the product $\overline{f(\mu)} f(\overline{\mu}) = 1$, (7.3.8) is obtained.

Note also that if $f(\lambda)$ and $R(\lambda)$ are analytic on a domain Ω as above and $f(\lambda) = e^{iR(\lambda)}$, then $f(\lambda)$ satisfies condition (7.3.7) on Ω, and hence (7.3.8), if $R(\lambda)$ is real symmetric on Ω. For

$$\overline{f(\lambda)} f(\overline{\lambda}) = e^{-i\overline{R(\lambda)}} e^{iR(\overline{\lambda})} = 1$$

provided $\overline{R(\lambda)} = R(\overline{\lambda})$, i.e., if $R(\lambda)$ is real symmetric on Ω. In particular, e^{iA} is H-unitary, as we have already observed in Theorem 7.3.1.

Proof of Theorem 7.3.3. Again, we can assume that (A, H) is in the canonical form of Theorem 5.1.1. So we have to consider the two cases (as in the proof of Theorem 7.3.2): (a) A is a Jordan block with real eigenvalue λ and size k, and

$\pm H$ is the $k \times k$ sip matrix; (b) $A = J \oplus \overline{J}$, where J is a Jordan block with nonreal eigenvalue λ and size k, and H is the $2k \times 2k$ sip matrix. In either case the equality $f(A)^* H f(A) = H$ expressing the property that $f(A)$ is H-unitary holds if and only if $f(\lambda)$ satisfies (7.3.7) and (7.3.8), as one checks easily using equation (7.1.2).   □

7.4   The Canonical Form and Sign Characteristic for a Function of an H-Selfadjoint Matrix

Let A be H-selfadjoint and $f(\lambda)$ be defined on $\sigma(A)$. If $f(A)$ is also H-selfadjoint, i.e., $f(\lambda)$ is real symmetric on $\sigma(A)$ (ref. Theorem 7.3.2), the problem arises of defining the canonical form and sign characteristic for the pair $(f(A), H)$ in terms of corresponding properties of the pair (A, H). An intermediate step is, of course, the determination of the Jordan form for $f(A)$ in terms of that of A and this has already been examined in Proposition 7.1.1.

Let λ_0 be a real eigenvalue of $f(A)$. We first find the canonical form and sign characteristic of $(f(A), H)$ at λ_0. Let $\lambda_1, \ldots, \lambda_k, \lambda_{k+1}, \ldots, \lambda_{k+\ell}$ be all the different eigenvalues of A such that $f(\lambda_i) = \lambda_0$, where $\lambda_1, \ldots, \lambda_k$ are real and $\lambda_{k+1}, \ldots, \lambda_{k+\ell}$ are nonreal. Note that the integer ℓ is necessarily even. The cases $k = 0$ or $\ell = 0$ are not excluded.

Let $s_{i1}, \ldots, s_{i,p_i}$ $(1 \leq i \leq k+\ell)$ be the sizes of Jordan blocks with the eigenvalue λ_i in the Jordan form of A, and let $\varepsilon_{i1}, \ldots, \varepsilon_{i,p_i}$ $(1 \leq i \leq k)$ be the corresponding signs in the sign characteristic of (A, H) (which exist, of course, only for real eigenvalues of A). Denote $m_i = \max \{s_{i1}, \ldots, s_{i,p_i}\}$ for $1 \leq i \leq k+\ell$, i.e., m_i is the index of λ_i.

In the next theorem it is assumed that $f(\lambda)$ is real symmetric on $\sigma(A)$.

Theorem 7.4.1. *For $i = 1, \ldots, k+\ell$ let r_i be the minimal integer $(1 \leq r_i \leq m_i - 1)$ such that $f^{(r_i)}(\lambda_i) \neq 0$ (if $f^{(j)}(\lambda_i) = 0$ for $j = 1, \ldots, m_i - 1$, put $r_i = m_i$). For a fixed positive integer q, put*

$$\gamma(q) = \sum_{i=1}^{k} \sum_{j=1}^{p_i} \varepsilon_{ij} \delta_{ij}(q), \qquad (7.4.9)$$

where

$$\delta_{ij}(q) = \begin{cases} 1, & \text{if } r_i(q+1) - s_{ij} \text{ is odd and either} \\ & q = \left[\frac{s_{ij}}{r_i}\right] + 1 \text{ or } q = \left[\frac{s_{ij}}{r_i}\right]; \\ 0, & \text{otherwise;} \end{cases}$$

put also

$$\eta(q) = \sum_{i=1}^{\ell} \sum_{j=1}^{p_i} \eta_{ij}(q). \qquad (7.4.10)$$

where

$$\eta_{ij}(q) = \begin{cases} r_i\left(\left[\frac{s_{ij}}{r_i}\right]+1\right) - s_{ij} & if \quad q = \left[\frac{s_{ij}}{r_i}\right]; \\ s_{ij} - r_i\left[\frac{s_{ij}}{r_i}\right] & if \quad q = \left[\frac{s_{ij}}{r_i}\right]+1; \\ 0 & otherwise. \end{cases}$$

Here, $[x]$ denotes the greatest integer less than or equal to x.

Then $\eta(q)$ is the number of Jordan blocks of $f(A)$ with eigenvalue λ_0 and size q. Furthermore, except in the case when $f^{(r_i)}(\lambda_i) < 0$ and q is even, $\frac{1}{2}(\eta(q)+\gamma(q))$ (resp. $\frac{1}{2}(\eta(q)-\gamma(q))$) is the number of signs $+1$ (resp. -1) in the sign characteristic of $(f(A), H)$ which correspond to these Jordan blocks. In the case when $f^{(r_i)}(\lambda_i) < 0$ and q is even, $\frac{1}{2}(\eta(q) - \gamma(q))$ (resp. $\frac{1}{2}(\eta(q) + \gamma(q))$) is the number of signs $+1$ (resp. -1) in the sign characteristic of $(f(A), H)$ which correspond to Jordan blocks of size q with eigenvalue λ_0.

It will be seen from the proof of this theorem that the numbers $\eta(q) \pm \gamma(q)$ are always even. In particular, Theorem 7.4.1 asserts that $\gamma(q)$ is the sum of signs corresponding to the Jordan blocks of $f(A)$ with eigenvalue λ_0 and size q (unless $f^{(r_i)}(\lambda_i) < 0$ and q is even, in which case $\gamma(q)$ is the negative of the sum of signs corresponding to the Jordan blocks of $f(A)$ with eigenvalue λ_0 and size q). By (7.4.9), this sum does not depend on the nonreal eigenvalues λ_i ($k+1 \le i \le k+\ell$).

Proof. By Proposition 7.1.1 we see that $\eta(q)$ given by (7.4.10) is just the number of Jordan blocks of $f(A)$ with eigenvalue λ_0 and size q. So it remains to prove that $\gamma(q)$ is the sum of signs in the sign characteristic of $(f(A), H)$ corresponding to these blocks.

In this proof we shall use a perturbation argument based on Theorem 5.9.1. This concerns the stability of the sign characteristic of pairs (B, G), where B is G-selfadjoint.

We can assume that the pair (A, H) is in the canonical form of (Theorem 5.1.1). So we have to consider only two cases: (a) A is the Jordan block with real eigenvalue λ_1 and size s, and $\pm H$ is the $s \times s$ sip matrix; (b) $A = J \oplus \overline{J}$, where J is the Jordan block with nonreal eigenvalue λ_1 and size s, and H is the $2s \times 2s$ sip matrix.

Consider the case (a). Let r ($= r_1$ in the notation of the theorem) be the least integer $1 \le r \le s - 1$ such that $f^{(r)}(\lambda_1) \ne 0$ (and $r = s$ if $f^{(i)}(\lambda_1) = 0$ for $1 \le i \le s - 1$). Let $f_t(\lambda)$, $t \in [0,1]$ be a family of analytic functions in a neighborhood of λ_1 with the following properties:

$$f_t(\lambda_1) = f(\lambda_1)(= \lambda_0), \quad f_t^{(i)}(\lambda_1) = 0 \quad for \quad 1 \le i < r;$$

$$f_t^{(r)}(\lambda_1) = (1 - t)f^{(r)}(\lambda_1) \pm r!t; \tag{7.4.11}$$

and

$$f_t^{(i)}(\lambda_1) = (1 - t)f^{(i)}(\lambda_1) \quad for \quad r < i \le s - 1.$$

The sign $\pm$ in (7.4.11) is chosen to guarantee that $f_t^{(r)}(\lambda_1) \neq 0$ for all $t \in [0,1]$. We assume in the subsequent argument that

$$f_t^{(r)}(\lambda_1) > 0, \qquad\qquad (7.4.12)$$

thus the sign is $+$. The situation when $f_t^{(r)}(\lambda_1) < 0$ is easily reduced to (7.4.12), by considering $-f(A)$ instead of $f(A)$ and using Corollary 5.8.2.

Recall that $f(\lambda)$ is symmetric on $\sigma(A)$, so the numbers $f^{(i)}(\lambda_1)$, $i = 0, \ldots, s-1$ are real. Formula (7.1.2) shows that $f_t(A)$ is a continuous function of $t \in [0,1]$, and $f_0(A) = f(A)$. Furthermore, Proposition 7.1.1 shows that the eigenvalue of $f_t(A)$ and the sizes of the Jordan blocks of $f_t(A)$ are independent of t on the interval $[0,1]$. As $f_t(A)$ is H-selfadjoint for $t \in [0,1]$, Theorem 5.9.1 implies that the sign characteristic of $(f_t(A), H)$ is also independent of t. So the sign characteristics of $(f(A), H)$ and $(f_1(A), H)$ are the same, and we shall determine the latter.

It follows from (7.1.2) that

$$f_1(A) = \begin{bmatrix} \lambda_0 & \cdots & 0 & 1 & 0 & \cdots & 0 \\ & \lambda_0 & \cdots & 0 & 1 & & \vdots \\ & & \lambda_0 & & \ddots & \ddots & 0 \\ & & & \ddots & & \ddots & 1 \\ & & & & \ddots & & 0 \\ & & & & & \ddots & \\ 0 & & & & & & \lambda_0 \end{bmatrix}, \qquad (7.4.13)$$

where 1's appear in the entries $(1, r+1), (2, r+2), \ldots, (r+1, s)$. It is easily seen that, denoting $v = \left[\frac{s}{r}\right]$ and letting e_j be the vector $\langle 0, \ldots, 0, 1, 0, \ldots, 0 \rangle$ with 1 in the jth place, the chains of vectors

$$\begin{aligned} e_i, e_{i+r}, \ldots, e_{i+vr}; \quad & i = 1, 2, \ldots, s - rv \\ e_i, e_{i+r}, \ldots, e_{i+(v-1)r}; \quad & i = s - rv + 1, \ldots, r \end{aligned} \qquad (7.4.14)$$

are Jordan chains of $f_1(A)$ and form a basis in $\mathbf{C}^s$. Now use the second description of the sign characteristic (Theorem 5.8.1) to compute the sign characteristic of $(f_1(A), H)$. Indeed, in the notation of Theorem 5.8.1, the linear transformation G_{v+1} in the basis $e_1, \ldots, e_{s-rv}$ is given by the right upper $(s - rv) \times (s - rv)$ corner of H. The linear transformation G_v in the basis $e_1, \ldots, e_r$ is given by the submatrix of H formed by its first r rows and columns $r(v-1)+1, \ldots, r(v-1)+r$. (In case $v = 0$ this description should be properly modified.) Since εH ($\varepsilon = \pm 1$) is the sip matrix, it is found with the help of Theorem 5.8.1 that, indeed, the sum of signs in the sign characteristic of $(f(A), H)$ which correspond to blocks of size q, is given by formula (7.4.9).

Consider case (b). A perturbation argument analogous to the proof of case (a) is used. Recall that

$$f^{(k)}(\lambda_1) = \overline{f^{(k)}(\overline{\lambda_1})}, \quad k = 0, 1, \ldots, s - 1.$$

Let r be the least integer $1 \le r \le s - 1$ such that $f^{(r)}(\lambda_1) \ne 0$ (and $r = s$ if $f^{(i)}(\lambda_1) = 0$ for $1 \le i \le s - 1$). Select a continuous function

$$\phi : [0, 1] \longrightarrow \mathbb{C}$$

such that

$$\phi(0) = f^{(r)}(\lambda_1), \quad \phi(1) = r!, \quad \phi(t) \ne 0 \text{ for all } t \in [0, 1].$$

Let $f_t(\lambda)$, $t \in [0, 1]$ be a family of analytic functions in a neighborhood of $\{\lambda_1, \overline{\lambda_1}\}$ with the following properties:

$$f_t(\lambda_1) = f(\lambda_1)(= \lambda_0), \quad f_t^{(i)}(\lambda_1) = 0 \quad \text{for} \quad 1 \le i < r;$$

$$f_t^{(r)}(\lambda_1) = \phi(t);$$

$$f_t^{(i)}(\lambda_1) = (1 - t)f^{(i)}(\lambda_1) \quad \text{for} \quad r < i \le s - 1,$$

and

$$f_t^{(i)}(\overline{\lambda_1}) = \overline{f_t^{(i)}(\lambda_1)}, \quad i = 0, 1, \ldots, s - 1, \quad 0 \le t \le 1.$$

Then, just as in the proof of the case (a), it follows that the sign characteristics of $(f(A), H)$ and $(f_1(A), H)$ are the same. Thus, we can assume that $f(A)$ is a direct sum of two equal blocks (7.4.13). The chains of vectors

$$e_i, e_{i+r}, \ldots, e_{i+vr}; \quad i = 1, 2, \ldots, s - rv, s + 1, \ldots, 2s - rv;$$
$$e_i, e_{i+r}, \ldots, e_{i+(v-1)r}; \quad i = s - rv + 1, \ldots, r, 2s - rv + 1, 2s - rv + 2, \ldots, s + r;$$

where $v = \left[\frac{s}{r}\right]$, are Jordan chains of $f(A)$ and form a basis of $\mathbb{C}^{2s}$. Now, as in case (a), the sign characteristic of $(f(A), H)$ is computed by applying Theorem 5.8.1. It turns out that the sum of signs in the sign characteristic of $(f(A), H)$ corresponding to equal Jordan blocks is zero. □

Example 7.4.2. Let

$$A = \begin{bmatrix} i & 1 \\ 0 & i \end{bmatrix} \oplus \begin{bmatrix} -i & 1 \\ 0 & -i \end{bmatrix}; \quad H = \begin{bmatrix} 0 & 0 & 0 & 1 \\ 0 & 0 & 1 & 0 \\ 0 & 1 & 0 & 0 \\ 1 & 0 & 0 & 0 \end{bmatrix}.$$

Then

$$A^2 = \begin{bmatrix} -1 & 2i & 0 & 0 \\ 0 & -1 & 0 & 0 \\ 0 & 0 & -1 & -2i \\ 0 & 0 & 0 & -1 \end{bmatrix}$$

is H-selfadjoint. According to Theorem 7.4.1 the sign characteristic of (A^2, H) at the eigenvalue -1 consists of $\{+1, -1\}$. Indeed, with

$$
T = \begin{bmatrix} 2ic & 0 & 2i & 0 \\ 0 & c & 0 & 1 \\ -2i & 0 & -2ic & 0 \\ 0 & 1 & 0 & c \end{bmatrix}, \qquad c = \frac{\sqrt{3}}{2} - \frac{i}{2}
$$

we have

$$
T^{-1}A^2 T = \begin{bmatrix} -1 & 1 \\ 0 & -1 \end{bmatrix} \oplus \begin{bmatrix} -1 & 1 \\ 0 & -1 \end{bmatrix}; \qquad T^*HT = \begin{bmatrix} 0 & 2 \\ 2 & 0 \end{bmatrix} \oplus \begin{bmatrix} 0 & -2 \\ -2 & 0 \end{bmatrix}.
$$

$\square$

Example 7.4.3. Let A and $f(\lambda)$ be as in Example 7.1.2, and let

$$
H = \varepsilon_1 S_5 \oplus \varepsilon_2 S_7, \qquad \varepsilon_1, \varepsilon_2 \in \{1, -1\},
$$

where S_5 and S_7 are the sip matrices of sizes 5×5 and 7×7, respectively. Using Theorem 7.4.1 we compute the sign characteristic of $I + \sin(X^3)$ as an H-selfadjoint matrix. In the notation of that theorem, we have

$$
\gamma(3) = \varepsilon_2, \quad \eta(3) = 1, \qquad \gamma(2) = 0, \quad \eta(2) = 4, \qquad \gamma(1) = \varepsilon_1, \quad \eta(1) = 1.
$$

Thus, the 3×3 Jordan block of $I + \sin(X^3)$ has sign ε_2, the four 2×2 Jordan blocks of $I + \sin(X^3)$ have signs 1 (two of them) and -1 (the other two), and the 1×1 Jordan block of $I + \sin(X^3)$ has sign ε_1. $\square$

7.5 Functions of H-Selfadjoint Matrices which are Selfadjoint in another Indefinite Inner Product

In this section we study the sign characteristic of a function of an H-selfadjoint matrix, but with respect to a different (but closely related) indefinite inner product. To start with, let A be an invertible H-selfadjoint matrix, and let $\widehat{H} = HA$. Obviously, $\widehat{H}$ is invertible and hermitian and, therefore, generates another inner product on $\mathbb{C}^n$. Moreover, A is $\widehat{H}$-selfadjoint. Indeed

$$
\widehat{H}A = HA^2 = A^*HA = A^{*2}H = A^*\widehat{H}.
$$

We compute the sign characteristic of $(A, \widehat{H})$ in terms of the sign characteristic of (A, H).

Proposition 7.5.1. Let $\lambda_0 \in \sigma(A) \cap \mathbb{R}$, and let $\varepsilon_1, \ldots, \varepsilon_r$ be the signs in the sign characteristic of (A, H) corresponding to the Jordan blocks of A with eigenvalue λ_0. Then the signs in the sign characteristic of $(A, \widehat{H})$ corresponding to these Jordan blocks are $\varepsilon_1 \cdot \operatorname{sgn} \lambda_0, \ldots, \varepsilon_r \cdot \operatorname{sgn} \lambda_0$, where $\operatorname{sgn} \lambda_0$ is 1 if $\lambda_0 > 0$ and -1 if $\lambda_0 < 0$.

(Recall that $\lambda_0 \neq 0$ since A is assumed invertible.)

Proof. We can assume that (A, H) is in the canonical form. Let J be a Jordan block in A with eigenvalue λ_0, and let εP be the corresponding part of H, where P is a sip matrix. Using the second description of the sign characteristic (of Theorem 5.8.1), it is easily seen that the canonical form of (J, PJ) is $(J, \operatorname{sgn} \lambda \cdot P)$. □

More generally, let A be an H-selfadjoint matrix, and let $f(\lambda)$ be a function defined on the spectrum of A such that $f(A)$ is H-selfadjoint and invertible (by Theorem 7.3.2 this means that $f(\lambda)$ is real symmetric on $\sigma(A)$ and $f(\lambda_0) \neq 0$ for every $\lambda_0 \in \sigma(A)$). Then $\widehat{H} = Hf(A)$ is invertible and hermitian, and $f(A)$ is $\widehat{H}$-selfadjoint. Combining Proposition 7.5.1 and the description of the sign characteristic of $(f(A), H)$ given in Theorem 7.4.1, we obtain the following description of the sign characteristic of $(f(A), \widehat{H})$. The notation introduced before and in the statement of Theorem 7.4.1 will be used.

Theorem 7.5.2. *Let A be H-selfadjoint, and let $f(A)$ be H-selfadjoint and invertible. For a fixed positive integer q, $\eta(q)$ is the number of Jordan blocks of $f(A)$ with eigenvalue λ_0 and size q and, except for the case when $f^{(r_i)}(\lambda_i) < 0$ and q is even, $\frac{1}{2}(\eta(q) + (\operatorname{sgn} \lambda_0) \cdot \gamma(q))$ (resp. $\frac{1}{2}(\eta(q) - (\operatorname{sgn} \lambda_0) \cdot \gamma(q)))$ is the number of signs $+1$ (resp. -1) in the sign characteristic of $(f(A), \widehat{H})$ corresponding to these Jordan blocks (here $\operatorname{sgn} \lambda_0 = 1$ if $\lambda_0 > 0$ and $\operatorname{sgn} \lambda_0 = -1$ if $\lambda_0 < 0$). In the exceptional case, $\frac{1}{2}(\eta(q) - (\operatorname{sgn} \lambda_0) \cdot \gamma(q))$ (resp. $\frac{1}{2}(\eta(q) + (\operatorname{sgn} \lambda_0) \cdot \gamma(q)))$ is the number of $+1$'s (resp. -1's) in the sign characteristic of $(f(A), \widehat{H})$ corresponding to these Jordan blocks.*

We note also the following fact concerning neutrality of invariant subspaces.

Theorem 7.5.3. *Let A be H-selfadjoint and let $f(A)$ be a function of A which is H-selfadjoint and invertible. Then an $f(A)$-invariant subspace $\mathcal{M} \subseteq \mathbb{C}^n$ (where n is the size of A) is H-neutral if and only if $\mathcal{M}$ is $\widehat{H}$-neutral, where $\widehat{H} = Hf(A)$.*

Proof. Let $\widehat{A} = f(A)$ and observe that the subspace $\mathcal{M}$ is $\widehat{A}$-invariant. Assuming $\mathcal{M}$ is H-neutral (so $(Hx, y) = 0$ for all $x, y \in \mathcal{M}$), for every $x, y \in \mathcal{M}$ we have

$$(\widehat{H}x, y) = (H\widehat{A}x, y) = 0.$$

Conversely, write $z = \widehat{A}^{-1}x \in \mathcal{M}$ and, if $\mathcal{M}$ is $\widehat{H}$-neutral, then for every $x, y \in \mathcal{M}$,

$$(Hx, y) = (\widehat{H}z, y) = 0.$$ □

Note that Theorem 7.5.3 does not hold in general for A-invariant subspaces which are H-nonnegative or H-nonpositive. For example, if $H = I$ and $f(A) = -I$, then every $f(A)$-invariant subspace is simultaneously H-positive and $Hf(A)$-negative.

7.6   Exercises

The exercises in this section pertain to the following H_j-selfadjoint matrices A_j:

(a) $A_1 = \begin{bmatrix} \lambda & \alpha \\ 0 & \bar{\lambda} \end{bmatrix}$, $\quad H_1 = \begin{bmatrix} 0 & 1 \\ 1 & 0 \end{bmatrix}$, $\quad \lambda \in C$, $\alpha \in R$.

(b) $A_2 = \begin{bmatrix} \lambda & \alpha & \gamma \\ 0 & \mu & \bar{\alpha} \\ 0 & 0 & \bar{\lambda} \end{bmatrix}$, $\quad H_2 = \begin{bmatrix} 0 & 0 & 1 \\ 0 & 1 & 0 \\ 1 & 0 & 0 \end{bmatrix}$, $\quad \mu, \gamma \in R$, $\lambda, \alpha \in C$.

(c) $A_3 = \begin{bmatrix} \mu & 0 & 0 \\ 0 & \lambda & \alpha \\ 0 & 0 & \bar{\lambda} \end{bmatrix}$, $\quad H_3 = \begin{bmatrix} 1 & 0 & 0 \\ 0 & 0 & 1 \\ 0 & 1 & 0 \end{bmatrix}$, $\quad \mu, \alpha \in R$, $\lambda \in C$.

(d) $A_4 = \begin{bmatrix} 0 & 1 & 0 & \cdots & 0 \\ 0 & 0 & 1 & \cdots & 0 \\ & & \cdots\cdots\cdots & & \\ a & 0 & 0 & \cdots & 0 \end{bmatrix}$, $\quad H_4 = \begin{bmatrix} 0 & 0 & \cdots & 0 & 1 \\ 0 & 0 & \cdots & 1 & 0 \\ & & \cdots\cdots & & \\ 1 & 0 & \cdots & 0 & 0 \end{bmatrix} \in R^{n \times n}$,

$a \in R$.

(e) $A_5 = \begin{bmatrix} 0 & 1 \\ -a_0 & -a_1 \end{bmatrix}$, $\quad H_5 = \begin{bmatrix} a_1 & 1 \\ 1 & 0 \end{bmatrix}$, $\quad a_0, a_1 \in R$.

(f) $A_6 = \begin{bmatrix} 0 & 1 & 0 \\ 0 & 0 & 1 \\ -a_0 & -a_1 & -a_2 \end{bmatrix}$, $\quad H_6 = \begin{bmatrix} a_2 & a_1 & 1 \\ a_1 & 1 & 0 \\ 1 & 0 & 0 \end{bmatrix}$, $\quad a_0, a_1, a_2 \in R$.

(g) $A_7 = \text{diag}\,(a_1, a_2, \ldots, a_n)$, $\quad H_7 = \begin{bmatrix} I_p & 0 \\ 0 & -I_q \end{bmatrix} \in R^{n \times n}$, $\quad a_1, \ldots, a_n \in R$,

$p + q = n$.

(h) $A_8 = \begin{bmatrix} 0 & 0 & \cdots & 0 & \lambda_1 \\ 0 & 0 & \vdots & \lambda_2 & 0 \\ \vdots & \vdots & \ddots & \vdots & \vdots \\ 0 & \lambda_{n-1} & \cdots & 0 & 0 \\ \lambda_n & 0 & \cdots & 0 & 0 \end{bmatrix}$, $\quad H_8 = \begin{bmatrix} 0 & 0 & \cdots & 0 & 1 \\ 0 & 0 & \vdots & 1 & 0 \\ \vdots & \vdots & \ddots & \vdots & \vdots \\ 0 & 1 & \cdots & 0 & 0 \\ 1 & 0 & \cdots & 0 & 0 \end{bmatrix} \in R^{n \times n}$,

$\lambda_1, \ldots, \lambda_n \in R$.

1. For each pair (A_j, H_j), $j = 1, \ldots, 8$, compute the canonical form, including the sign characteristic, of A_j^2 as an H_j-selfadjoint matrix.

2. The same as in Exercise 1, but for A_j^3.

3. Compute the canonical form, including the sign characteristic, of the H_j-unitary matrix e^{iA_j}, for $j = 1, 2, \ldots, 8$.

4. Show by example that not every invertible H-selfadjoint matrix has an H-selfadjoint square root. Recall that a matrix $Y \in C^{n \times n}$ is called a *square root* of a matrix $X \in C^{n \times n}$ if $Y^2 = X$.

5. Prove that every invertible H-selfadjoint matrix with no negative eigenvalues has an H-selfadjoint square root.

6. For the invertible matrices A_j, $j = 1, \ldots, 8$, find whether there exists an H_j-selfadjoint square root of A_j. If it does exist, determine whether or not it is unique, and find the canonical form of an H_j-selfadjoint square root of A_j.

7. Does there exist a cube root of every invertible H-selfadjoint matrix?

8. For the invertible matrices A_j, $j = 1, \ldots, 8$, find the canonical form, including the sign characteristic, of A_j as an $H_j A_j$-selfadjoint matrix.

9. Prove that every invertible H-selfadjoint matrix A with no negative eigenvalues has an H-selfadjoint logarithm, i.e., an H-selfadjoint matrix B such that $e^B = A$.

10. The condition in the previous exercise is not necessary: Show that $-I_{2m}$ has an S_{2m}-selfadjoint logarithm, where S_{2m} is the $2m \times 2m$ sip matrix. On the other hand, show that $-I_n$, where n is odd, has no H-selfadjoint logarithms, for any invertible hermitian $n \times n$ matrix H.

11. Let

$$
A = \begin{bmatrix} a_0 & a_1 & a_2 & \cdots & a_n \\ 0 & a_0 & a_1 & \cdots & a_{n-1} \\ & & \cdots\cdots\cdots & & \\ 0 & 0 & \cdots & 0 & a_0 \end{bmatrix}
$$

be an upper triangular Toeplitz matrix. Find the Jordan form of $f(A)$, for the following scalar functions:
(a) $f(\lambda) = \lambda^2 + \lambda$,   (b) $f(\lambda) = e^{2\lambda}$,   (c) $f(\lambda) = \sqrt[m]{\lambda}$,
(d) $f(\lambda) = \sin \lambda$,   (e) $f(\lambda) = -\cos(3\lambda + 1)$,   (f) $f(\lambda) = \ln(\lambda)$.

For the functions $\sqrt[m]{\lambda}$ and $\ln(\lambda)$ assume that $a_0 > 0$ and the branches of the mth root and the logarithmic functions are chosen so that $\sqrt[m]{\lambda} > 0$ for $\lambda > 0$ and $\ln(\lambda)$ is real for $\lambda > 0$.

12. Consider a matrix A which is a direct sum of two upper triangular Toeplitz matrices:

$$
A = \begin{bmatrix} a_0 & a_1 & a_2 & \cdots & a_n \\ 0 & a_0 & a_1 & \cdots & a_{n-1} \\ & & \cdots\cdots\cdots & & \\ 0 & 0 & \cdots & 0 & a_0 \end{bmatrix} \oplus \begin{bmatrix} b_0 & b_1 & b_2 & \cdots & b_m \\ 0 & b_0 & b_1 & \cdots & b_{m-1} \\ & & \cdots\cdots\cdots & & \\ 0 & 0 & \cdots & 0 & b_0 \end{bmatrix}.
$$

For the functions $f(\lambda)$ of the preceding exercise, determine whether $f(A)$ is:

(1) nonderogatory, i.e., has only one Jordan block in the Jordan form for every eigenvalue;

(2) diagonalizable;

(3) diagonalizable with all eigenvalues distinct.

7.7   Notes

The material of this chapter is based on [40, Chapter 6]. A stronger version of Theorem 7.3.1 has recently been proved by Higham et al. [48].

Chapter 8

H-Normal Matrices

The main ideas developed in this chapter are, first, a classification (up to unitary similarity) of normal matrices in an indefinite inner product space and, second, the nature of canonical forms in this classification. In full generality, this problem area remains unsolved. We will show that it is not less complex than the problem of classification of pairwise $v \times v$ commuting matrices (up to simultaneous similarity), where $v = \min(v_+, v_-)$ and $v_+(v_-)$ is the number of positive (negative) eigenvalues of H. In the research literature the latter problem is said to be "wild". In contrast, it is well known that for the case when H is positive definite the situation is clear and can be reduced to the problem of simultaneous diagonalization of pairwise commuting selfadjoint matrices.

A complete theory is presented here for the special case when the indefinite inner product is defined by

$$[x, y] = (Hx, y),$$

where H is invertible and hermitian *with only one negative or positive eigenvalue*. This includes the description of canonical form and invariants.

In this chapter we use consistently the language of linear transformations, i.e., a linear transformation on $\mathbb{C}^n$ is identified with its matrix representation in the standard basis $e_1, \ldots, e_n$. On the other hand, it will also be convenient to work with matrix representations of the same linear transformation with respect to different bases.

8.1 Decomposability: First Remarks

First recall some basic definitions. If $H \in \mathbb{C}^{n \times n}$ is hermitian and nonsingular, then it generates an inner product

$$[x, y] = (Hx, y), \quad x, y \in \mathbb{C}^n.$$

which, in general, is indefinite. In this inner product the adjoint linear transformation is denoted $A^{[*]}$ and satisfies

$$[Ax, y] = \left[x, A^{[*]}y\right] = \left[x, H^{-1}A^*Hy\right].$$

The matrix A is said to be H-normal, or normal in the inner product $[x, y]$, if

$$AA^{[*]} = A^{[*]}A, \quad \text{or} \quad AH^{-1}A^*H = H^{-1}A^{[*]}HA.$$

We remark that if H is positive definite and

$$(AH^{-1}A^*)H = (H^{-1}A^*H)A,$$

then A and $A^{[*]}$ are simultaneously diagonalizable in an H-orthogonal basis.

Example 8.1.1. *Let J be an $n \times n$ Jordan block with eigenvalue $\lambda_0 \in \mathbb{C}$ and let $H = P$ be the $n \times n$ sip matrix. It is easily verified that J is P-normal for any λ_0.*

$\square$

Proposition 8.1.2. *Every square matrix $A \in \mathbb{C}^{n \times n}$ is H-normal for some hermitian nonsingular H.*

Proof. Let $J = \text{diag}\,(J_1, \ldots, J_k)$ be a Jordan form of the matrix A, with $J_1, J_2, \ldots,$ J_k all Jordan blocks and let $A = S^{-1}JS$. Let P_j be the sip matrix of the same size as J_j $(j = 1, 2, \ldots, k)$. Form $P = \text{diag}\,(P_1, P_2, \ldots, P_k)$, and it is obvious that J is P-normal. Hence A is S^*PS-normal. $\square$

Let $A \in \mathbb{C}^{n \times n}$ and $\mathcal{V}$ be a subspace of $\mathbb{C}^n$. Recall that, if $A\mathcal{V} \subseteq \mathcal{V}$, then $\mathcal{V}$ is said to be an *invariant subspace* of A, and the restriction of A to $\mathcal{V}$ is denoted by $A\,|_{\mathcal{V}}$.

A linear transformation A acting in $\mathbb{C}^n$ is called *decomposable* if there exists a nondegenerate subspace $\mathcal{V} \subseteq \mathbb{C}^n$, $\mathcal{V} \neq \{0\}$, $\mathcal{V} \neq \mathbb{C}^n$, such that both $\mathcal{V}$ and $\mathcal{V}^{[\perp]}$ are invariant for A. If $\mathcal{V}$ and $\mathcal{V}^{[\perp]}$ are nondegenerate nonzero subspaces which are invariant for A, and $A_1 := A\,|_{\mathcal{V}}$, $A_2 := A\,|_{\mathcal{V}^{[\perp]}}$, then A is called *the orthogonal sum of A_1 and A_2*, or the *H-orthogonal sum* if the indefinite inner product generated by H is to be emphasized. In the above decomposition A and H can be represented as

$$A = \begin{bmatrix} A_1 & 0 \\ 0 & A_2 \end{bmatrix}, \quad H = \begin{bmatrix} H_1 & 0 \\ 0 & H_2 \end{bmatrix}. \tag{8.1.1}$$

If a linear transformation is not decomposable into an H-orthogonal sum, it is called *indecomposable*. It is easy to see that A is decomposable if and only if there exists a nondegenerate subspace $\mathcal{V} \in \mathbb{C}^n$ different from $\{0\}$ and $\mathbb{C}^n$, such that $\mathcal{V}$ is invariant for both A and $A^{[*]}$. Indeed, if $v \in \mathcal{V}$, $w \in \mathcal{V}^{[\perp]}$, then

$$[Aw, v] = \left[w, A^{[*]}v\right] \subseteq [w, \mathcal{V}] = 0,$$

so $Aw \in \mathcal{V}^{[\perp]}$. The converse is trivial.

It is clear that the definition of orthogonal sum can be generalized to admit any number, k, of terms and then it is easily proved that any linear transformation can be decomposed into an orthogonal sum of k indecomposable linear transformations.

In the case when H is positive definite it is known that if a subspace $V \subseteq \mathbf{C}^n$ is invariant for a normal linear transformation N, then $V^{[\perp]}$ is also invariant for N and, therefore, N is decomposable (if $n > 1$). This property does not hold in general when H is indefinite. Indeed, let N and H be the linear transformations in $\mathbf{C}^4$ having the matrix representation

$$ N = \begin{bmatrix} 0 & 0 & 0 & 0 \\ 0 & 0 & 0 & 0 \\ 0 & 0 & 0 & 1 \\ 0 & 0 & 0 & 0 \end{bmatrix}, \quad H = \begin{bmatrix} 0 & 0 & 0 & 1 \\ 0 & 0 & 1 & 0 \\ 0 & 1 & 0 & 0 \\ 1 & 0 & 0 & 0 \end{bmatrix} $$

in some basis $\{v_1, v_2, v_3, v_4\}$ of $\mathbf{C}^4$. Then

$$ N^{[*]} = \begin{bmatrix} 0 & 1 & 0 & 0 \\ 0 & 0 & 0 & 0 \\ 0 & 0 & 0 & 0 \\ 0 & 0 & 0 & 0 \end{bmatrix} $$

and a simple computation shows that $NN^{[*]} = N^{[*]}N = 0$.

Let $V_1 = \mathrm{Span}\,\{v_2, v_3\}$ and $V_2 = \mathrm{Span}\,\{v_1, v_4\}$. Each subspace V_i is nondegenerate. It is easy to see that $V_1^{[\perp]} = V_2$, V_1 is invariant for N and V_2 is invariant for $N^{[*]}$. However, N is indecomposable. To prove this we assume that, on the contrary, there is a nondegenerate nonzero subspace V, $V^{[\perp]} \subseteq \mathbf{C}^4$ for which

$$ N = \begin{bmatrix} N_1 & 0 \\ 0 & N_2 \end{bmatrix} $$

where $N_1 : V \to V$, $N_2 : V^{[\perp]} \to V^{[\perp]}$.

We first show that $\dim V \neq 1$. Indeed, if a vector $v \in \mathbf{C}^4$ spans V, then

$$ Nv = 0, \quad N^{[*]}v = 0, \quad \text{and} \quad [v, v] \neq 0. $$

But the first two relations imply that $v \in \mathrm{Span}\,\{v_1, v_3\}$ and, hence, $[v, v] = 0$. The equality $\dim V = 2$ is also impossible because then $N_1 = 0$ or $N_2 = 0$. If $N_1 = 0$, then $N_1^{[*]} = 0$, $V = \mathrm{Span}\,\{v_1, v_3\}$, and $[V, V] = 0$. If $N_2 = 0$, then, similarly, $[V^{[\perp]}, V^{[\perp]}] = 0$. So N is indecomposable.

Lemma 8.1.3. *Any H-normal linear transformation $N : \mathbf{C}^n \to \mathbf{C}^n$ is an orthogonal sum of normal linear transformations each of which has one or two distinct eigenvalues.*

In the proof we will take advantage of the following well-known result regarding two commuting linear transformations acting in the vector space $\mathbf{C}^n$ (see [41, Chapter 9]):

Proposition 8.1.4. *Let* $A, B : C^n \to C^n$ *be linear transformations such that* $AB = BA$, *let* $\{\lambda_1, \lambda_2, \ldots, \lambda_\ell\}$ *be the set of all distinct eigenvalues of* A, *and let* $\{\mu_1, \mu_2, \ldots, \mu_m\}$ *be the set of all distinct eigenvalues of* B. *For each pair* (i, j), $(i = 1, 2, \ldots, \ell; \ j = 1, 2, \ldots, m)$ *we define the subspace* $\mathcal{Q}_{ij} \subseteq C^n$:

$$\mathcal{Q}_{ij} := \{x \in C^n \ : \ (A - \lambda_i I)^n x = (B - \mu_j I)^n x = 0\}. \tag{8.1.2}$$

Let $\Omega = \{(i, j) : \mathcal{Q}_{ij} \neq 0\}$. *Then the subspaces* $\mathcal{Q}_{ij}$ *have the following properties:*

(a) $\mathcal{Q}_{ij} \cap \text{Span} \{\mathcal{Q}_{rs} : (r, s) \in \Omega, \ (r, s) \neq (i, j)\} = \{0\}$ *for any* $(i, j) \in \Omega$;

(b) $\text{Span} \{\mathcal{Q}_{ij} : (i, j) \in \Omega\} = C^n$;

(c) $\mathcal{Q}_{ij}$ *is an invariant subspace for both* A *and* B *for all* $(i, j) \in \Omega$;

(d) λ_i *is the only eigenvalue of the linear transformation* $A \mid_{\mathcal{Q}_{ij}}$ *and* μ_j *is the only eigenvalue of the linear transformation* $B \mid_{\mathcal{Q}_{ij}}$.

Proof of Lemma 8.1.3. We use Proposition 8.1.4 with $A = N$, $B = N^{[*]}$. In this case, in the notation of Proposition 8.1.4, we have $\ell = m$ and it can be assumed that $\mu_i = \overline{\lambda_i}$ $(i = 1, 2, \ldots, \ell)$. We claim that, for any pair of indices (i, j), $(r, s) \in \Omega$ we have

$$[\mathcal{Q}_{ij}, \mathcal{Q}_{rs}] = 0 \quad \text{unless} \quad i = s \quad \text{and} \quad j = r. \tag{8.1.3}$$

Suppose first that $i \neq s$. It is clear that λ_s is not an eigenvalue of $N \mid_{\mathcal{Q}_{ij}}$ and, hence, the linear transformation $(N - \lambda_s I) \mid_{\mathcal{Q}_{ij}}$ is nonsingular. Let $x \in \mathcal{Q}_{ij}$. Then there exists a vector $z \in \mathcal{Q}_{ij}$ such that $(N - \lambda_s I)^n z = x$. Now for any $y \in \mathcal{Q}_{rs}$ we have

$$[x, y] = [(N - \lambda_s I)^n z, y] = \left[z, \left(N^{[*]} - \overline{\lambda_s} I\right)^n y\right] = [z, 0] = 0.$$

The case $i = s$, $j \neq r$ is treated similarly. So, (8.1.3) holds. Now we can construct a desired decomposition of N. Let

$$\mathcal{V}_i = \mathcal{Q}_{ii} \quad \text{for} \quad (i, i) \in \Omega, \qquad \mathcal{V}_{jk} = \text{Span} \{\mathcal{Q}_{jk}, \mathcal{Q}_{kj}\} \quad ((j, k) \in \Omega, j < k). \tag{8.1.4}$$

The subspaces (8.1.4) are mutually orthogonal (due to (8.1.3)), the intersection of any one of them with the sum of the other subspaces is zero, and they span C^n. By Proposition 2.2.2, each of the subspaces $\mathcal{V}_i$ and $\mathcal{V}_{jk}$ is nondegenerate. Therefore, the linear transformation N is the orthogonal sum of linear transformations N_i and N_{jk} where

$$N_i = N \mid_{\mathcal{V}_i}, \quad ((i, i) \in \Omega), \qquad N_{jk} = N \mid_{\mathcal{V}_{jk}}, \quad ((j, k) \in \Omega, j < k).$$

It follows from (8.1.2) and (8.1.4) that each N_i has only the eigenvalue λ_i and that each N_{jk} has exactly two distinct eigenvalues λ_j and λ_k. This concludes the proof. $\qquad \square$

Note that if $(j, k) \in \Omega$, (in the notation of the proof of Lemma 8.1.3) then $(k, j) \in \Omega$ and the characteristic polynomial $\varphi(\lambda)$ of N_{jk} has zeros λ_j and λ_k.

Example 8.1.5. *Assume that all the Jordan blocks of an H-normal linear transformation N are one-dimensional, i.e.,, N is diagonalizable. In this case the subspaces $\mathcal{V}_i$ appearing in Lemma 8.1.3 consist of all eigenvectors common to N and $N^{[*]}$ with eigenvalues λ_i and $\overline{\lambda}_i$ respectively. Each of the subspaces $\mathcal{Q}_{jk}$, $(j \neq k)$ consists of all eigenvectors common to N and $N^{[*]}$ with eigenvalues λ_j and $\overline{\lambda}_k$. There exists a basis*

$$\left\{ v_{i_1}, v_{i_2}, \ldots, v_{i_p}; v_{j_1}, v_{k_1}, v_{j_2}, v_{k_2}, \ldots, v_{j_q}, v_{k_q} \right\}, \qquad (p + 2q = n)$$

of $\mathbb{C}^n$ such that

$$N v_{i_s} = \lambda_{i_s} v_{i_s}, \quad (s = 1, 2, \ldots, p),$$

$$N v_{j_t} = \lambda_{k_t} v_{j_t}, \quad N v_{k_t} = \lambda_{j_t} v_{k_t} \quad (t = 1, 2, \ldots, q),$$

$$[v_{i_s}, v_{i_s}] = \pm 1, \quad (s = 1, 2, \ldots, p), \quad [v_{j_t}, v_{k_t}] = \pm 1, \quad (t = 1, 2, \ldots, q);$$

all the remaining indefinite inner products of the vectors v_m are equal to zero. Therefore, in an appropriate basis the matrices of N and H are block-diagonal where each block is either one-dimensional

$$N_\ell = [\lambda_\ell], \quad H_\ell = [\pm 1]$$

or two-dimensional

$$N_{uv} = \begin{bmatrix} \lambda_u & 0 \\ 0 & \lambda_v \end{bmatrix}, \quad H_{uv} = \pm \begin{bmatrix} 0 & 1 \\ 1 & 0 \end{bmatrix}.$$

$\square$

8.2   *H*-Normal Linear Transformations and Pairs of Commuting Matrices

We are to describe all classes of H-unitarily similar normal linear transformations and to find the canonical representation of each class. From Section 8.1 it follows that it is sufficient to solve the problem for the classes consisting of indecomposable linear transformations. Moreover, as a result of Lemma 8.1.3, we can even limit discussion to the linear transformations that have only one or two distinct eigenvalues. However, in spite of this promising start, the general problem proves to be in a certain sense unsolvable. In fact, it will be shown that: *In the general case, the problem of classification of H-normal linear transformations up to H-unitary similarity is at least as complicated as the problem of classification up to simultaneous similarity of pairs of commuting matrices of size $\min(\nu_+, \nu_-)$, where ν_+ (resp., ν_-) is the number of positive (negative) eigenvalues of H (counted with multiplicities).*

Indeed, let H have ν_+ positive and ν_- negative eigenvalues $(\nu_+ + \nu_- = n)$. Without loss of generality we can assume that $\nu_+ \geq \nu_-$ and, hence

$$\nu_0 := \nu_+ - \nu_- \geq 0.$$

Let $\mathbf{M}_{\nu_-}$ be the set of all pairs of commuting $\nu_- \times \nu_-$-matrices. We consider the case $\nu_0 > 0$; it will be obvious how to treat the case $\nu_0 = 0$. For each pair of $\nu_- \times \nu_-$ commuting matrices P, Q we define two nonnegative integers ℓ and m as follows: ℓ is the smallest nonnegative integer such that the linear transformations $P - \ell I$ and Q do not have a common eigenvalue and m is the smallest nonnegative integer which is not an eigenvalue of $P - \ell I$ and not an eigenvalue of Q^*. Here, as usual, W^* denotes the conjugate transpose of the matrix W.

There exists a direct sum decomposition of $\mathbf{C}^n$ into subspaces:

$$\mathbf{C}^n = \mathcal{C}_+ + \mathcal{C}_- + \mathcal{C}_0 \tag{8.2.5}$$

with $\dim \mathcal{C}_+ = \dim \mathcal{C}_- = \nu_-$, $\dim \mathcal{C}_0 = \nu_0$ such that, up to a congruence, H has the form

$$H = \begin{bmatrix} 0 & I & 0 \\ I & 0 & 0 \\ 0 & 0 & I \end{bmatrix}.$$

Applying, if necessary, a congruence $H \mapsto S^*HS$ for a suitable invertible S, we can assume without loss of generality that the decomposition (8.2.5) is orthogonal with respect to the standard euclidean inner product. Given the pair $\{P, Q\} \in \mathbf{M}_{\nu_-}$, introduce the linear transformation

$$N = \begin{bmatrix} P - \ell I & 0 & 0 \\ 0 & Q^* & 0 \\ 0 & 0 & mI \end{bmatrix}. \tag{8.2.6}$$

The decomposition (8.2.6) corresponds to the decomposition (8.2.5) of $\mathbf{C}^n$. It is easy to see that the linear transformation N defined in (8.2.6) is H-normal.

Let the H-normal linear transformations N_1 and N_2 be produced by two pairs

$$\{P_1, Q_1\}, \{P_2, Q_2\} \in \mathbf{M}_{\nu_-} \tag{8.2.7}$$

where P_1 is similar to P_2 and Q_1 is similar to Q_2. We will prove that N_1 and N_2 are H-unitarily similar if and only if the pairs of matrices (8.2.7) are simultaneously similar. The latter means that there exists a nonsingular $\nu_- \times \nu_-$-matrix R such that

$$P_2 = RP_1R^{-1}, \quad Q_2 = RQ_1R^{-1}. \tag{8.2.8}$$

Indeed, if (8.2.8) holds, the pairs (8.2.7) have the same ℓ and m, the linear transformation

$$U = \begin{bmatrix} R & 0 & 0 \\ 0 & (R^*)^{-1} & 0 \\ 0 & 0 & I \end{bmatrix} \tag{8.2.9}$$

is H-unitary and

$$N_2 = U N_1 U^{-1}. \tag{8.2.10}$$

Conversely, suppose that (8.2.10) holds for an H-unitary linear transformation U. Representing U in the block form corresponding to the decomposition (8.2.5) and equating the corresponding blocks on both sides of the identity $U N_1 = N_2 U$, we show first that all the off-diagonal blocks of U are zero. (In the process we take advantage of the well-known fact that if square matrices A, B, C satisfy the equation $AC = CB$ and A and B do not have a common eigenvalue, then $C = 0$; see Theorem A.4.1.) Therefore U can be written in the form

$$U = \begin{bmatrix} R & 0 & 0 \\ 0 & (R^*)^{-1} & 0 \\ 0 & 0 & U_0 \end{bmatrix}, \tag{8.2.11}$$

where $U_0 \in \mathsf{C}^{\nu_0 \times \nu_0}$ is unitary, because U is H-unitary, and the linear transformation R that appears in (8.2.11) realizes the simultaneous similarity (8.2.8).

8.3 On Unitary Similarity in an Indefinite Inner Product

Let H_1 and H_2 be invertible hermitian linear transformations and recall that a linear transformation U for which

$$H_1 = U^* H_2 U. \tag{8.3.12}$$

is said to be (H_1, H_2)-*unitary*.

Linear transformations A_1 and A_2 from $\mathsf{C}^n(H_1)$ to $\mathsf{C}^n(H_2)$ are called (H_1, H_2)-*unitarily similar* (or unitarily similar if H_1 and H_2 are understood from the context) if there exists an (H_1, H_2)-unitary linear transformation U such that $A_1 = U^{-1} A_2 U$. It is readily seen that A_1 and A_2 are unitarily similar if and only if there exists an invertible linear transformation $T : \mathsf{C}^n \to \mathsf{C}^n$ such that

$$H_1 = T^* H_2 T, \quad A_1 = T^{-1} A_2 T. \tag{8.3.13}$$

Lemma 8.3.1. *Let A_i be a linear transformation acting in the space $\mathsf{C}^n(H_i)$, $i = 1, 2$. Let $\{u_1, u_2, \ldots, u_n\}$ be a basis of $\mathsf{C}^n(H_1)$ and $\{w_1, w_2, \ldots, w_n\}$ a basis of $\mathsf{C}^n(H_2)$ such that A_1 and A_2 have the same matrix A in these bases. Then A_1 and A_2 are (H_1, H_2)-unitarily similar if and only if*

$$[u_i, u_j] = [w_i, w_j], \quad \text{for } i, j = 1, 2, \ldots, n. \tag{8.3.14}$$

Proof. Assume that (8.3.14) holds. Define a linear transformation $T : \mathsf{C}^n(H_1) \to \mathsf{C}^n(H_2)$ by the equalities $T u_i = w_i$, $i = 1, 2, \ldots, n$. Since

$$[u_i, u_j] = (H_1 u_i, u_j), \quad [w_i, w_j] = (H_2 w_i, w_j),$$

the equality $[u_i, u_j] = [w_i, w_j]$ can be rewritten as

$$(H_1 u_i, u_j) = (H_2 T u_i, T u_j) = (T^* H_2 T u_i, u_j).$$

Hence $H_1 = T^* H_2 T$. Next, if a_{ij} is the (i, j)-entry of the matrix A, we have

$$T A_1 u_i = T \left(\sum_{k=1}^n a_{ki} u_k \right) = \sum_{k=1}^n a_{ki} T u_k = \sum_{k=1}^n a_{ki} w_k = A_2 w_i = A_2 T u_i,$$

and, therefore, $A_1 = T^{-1} A_2 T$. Thus, both of the identities (8.3.13) are satisfied, and A_1 and A_2 are (H_1, H_2)-unitarily similar.

Conversely, if A_1 and A_2 are (H_1, H_2)-unitarily similar, then equalities (8.3.14) are easily verified using (8.3.13).   □

Corollary 8.3.2. *Let* $\mathbb{C}^n(H_1)$ *be an* n-*dimensional space with indefinite inner product defined by an invertible hermitian matrix* H_1, A_1 *a linear transformation acting in* $\mathbb{C}^n(H_1)$, *and* A *its matrix in a basis* $\{u_1, u_2, \ldots, u_n\}$. *Let* $\{w_1, w_2, \ldots, w_n\}$ *be an orthonormal basis in* $\mathbb{C}^n$ *with respect to the standard euclidean inner product. Define two linear transformations* A_2 *and* H_2 *acting in* $\mathbb{C}^n$ *via their matrices in the basis* $\{w_1, w_2, \ldots, w_n\}$ *as follows: the matrix of* A_2 *is* A, *the* (i, j)-*entry of* H_2 *is* $[u_j, u_i]$, $i, j = 1, 2, \ldots, n$. *Then* A_1 *and* A_2 *are* (H_1, H_2)-*unitarily similar.*

Proof. For any $i, j = 1, 2, \ldots, n$ we have

$$[w_i, w_j] = (H_2 w_i, w_j) = \left(\sum_{k=1}^n [u_i, u_k] w_k, w_j \right) = [u_i, u_j].$$

So, the conditions of Lemma 8.1.3 are satisfied and A_1 and A_2 are (H_1, H_2)-unitarily similar.   □

8.4   The Case of Only One Negative Eigenvalue of H

The result of Section 8.2 shows that a classification of H-normal linear transformations can be expected only when ν_- is small. Such a classification is established in this section for $\nu_- = 1$. The (rather long) proof occupies the rest of the section.

Theorem 8.4.1. *Let* H *be an invertible hermitian matrix with only one negative eigenvalue. Any* H-*normal linear transformation acting in* $\mathbb{C}^n$ *with* $n > 4$ *is decomposable. If* $n = 1$, *any linear transformation is* H-*normal and indecomposable. For* $n = 2, 3, 4$ *any indecomposable* H-*normal linear transformation is unitarily similar to one and only one of the following canonical linear transformations:*

$$N = \begin{bmatrix} \lambda_1 & 0 \\ 0 & \lambda_2 \end{bmatrix}, \quad H = \begin{bmatrix} 0 & 1 \\ 1 & 0 \end{bmatrix}, \quad \lambda_1 \neq \lambda_2; \qquad (8.4.15)$$

$$N = \begin{bmatrix} \lambda & z \\ 0 & \lambda \end{bmatrix}, \quad H = \begin{bmatrix} 0 & 1 \\ 1 & 0 \end{bmatrix}, \quad |z| = 1; \qquad (8.4.16)$$

$$N = \begin{bmatrix} \lambda & z & r \\ 0 & \lambda & z \\ 0 & 0 & \lambda \end{bmatrix}, \quad H = \begin{bmatrix} 0 & 0 & 1 \\ 0 & 1 & 0 \\ 1 & 0 & 0 \end{bmatrix}, \quad r \in \mathsf{R}, \quad |z| = 1, \quad 0 < \arg z < \pi;$$
$$(8.4.17)$$

$$N = \begin{bmatrix} \lambda & 1 & ir \\ 0 & \lambda & 1 \\ 0 & 0 & \lambda \end{bmatrix}, \quad H = \begin{bmatrix} 0 & 0 & 1 \\ 0 & 1 & 0 \\ 1 & 0 & 0 \end{bmatrix}, \quad r \in \mathsf{R}; \qquad (8.4.18)$$

$$N = \begin{bmatrix} \lambda & \cos\alpha & \sin\alpha & 0 \\ 0 & \lambda & 0 & 1 \\ 0 & 0 & \lambda & 0 \\ 0 & 0 & 0 & \lambda \end{bmatrix}, \quad H = \begin{bmatrix} 0 & 0 & 0 & 1 \\ 0 & 1 & 0 & 0 \\ 0 & 0 & 1 & 0 \\ 1 & 0 & 0 & 0 \end{bmatrix}, \quad 0 < \alpha \le \frac{\pi}{2}. \quad (8.4.19)$$

Proof. Throughout the proof we assume that N is indecomposable. Then, according to Lemma 8.1.3, N has only one or two distinct eigenvalues. In the latter case, as the proof of Lemma 8.1.3 shows, n is even:

$$n = 2m, \quad \mathsf{C}^n = \mathrm{Span}\,(\mathcal{Q}_1, \mathcal{Q}_2), \quad \mathcal{Q}_1 \cap \mathcal{Q}_2 = \{0\}$$

with

$$\dim \mathcal{Q}_1 = \dim \mathcal{Q}_2 = m, \quad [\mathcal{Q}_1, \mathcal{Q}_1] = [\mathcal{Q}_2, \mathcal{Q}_2] = 0;$$

the linear transformation $N|_{\mathcal{Q}_1}$ (resp., $N|_{\mathcal{Q}_2}$) has the only eigenvalue λ_1 (resp., λ_2). Since in a space with an indefinite inner product no neutral subspace can be of dimension larger than $\min(\nu_+, \nu_-)$ (Theorem 2.3.4), we conclude that $m = 1$ and $n = 2$. The vectors u_1, u_2 that span $\mathcal{Q}_1$ and $\mathcal{Q}_2$ correspondingly can be normed so that

$$[u_1, u_1] = [u_2, u_2] = 0, \quad [u_1, u_2] = 1. \qquad (8.4.20)$$

Thus, we found a basis $\{u_1, u_2\}$ of C^2 with the properties (8.4.20) such that the matrix of N is as in (8.4.15). According to Corollary 8.3.2, the linear transformation N is unitarily similar to the canonical form (8.4.15).

Now it remains to consider the case when N has only one eigenvalue λ. Then $N^{[*]}$ has only one eigenvalue $\overline{\lambda}$. Define the subspace $\mathcal{S}_0 \subseteq \mathsf{C}^n$ as

$$\mathcal{S}_0 = \left\{ x \in \mathsf{C}^n \,:\, (N - \lambda I)x = (N^{[*]} - \overline{\lambda}I)x = \{0\} \right\}.$$

In other words, $\mathcal{S}_0$ is the subspace spanned by all eigenvectors common to N and $N^{[*]}$. Since N and $N^{[*]}$ commute, we have $\dim \mathcal{S}_0 > 0$. We consider the following cases and subcases, depending on the properties of $\mathcal{S}_0$.

Case 1. $\mathcal{S}_0$ is not neutral. Then there exists a vector $v \in \mathsf{C}^n$ such that

$$Nv = \lambda v, \quad N^{[*]}v = \overline{\lambda}v, \quad [v, v] \ne 0.$$

The one-dimensional subspace $\mathcal{V} := \mathrm{Span}\,\{v\}$ is nondegenerate and invariant for both N and $N^{[*]}$. Hence, N is indecomposable if and only if $\mathcal{S}_0$ is equal to the space C^n $(n = 1)$ and the case is one-dimensional.

Case 2. $\mathcal{S}_0$ is neutral. Then

$$\dim \mathcal{S}_0 = 1 \qquad\qquad (8.4.21)$$

because H has only one negative eigenvalue. The subspace $\mathcal{S}_0^{[\perp]}$ is of dimension $n - 1$ and $\mathcal{S}_0 \subseteq \mathcal{S}_0^{[\perp]}$.

Case 3. (Subcase of Case 2). $\mathcal{S}_0 = \mathcal{S}_0^{[\perp]}$. We have $n = 2$. Let $\{v_1, v_2\}$ be a basis of C^2 such that v_1 spans $\mathcal{S}_0$ and $[v_1, v_2] \neq 0$. It is easy to find a vector $\tilde{v}_2 = \alpha_1 v_1 + \alpha_2 v_2$ such that

$$[v_1, v_1] = [\tilde{v}_2, \tilde{v}_2] = 0, \quad [v_1, \tilde{v}_2] = 1 \quad (v_1 \in \mathcal{S}_0).$$

Then in the basis $\{v_1, \tilde{v}_2\}$ the matrix of N is

$$N = \begin{bmatrix} \lambda & \mu \\ 0 & \lambda \end{bmatrix}$$

where $0 \neq \mu \in \mathsf{C}$. Let $u_1 = \sqrt{|\mu|}v_1, u_2 = \frac{1}{\sqrt{|\mu|}}\tilde{v}_2$. Then we have

$$[u_1, u_1] = [u_2, u_2] = 0, \quad [u_1, u_2] = 1.$$

In the basis $\{u_1, u_2\}$ the matrix of N is as in (8.4.16) with $z = \mu/|\mu|$.

According to Corollary 8.3.2, the linear transformation N is unitarily similar to the canonical form (8.4.16). Note that the value of z in (8.4.16) is uniquely defined, i.e., it is an H-unitary invariant of N. To prove this it is sufficient to prove that if

$$N_1 = \begin{bmatrix} \lambda & \tilde{z} \\ 0 & \lambda \end{bmatrix}, \quad |\tilde{z}| = 1$$

and T is a nonsingular 2×2-matrix such that

$$N_1 = T^{-1}NT \qquad\qquad (8.4.22)$$

and

$$H = T^*HT, \qquad\qquad (8.4.23)$$

then $\tilde{z} = z$. To prove this, note first of all that it follows from (8.4.22) that T is upper triangular,

$$T = \begin{bmatrix} t_1 & t_3 \\ 0 & t_2 \end{bmatrix},$$

and

$$zt_2 = \tilde{z}t_1. \qquad\qquad (8.4.24)$$

From (8.4.23) it follows that $t_1\overline{t_2} = 1$. Thus, (8.4.24) can be rewritten as $z = \tilde{z}t_1\overline{t_1}$. The last equality along with $|\tilde{z}| = |z| = 1$ yields $\tilde{z} = z$.

Case 4. (Subcase of Case 2). $\mathcal{S}_0 \subseteq \mathcal{S}_0^{[\perp]}$, $\mathcal{S}_0 \neq \mathcal{S}_0^{[\perp]}$. As above, we can find the vectors $v_1, w \in \mathbb{C}^n$ such that

$$[v_1, v_1] = [w, w] = 0, \quad [v_1, w] = 1, \quad v_1 \in \mathcal{S}_0. \tag{8.4.25}$$

The subspace Span $\{v_1, w\}$ is nondegenerate and so is the subspace

$$\mathcal{S} := (\text{Span } \{v_1, w\})^{[\perp]}.$$

It is easy to check that

$$\text{Range}(N - \lambda I) \subseteq \mathcal{S}_0^{[\perp]}, \quad \text{Range}\left(N^{[*]} - \lambda I\right) \subseteq \mathcal{S}_0^{[\perp]}.$$

Hence, $\mathcal{S}_0^{[\perp]}$ is invariant for both N and $N^{[*]}$, and

$$Nw = \lambda w + v_0, \quad \text{for some} \quad v_0 \in \mathcal{S}_0^{[\perp]}.$$

We have a direct sum decomposition $\mathbb{C}^n = \mathcal{S}_0 + \mathcal{S} + \mathcal{S}_1$, where $\mathcal{S}_1 = \text{Span } \{w\}$, and the corresponding decompositions of N and $N^{[*]}$ are

$$N = \begin{bmatrix} \lambda & * & * \\ 0 & N_1 & * \\ 0 & 0 & \lambda \end{bmatrix}, \quad N^{[*]} = \begin{bmatrix} \bar{\lambda} & * & * \\ 0 & N_2 & * \\ 0 & 0 & \bar{\lambda} \end{bmatrix}. \tag{8.4.26}$$

Note that the subspace $\mathcal{S} + \mathcal{S}_0 = \mathcal{S}_0^{[\perp]}$ is both N-invariant and $N^{[*]}$-invariant. It turns out that $N_1^{[*]} = N_2$. Indeed, let $x, y \in \mathcal{S}$. Then for some complex numbers α and β, we have

$$Nx = N_1 x + \alpha v_1, \quad N^{[*]}y = N_2 y + \beta v_1$$

and hence,

$$[N_1 x, y] = [Nx - \alpha v_1, y] = [Nx, y] = \left[x, N^{[*]}y\right] = [x, N_2 y + \beta v_1] = [x, N_2 y]$$

which means that $N_1^{[*]} = N_2$. Moreover, N_1 is normal (with respect to the indefinite inner product) in $\mathcal{S}$, which follows from the block structure (8.4.26) of N and $N^{[*]}$. Next, since H has only one negative eigenvalue, the inner product on $\mathcal{S}$ inherited from $\mathbb{C}^n(H)$ is positive definite. The linear transformation N_1 is normal and has only one eigenvalue. Hence, $N_1 = \lambda I$, $N_1^{[*]} = \bar{\lambda} I$.

Case 5. (Subcase of Case 4) $\mathcal{S}_0 \neq \mathcal{S}_0^{[\perp]}$, $n = 3$. Denote the vector w by v_3. We can find a vector $v_2 \in \mathcal{S}$ such that $[v_2, v_2] = 1$. So, we have a basis

$$\{v_1, v_2, v_3\} \tag{8.4.27}$$

of C^3 with

$$[v_i, v_j] = \begin{cases} 1 & \text{if } i+j = 4 \\ 0 & \text{if } i+j \neq 4. \end{cases}$$

Because of (8.4.26), the matrices of N and $N^{[*]}$ in the basis (8.4.27) are

$$N = \begin{bmatrix} \lambda & \mu_1 & \mu_3 \\ 0 & \lambda & \mu_2 \\ 0 & 0 & \lambda \end{bmatrix}, \quad N^{[*]} = \begin{bmatrix} \bar{\lambda} & \bar{\mu}_2 & \bar{\mu}_3 \\ 0 & \bar{\lambda} & \bar{\mu}_1 \\ 0 & 0 & \bar{\lambda} \end{bmatrix}.$$

The linear transformation N is H-normal if and only if $\mu_1\bar{\mu}_1 = \mu_2\bar{\mu}_2$. Note that $\mu_1\mu_2 \neq 0$; otherwise we would have

$$(N - \lambda I)v_2 = (N^{[*]} - \bar{\lambda}I)v_2 = 0, \quad v_2 \in S_0, \quad \text{and} \quad \dim S_0 > 1,$$

which contradicts (8.4.21). Thus,

$$\mu_1 = \rho e^{i\theta_1}, \quad \mu_2 = \rho e^{i\theta_2}, \quad \rho \neq 0, \quad 0 \leq \theta_1, \theta_2 < 2\pi.$$

Let

$$w_1 = \rho v_1, \quad w_2 = \pm e^{\frac{i(\theta_2-\theta_1)}{2}} v_2, \quad w_3 = \frac{v_3}{\rho},$$

where the $+$ or $-$ sign is chosen so that the argument of the complex number

$$\pm e^{\frac{i(\theta_2-\theta_1)}{2}} e^{i\theta_1} = \pm e^{\frac{i(\theta_2+\theta_1)}{2}}$$

is in the interval $[0, \pi)$. Obviously, we have

$$[w_i, w_j] = \begin{cases} 1 & \text{if } i+j = 4 \\ 0 & \text{if } i+j \neq 4 \end{cases} \tag{8.4.28}$$

and the matrix of N in the basis $\{w_1, w_2, w_3\}$ is

$$N = \begin{bmatrix} \lambda & z & \mu \\ 0 & \lambda & z \\ 0 & 0 & \lambda \end{bmatrix}$$

where

$$\mu = \frac{\mu_3}{\rho^2} \quad \text{and} \quad z = \pm e^{\frac{i(\theta_1+\theta_2)}{2}}$$

with the $+$ or $-$ sign chosen as explained above. Thus, $|z| = 1$, $0 \leq \arg z < \pi$.
Next we introduce a new basis

$$\{u_1, u_2, u_3\} \tag{8.4.29}$$

of C^3 as follows:

$$u_1 = t_1 w_1, \quad u_2 = t_2 w_2 + t_4 w_1, \quad u_3 = t_3 w_3 + t_5 w_2 + t_6 w_1, \tag{8.4.30}$$

where $t_1, \ldots, t_6$ are parameters to be defined later. In order for the properties (8.4.28) of the indefinite inner products to be preserved, the following equalities have to be satisfied:

$$t_1\overline{t_3} = 1, \quad t_2\overline{t_2} = 1, \quad t_2\overline{t_5} + t_4\overline{t_3} = 0, \quad t_3\overline{t_6} + t_6\overline{t_3} + t_5\overline{t_5} = 0. \tag{8.4.31}$$

We set

$$t_1 = t_2 = t_3 = 1. \tag{8.4.32}$$

In the new basis (8.4.29), with (8.4.30)–(8.4.32) satisfied, the matrix of N becomes

$$N = \begin{bmatrix} \lambda & z & \nu \\ 0 & \lambda & z \\ 0 & 0 & \lambda \end{bmatrix}$$

where $\nu = \mu + (t_5 + \overline{t_5})z$. If $z \neq 1$, we can find the value of t_5 (and the values $t_4 = -\overline{t_5}$, $t_6 = -\frac{1}{2}|t_5|^2$) such that all the equalities in (8.4.31) will hold and $\nu = r \in \mathsf{R}$.

Because of Corollary 8.3.2, the linear transformation N is unitarily similar to the canonical form (8.4.17). If $z = 1$, we can find the value of t_5 (and the values of t_4 and t_6 as above) such that again all the equalities in (8.4.31) will hold and $\nu = ir$, $r \in \mathsf{R}$. In this case, the linear transformation N is similar to the canonical form (8.4.18).

Now we will show that the numbers z and r in the canonical form (8.4.17) are invariants. To prove this, it is sufficient to prove that if

$$\widetilde{N} = \begin{bmatrix} \lambda & \widetilde{z} & \widetilde{r} \\ 0 & \lambda & \widetilde{z} \\ 0 & 0 & \lambda \end{bmatrix}, \quad |\widetilde{z}| = 1, \quad 0 < \arg \widetilde{z} < \pi, \quad \widetilde{r} \in \mathsf{R}$$

and if T is an invertible 3×3-matrix such that

$$\widetilde{N} = T^{-1}NT \tag{8.4.33}$$

and

$$T^*HT = H, \tag{8.4.34}$$

then $\widetilde{z} = z$ and $\widetilde{r} = r$. To show this, we first take into account that from (8.4.33) it follows that the matrix T is upper triangular, say

$$T = \begin{bmatrix} t_1 & t_4 & t_6 \\ 0 & t_2 & t_5 \\ 0 & 0 & t_3 \end{bmatrix},$$

and that

$$\widetilde{z}t_1 - zt_2 = 0, \quad \widetilde{z}t_2 - zt_3 = 0, \quad \widetilde{z}t_4 + \widetilde{r}t_1 - zt_5 - rt_3 = 0. \tag{8.4.35}$$

The matrix equation (8.4.34) is now equivalent to the set of scalar equations in (8.4.31). From the first two equations in (8.4.35) it follows that

$$\frac{\tilde{z}}{z} = \frac{t_2}{t_1} = \frac{t_3}{t_2}$$

which, along with two first equations (8.4.31), yields

$$\frac{\tilde{z}}{z} = \frac{t_2}{t_1} = \frac{\overline{t_2}}{\overline{t_1}}.$$

So, t_2/t_1 is real and, since $|\tilde{z}| = |z| = 1$ and $0 < \arg \tilde{z}$, $\arg z < \pi$, we obtain $t_1 = t_2$. Thus,

$$\tilde{z} = z, \qquad\qquad t_1 = t_2 = t_3.$$

Now the last equation in (8.4.35) can be rewritten as

$$r - \tilde{r} = (t_4 \overline{t_3} - t_5 \overline{t_2})z,$$

or, after the third equation in (8.4.31) is taken into account, as

$$r - \tilde{r} = (t_4 \overline{t_3} + t_3 \overline{t_4})z.$$

Since $\tilde{r} - r$ and $(t_4 \overline{t_3} + t_3 \overline{t_4})$ are both real and z is not, the last equation yields $\tilde{r} = r$.

The proof that r is an invariant in the case when N has the canonical form (8.4.18) is similar.

Case 6. (Subcase of Case 4). $S_0 \neq S_0^{[\perp]}$, $n > 3$. Let v_1 and w be as in (8.4.25). Recall that

$$S_0 = \text{Span } \{v_1\}, \quad S = \text{Span } \{v_1, w\}^{[\perp]}, \quad S_1 = \text{Span } \{w\}.$$

The direct sum decomposition $C^n = S_0 \dotplus S \dotplus S_1$ gives rise to the following block structure of the matrices of N and $N^{[*]}$ (see (8.4.26)):

$$N = \begin{bmatrix} \lambda & M_1 & \beta \\ 0 & \lambda I & M_3 \\ 0 & 0 & \lambda \end{bmatrix}, \quad N^{[*]} = \begin{bmatrix} \overline{\lambda} & M_2 & \tilde{\beta} \\ 0 & \overline{\lambda} I & M_4 \\ 0 & 0 & \overline{\lambda} \end{bmatrix}, \quad \beta, \tilde{\beta} \in C.$$

Note that H is positive definite on S. If $n > 4$, then $\dim S > 2$, and therefore the subspace

$$S_0 \dotplus S_1 \dotplus (\text{Range } M_3 + \text{Range } M_4) \tag{8.4.36}$$

has dimension less than n; however, it is easy to see that the subspace (8.4.36) is H-nondegenerate and invariant for both N and $N^{[*]}$, a contradiction with the indecomposability of N. Thus, we must have $n = 4$.

Next, consider $v_2 := M_3 w \in \mathcal{S}$. The vector v_2 is nonzero (otherwise, the subspace (8.4.36) is 3-dimensional, a contradiction with the irreducibility of N). By scaling v_1 and w, if necessary, we may assume that $[v_2, v_2] = 1$. Let $v_3 \in \mathcal{S}$ be such that $[v_2, v_3] = 0$ and $[v_3, v_3] = 1$. Then, in the basis $\{v_1, v_2, v_3, v_4\}$, where $v_4 = w$, the matrices of N and H are

$$
N = \begin{bmatrix} \lambda & \mu_2 & \mu_3 & \beta \\ 0 & \lambda & 0 & 1 \\ 0 & 0 & \lambda & 0 \\ 0 & 0 & 0 & \lambda \end{bmatrix}, \quad H = \begin{bmatrix} 0 & 0 & 0 & 1 \\ 0 & 1 & 0 & 0 \\ 0 & 0 & 1 & 0 \\ 1 & 0 & 0 & 0 \end{bmatrix}
$$

where $\mu_2, \mu_3, \beta \in \mathbb{C}$. Consequently, $N^{[*]}$ in the same basis takes the form

$$
N^{[*]} = \begin{bmatrix} \overline{\lambda} & 1 & 0 & \overline{\beta} \\ 0 & \overline{\lambda} & 0 & \overline{\mu_2} \\ 0 & 0 & \overline{\lambda} & \overline{\mu_3} \\ 0 & 0 & 0 & \overline{\lambda} \end{bmatrix}.
$$

Next we show that by a suitable transformation we can replace β by zero. Indeed, let

$$
T := \begin{bmatrix} 1 & \beta & 0 & y \\ 0 & 1 & 0 & -\overline{\beta} \\ 0 & 0 & 1 & 0 \\ 0 & 0 & 0 & 1 \end{bmatrix}, \quad y \in \mathbb{C} \quad \text{satisfies} \quad 2\Re y + |\beta|^2 = 0.
$$

One verifies by a straightforward computation that $T^* H T = H$ (i.e., T is H-unitary), and

$$
T^{-1} \begin{bmatrix} \lambda & \mu_2 & \mu_3 & \beta \\ 0 & \lambda & 0 & 1 \\ 0 & 0 & \lambda & 0 \\ 0 & 0 & 0 & \lambda \end{bmatrix} T = \begin{bmatrix} \lambda & \mu_2 & \mu_3 & 0 \\ 0 & \lambda & 0 & 1 \\ 0 & 0 & \lambda & 0 \\ 0 & 0 & 0 & \lambda \end{bmatrix}.
$$

Thus, we assume

$$
N = \begin{bmatrix} \lambda & \mu_2 & \mu_3 & 0 \\ 0 & \lambda & 0 & 1 \\ 0 & 0 & \lambda & 0 \\ 0 & 0 & 0 & \lambda \end{bmatrix}, \quad N^{[*]} = \begin{bmatrix} \overline{\lambda} & 1 & 0 & 0 \\ 0 & \overline{\lambda} & 0 & \overline{\mu_2} \\ 0 & 0 & \overline{\lambda} & \overline{\mu_3} \\ 0 & 0 & 0 & \overline{\lambda} \end{bmatrix}, \quad H = \begin{bmatrix} 0 & 0 & 0 & 1 \\ 0 & 1 & 0 & 0 \\ 0 & 0 & 1 & 0 \\ 1 & 0 & 0 & 0 \end{bmatrix}
$$

in some basis $\{\tilde{v}_1, \tilde{v}_2, \tilde{v}_3, \tilde{v}_4\}$.

Since N is H-normal, we have $\mu_2 \overline{\mu_2} + \mu_3 \overline{\mu_3} = 1$ and, therefore,

$$
\mu_2 = \cos \alpha \, e^{i\phi_2}, \quad \mu_3 = \sin \alpha \, e^{i\phi_3}
$$

for appropriate $0 < \alpha \leq \pi/2$, $\phi_2, \phi_3 \in \mathbb{R}$. (If $\alpha = 0$, the linear transformation N is decomposable because v_3 is an eigenvector of both N and $N^{[*]}$ and $\dim \mathcal{S}_0 > 1$.) The vectors

$$u_1 = e^{\frac{i\phi_2}{2}}\widetilde{v}_1, \quad u_2 = e^{\frac{-i\phi_2}{2}}\widetilde{v}_2, \quad u_3 = e^{i\left(\frac{\phi_2}{2}-\phi_3\right)}\widetilde{v}_3, \quad u_4 = e^{i\frac{\phi_2}{2}}\widetilde{v}_4$$

satisfy

$$[u_i, u_j] = \begin{cases} 1 & \text{if } (i,j) = (1,4) \text{ or } (4,1) \text{ or } (2,2) \text{ or } (3,3) \\ 0 & \text{otherwise.} \end{cases}$$

Moreover, in the basis consisting of these vectors, the matrix of N is as in (8.4.19). Applying Corollary 8.3.2, we conclude that N is unitarily similar to the canonical form (8.4.19).

Now we will prove that the number α in (8.4.19) is an invariant. To prove this, it is sufficient to prove that if

$$N_1 = \begin{bmatrix} \lambda & \cos\widetilde{\alpha} & \sin\widetilde{\alpha} & 0 \\ 0 & \lambda & 0 & 1 \\ 0 & 0 & \lambda & 0 \\ 0 & 0 & 0 & \lambda \end{bmatrix}, \quad 0 < \widetilde{\alpha} \leq \frac{\pi}{2} \tag{8.4.37}$$

and $T = (t_{jk})$ is a nonsingular matrix such that

$$N_1 = T^{-1}NT \tag{8.4.38}$$

and

$$T^*HT = H, \tag{8.4.39}$$

then $\widetilde{\alpha} = \alpha$. To prove this, we first observe that from (8.4.38) it follows that T is upper triangular with $t_{44} = t_{22}$. Comparing the $(1,4)$- and $(2,2)$- entries of the matrices that appear in both sides of (8.4.39), we conclude that $t_{11} = t_{22} \neq 0$. Since the $(1,2)$-entry of the matrix $0 = TN_1 - NT$ (see (8.4.38)) is $t_{11}(\cos\widetilde{\alpha} - \cos\alpha)$ and $0 < \widetilde{\alpha}$, $\alpha \leq \frac{\pi}{2}$, it follows that $\cos\widetilde{\alpha} = \cos\alpha$ and, hence, $\widetilde{\alpha} = \alpha$.

This completes the proof of Theorem 8.4.1.   □

It is interesting to note that the Jordan form of N in (8.4.19) consists of one three-dimensional block and one one-dimensional block, if $0 < \alpha < \frac{\pi}{2}$, and consists of two two-dimensional blocks, if $\alpha = \frac{\pi}{2}$.

8.5   Exercises

1. Let $A \in \mathbb{C}^{n \times n}$ be a normal linear transformation (with respect to the *standard* inner product $(\cdot, \cdot)$). Which of the following statements are correct?

   (a) There exists an orthogonal basis such that A is diagonal.

(b) There exists a basis in which A is diagonal.

2. Solve the preceding exercise for a linear transformation which is normal with respect to an indefinite inner product.

3. (a) If $A \in \mathbb{C}^{n \times n}$ is diagonal in a suitable basis of $\mathbb{C}^n$, how can the standard inner product in $\mathbb{C}^n$ be changed so that A is normal with respect to the new inner product?

   (b) Can the new inner product be positive definite?

4. Let $A \in \mathbb{C}^{n \times n}$ have a block diagonal form. In each of the following cases, does there exist an indefinite inner product in $\mathbb{C}^n$ such that A is normal with respect to this inner product?

   (a) All diagonal blocks are of size at most 2.

   (b) All diagonal blocks are of size 2.

   (c) What is the smallest possible $\nu_-(H)$ in both cases (a) and (b), where the invertible hermitian matrix H induces the inner product?

   (d) If all diagonal blocks are of size at most two, except one which is a Jordan block of size 3, what is the smallest possible $\nu_-(H)$?

   (e) If there are only two Jordan blocks both of size 4 on the diagonal of A, whereas all other diagonal blocks are of size 1, what is the smallest possible $\nu_-(H)$?

5. Let A be an H-normal linear transformation in $\mathbb{C}^n$ and let the Jordan form of A consist of one Jordan block.

   (a) Prove that there exists a basis of $\mathbb{C}^n$ in which H and A have the following representation:

   $$H = \varepsilon \begin{bmatrix} 0 & 0 & \cdots & 0 & 1 \\ 0 & 0 & \cdots & 1 & 0 \\ \cdots & \cdots & \cdots & \cdots & \cdots \\ 1 & 0 & \cdots & 0 & 0 \end{bmatrix}, \quad \varepsilon = \pm 1,$$

   $$A = \lambda I + e^{i\theta} \begin{bmatrix} 0 & -i & c_2 & \cdots & c_{n-1} \\ 0 & 0 & -i & \cdots & c_{n-2} \\ \cdots & \cdots & \cdots & \cdots & \cdots \\ 0 & 0 & \cdots & \cdots & 0 \end{bmatrix}$$

   with $c_1, c_2, \ldots, c_{n-1} \in \mathbb{R}$; $0 \le \theta < \pi$.

   (b) Show that this representation is unique.

   (c) When is A H-selfadjoint?

   (d) When is A H-unitary?

6. Let the Jordan form of a decomposable H-normal linear transformation on $\mathbb{C}^{2n}$ consist of two equal blocks with different eigenvalues $\lambda \in \mathbb{C}$ and $\mu \in \mathbb{C}$. Assume that either

$$|\lambda| < |\mu|$$

or

$$|\lambda| = |\mu| \quad \text{and} \quad \arg \lambda < \arg \mu \quad (0 \leq \arg \lambda, \; \arg \mu < 2\pi).$$

(a) Prove that there exists a basis in $\mathbb{C}^{2n}$ in which the matrices H and A have the following forms:

$$H = \begin{bmatrix} 0 & S_n \\ S_n & 0 \end{bmatrix}, \quad N = \begin{bmatrix} C_1 & 0 \\ 0 & C_2 \end{bmatrix},$$

where S_n is the $n \times n$ sip matrix, and either

$$C_1 = \lambda I + e^{i\theta} \begin{bmatrix} 0 & -i & c_2 & \cdots & c_{n-1} \\ 0 & 0 & -i & \cdots & c_{n-2} \\ \cdots & \cdots & \cdots & \cdots & \cdots \\ 0 & 0 & 0 & \cdots & 0 \end{bmatrix},$$

$$C_2 = \mu I + e^{i\theta} \begin{bmatrix} 0 & -ri & \overline{c_2} & \cdots & \overline{c_{n-1}} \\ 0 & 0 & -ri & \cdots & \overline{c_{n-2}} \\ \cdots & \cdots & \cdots & \cdots & \cdots \\ 0 & 0 & 0 & \cdots & 0 \end{bmatrix},$$

where

$$0 \leq \theta < \pi, \quad r > 0, \quad r \neq ie^{-i\theta}, \quad c_2, \ldots, c_{n-1} \in \mathbb{C},$$

or

$$C_1 = \lambda I + \begin{bmatrix} 0 & 1 & c_2 & \cdots & c_{n-1} \\ 0 & 0 & 1 & \cdots & c_{n-2} \\ \cdots & \cdots & \cdots & \cdots & \cdots \\ 0 & 0 & 0 & \cdots & 0 \end{bmatrix},$$

$$C_2 = \mu I + \begin{bmatrix} 0 & 1 & -\overline{c_2} & \cdots & -\overline{c_{n-1}} \\ 0 & 0 & 1 & \cdots & -\overline{c_{n-2}} \\ \cdots & \cdots & \cdots & \cdots & \cdots \\ 0 & 0 & 0 & \cdots & 0 \end{bmatrix},$$

where $c_2, c_3, \ldots, c_{n-1} \in \mathbb{C}$.

(b) Prove that the representation of A and H as in part (a) is unique.

7. Let an $n \times n$ matrix A have the form

$$A = \lambda \begin{bmatrix} 1 & c_1 & c_2 & \cdots & c_{n-1} \\ 0 & 1 & c_1 & \cdots & c_{n-2} \\ \cdots & \cdots & \cdots & \cdots & \cdots \\ 0 & 0 & 0 & \cdots & 1 \end{bmatrix}, \quad \lambda \in \mathbb{C}, \quad |\lambda| = 1.$$

Verify that A is H-unitary, where H is the $n \times n$ sip matrix, if and only if the following equalities hold:

$$c_1 + \overline{c_1} = 0$$

$$c_2 + \overline{c_2} + c_1\overline{c_1} = 0$$

$$\cdots\cdots\cdots$$

$$c_{n-1} + \overline{c_{n-1}} + c_1\overline{c_{n-2}} + \cdots + c_{n-2}\overline{c_1} = 0.$$

8. Consider matrices $A, B \in \mathbb{C}^{n \times n}$ for which $B = \lambda I + \mu A$, where $\lambda \in \mathbb{C}$, $\mu \in \mathbb{C} \setminus \{0\}$, and let H be an invertible hermitian matrix.

   (a) Show that A is H-normal if and only if B is H-normal.

   (b) Show that A is indecomposable and H-normal if and only if B has the same properties.

   (c) Assuming that H has only one negative eigenvalue, and that A is indecomposable and H-normal - having the canonical form of Theorem 8.4.1, find the canonical form of B.

9. Construct an example of an H_0-normal matrix which is neither H-selfadjoint nor H-unitary, for any invertible hermitian matrix H.

8.6 Notes

In this chapter we follow the paper [43]. See also [44], [45]. For further developments on classification of indecomposable H-normal matrices, see [51], [52], [50]. For other recently developed aspects of the theory of H-normal matrices, see [79], [85], [84].

As shown in [28], the problem of classification up to simultaneous similarity of pairs of commuting matrices discussed in Section 8.2 is equivalent to the problem of classification of k-tuples of arbitrary square matrices of appropriate size up to simultaneous similarity. For discussion of complexity of the latter problem, using methods of linear algebra, see [8].

Chapter 9

General Perturbations. Stability of Diagonalizable Matrices

If A is an H-selfadjoint matrix, a "general perturbation" of the pair (A, H) results in a pair (B, G), in which B is G-selfadjoint and is close to the unperturbed pair (A, H) in an appropriate sense. A similar convention applies to the perturbations of H-unitary matrices considered here.

Identification of a quantity which is invariant under such perturbations is one of the main results of the chapter. This general theorem will admit the characterization of all diagonalizable H-selfadjoint matrices with real spectrum which retain these properties after a general perturbation. Also a description of those cases in which analytic perturbations of H-selfadjoint matrices retain spectral properties which are familiar from the classical hermitian case is obtained. Analogous results for perturbations of H-unitary matrices are also discussed.

9.1 General Perturbations of H-Selfadjoint Matrices

Recall that the signature, sig H, of a hermitian matrix H is defined as the difference between the number of positive eigenvalues of H and the number of negative eigenvalues of H (in both cases counting multiplicities). Zero eigenvalues of H, if any, are not counted.

For a given $n \times n$ matrix A and $\lambda \in \mathsf{C}$, let

$$\mathcal{R}_\lambda(A) = \{x \in \mathsf{C}^n \mid (A - \lambda I)^n x = 0\}.$$

So $\mathcal{R}_\lambda(A) \neq \{0\}$ if and only if λ is an eigenvalue of A, and in this case $\mathcal{R}_\lambda(A)$ is the root subspace of A corresponding to λ. The orthogonal projection onto $\mathcal{R}_\lambda(A)$ is denoted by $P_{\mathcal{R}_\lambda(A)}$.

Observe that if $\lambda \in \sigma(A)$ is real and A is H-selfadjoint, then $P_{\mathcal{R}_\lambda(A)} H P_{\mathcal{R}_\lambda(A)}$ determines the quadratic form on $\mathcal{R}_\lambda(A)$ associated with a restriction of H. The main theorem shows that an invariant of general perturbations of the pair (A, H) is determined by the signatures of quadratic forms of this kind.

Theorem 9.1.1. *Let A be H-selfadjoint and $\Omega \subseteq \mathrm{R}$ be any open set such that the boundary of Ω does not intersect $\sigma(A)$. Then for some sufficiently small neighborhoods U_A of A and U_H of H,*

$$\sum_{\lambda \in \Omega} \mathrm{sig}\,\left(P_{\mathcal{R}_\lambda(A)} H P_{\mathcal{R}_\lambda(A)}\right) = \sum_{\mu \in \Omega} \mathrm{sig}\,\left(P_{\mathcal{R}_\mu(B)} G P_{\mathcal{R}_\mu(B)}\right) \qquad (9.1.1)$$

for every $B \in U_A$ which is G-selfadjoint for an invertible selfadjoint $G \in U_H$.

Moreover, for such a B, the number $\nu_\Omega(B)$ of eigenvalues in Ω (counting multiplicities), satisfies the inequality

$$\nu_\Omega(B) \ge \sum_{\lambda \in \Omega} \left|\mathrm{sig}\,\left(P_{\mathcal{R}_\lambda(A)} H P_{\mathcal{R}_\lambda(A)}\right)\right|, \qquad (9.1.2)$$

and in every neighborhood $U \subseteq U_A$ of A there exists an H-selfadjoint matrix B for which equality holds in (9.1.2).

Proof. We first prove (9.1.1). Evidently, it is sufficient to consider the case

$$\Omega = \{\mu \in \mathrm{R} \mid \mu_1 < \mu < \mu_2\},$$

where μ_1, $\mu_2 \notin \sigma(A)$.

Let us compute the signature of the hermitian matrix $\mu H - HA$ where $\mu \in \mathrm{R} \setminus \sigma(A)$. Passing to the canonical form $(J, P_{\varepsilon,J})$ of (A, H) (Theorem 5.1.1) one sees easily that

$$\mathrm{sig}(\mu H - HA) = \sum_{\lambda \in \sigma(A) \cap \mathrm{R}} \mathrm{sgn}(\mu - \lambda) \sum_{i=1}^{k(\lambda)} \frac{\varepsilon_i(\lambda)}{2}[1 - (-1)^{m_i(\lambda)}], \qquad (9.1.3)$$

where $m_1(\lambda), \ldots, m_{k(\lambda)}(\lambda)$ are the sizes of Jordan blocks in J with eigenvalue λ, and $\varepsilon_1(\lambda), \ldots, \varepsilon_{k(\lambda)}(\lambda)$ are the corresponding signs in the sign characteristic of (A, H). As usual, $\mathrm{sgn}(\mu - \lambda) = 1$ if $\mu - \lambda > 0$ and $\mathrm{sgn}(\mu - \lambda) = -1$ if $\mu - \lambda < 0$. From (9.1.3) we find that

$$\mathrm{sig}(\mu_2 H - HA) - \mathrm{sig}(\mu_1 H - HA) = \sum_{\mu_1 < \lambda < \mu_2,\ \lambda \in \sigma(A)} \sum_{i=1}^{k(\lambda)} \varepsilon_i(\lambda)[1 - (-1)^{m_i(\lambda)}],$$

$$(9.1.4)$$

which is equal to $\sum_{\mu_1 < \lambda < \mu_2} \mathrm{sig}(P_{\mathcal{R}_\lambda(A)} H P_{\mathcal{R}_\lambda(A)})$. Since the signature of an invertible hermitian matrix is constant under small perturbations (Theorem A.1.2(b)), there exist neighborhoods U_A of A and U_H of H such that, for every G-selfadjoint $B \in U_A$ with $G \in U_H$, we have:

$$\mathrm{sig}(\mu_i H - HA) = \mathrm{sig}(\mu_i G - GB), \quad i = 1, 2.$$

For such B and G (9.1.1) follows from (9.1.4).

Inequality (9.1.2) is a direct consequence of (9.1.1). Indeed, the canonical form of B shows that

$$\nu_\Omega(B) \geq \sum_{\mu \in \Omega} \left| \mathrm{sig}(P_{\mathcal{R}_\mu(B)} G P_{\mathcal{R}_\mu(B)}) \right|.$$

To prove the last part of the theorem let $(J, P_{\varepsilon,J})$ be the canonical form of (A, H). Consider the part J_0 of the Jordan matrix J which corresponds to a fixed real eigenvalue λ_0 of A and let P_{ε,J_0} be the corresponding part of $P_{\varepsilon,J}$. Clearly, it is sufficient to find a P_{ε,J_0}-selfadjoint matrix K in every neighborhood of J_0 with the property that the number of real eigenvalues of K is exactly $|\mathrm{sig}P_0 H P_0|$, where $P_0 = P_{\mathcal{R}_{\lambda_0}}(J)$. To simplify notation assume that $\lambda_0 = 0$.

Consider first the construction of K in three particular cases:

(i) J_0 is a Jordan block of even size α;

(ii) J_0 is a Jordan block of odd size α;

(iii) $J_0 = J_1 \oplus J_2$ consists of two Jordan blocks J_1 and J_2 of odd sizes α_1 and α_2, respectively, and with opposite signs in the sign characteristic.

Denote by $J(\pm i, \beta)$ the Jordan block of size β with eigenvalue $\pm i$, and let ξ be a small positive number. In case (i) put $K = J_0 + \xi \left(\mathrm{diag} \left(J \left(i, \frac{\alpha}{2} \right), J \left(-i, \frac{\alpha}{2} \right) \right) \right)$. It is easy to check that K is P_{ε,J_0}-selfadjoint with all eigenvalues nonreal.

In case (ii) put

$$K = J_0 + \xi \left(\mathrm{diag} \left(J \left(i, \frac{\alpha - 1}{2} \right), 0, J \left(-i, \frac{\alpha - 1}{2} \right) \right) \right);$$

then K is P_{ε,J_0}-selfadjoint with exactly $\alpha - 1$ nonreal eigenvalues.

In case (iii) put

$$K = J_0 + \xi \left(\mathrm{diag} \left(J \left(i, \frac{\alpha_1 - 1}{2} \right), \begin{bmatrix} 1 & 0 & 0 & 1 \\ 0 & J\left(-i, \frac{\alpha_1-1}{2}\right) & 0 & 0 \\ 0 & 0 & J\left(i, \frac{\alpha_2-1}{2}\right) & 0 \\ -1 & 0 & 0 & 1 \end{bmatrix}, J \left(-i, \frac{\alpha_2 - 1}{2} \right) \right) \right),$$

then K is P_{ε,J_0}-selfadjoint, and

$$\det(\lambda I - K) = \left(\lambda^2 + \xi^2 \right)^{\frac{1}{2}(\alpha_1 + \alpha_2) - 1} \left[(\lambda - \xi)^2 + \xi^2 \right],$$

so that all eigenvalues of K are nonreal.

In the general case we apply the construction of case (i) to each Jordan block of J_0 of even size, the construction of case (iii) to each pair of Jordan blocks of J_0 of odd size and different signs, and if Jordan blocks of odd size are left, we apply the construction of case (ii) to each of them. It is easily seen that in this way we produce a P_{ε,J_0}-selfadjoint matrix K in every neighborhood of J_0 such that $\nu_\mathbb{R}(K) = |\mathrm{sig}P_0 H P_0|$. This completes the proof. □

The following special case of Theorem 9.1.1 is noteworthy. For a real eigenvalue λ_0 there is an associated set of signs $\varepsilon(\lambda_0) \subseteq \varepsilon$, the sign characteristic of (A, H). The statement of the corollary concerns the number $k(\lambda_0)$ which is, by definition, the minimum of the number of positive signs in $\varepsilon(\lambda_0)$ and the number of negative signs in $\varepsilon(\lambda_0)$.

Corollary 9.1.2. *Let A be H-selfadjoint and assume that all partial multiplicities of A corresponding to λ_0 are equal to 1. Then for every $\delta > 0$ there exist neighborhoods U_A of A and U_H of H such that, for every pair $(B, G) \in U_A \times U_H$ with B G-selfadjoint, the number $s_c(B)$ of nonreal eigenvalues of B in the disc $\{\lambda \mid |\lambda - \lambda_0| < \delta\}$ does not exceed $2k(\lambda_0)$, counting multiplicities.*
  Moreover, in every neighborhood U of A contained in U_A there exists an H-selfadjoint matrix B such that $s_c(B) = 2k(\lambda_0)$.

We conclude this section with two illustrative examples.

Example 9.1.3. *Let $A = J_n(0)$ be the $n \times n$ nilpotent Jordan block $(n \geq 2)$, and let $H = \varepsilon S_n$, the $n \times n$ sip matrix. Consider a small εS_n-selfadjoint perturbation $B = A + \delta e_n e_1^T$, where δ is a positive number close to zero. Thus, B is obtained from A by adding δ in the left bottom corner. It is easy to see that B has eigenvalues*

$$\sqrt[n]{\delta} e^{2m\pi i/n}, \quad m = 0, 1, \ldots, n-1,$$

with the corresponding eigenvectors

$$\langle 1, \sqrt[n]{\delta} e^{2m\pi i/n}, (\sqrt[n]{\delta})^2 e^{2(2m)\pi i/n}, \cdots, (\sqrt[n]{\delta})^{n-1} e^{2((n-1)m\pi i/n)}\rangle,$$

for $m = 0, 1, \ldots, n-1$. If Ω is an open set on the real line that contains zero, we have

$$\sum_{\mu \in \Omega} \text{sig}\, \left(P_{\mathcal{R}_\mu(B)}(\varepsilon S_n)P_{\mathcal{R}_\mu(B)}\right) = \begin{cases} \varepsilon & \text{if}\quad n \quad \text{is odd} \\ 0 & \text{if}\quad n \quad \text{is even}. \end{cases}$$

This is equal to $\text{sig}\, \varepsilon S_n$, as required by Theorem 9.1.1. □

Example 9.1.4. *Let*

$$A = 0, \quad H = \begin{bmatrix} I_p & 0 \\ 0 & -I_q \end{bmatrix} \in \mathbb{C}^{n \times n}.$$

For a fixed matrix $C \in \mathbb{C}^{p \times q}$, consider the H-selfadjoint perturbation of A:

$$B = \begin{bmatrix} 0 & \varepsilon C \\ -\varepsilon C^* & 0 \end{bmatrix} \in \mathbb{C}^{n \times n}, \quad \varepsilon > 0 \quad \text{small}.$$

Since B is skew-hermitian, zero is the only possible real eigenvalue of B, and then

$$\mathcal{R}_0(B) = \begin{bmatrix} \text{Ker}\, C^* \\ \text{Ker}\, C \end{bmatrix}.$$

Let $k = \operatorname{rank} C = \operatorname{rank} C^*$. Then $\mathcal{R}_0(B)$ has dimension $(p - k) + (q - k)$, the $(p - k)$-dimensional subspace $\operatorname{Ker} C^*$ is H-positive, and the $(q - k)$-dimensional subspace $\operatorname{Ker} C$ is H-negative. Thus,

$$\operatorname{sig}\left(P_{\mathcal{R}_0(B)} H P_{\mathcal{R}_0(B)}\right) = p - q = \operatorname{sig} H,$$

as asserted by Theorem 9.1.1. □

9.2 Stably Diagonalizable H-Selfadjoint Matrices

Let A be an H-selfadjoint matrix, and let Ω be an open subset of the real line. We say that A is Ω-*diagonalizable* if for every $\lambda_0 \in \Omega \cap \sigma(A)$ the multiplicity of λ_0 as a zero of $\det(I\lambda - A)$ coincides with $\dim \operatorname{Ker}(\lambda_0 I - A)$. In other words, the restriction of A to the spectral subspace corresponding to the eigenvalues of A in Ω is similar to a diagonal matrix.

Next, we need to consider matrices for which all neighboring matrices (with similar symmetries) are also Ω-diagonalizable. More formally, we call matrix A *stably* Ω-*diagonalizable* if there exist neighborhoods U_A of A and U_H of H such that, whenever B is G-selfadjoint and $(B, G) \in U_A \times U_H$ it follows that B has the same number of eigenvalues as A in Ω (counting multiplicities) and is Ω-diagonalizable. Note that, in particular, A must be Ω-diagonalizable.

We will also use the corresponding notion in which the matrix H is kept fixed. Thus, an H-selfadjoint matrix A is called H-*stably* Ω-*diagonalizable* if there exists a neighborhood U_A of A such that every H-selfadjoint matrix B in U_A is Ω-diagonalizable. In the next theorem we assume that Ω is an open subset of the real line such that its boundary does not contain eigenvalues of A.

Theorem 9.2.1. Let A be H-selfadjoint. Then the following statements are equivalent:

(i) A is stably Ω-diagonalizable;

(ii) A is H-stably Ω-diagonalizable;

(iii) the quadratic form (Hx, x) is either positive definite or negative definite on the subspace $\operatorname{Ker}(\lambda_0 I - A)$, for every $\lambda_0 \in \sigma(A) \cap \Omega$.

We shall call the real eigenvalue λ_0 of an H-selfadjoint matrix A *definite* if the quadratic form (Hx, x) is either positive definite or negative definite on the root subspace of A corresponding to λ_0. The canonical form (Theorem 5.1.1) shows that the real eigenvalue λ_0 is definite if and only if the Jordan blocks of A corresponding to λ_0 all have size 1, and the signs in the H-sign characteristic of A corresponding to λ_0 are either all equal to $+1$, or all equal to -1. Thus, statement (iii) above is equivalent to the definiteness of each eigenvalue of A in Ω.

The proof of Theorem 9.2.1 will indicate some additional properties of an H-selfadjoint stably Ω-diagonalizable matrix A. First, there exist neighborhoods

U_A of A, U_H of H such that if B is G-selfadjoint and $(B, G) \in U_A \times U_H$, then B is *stably* Ω-diagonalizable (and not only Ω-diagonalizable as the definition of "stably Ω-diagonalizable" required). Second, if a real eigenvalue λ_0 of A "splits" under the perturbation and produces eigenvalues $\mu_1, \ldots, \mu_r$ of B (all of them real), then the sign of μ_j in the G-sign characteristic of B ($j = 1, \ldots, r$) is just the sign of λ_0 in the H-sign characteristic of A.

Note also that in every neighborhood of an H-selfadjoint matrix which is Ω-diagonalizable but not stably Ω-diagonalizable, there exists an H-selfadjoint B with nonreal eigenvalues.

Proof of Theorem 9.2.1. We start with the part (iii)$\Rightarrow$(i). We are given that each eigenvalue of A in Ω is definite so that, in particular, A is Ω-diagonalizable. Let $\lambda_1 < \cdots < \lambda_r$ be the eigenvalues of A in Ω and let

$$\delta = \min \left\{ \frac{1}{3}(\lambda_2 - \lambda_1), \ldots, \frac{1}{3}(\lambda_r - \lambda_{r-1}), \widetilde{\lambda}_1, \ldots, \widetilde{\lambda}_r \right\}$$

where $\widetilde{\lambda}_i$ is the distance from λ_i to the boundary of Ω.

If B is a perturbation of A, we write P_i (resp., Q_μ) for the orthogonal projection onto the root subspace of A (resp., of B) associated with λ_i (resp., with $\mu \in \mathbb{R}$), and $\nu_i(B)$ for the number of real eigenvalues (counting multiplicities) of B whose distance from λ_i is less than δ.

Using Theorem 9.1.1 neighborhoods U_A of A and U_H of H can be found so that, if $(B, G) \in U_A \times U_H$ and B is G-selfadjoint then, for $i = 1, 2, \ldots, r$,

$$\nu_i(B) \geq \left| \sum \mathrm{sig} \left(Q_\mu G Q_\mu \right) \right| = \left| \mathrm{sig} \left(P_i H P_i \right) \right|, \qquad (9.2.5)$$

and the summation is over all *real* μ whose distance from λ_i is less than δ. Since λ_i is a definite eigenvalue of A, the last term in (9.2.5) is just the dimension of the root subspace $\mathcal{R}_{\lambda_i}(A)$ of λ_i. But for B close to A, since the eigenvalues of B are continuous functions of the entries of B, we obviously have $\nu_i(B) \leq \dim \mathcal{R}_{\lambda_i}(A)$ and so, taking U_A smaller, if necessary, the inequality in (9.2.5) is in fact an equality which means that B is Ω-diagonalizable and all its eigenvalues are definite. Further, the relation $\nu_i(B) = \dim \mathcal{R}_{\lambda_i}(A)$ shows that B has the same number of eigenvalues as A in Ω (counting multiplicities). So A is stably Ω-diagonalizable.

(i)$\Rightarrow$(ii) is evident.

(ii)$\Rightarrow$(iii) Assume that A is Ω-diagonalizable, but its eigenvalue $\lambda_0 \in \Omega$ is not definite. By Theorem 9.1.1, in every neighborhood of A there exists an H-selfadjoint matrix B such that the number of real eigenvalues of B in a neighborhood of λ_0 is less than the dimension of the root subspace $\mathcal{R}_{\lambda_0}(A)$. This means that B has nonreal eigenvalues, a contradiction to (ii).   □

The case $\Omega = \mathbb{R}$ in Theorem 9.2.1 will be of particular interest for us. An H-selfadjoint matrix A is called *diagonalizable with real eigenvalues* (in short, *r-diagonalizable*) if A is similar to a diagonal matrix with real eigenvalues, i.e., A is

Ω-diagonalizable with $\Omega = \mathbb{R}$ and all eigenvalues of A are real. The definition of *stably r-diagonalizable* matrices, and of *H-stably r-diagonalizable* matrices are now evident. The following result is proved in the same way as Theorem 9.2.1.

Theorem 9.2.2. *Let A be H-selfadjoint. Then the following statements are equivalent:*

(i) *A is stably r-diagonalizable;*

(ii) *A is H-stably r-diagonalizable;*

(iii) *all eigenvalues of A are real and definite.*

The remarks concerning additional properties of stably Ω-diagonalizable matrices (stated before the proof of Theorem 9.2.1) apply for the stably r-diagonalizable matrices as well.

9.3 Analytic Perturbations and Eigenvalues

In Section 9.2 we have studied H-selfadjoint matrices A which are stably r-diagonalizable. Observe that in the classical case, when H is positive definite, every H-selfadjoint matrix is stably r-diagonalizable. So stably r-diagonalizable matrices can be viewed as H-selfadjoint matrices which behave like hermitian ones with respect to small perturbations.

Hermitian matrices are noted also for their special properties with respect to analytic perturbations. Namely, if $A(\tau) = \sum_{j=0}^{\infty} \tau^j A_j$ is an analytic matrix function of a real parameter τ with hermitian coefficients A_j, then the eigenvalues of $A(\tau)$ are analytic functions of τ (see, e.g. Theorem 5.11.1 for a particular case of $A(\tau)$ and [58, Section 2.6] for the general situation). Note that in general (i.e., without the assumption of the hermitian property), the eigenvalues of $A(\tau)$ need not be analytic; one can claim only their continuity.

This point of view of analytic perturbations leads naturally to the following definition. Let A_0 be H_0-selfadjoint and let λ_0 be a real eigenvalue of A_0. We say that λ_0 is *analytically extendable* by analytic perturbations if for any pair of matrix functions $A(\tau)$, $H(\tau)$ which are analytic in the real variable τ on a neighborhood U of zero and such that $A(\tau)$ is $H(\tau)$-selfadjoint for all $\tau \in U$ and $A(0) = A_0$, $H(\tau) = H_0$, the eigenvalues of $A(\tau)$ which tend to λ_0 as $\tau \to 0$ can be chosen analytic functions on U.

When H_0 is positive definite, every eigenvalue λ_0 of an H_0-selfadjoint matrix A_0 is analytically extendable. Indeed, if H_0 is not perturbed (i.e., $H(\tau) \equiv H_0$) the result mentioned above ([58, Section 2.6]) applies. The general analytic perturbation $A(\tau), H(\tau)$ is easily reduced to this case by considering an analytic matrix function $S(\tau)$ such that $H(\tau) = S(\tau)^* S(\tau)$ and replacing $A(\tau)$ by $S(\tau)A(\tau)S(\tau)^{-1}$.

It will be clear from Theorem 9.3.1 below (in view of Theorem 9.2.1) that stably r-diagonalizable matrices, and only they, have all eigenvalues analytically

extendable. So, with respect to analytic perturbations as well, the stably r-diago-
nalizable matrices behave like hermitian ones.

Theorem 9.3.1. *Let A_0 be H_0-selfadjoint. Then a real eigenvalue λ_0 of A_0 is analytically extendable if and only if the quadratic form $(H_0 x, x)$ is either positive
definite or negative definite on the subspace $\mathrm{Ker}(\lambda_0 I - A_0)$.*

Proof. Let the quadratic form $(H_0 x, x)$ be either positive or negative definite on
$\mathrm{Ker}(\lambda_0 I - A)$ and let $\triangle$ be an open disc with center λ_0 whose closure intersects
$\sigma(A)$ only in $\{\lambda_0\}$. By Theorem 9.2.1 there exists an $\varepsilon > 0$ such that, for every
G-selfadjoint matrix B such that $||A - B|| + ||H - G|| < \varepsilon$, all the eigenvalues of
B in $\triangle$ are real. Now let $A(\tau), H(\tau)$ be $n \times n$ matrix functions with the properties
described in the definition of the analytic extendability of eigenvalues. Then the
eigenvalues $\lambda_1(\tau), \ldots, \lambda_\nu(\tau)$ of $A(\tau)$ which tend to λ_0 as $\tau \to 0$ are real for τ
sufficiently close to 0 (namely, those τ for which $\lambda_i(\tau) \in \triangle$ and $||A(\tau) - A|| +
||H(\tau) - H|| < \varepsilon$). This implies that, for $j = 1, 2, \ldots, n$, $\lambda_j(\tau)$ is analytic in τ on a
neighborhood of zero. Indeed, $\lambda_j(\tau)$ is a zero of the polynomial $\det(\lambda I - A(\tau))$ with
coefficients analytic in τ and, as such, admits expansion in a series of fractional
powers of τ. More exactly, there exist positive integers $\alpha_1, \ldots, \alpha_m$ such that

$$\alpha_1 + \cdots + \alpha_m = \dim \mathcal{R}_{\lambda_0}(A) =: \nu$$

and (admitting a reordering of $\lambda_j(\tau)$, if necessary)

$$\lambda_p(\tau) = \lambda_0 + \sum_{k=1}^{\infty} c_k^{(q)}(x_p)^k, \quad \alpha_1 + \cdots + \alpha_{q-1} < p \le \alpha_1 + \cdots + \alpha_q \qquad (9.3.6)$$

where

$$
\begin{aligned}
x_p &= x_p(\tau) \\
&= |\tau|^{1/(\alpha_q)} \left\{ \cos\left[\frac{1}{\alpha_q}\left(\arg \tau + 2\pi i \left(p - \sum_{i=1}^{q-1} \alpha_i \right) \right) \right] \right. \\
&\quad \left. + i \sin\left[\frac{1}{\alpha_q}\left(\arg \tau + 2\pi i \left(p - \sum_{i=1}^{q-1} \alpha_i \right) \right) \right] \right\},
\end{aligned}
$$

and where $c_k^{(q)}$ are complex numbers (so x_p is an α_q-th root of τ). By definition,
$\alpha_0 = 0$. For details see [7], for example. Consider

$$\lambda_{\alpha_1}(\tau) = \lambda_0 + \sum_{k=1}^{\infty} c_k^{(1)}(x_{\alpha_1})^k$$

and let k_1 be the smallest index such that $c_{k_1}^{(1)} \ne 0$ (if all $c_k^{(1)} = 0$, then $\lambda_1(\tau) \equiv \lambda_0$
is obviously analytic in τ for $p = 1, \ldots, \alpha_1$). Then

$$c_{k_1}^{(1)} = \lim_{\tau \to +0} \frac{\lambda_1(\tau) - \lambda_0}{(x_{\alpha_1})^{k_1}}. \qquad (9.3.7)$$

We find that, because $\lambda_1(\tau) - \lambda_0$ is real, so is $c_{k_1}^{(1)}$. A similar argument shows that all nonzero $c_k^{(1)}$ are real. As the imaginary part of $\lambda_{\alpha_1}(\tau)$ is zero, we obtain

$$\sum_{k=1}^{\infty} c_k^{(1)} |\tau|^{k/(\alpha_1)} \sin\left(\frac{k}{\alpha_1} \arg \tau\right) = 0$$

for all $|\tau| < \delta$. In particular, for $\tau < 0$ this implies that $c_k^{(1)} \sin\left(\frac{k\pi}{\alpha_1}\right) = 0$ for $k = 1, 2, \ldots$, and means that k is an integer multiple of α_1 if $c_k^{(1)} \neq 0$. So all $\lambda_p(\tau)$, $1 \leq p \leq \alpha_1$, are analytic on a neighborhood of $\tau = 0$. The same argument shows that all $\lambda_1(\tau), \ldots, \lambda_\nu(\tau)$ are analytic in τ on a neighborhood of zero. Hence λ_0 is analytically extendable.

Assume now that the form $(H_0 x, x)$, is neither positive definite nor negative definite on the subspace $\text{Ker}(\lambda_0 I - A_0)$. We shall construct an analytic matrix function $A(\tau)$, $-\infty < \tau < \infty$, which is H_0-selfadjoint and satisfies $A(0) = A_0$, but has a nonanalytic eigenvalue $\lambda_0(\tau)$ which is equal to λ_0 in the limit as $\tau \to 0$.

Without loss of generality we can assume that (A_0, H_0) is in the canonical form. From the condition on the quadratic form $(H_0 x, x)$ it follows that either A_0 has a Jordan block J_0 of size $m \geq 2$ corresponding to the eigenvalue λ_0, or A_0 has two Jordan blocks $J_0 \oplus J_0$, each of size 1 and with opposite signs in the sign characteristic of (A_0, H_0). Consider the first case:

$$A_0 = J_0 \oplus J_1; \quad H_0 = \pm P_0 \oplus P_1,$$

where P_0 is the $m \times m$ sip matrix, and (J_1, P_1) is the rest of (A_0, H_0). Then

$$A(\tau) = \begin{bmatrix} \lambda_0 & 1 & & & 0 \\ 0 & \lambda_0 & \ddots & & \\ \vdots & & \ddots & \ddots & \\ 0 & \cdots & & \ddots & 1 \\ \tau & 0 & & 0 & \lambda_0 \end{bmatrix} \oplus J_1$$

and $\lambda_0(\tau) = \lambda_0 + \tau^{1/m}$ will do, i.e., is a nonanalytic eigenvalue.

In the second case let

$$A_0 = \text{diag}\,(\lambda_0, \lambda_0, J_1); \quad H_0 = \text{diag}\,(1, -1, P_1).$$

Then

$$A(\tau) = \begin{bmatrix} \lambda_0 + 2\tau + \tau^2 & -\tau \\ \tau & \lambda_0 \end{bmatrix} \oplus J_1,$$

and $\lambda_0(\tau) = \lambda_0 + \frac{1}{2}(2\tau + \tau^2 + \tau(4\tau + \tau^2)^{1/2})$ is not analytic on a neighborhood of $\tau = 0$. $\qquad\square$

Several interesting remarks on this theorem are the subjects of Exercises 17, 18, 19, and 20 at the end of this chapter.

The following result concerns a case of nonanalytic behavior.

Theorem 9.3.2. *Let λ_0 be a real eigenvalue of an H_0-selfadjoint matrix A_0. Suppose that the quadratic form $(H_0 x, x)$ is not definite on $\mathrm{Ker}(\lambda_0 I - A_0)$, and let m_+ (resp., m_-) be the number of positive (resp. negative) squares in the canonical representation of the quadratic form $(H_0 x, x)$ on the root subspace $\mathcal{R}_{\lambda_0}(A_0) = \mathrm{Ker}(\lambda_0 I - A_0)^n$. Then a continuous eigenvalue $\lambda_0(\tau)$ of an $H(\tau)$-selfadjoint analytic matrix $A(\tau)$, where $H(0) = H_0$, $A(0) = A_0$ and τ belongs to some real neighborhood of zero, has a fractional power expansion*

$$\lambda_0(\tau) = \lambda_0 + \sum_{j=1}^{\infty} c_j \left(\tau^{1/p}\right)^j, \quad c_j \in \mathbb{C} \tag{9.3.8}$$

with

$$p \le 2 \min(m_+, m_-) + 1. \tag{9.3.9}$$

Observe that the quadratic form $(H_0 x, x)$ is nondegenerate on $\mathcal{R}_{\lambda_0}(A_0)$. (This follows readily on consideration of the canonical form of (A_0, H_0)).

Proof. It is well-known that $\lambda_0(\tau)$ has an expansion (9.3.8) in fractional powers of τ; (i.e., a *Puiseux expansion*, see [7], for example). It remains to prove the estimate (9.3.9) for p.

Assume $\lambda_0(\tau) \not\equiv \lambda_0$, and let φ be the number of nonreal functions $\lambda_0(\tau)$ given by the formula (9.3.8) (in this formula each of the p different values for $\tau^{1/p}$ is admitted). Using arguments employed in the proof of Theorem 9.3.1, it is not difficult to see that $\varphi \ge p - 1$ if p is odd and $\varphi = p$ if p is even. On the other hand, by Corollary 9.1.2, the number of nonreal eigenvalues of $A(\tau)$ (for τ sufficiently close to zero) does not exceed $2 \min(m_+, m_-)$. Hence the inequality (9.3.9) holds. If $(H_0 x, x)$ is either positive or negative definite on $\mathrm{Ker}(\lambda_0 I - A)$, then (9.3.9) gives $p = 1$, i.e., λ_0 is analytically extendable. This is the "if" part of Theorem 9.3.1. □

Example 9.3.3. *Let*

$$A = 0, \quad H = \begin{bmatrix} I_p & 0 \\ 0 & -I_q \end{bmatrix} \in \mathbb{C}^{n \times n}, \quad A(\tau) = \tau C,$$

where C is any fixed H-selfadjoint matrix, and $\tau \in \mathbb{R}$, with $|\tau|$ small. The eigenvalues of $A(\tau)$ are of the form $\tau \lambda_j$, where $\lambda_1, \ldots, \lambda_k$ are the distinct eigenvalues of C. Obviously, the eigenvalues of $A(\tau)$ are analytic functions of τ. On the other hand, by Theorem 9.3.2, there exist H-selfadjoint analytic perturbations of A whose eigenvalues are not analytic (if $p > 0$, $q > 0$). Letting (for simplicity) $p = q = 1$ (and hence $n = 2$), one such perturbation $A(\tau)$ is given by

$$A(\tau) = \begin{bmatrix} \tau & \dfrac{\tau}{2} \\ -\dfrac{\tau}{2} & -\tau^2 \end{bmatrix}.$$

Indeed, the eigenvalues of $A(\tau)$ are

$$\lambda_0(\tau) = \frac{-\tau^2 + \tau \pm \tau\sqrt{2\tau + \tau^2}}{2},$$

which is not analytic in a neighborhood of $\tau = 0$. In fact, $\lambda_0(\tau)$ admits a fractional power series (9.3.8) with $p = 2$. $\qquad\Box$

9.4 Analytic Perturbations and Eigenvectors

Now let λ_0 be a real eigenvalue of an H_0-selfadjoint matrix A_0. We say that the eigenvectors of A_0 corresponding to λ_0 are *analytically extendable* if the following conditions hold. Let $A(\tau)$, $H(\tau)$ be a pair of matrix functions which are analytic in τ on a real neighborhood U of 0, are such that $A(\tau)$ is $H(\tau)$-selfadjoint for all $t \in U$, and satisfy $A(0) = A_0$, $H(0) = H_0$. Also let Γ be a circle with the center λ_0 and radius so small that λ_0 is the only eigenvalue of A_0 inside or on Γ. Then for each real τ sufficiently close to zero there exists an $H(\tau)$-orthonormal basis $x_1(\tau), \ldots, x_k(\tau)$ of eigenvectors of $A(\tau)$ in the subspace $K(\tau)$ defined by $\mathrm{Ker}(\lambda I - A(\tau))$, where the sum is taken over all eigenvalues λ of $A(\tau)$ inside Γ, and the vector functions $x_1(\tau), \ldots, x_k(\tau)$ are analytic functions of τ. Recall that $H(\tau)$-orthonormality of $x_1(\tau), \ldots, x_k(\tau)$ means that $(H(\tau)x_i(\tau), x_j(\tau))$ is equal to 0 if $i \neq j$, and equal to ± 1 if $i = j$. As we shall see shortly, $K(\tau)$ is, in fact, the sum of the root subspaces of $A(\tau)$ corresponding to the eigenvalues of $A(\tau)$ inside Γ.

If the eigenvectors of A_0 corresponding to λ_0 are analytically extendable, then the eigenvalue λ_0 is necessarily analytically extendable. Indeed, assuming the contrary, Theorem 9.3.1 shows that the form (H_0x, x) is neither positive definite nor negative definite on $\mathrm{Ker}(\lambda_0 I - A_0)$. Then, arguing as in the second part of the proof of Theorem 9.3.1, we find that the eigenvectors of A_0 corresponding to λ_0 are not analytically extendable. In particular, the analytic extendability of the eigenvectors of A_0 at λ_0 implies that the number k coincides with the multiplicity of λ_0 as a zero of $\det(\lambda I - A)$. Consequently, all Jordan blocks in the Jordan form of $A(\tau)$ corresponding to the eigenvalues of $A(\tau)$ inside Γ have size 1 (in particular, all Jordan blocks in the Jordan form of A_0 corresponding to λ_0 have size 1).

When H_0 is positive (or negative) definite, it is a well-known fact that the eigenvectors of A_0 are always analytically extendable. If H_0 is not definite then, in general, the analytic extendability of eigenvectors fails. The next theorem gives necessary and sufficient conditions for analytic extendability of eigenvectors. It will be seen from this theorem that the eigenvectors of A_0 corresponding to each eigenvalue are analytically extendable if and only if A_0 is stably r-diagonalizable. So, once again, stably r-diagonalizable matrices behave like hermitian ones.

In fact, it turns out that analytic extendability of λ_0 and that of the eigenvectors corresponding to λ_0 are equivalent.

Theorem 9.4.1. *Let A_0 be H_0-selfadjoint. Then the eigenvectors of A_0 corresponding to $\lambda_0 \in \sigma(A_0) \cap \mathbb{R}$ are analytically extendable if and only if the quadratic form $(H_0 x, x)$ is either positive definite or negative definite on the subspace $\mathrm{Ker}(\lambda_0 I - A_0)$.*

For the proof of Theorem 9.4.1 we need a technical result:

Theorem 9.4.2. *Let $W(t)$ be an analytic $m \times n$ complex matrix-valued function of the real variable t, $a < t < b$. Let*

$$r = \max\{\mathrm{rank}\, W(t) : a < t < b\}.$$

(a) *Assume that $r > 0$. Then there exist vector-valued functions $x_1(t), \ldots, x_r(t) \in \mathbb{C}^m$ which are analytic on (a, b) and have the following properties:*

  (1) *$x_1(t), \ldots, x_r(t)$ are linearly independent for every $t \in (a, b)$;*

  (2) $\mathrm{Span}\,\{x_1(t), \ldots, x_r(t)\} = \mathrm{Range}\, W(t)$
     for every $t \in (a, b)$ except for the set Ξ of isolated points, possibly empty, that consists of exactly those $t \in (a, b)$ for which $\mathrm{rank}\, W(t) < r$;

  (3) *for every $t_0 \in \Xi$,*

$$\mathrm{Span}\,\{x_1(t_0), \ldots, x_r(t_0)\} \supseteq \mathrm{Range}\, W(t_0).$$

(b) *Assume that $r < n$. Then there exist vector-valued functions $y_{r+1}(t), \ldots, y_n(t) \in \mathbb{C}^n$ which are analytic on (a, b) and have the following properties:*

  (1) *$y_{r+1}(t), \ldots, y_n(t)$ are linearly independent for every $t \in (a, b)$;*

  (2) $\mathrm{Span}\,\{y_{r+1}(t), \ldots, y_n(t)\} = \mathrm{Ker}\, W(t)$
     for every $t \in (a, b)$ except for the set Ξ of isolated points, possibly empty, that consists of exactly those $t \in (a, b)$ for which $\mathrm{rank}\, W(t) < r$;

  (3) *for every $t_0 \in \Xi$,*

$$\mathrm{Span}\,\{y_{r+1}(t), \ldots, y_n(t)\} \subseteq \mathrm{Ker}\, W(t_0).$$

Theorem 9.4.2 was proved in [95] in the context of complex analytic operator functions. A proof of the theorem as stated can be found in [39, Chapter S6] or [41]. To illustrate Theorem 9.4.2, consider the following situation:

Example 9.4.3. *Assume that*

$$W(t) = E(t)\,(\mathrm{diag}\,(w_1(t), w_2(t), \ldots, w_r(t), 0, \ldots, 0))\, F(t), \quad a < t < b, \quad (9.4.10)$$

where $E(t)$ and $F(t)$ are analytic matrix functions of sizes $m \times m$ and $n \times n$ respectively, such that

$$\det E(t) \neq 0, \quad \det F(t) \neq 0, \quad \text{for all } t \in (a, b),$$

and $w_1(t), \ldots, w_r(t)$ are analytic scalar functions not identically equal to zero. Then the vectors $x_1(t), \ldots, x_r(t)$ of Theorem 9.4.2 can be taken as the first r columns of the matrix function $E(t)$, and the vectors $y_{r+1}(t), \ldots, y_n(t)$ of Theorem 9.4.2 can be taken as the $n - r$ right most columns of the matrix function $F(t)^{-1}$.

We note in passing that every $m \times n$ analytic matrix function $W(t)$ can be written in the form (9.4.10), although we will not use this fact in the sequel.

Proof of Theorem 9.4.1. In view of the remark preceding the statement of this theorem, and of Theorem 9.4.1, we have only to show that if λ_0 is analytically extendable (or, equivalently, if the quadratic form $(H_0 x, x)$ is definite on $\mathrm{Ker}(\lambda_0 I - A_0)$), then the eigenvectors corresponding to λ_0 are analytically extendable as well.

Let $A(\tau), H(\tau)$ be a pair of matrix functions as in the definition of analytically extendable eigenvectors defined for $|\tau| < \delta$ (δ is a positive number), and let Γ be a small circle with center λ_0. By Theorem 9.2.1 there is a $\delta_1 \in (0, \delta]$ such that for $|\tau| < \delta_1$ all eigenvalues of $A(\tau)$ inside Γ are real, all Jordan blocks of $A(\tau)$ corresponding to these eigenvalues are of size 1, and the quadratic form $(H(\tau)x, x)$ is definite on $\mathrm{Ker}(\lambda(\tau)I - A(\tau))$ for every eigenvalue $\lambda(\tau)$ of $A(\tau)$ inside Γ. In particular, $(H(\tau)x, x) \neq 0$ for every eigenvector x corresponding to $\lambda(\tau)$ ($|\tau| < \delta_1$).

Let $\lambda(\tau)$ be an eigenvalue of $A(\tau)$ which is analytic for $|\tau| < \delta$ and such that $\lambda(0) = \lambda_0$. Choose a nonzero analytic vector function $x(\tau) \in \mathrm{Ker}(\lambda(\tau)I - A(\tau))$, $|\tau| < \delta$ (such an $x(\tau)$ exists in view of Theorem 9.4.2). As we have seen in the preceding paragraph, $(H(\tau)x(\tau), x(\tau)) \neq 0$ for $|\tau| < \delta_1$. Put $x_1(\tau) = |(H(\tau)x(\tau), x(\tau))|^{-1/2} x(\tau)$; then $x_1(\tau)$ is an analytic eigenvector of $A(\tau)$ with $(H(\tau)x_1(\tau), x_1(\tau)) = \pm 1$.

Now consider the $H(\tau)$-orthogonal companion $\mathcal{M}(\tau)$ of Span $\{x_1(\tau)\}$ ($|\tau| < \delta_1$). Since the subspace Span $\{x_1(\tau)\}$ is $H(\tau)$-nondegenerate, $\mathcal{M}(\tau)$ is, in fact, a direct complement to Span $\{x_1(\tau)\}$ in C^n (n is the size of A_0). Moreover, we have

$$\mathcal{M}(\tau) = H(\tau)^{-1}(\mathrm{Span}\ \{x_1(\tau)\}^{\perp}) = H(\tau)^{-1}\left(\mathrm{Ker}\ x_1(\tau)^*\right) = \mathrm{Ker}\ (x_1(\tau)^* H(\tau)).$$

By Theorem 9.4.2 there exists an analytic basis in $\mathcal{M}(\tau)$, and applying the Gram–Schmidt orthogonalization, we obtain an analytic orthonormal basis $y_1(\tau), \ldots, y_{n-1}(\tau)$ in $\mathcal{M}(\tau)$ ($|\tau| < \delta_1$). Consider the linear transformation $A(\tau)\ |_{\mathcal{M}(\tau)}$: $\mathcal{M}(\tau) \to \mathcal{M}(\tau)$ and the quadratic form determined by $P_{\mathcal{M}(\tau)} H(\tau)\ |_{\mathcal{M}(\tau)}$: $\mathcal{M}(\tau) \to \mathcal{M}(\tau)$, where $P_{\mathcal{M}(\tau)}$ is the orthogonal projection on $\mathcal{M}(\tau)$ (note that since $A(\tau)$ is $H(\tau)$-selfadjoint, the subspace $\mathcal{M}(\tau)$ is $A(\tau)$-invariant). Writing these linear transformations in the basis $y_1(\tau), \ldots, y_{n-1}(\tau)$ we obtain $(n-1) \times (n-1)$ matrices $A_1(\tau)$ and $H_1(\tau)$ ($|\tau| < \delta_1$) such that $H_1(\tau)$ is hermitian and invertible, $A_1(\tau)$ is $H_1(\tau)$-selfadjoint, and $A_1(\tau)$ and $H_1(\tau)$ are analytic on τ. (The analyticity of $A_1(\tau)$ follows from the analyticity of the *unique* solution $\{\alpha_{ij}\}_{i,j=1}^{n-1}$ of the system of linear equations

$$A(\tau)y_i(\tau) = \sum_{j=1}^{n-1} \alpha_{ji} y_j(\tau), \quad i = 1, \ldots, n-1$$

with analytic coefficients.) Apply the argument employed in the first part of the proof to produce an analytic eigenvector $x_2(\tau)$ of $A(\tau) |_{\mathcal{M}(\tau)}$ for $|\tau| < \delta_2 \leq \delta_1$ such that $(H(\tau)x_2(\tau), x_2(\tau)) = \pm 1$, and so on. Eventually we obtain the analytic (for $|\tau| < \delta_k < \delta$, where δ_k is positive) $H(\tau)$-orthonormal basis $x_1(\tau), \ldots, x_k(\tau)$ of eigenvectors of $A(\tau)$ in the sum of the root subspaces of $A(\tau)$ corresponding to the eigenvalues inside Γ, i.e., in the subspace Range $\left[\frac{1}{2\pi i} \int_\Gamma (\lambda I - A(\tau))^{-1} d\lambda \right]$. We remark that this subspace is the range of the Riesz projection corresponding to the eigenvalues of $A(\tau)$ inside Γ; see Section A.3 and in particular Proposition A.3.1. Hence the eigenvectors of A_0 corresponding to λ_0 are analytically extendable.   □

The proof of Theorem 9.4.1 also shows the validity of the following statement (and this also follows from Theorem 9.1.1): Let the eigenvectors of an H_0-selfadjoint matrix A_0 corresponding to $\lambda_0 \in \sigma(A) \cap \mathbb{R}$ be analytically extendable, let $A(\tau), H(\tau)$ be as in the definition of analytically extendable eigenvectors, and let $x_1(\tau), \ldots, x_k(\tau)$ be an analytic $H(\tau)$-orthonormal basis of eigenvectors of $A(\tau)$ in the range of the Riesz projection Range $\left[\frac{1}{2\pi i} \int_\Gamma (\lambda I - A(\tau))^{-1} d\lambda \right]$ (such a basis exists by the analytical extendability). Then $(H(\tau)x_i(\tau), x_i(\tau))$ is $+1$ (resp. -1) if the form $(H_0 x, x)$ is positive (resp. negative) definite on $\mathrm{Ker}(\lambda_0 I - A_0)$.

9.5   The Real Case

Consider the important case of an H-selfadjoint matrix A where both A and H are real and pairs (B, G) obtained from perturbations of (A, H) are also confined to real matrices. It is not difficult to see that the results of Sections 9.1 and 9.2 have precise analogues in this context. The proofs are also the same with one exception.

The exception concerns the construction of matrices with nonreal eigenvalues developed in three cases in the final part of the proof of Theorem 9.1.1, and needed to establish the case of equality in the relation (9.1.2). For case (i) the role played by diag $\left(J\left(i, \frac{\alpha}{2}\right), J\left(-i, \frac{\alpha}{2}\right) \right)$ is now played by a block of the real Jordan form:

$$
\begin{bmatrix}
0 & 1 & 1 & 0 & 0 & 0 & & & \\
-1 & 0 & 0 & 1 & 0 & 0 & & & \\
& & 0 & 1 & 1 & 0 & & & \\
& & -1 & 0 & 0 & 1 & & & \\
& & & & & & \ddots & & \\
& & & & & & & 0 & 1 \\
& & & & & & & -1 & 0
\end{bmatrix}
$$

with similar modifications for cases (ii) and (iii).

Using the real analogue of Theorem 9.1.1 the proofs of "real" versions of Theorems 9.2.1 and 9.2.2 are essentially the same.

9.6 Positive Perturbations of H-Selfadjoint Matrices

Let H be an $n \times n$ invertible hermitian matrix, and let A be H-selfadjoint. An eigenvalue λ_0 of A will be called *semi-simple* if $\dim \operatorname{Ker}(\lambda_0 I - A)$ coincides with the multiplicity of λ_0 as a zero of $\det(\lambda I - A)$, or, in other words, if the Jordan blocks of A corresponding to λ_0 all have size 1. If λ_0 is a semi-simple real eigenvalue of A, then the quadratic form (Hx, x) is nondegenerate on $\operatorname{Ker}(\lambda_0 I - A)$, i.e., zero is the only vector $x_0 \in \operatorname{Ker}(\lambda_0 I - A)$ with the property that $(Hx_0, y) = 0$ for all $y \in \operatorname{Ker}(\lambda_0 I - A)$. This follows easily from the canonical form for pairs (A, H) in which A is H-selfadjoint.

Let $r_+(\lambda_0)$ (resp. $r_-(\lambda_0)$) be the number of positive (resp. negative) squares in the canonical representation of the quadratic form (Hx, x) on $\operatorname{Ker}(\lambda_0 I - A)$, where λ_0 is semi-simple. So, in particular,

$$r_+(\lambda_0) + r_-(\lambda_0) = \dim \operatorname{Ker}(\lambda_0 I - A),$$

and is also the algebraic multiplicity of λ_0 as an eigenvalue of A. If $r_-(\lambda_0) = 0$, $(r_+(\lambda_0) = 0)$, we say that λ_0 is a *positive definite (negative definite)* eigenvalue of A. Recall that a matrix A is said to be H-*positive* if $[Ax, x] > 0$ for all $x \neq 0$, where $[x, y] = (Hx, y)$. In other words, A is H-positive if and only if the matrix HA is positive definite.

Theorem 9.6.1. *Let A be H-selfadjoint, and let λ_0 be a semi-simple real eigenvalue of A. Let Γ be any contour such that λ_0 is the only eigenvalue of A inside or on Γ. Then there exists an $\varepsilon > 0$ with the following properties:*

(1) *For every H-positive matrix A_0 with $\|A_0\| < \varepsilon$, the H-selfadjoint matrix $A + A_0$ has exactly $r \overset{\text{def}}{=} r_+(\lambda_0) + r_-(\lambda_0)$ eigenvalues inside Γ (counting multiplicities), and all of them are real.*

(2) *$r_+(\lambda_0)$ of these eigenvalues (counting multiplicities) are greater than λ_0, and $r_-(\lambda_0)$ of them are smaller than λ_0.*

(3) *Every eigenvalue $\widetilde{\lambda}$ of $A + A_0$ inside Γ is semi-simple and positive definite if $\widetilde{\lambda} > \lambda_0$, or negative definite if $\widetilde{\lambda} < \lambda_0$.*

Proof. First observe that λ is an eigenvalue of A if and only if it is an eigenvalue of the hermitian pencil $\lambda H - HA$, in the sense that $\det(\lambda H - HA) = 0$. It will be convenient to prove the theorem in this context of hermitian pencils. Thus, writing $B = HA$ and $B_0 = HA_0$ it will be proved that if B_0 is positive definite (in the classical sense) and $\|B_0\| < \varepsilon$, then the eigenvalues of the linear polynomial $\lambda H - (B + B_0)$ have the properties indicated in the theorem. The conclusions of the theorem will then hold on replacing ε by $\|H^{-1}\|^{-1} \varepsilon$, i.e., $\|B_0\| < \|H^{-1}\|^{-1} \varepsilon$ will imply $\|A_0\| < \varepsilon$.

Since B_0 is positive definite there is an invertible matrix S such that $B_0 = S^* S$ and then we may write

$$\lambda H - (B + B_0) = S^* \{\lambda H' - (B' + I)\} S,$$

where $H' = (S^*)^{-1}HS^{-1}$ and $B' = (S^*)^{-1}BS^{-1}$. This congruence implies that when λ_0 is an eigenvalue of $\lambda H - B$ with associated parameters $r_+(\lambda_0)$, $r_-(\lambda_0)$ as in the statement of the theorem, then it is also an eigenvalue of $\lambda H' - B'$ with the same parameters $r_+(\lambda_0)$, $r_-(\lambda_0)$. Furthermore, with Γ defined as in the theorem, λ_0 is the only eigenvalue of $\lambda H' - B'$, inside or on Γ.

Theorem 5.11.1 shows that $\lambda H' - B'$ admits a representation

$$\lambda H' - B' = U(\lambda)\text{diag } (\mu_1(\lambda), \dots, \mu_n(\lambda)) \, U(\lambda)^*$$

for all $\lambda \in \mathsf{R}$, where $U(\lambda)$ is an analytic matrix function with unitary values $(U(\lambda)U(\lambda)^* = I)$ for $\lambda \in \mathsf{R}$; and $\mu_1(\lambda), \dots, \mu_n(\lambda)$ are real-valued analytic functions of the real variable λ. Since λ_0 is a simple eigenvalue of $\lambda H - B$, and hence of $\lambda H' - B'$, it follows that λ_0 is a simple eigenvalue of $\lambda H - B$, and hence $\lambda H' - B'$, it follows that λ_0 is a simple zero of exactly r functions $\mu_{i_k}(\lambda)$, $k = 1, 2, \dots, r$ among $\mu_1(\lambda), \dots, \mu_n(\lambda)$ and, moreover, exactly $r_+(\lambda_0)$ and $r_-(\lambda_0)$ of the r derivatives $\mu'_{i_k}(\lambda_0)$ are positive, and negative, respectively. It is supposed, for simplicity, that $\mu_1(\lambda) = \cdots = \mu_r(\lambda_0) = 0$ and $\mu'_j(\lambda_0)$ is positive for $j = 1, 2, \dots, r_+(\lambda_0)$, and negative for $j = r_+(\lambda_0) + 1, \dots, r$.

Choose an $\varepsilon > 0$ so that, for $\|B_0\| < \varepsilon$, $\lambda H - (B + B_0)$, and hence $\lambda H' - (B' + I)$, has exactly r eigenvalues inside Γ. By continuity of the eigenvalues as functions of δ the same is true for $\lambda H' - (B' + \delta I)$ for any $\delta \in [0, 1]$. Furthermore, defining $\mu_j(\lambda, \delta) = \mu_j(\lambda) - \delta$, we have

$$\lambda H' - (B' + \delta I) = U(\lambda)\text{diag } (\mu_1(\lambda, \delta), \dots, \mu_n(\lambda, \delta)) \, U(\lambda)^*.$$

Let $[\alpha, \beta]$ be the interval of R consisting of points inside and on Γ for which $\lambda_0 \in (\alpha, \beta)$. Fix j and consider the family of functions $\mu_j(\lambda, \delta)$ defined on $[\alpha, \beta]$; one for each $\delta \in [0, 1]$. Clearly, $\mu_j(\lambda, 0)$ has a simple zero at $\lambda = \lambda_0$ and for any $\delta \in [0, 1]$, $\mu_j(\alpha, \delta) \neq 0$ and $\mu_j(\beta, \delta) \neq 0$. As $\mu_j(\alpha, \delta)$, $\mu_j(\beta, \delta)$ are continuous nonzero functions of $\delta \in [0, 1]$ and

$$\mu_j(\alpha, 0)\mu_j(\beta, 0) < 0 \quad \text{for} \quad j = 1, \dots, r,$$

also

$$\mu_j(\alpha, \delta)\mu_j(\beta, \delta) < 0 \quad \text{for} \quad j = 1, \dots, r \quad \text{and for} \quad \delta \in [0, 1].$$

Thus, for each $\delta \in [0, 1]$, $\mu_j(\lambda, \delta)$ has at least one zero in (α, β). The same applies for each j from 1 to r. Since the total number of zeros (counting multiplicities) of $\mu_j(\lambda, \delta)$, $j = 1, \dots, r$ in the interval $[\alpha, \beta]$ does not exceed r, each function $\mu_j(\lambda, \delta)$ for a fixed j between 1 and r and fixed $\delta \in [0, 1]$ has exactly one simple zero $\lambda_j(\delta)$ in (α, β). In particular, it follows that for any $\delta \in [0, 1]$ the r eigenvalues of $\lambda H - (B + \delta B_0)$ which are inside Γ are all real.

Since $\lambda_j(\delta)$ is a simple zero of $\mu_j(\lambda, \delta)$, $j = 1, 2, \dots, r$, which depends continuously on $\delta \in [0, 1]$, the derivative

$$\frac{d\mu_j}{d\lambda}(\lambda, \delta)\big|_{\lambda_j(\delta)} := \nu_j(\delta)$$

is either positive for all $\delta \in [0,1]$ or negative for all $\delta \in [0,1]$. Since $\nu_j(0) > 0$ for $j = 1,\ldots,r_+(\lambda_0)$ and $\nu_j(0) < 0$ for $j = r_+(\lambda_0)+1,\ldots,r$, we conclude that for any $\delta \in [0,1]$ $\nu_j(\delta) > 0$ for $j = 1,\ldots,r_+(\lambda_0)$ and $\nu_j(\delta) < 0$ for $j = r_+(\lambda_0)+1,\ldots,r$.

Now let $\delta \in (0,1]$ and, since $\mu_j(\lambda_0,\delta) = \mu_j(\lambda_0) - \delta < 0$, it follows that $\lambda_j(\delta) > \lambda_0$ for $j = 1,2,\ldots,r_+(\lambda_0)$ and $\lambda_j(\delta) < \lambda_0$ for $j = r_+(\lambda_0) + 1,\ldots,n$. Putting $\delta = 1$ the theorem is proved. □

Note that the ε appearing in the statement of Theorem 9.6.1 can be estimated. Using [46, Theorem 2.2] it is found that one can use any ε satisfying

$$0 < \varepsilon < (\sup_{\lambda \in \Gamma} ||\lambda H - HA||)^{-1} \cdot ||H^{-1}||^{-1}.$$

A result similar to Theorem 9.6.1 holds for H-negative perturbations of A; in this case the words "greater than" and "less than" in (2), as well as $>$ and $<$ in (3), in the statement of the theorem must be interchanged.

For positive definite H the statement of Theorem 9.6.1 reduces to the following well-known fact (which is not difficult to prove using variational properties of the eigenvalues of a hermitian matrix): Let A be an hermitian $n \times n$ matrix with eigenvalues $\lambda_1 \le \cdots \le \lambda_n$, then the eigenvalues $\mu_1 \le \cdots \le \mu_n$ of the hermitian matrix $A + A_0$ with positive definite A_0 satisfy the inequalities

$$\mu_i > \lambda_i, \quad i = 1,\ldots,n.$$

Example 9.6.2. *Let* $A = 0$, $H = \begin{bmatrix} I_p & 0 \\ 0 & -I_q \end{bmatrix} \in \mathbb{C}^{n \times n}$, $(p + q = n)$. *Then an H-positive matrix has the form*

$$A_0 = \begin{bmatrix} B & C \\ -C^* & -D \end{bmatrix},$$

where $B \in \mathbb{C}^{p \times p}$, $C \in \mathbb{C}^{p \times q}$, *and* $D \in \mathbb{C}^{q \times q}$ *are such that the matrix* $\begin{bmatrix} B & C \\ C^* & D \end{bmatrix}$ *is positive definite. Theorem 9.6.1 asserts that the matrix A_0 is diagonalizable, all eigenvalues of A_0 are real, and p of them are positive, whereas q of them are negative.*

9.7 *H*-Selfadjoint Stably *r*-Diagonalizable Matrices

Let $H = H^*$ be an invertible complex $n \times n$ matrix. Recall that an H-selfadjoint matrix A is called stably r-diagonalizable if A is similar to a real diagonal matrix and this property holds also for every matrix A' which is sufficiently close to A and which is H'-selfadjoint for some hermitian matrix H' sufficiently close to H. Denote by $S_r(H)$ the class of all H-selfadjoint stably r-diagonalizable matrices. We know by Theorem 9.2.2 that $A \in S_r(H)$ if and only if all eigenvalues of A are real and definite. The latter means that the quadratic form (Hx, x) is either positive definite or negative definite on the subspace $\mathrm{Ker}(\lambda_0 I - A)$ for every eigenvalue λ_0 of A. So there is a unique sign associated with each eigenvalue λ_0 of a matrix $A \in S_r(H)$ (which coincides with the sign of the quadratic form (Hx, x), $x \in \mathrm{Ker}(\lambda_0 I - A)$).

Let A be an H-selfadjoint stably r-diagonalizable matrix. We now define the *index* of A as follows. Let (α_0, α_1), (α_1, α_2), ..., (α_{p-1}, α_p), $\alpha_0 = -\infty$, $\alpha_p = \infty$, be consecutive intervals on the real line such that every interval (α_i, α_{1+i}) contains the largest possible number of eigenvalues of A having the same sign in the sign characteristic (so adjacent intervals contain eigenvalues with opposite signs). Let n_i be the sum of multiplicities of the eigenvalues of A lying in (α_{i-1}, α_i), multiplied by (-1) if the sign of these eigenvalues is negative. Thus, the sign of n_i coincides with the sign (in the sign characteristic of (A, H)) of eigenvalues belonging to (α_{i-1}, α_i). The sequence $\{n_1, \ldots, n_p\}$ will be called the *index* of A and will be denoted $\mathrm{ind}_r (A, H)$. It is easily seen that the index does not depend on the choice of α_i (subject to the condition mentioned above). Observe the following properties of $\mathrm{ind}_r (A, H)$:

$$n_i n_{i+1} < 0, \quad i = 1, \ldots, p-1; \tag{9.7.11}$$

$$\sum_{i=1}^{p} |n_i| = n; \tag{9.7.12}$$

$$\sum_{i=1}^{p} n_i = \mathrm{sig}\, H. \tag{9.7.13}$$

Note that if $n_1, \ldots, n_p$ are integers with the properties (9.7.11)–(9.7.13), then there is a matrix $A \in S_r(H)$ such that $\{n_1, \ldots, n_p\}$ is the index of A. Indeed, put

$$K = \mathrm{diag}\left(I_{n_1}, 2I_{n_2}, \ldots, pI_{n_p}\right),$$

and

$$Q = \mathrm{diag}\left(\varepsilon_1 I_{n_1}, \varepsilon_2 I_{n_2}, \ldots, \varepsilon_p I_{n_p}\right),$$

where I_j is the $j \times j$ unit matrix and $\varepsilon_j = \mathrm{sgn}\, n_j$. Clearly, K is Q-selfadjoint. Moreover, because of (9.7.13), $\mathrm{sig}\, Q = \mathrm{sig}\, H$. So there exists an invertible S such that $H = S^* Q S$. Now $A = S^{-1} K S$ is H-selfadjoint, stably r-diagonalizable and

$$\mathrm{ind}_r (A, H) = \{n_1, \ldots, n_p\}.$$

The set $S_r(H)$ of all H-selfadjoint stably r-diagonalizable matrices is open, and therefore splits into open connected components. All such components may be described as follows.

Theorem 9.7.1. *All matrices from $S_r(H)$ with the same index form a connected component in $S_r(H)$, and each connected component in $S_r(H)$ has such a form.*

Proof. Let $A, B \in S_r(H)$ with canonical forms

$$J_A = \text{diag}\,(\alpha_1, \ldots, \alpha_n)\,, \quad P_{\varepsilon, J_A} = \text{diag}\,(\delta_1, \ldots, \delta_n)$$

and

$$J_B = \text{diag}\,(\beta_1, \ldots, \beta_n)\,, \quad P_{\varepsilon, J_B} = \text{diag}\,(\zeta_1, \ldots, \zeta_n)\,,$$

respectively (so $\alpha_i, \beta_i \in \mathbb{R}$ and $\delta_i, \zeta_i = \pm 1$). We assume that $\alpha_1 \leq \cdots \leq \alpha_n$; $\beta_1 \leq \cdots \leq \beta_n$.

Suppose first that

$$\text{ind}_r\,(A, H) = \text{ind}_r\,(B, H).$$

Then $P_{\varepsilon, J_A} = P_{\varepsilon, J_B}$, and

$$J(t) = \text{diag}\,(t\alpha_1 + (1 - t)\beta_1, \ldots, t\alpha_n + (1 - t)\beta_n)\,, \quad t \in [0, 1]$$

is a continuous path of P_{ε, J_A}-selfadjoint stably r-diagonalizable matrices connecting J_A and J_B. Let $A = S^{-1}J_A S$, $H = S^* P_{\varepsilon, A} S$ be the unitary similarity relations. Then $A(t) = S^{-1}J_A(t)S$, $t \in [0, 1]$ is a continuous path of matrices from $S_r(H)$, and $A(1)$ is similar to B and has the same sign characteristic. By Theorem 5.4.1 $A(1)$ and B are H-unitarily similar. Since the classes of H-unitary similarity are connected (see Theorem 5.4.4), $A(1)$ and B belong to the same connected component of $S_r(H)$. So the same is true for A and B.

Suppose now that $\text{ind}_r\,(A, H)$ is not equal to $\text{ind}_r\,(B, H)$. We shall prove that A and B belong to different connected components of $S_r(H)$. Assume the contrary. Then there exists a continuous path $A(t)$, $t \in [0, 1]$ from A to B in $S_r(H)$. Let

$$t_0 = \inf\,\{t \in [0, 1] \mid \text{ind}_r(A(t), H) \neq \text{ind}_r(A, H)\}\,.$$

Let $\mu_1 < \cdots < \mu_r$ be the different eigenvalues of $A(t_0)$. We claim that for some μ_j, the form (Hx, x), $x \in \text{Ker}(\mu_j I - A(t_0))$ is not definite. Indeed, if all the forms (Hx, x), $x \in \text{Ker}(\mu_j I - A(t_0))$ were definite, then the same is true for the forms

$$(Hx, x)\,, \quad x \in \text{Range}\,\left[\frac{1}{2\pi i}\int_{\Gamma_j}(\lambda I - A(t))^{-1}d\lambda\right]\,, \quad j = 1, 2, \cdots, r\,, \quad (9.7.14)$$

where Γ_j is a small contour around μ_j, and t belongs to some neighborhood U of t_0 (see Section A.3 for some information about Riesz projections and their continuity, Theorem A.3.2). Indeed, use Theorem 9.2.1 and remarks thereafter to see the definiteness of the forms (9.7.14) with $t = t_0$; then by continuity of $A(t)$ and of the corresponding Riesz projections, the forms (9.7.14) are definite for all $t \in U$ and for $j = 1, 2, \ldots, r$. (Recall that $\frac{1}{2\pi i}\int_{\Gamma_j}(\lambda I - A(t))^{-1}d\lambda$ is the Riesz projection corresponding to the eigenvalues of $A(t)$ inside Γ_j, see Section A.3.) But then, by definition of the index, we have

$$\text{ind}_r\,(A(t), H) = \text{ind}_r\,(A(t_0), H)\,, \quad t \in U,$$

a contradiction with the choice of t_0. By Theorem 9.2.2, $A(t_0)$ is not stably r-diagonalizable, which contradicts the choice of the path $A(t)$, $t \in [0,1]$. $\square$

More generally, consider the connected components of the set S_r of pairs (A, H), where A is an H-selfadjoint stably r-diagonalizable $n \times n$ matrix. Given integers $n_1, \ldots, n_p$ such that $n_i n_{i+1} < 0$, $i = 1, \ldots, p$, and $\sum_{i=1}^{p} |n_i| = n$, it is easily seen that the set of all pairs $(A, H) \in S_r$ for which the index $\mathrm{ind}_r (A, H)$ is equal to $\{n_1, \ldots, n_p\}$ (then necessarily $\sum_{i=1}^{p} n_i = \mathrm{sig}\, H$) is not empty. As in Theorem 9.7.1 such sets are precisely the connected components of S_r:

Theorem 9.7.2. *All pairs $(A, H) \in S_r$ with the same index $\mathrm{ind}_r(A, H)$ form a connected component in S_r, and each connected component in S_r has such a form.*

Proof. Let H_i-selfadjoint matrices A_i, $i = 1, 2$, be such that A_i is stably r-diagonalizable and $\mathrm{ind}_r(A_1, H_1) = \mathrm{ind}_r(A_2, H_2) = \{n_1, \ldots, n_p\}$. Then, in particular, $\mathrm{sig}\, H_1 = \mathrm{sig}\, H_2$. So there exists an invertible matrix S such that $H_2 = S^* H_1 S$. Put $A_2' = S^{-1} A_1 S$. Then A_2' is H_2-selfadjoint, and since $\sigma(A_2') = \sigma(A_1)$, also $\mathrm{ind}_r(A_2', H_2) = \mathrm{ind}_r(A_1, H_1)$. By Theorem 9.7.1 there is a continuous path of matrices from $S_r(H_2)$ connecting A_2 and A_2'. Also there is a continuous path $F(t)$, $t \in [0,1]$ of invertible matrices connecting S and I. Put $A(t) = F(t)^{-1} A_1 F(t)$, $H(t) = F(t)^* H_1 F(t)$ to obtain a continuous path of matrices $A(t) \in S_r(H(t))$ such that $A(0) = A_2'$; $A(1) = A_1$. So the set of all pairs $(A, H) \in S_r$ with $\mathrm{ind}_r (A, H) = \{n_1, \ldots, n_p\}$ is connected.

Now let $(A_1, H_1), (A_2, H_2) \in S_r$, and assume that there exists a continuous path $(A(t), H(t)) \in S_r$ such that $A(0) = A_1$; $H(0) = H_1$; $A(1) = A_2$; $H(1) = H_2$. Then clearly, $\mathrm{sig}\, H_1 = \mathrm{sig}\, H_2$. Also, as in the proof of Theorem 9.7.1, $\mathrm{ind}_r (A_1, H_1) = \mathrm{ind}_r (A_2, H_2)$. $\square$

9.8   General Perturbations and Stably Diagonalizable H-Unitary Matrices

Results on the perturbation of H-unitary matrices are discussed in this section. They are derived from, and are similar to, those obtained in Sections 9.1 and 9.2 for H-selfadjoint matrices.

We continue to use the notation $P_{\mathcal{R}_\lambda(A)}$ for the orthogonal projection on the root subspace for A associated with λ. The unit circle is denoted T.

Theorem 9.8.1. *Let U be H-unitary, and let $\Omega \subseteq \mathsf{T}$ be an open set (relative to T) whose boundary does not intersect $\sigma(U)$. Then for some neighborhoods $\mathcal{U}_U$ of U and $\mathcal{U}_H$ of H we have*

$$\sum_{\lambda \in \Omega} \mathrm{sig}\, \left(P_{\mathcal{R}_\lambda(U)}\, H\, P_{\mathcal{R}_\lambda(U)} \right) = \sum_{\mu \in \Omega} \mathrm{sig}\, \left(P_{\mathcal{R}_\mu(V)}\, G\, P_{\mathcal{R}_\mu(V)} \right)$$

for every $V \in \mathcal{U}_U$ which is G-unitary for some hermitian $G \in \mathcal{U}_H$. Moreover, the number $v_\Omega(V)$ of eigenvalues of such a matrix V in Ω (counting multiplicities) satisfies the inequality

$$v_\Omega(V) \geq \sum_{\lambda \in \Omega} |\text{sig}\left(P_{\mathcal{R}_\lambda(U)} \, H \, P_{\mathcal{R}_\lambda(U)}\right)|, \qquad (9.8.15)$$

and in every (open) neighborhood $\mathcal{U} \subset \mathcal{U}_U$ of U there exists an H-unitary V for which the equality holds in (9.8.15).

Note that by taking $\mathcal{U}_H$ sufficiently small in this statement, the invertibility of the matrix G is guaranteed.

Theorem 9.8.1 can be obtained from Theorem 9.1.1 by using the Cayley transform.

Let U be an H-unitary matrix and let Ω be an open subset of the unit circle. The matrix U is called Ω-*diagonalizable* if, for every $\lambda_0 \in \Omega \cap \sigma(A)$, the multiplicity of λ_0 as a zero of $\det(\lambda I - A)$ is just $\dim \text{Ker}\,(\lambda_0 I - A)$. The matrix U is H-*stably* Ω-*diagonalizable* if every H-unitary matrix V sufficiently close to U has the same number of eigenvalues in Ω as U, and is Ω-diagonalizable. The matrix U is *stably* Ω-*diagonalizable* if this property holds for every G-unitary V such that G (resp. V) is sufficiently close to H (resp. U).

In the following theorem we assume that the boundary of Ω does not intersect $\sigma(U)$.

Theorem 9.8.2. *Let U be an H-unitary matrix. The following statements are equivalent:*

(i) *U is stably Ω-diagonalizable;*

(ii) *U is H-stably Ω-diagonalizable;*

(iii) *the quadratic form (Hx, x) is either positive or negative definite on the subspace $\text{Ker}\,(\lambda_0 I - U)$, for every $\lambda_0 \in \sigma(U) \cap \Omega$.*

It is easy to see that Theorems 9.2.1 and 9.8.2 can be obtained one from another by using the Cayley transform and its inverse. A direct proof of Theorem 9.8.2 can also be obtained from the general perturbation Theorem 9.8.1.

An important particular case of Theorem 9.8.2 arises when Ω is the whole unit circle T.

An H-unitary matrix U is called *diagonalizable with unimodular eigenvalues* (in short, *u-diagonalizable*) if U is similar to a diagonal matrix with unimodular entries on the diagonal, i. e. U is Ω-diagonalizable with $\Omega = \mathsf{T}$, and all eigenvalues of U lie on the unit circle T. The meaning of the notions of *stably u-diagonalizable* matrices and *H-stably u-diagonalizable* matrices is clear. They are direct analogues of definitions made in Section 9.2. The analogue of Theorem 9.2.2 is:

Theorem 9.8.3. *Let U be an H-unitary matrix. The following statements are equivalent:*

(i) *U is stably u-diagonalizable;*

(ii) U *is* H-*stably* u-*diagonalizable;*

(iii) *all eigenvalues of* U *lie on the unit circle, and the quadratic form* (Hx, x)
is either positive definite or negative definite on $\operatorname{Ker}(\lambda_0 I - U)$, *for every*
$\lambda_0 \in \sigma(U)$.

9.9 H-Unitarily Stably u-Diagonalizable Matrices

Let $H = H^*$ be an invertible complex $n \times n$ matrix. Recall from the preceding
section that an H-unitary matrix U is called stably u-diagonalizable if in some
basis in $\mathbf{C}^n$, U is a diagonal matrix with unimodular eigenvalues and this property
holds also for every matrix U' which is sufficiently close to U and which is G-
unitary for some hermitian invertible matrix G sufficiently close to H.

Denoting by $S_u(H)$ the set of all H-unitary stably u-diagonalizable matrices,
it follows from Theorem 9.8.3, that $U \in S_u(H)$ if and only if U is similar to a
diagonal matrix with unimodular eigenvalues and the eigenvalues of U are definite.
The latter statement means that the signs in the sign characteristic of the pair
(U, H) which correspond to Jordan blocks with the same eigenvalue (as described
in Section 5.15), are either all $+1$'s or -1's. So there is a sign corresponding to each
eigenvalue of $U \in S_u(H)$ (this sign coincides with the sign in the sign characteristic
of (U, H) corresponding to this eigenvalue).

We will describe the structure of the connected components in $S_u(H)$. To
this end we introduce the *index* for a stably u-diagonalizable H-unitary matrix U
(cf. the definition of the index for stably r-diagonalizable H-selfadjoint matrices).
Let

$$(\alpha_0, \alpha_1), (\alpha_1, \alpha_2), \ldots, (\alpha_{p-1}, \alpha_p), \quad \alpha_p = \alpha_0,$$

be consecutive intervals on the unit circle such that every interval (α_i, α_{i+1}) contains the largest possible number of eigenvalues of U of the same sign (so adjacent
intervals contain eigenvalues of opposite sign). In particular, either p is even or
$p = 1$, and the latter case may occur only if H is positive definite or negative
definite.

Let v_i $(i = 1, \ldots, p)$ be the sum of multiplicities of eigenvalues of U lying in
(α_{i-1}, α_i), multiplied by (-1) if the sign of these eigenvalues is negative (so that
the sign of v_i coincides with the sign of eigenvalues belonging to (α_{i-1}, α_i)). The
sequence $\{v_1, \ldots, v_p\}$, as well as any sequence

$$\{v_i, v_{i+1}, \ldots, v_p, v_1, \ldots, v_{i-1}\}, \quad i = 2, \ldots, p, \tag{9.9.16}$$

(obtained from $\{v_1, \ldots, v_p\}$ by a cyclic permutation) will be called the *index* of
U and denoted $\operatorname{ind}_u(U, H)$. It is seen that the index of U does not depend on
the choice of α_i (subject to the above conditions), provided that one takes into
account the possible cyclic permutation of $\operatorname{ind}_u(U, H)$.

Observe the following properties of $\operatorname{ind}_u(U, H)$:

If $p > 1$, then $v_i v_{i+1} < 0$, $\quad i = 1, \ldots, p$ $\quad (v_{p+1} = v_1$ by definition), $\tag{9.9.17}$

$$\sum_{i=1}^{p} |v_i| = n, \tag{9.9.18}$$

$$\sum_{i=1}^{p} v_i = \text{sig } H, \tag{9.9.19}$$

and note that (9.9.17) implies $p = 1$ or p is even.

Let $\{v_1, \ldots, v_p\}$ be any sequence of nonzero integers with properties (9.9.17)–(9.9.19). It is easily seen that the set of all matrices from $S_u(H)$ whose index is $\{v_1, \ldots, v_p\}$ is not empty (cf. the proof of the corresponding property for $\text{ind}_r(A, H)$ in Section 9.7).

Theorem 9.9.1. *All matrices from $S_u(H)$ whose indices are obtained from each other by cyclic permutation from a connected component in $S_u(H)$, and every connected component in $S_u(H)$ has this form.*

Proof. Let $U_1, U_2 \in S_u(H)$; let

$$K_1 = \text{diag}\left(e^{i\theta_1}, e^{i\theta_2}, \ldots, e^{i\theta_n}\right), \quad P_{\varepsilon_1, J} = \text{diag}\left(\zeta_1, \zeta_2, \ldots, \zeta_n\right), \quad \zeta_i = \pm 1,$$

be the canonical form of the pair (U_1, H), and let

$$K_2 = \text{diag}\left(e^{i\pi_1}, e^{i\pi_2}, \ldots, e^{i\pi_n}\right), \quad P_{\varepsilon_2, J} = \text{diag}\left(\eta_1, \eta_2, \ldots, \eta_n\right), \quad \eta_i = \pm 1,$$

be the canonical form of the pair (U_2, H). We assume that

$$\theta_1 \le \theta_2 \le \cdots \le \theta_n, \quad \pi_1 \le \pi_2 \le \cdots \le \pi_n, \quad \theta_n - \theta_1 < 2\pi, \quad \pi_n - \pi_1 < 2\pi. \tag{9.9.20}$$

Suppose now that $\text{ind}_u(U_1, H)$ is equal to $\text{ind}_u(U_2, H)$. This means that, after some cyclic permutation of terms on the main diagonal in K_2, and the same permutation of terms in $P_{\varepsilon_2, J}$, we obtain

$$K_3 = \text{diag}\left(e^{ip_1}, e^{ip_2}, \ldots, e^{ip_n}\right)$$

and $P_{\varepsilon_1, J}$, respectively. By adding 2π to some of the p_j's (if necessary), we ensure that

$$p_1 \le p_2 \le \cdots \le p_n, \quad p_n - p_1 < 2\pi.$$

Put

$$K(t) = \text{diag}\left[e^{i\sigma_1(t)}, e^{i\sigma_2(t)}, \ldots, e^{i\sigma_n(t)}\right], \quad \sigma_j(t) = (1-t)\theta_j + t p_j, \quad j = 1, 2, \ldots, n.$$

Now write $U_1 = S^{-1} K_1 S$, $H = S^* P_{\varepsilon_1, J} S$ for some invertible matrix S. Then

$$U(t) = S^{-1} K(t) S, \quad t \in [0, 1]$$

is a continuous path of H-unitary matrices from $S_u(H)$ connecting U_1 and $S^{-1} K_3 S$. Also the H-unitary matrices $S^{-1} K_3 S$ and U_2 are similar and have the same sign characteristic. By Theorem 5.17.2, $S^{-1}K_3S$ and U_2 are H-unitarily similar. Since the class of H-unitary similarity of H-unitary matrices is connected (Theorem 5.17.3), we can find a continuous path from $S^{-1}K_3S$ to U_2 in this class. Clearly, this path belongs to $S_u(H)$.

Now suppose that $\text{ind}_u(U_1, H)$ is not equal to $\text{ind}_u(U_2, H)$. Assume there exists a continuous path $U(t)$, $t \in [0,1]$, of H-unitary matrices in $S_u(H)$ connecting U_1 and U_2. As in the proof of Theorem 9.7.1, we pick

$$t_0 = \inf\left\{t \in [0,1] \mid \text{ind}_u(U(t), H) \text{ is not equal to } \text{ind}_u(U_1, H)\right\},$$

and show that for all t in a sufficiently small neighborhood of t_0, $\text{ind}_u(U(t), H)$ is equal to $\text{ind}_u(U(t_0), H)$; a contradiction. $\square$

Finally, we consider the connected components of the set S_u of all pairs (U, H), such that $H = H^* \in C^{n \times n}$ is invertible and U is an $n \times n$ matrix belonging to $S_u(H)$. Again, given integers $v_1, \ldots, v_p$ with properties (9.9.17) and (9.9.18), the set of all pairs $(U, H) \in S_u$ such that $\text{ind}_u(U, H)$ is equal to $\{v_1, \ldots, v_p\}$ is not empty.

Theorem 9.9.2. *All pairs $(U, H) \in S_u$ whose indices $\text{ind}_u(U, H)$ are obtained from each other by a cyclic permutation form a connected component in S_u. Every connected component in S_u has such a form.*

Proof. Assume $(U_1, H_1), (U_2, S_2) \in S_u$ are such that $\text{ind}_u(U_1, H_1) = \{v_1, \ldots, v_p\}$ is obtained from $\text{ind}_u(U_2, H_2)$ by a cyclic permutation. Then

$$\sum_{i=1}^{p} v_i = \text{sig } H_1 = \text{sig } H_2.$$

Let S be an invertible matrix such that $H_2 = S^* H_1 S$, and put $U_3 = S^{-1}U_1S$. Let $S(t)$, $t \in [0,1]$, be a continuous path of invertible matrices with $S(0) = S$, $S(1) = I$. Put

$$U_3(t) = S^{-1}(t)\, U_1\, S(t), \quad H_2(t) = S^*(t)\, H_1\, S(t).$$

Then $U_3(t)$ is $H_2(t)$-unitary, sig $H_2(t) = \text{sig } H_1$, and $U_3(t)$ is stably u-diagonalizable and the index $\text{ind}_u(U_3(t), H_2)$ is equal to $\text{ind}_u(U_1, H_1)$. Further,

$$\left(U_3(0), H_2(0)\right) = (U_3, H_2), \quad \left(U_3(1), H_2(1)\right) = (U_1, H_1).$$

By Theorem 9.9.1 there is a continuous path between (U_3, H_2) and (U_2, H_2) in the set $S_u(H_2)$. So (U_1, H_1) and (U_2, H_2) belong to the same connected component of the set S_u.

Now let (U_1, H_1) and (U_2, H_2) be in S_u. If sig $H_1 \neq$ sig H_2, then clearly (U_1, H_1) and (U_2, H_2) belong to different connected components of S_u (cf. Theorem A.1.2(b)). Suppose sig $H_1 =$ sig H_2, but the indices $\text{ind}_u(U_1, H_1)$ and $\text{ind}_u(U_2, H_2)$ are not equal. If it is assumed that there exists a continuous path between (U_1, H_1) and (U_2, H_2) in S_u then, arguing as in the proof of Theorem 9.7.1, we arrive at a contradiction. $\square$

9.10 Exercises

1. Give an example of a diagonalizable matrix which after a suitable arbitrarily small perturbation becomes nondiagonalizable.

2. When is a diagonalizable matrix A *stably diagonalizable*, that is all sufficiently small perturbations of A are also diagonalizable matrices?

3. Let A be a 2×2 H-selfadjoint matrix, where $H = \begin{bmatrix} 0 & 1 \\ 1 & 0 \end{bmatrix}$. What statement can be made about the canonical form of small H-selfadjoint perturbations of A, under each of the following three hypotheses?

    (a) A has no real eigenvalues;

    (b) A has two different eigenvalues;

    (c) A has one real eigenvalue of multiplicity 2.

4. Answer the question in Exercise 3 for H-unitary perturbations of an H-unitary matrix A, using unimodular rather than real eigenvalues.

5. Let there be given the following pairs of matrices (A_j, H_j), $j = 1, 2, 3, 4$, where A_j is H_j-selfadjoint:

    (a)
    $$A_1 = \begin{bmatrix} \lambda_0 & 1 & 0 \\ 0 & \lambda_0 & 1 \\ 0 & 0 & \lambda_0 \end{bmatrix}, \quad H_1 = \begin{bmatrix} 0 & 0 & 1 \\ 0 & 1 & 0 \\ 1 & 0 & 0 \end{bmatrix}, \quad \lambda_0 \in \mathsf{R}.$$

    (b)
    $$A_2 = \begin{bmatrix} \lambda_1 & 1 & 0 \\ 0 & \lambda_2 & 1 \\ 0 & 0 & \lambda_1 \end{bmatrix}, \quad H_2 = H_1, \quad \lambda_1, \lambda_2 \in \mathsf{R}.$$

    (c)
    $$A_3 = \begin{bmatrix} \lambda_0 & 1 & 0 \\ 0 & \lambda_0 & 0 \\ 1 & 0 & \lambda_0 \end{bmatrix}, \quad H_3 = \begin{bmatrix} 1 & 0 & 0 \\ 0 & 0 & 1 \\ 0 & 1 & 0 \end{bmatrix}, \quad \lambda_0 \in \mathsf{R}.$$

    (d)
    $$A_4 = \begin{bmatrix} \lambda_0 & 1 & 0 \\ 0 & \lambda_1 & 1 \\ 0 & 0 & \lambda_0 \end{bmatrix}, \quad H_4 = H_1, \quad \lambda_0 \in \mathsf{C}, \quad \lambda_1 \in \mathsf{R}.$$

    Describe the canonical forms of small H_j-selfadjoint perturbations of these pairs.

6. Let A be an H-selfadjoint $n \times n$ matrix. Consider the following statements:

   (a) There exist open neighborhoods $\mathcal{U}_A$ of A and $\mathcal{U}_H$ of H such that, for every pair $(B, G) \in \mathcal{U}_A \times \mathcal{U}_H$ with B G-selfadjoint, the numbers of real eigenvalues (counted with multiplicities) of A and B coincide.

   (b) For every pair (B, G) as in (a), the numbers of nonreal eigenvalues of A and B coincide.

   (c) For every pair (B, G) as in (a), all partial multiplicities of B are equal to 1.

   Verify whether or not the each of the statements (a), (b), and (c) holds when all partial multiplicities of A are equal to 1.

7. Solve the preceding exercise under each of the following hypotheses:

   (a) A has n distinct eigenvalues.

   (b) all real eigenvalues of A (if any) have algebraic multiplicity 1.

   (c) all partial multiplicities of A corresponding to real eigenvalues (if any) are equal to 1.

   (d) A has at least one partial multiplicity greater than 1 corresponding to a real eigenvalue.

8. Generalize Exercises 6 and 7 for H-unitary matrices A.

9. Let H be negative definite, and let A be H-selfadjoint. Find the canonical forms of all small H-selfadjoint perturbations of A.

10. Are the following matrices A_j, $j = 1, 2$, stably r-diagonalizable with respect to H_j?

   (a)
   $$A_1 = \begin{bmatrix} I_n & 0 \\ 0 & -I_n \end{bmatrix} \in \mathbb{R}^{2n \times 2n}, \quad H_1 = \begin{bmatrix} S_n & 0 \\ 0 & S_n \end{bmatrix}.$$

   (b) $A_2 = \begin{bmatrix} 0 & 1 & 0 & \cdots & 0 \\ 0 & 0 & 1 & \cdots & 0 \\ & & \cdots\cdots\cdots & & \\ -a_0 & -a_1 & -a_2 & \cdots & -a_{n-1} \end{bmatrix} \in \mathbb{R}^{n \times n};$

   $$H_2 = \begin{bmatrix} a_1 & a_2 & \cdots & a_{n-1} & 1 \\ a_2 & & & \cdot & \\ \vdots & & \cdot & \cdot & 0 \\ 1 & & & & \end{bmatrix}, \text{ where } a_0, \ldots, a_{n-1} \in \mathbb{R}.$$

11. Let A be H-selfadjoint with a real eigenvalue λ_0 for which the form (Hx, x) is indefinite on the subspace $\mathrm{Ker}\,(\lambda_0 I - A)$. Prove that there exists an H-selfadjoint matrix B which is arbitrarily close to A and such that B has eigenvalues in the open upper and lower halfplanes.

12. Establish an analogue of the preceding exercise for H-unitary matrices.

13. Let $A_3 = J_n(\lambda_0)$ be the $n \times n$ Jordan block with real eigenvalue λ_0, and let $H_3 = S_n$, the $n \times n$ sip matrix. Describe the canonical forms of all H_3-selfadjoint small perturbations of A_3.

14. Let A_j and H_j, $j = 1, 2, 3$, be as in Exercises 10 and 13, and let B be any small H-selfadjoint perturbation of A. Show that:

    (a) If $n = 2k + 1$ is odd, then B always has at least one real eigenvalue.

    (b) If $n = 2k$ is even, then B does not necessarily have a real eigenvalue.

15. Establish an analogue of Exercise 13 for H-unitary matrices.

16. Establish an analogue of Exercise 14 for H-unitary matrices.

17. Let λ_0 be a real eigenvalue of an H_0-selfadjoint $n \times n$ matrix A_0. Prove that if the quadratic form $(H_0 x, x)$ is not definite on the subspace $\mathrm{Ker}(\lambda_0 I - A_0)$, then there exists a quadratic polynomial $A_0 + \tau A_1 + \tau^2 A_2$ with H_0-selfadjoint coefficients which has a continuous nonanalytic eigenvalue tending to λ_0 as $\tau \to 0$.

18. Show that in the preceding exercise one can take A_1 and A_2 such that $\mathrm{rank} A_1 \leq 2$, $\mathrm{rank} A_2 \leq 1$ and $\mathrm{Range}\,A_2 \subseteq \mathrm{Range}\,A_1$.

19. Give an example to demonstrate that the statement in Exercise 17 is not generally true if we replace the quadratic polynomial by a linear polynomial $A_0 + \tau A_1$.

20. Let λ_0 be a real eigenvalue of an H_0-selfadjoint $n \times n$ matrix A. Assume that the quadratic form $(H_0 x, x)$ is degenerate on $\mathrm{Ker}(\lambda_0 I - A)$, meaning that there is a nonzero vector $x \in \mathrm{Ker}(\lambda_0 I - A)$ such that $(H_0 x, y) = 0$ for every $y \in \mathrm{Ker}(\lambda_0 I - A)$. Prove that there exists an H_0-selfadjoint matrix A_1 of rank 1 such that $A_0 + \tau A_1$ has a nonanalytic eigenvalue tending to λ_0 as $\tau \to 0$.

9.11 Notes

The material of Section 9.1 is taken from [38]. The results of Section 9.2 are due mainly to Krein [61]; see also [27]. The results of Sections 9.3 and 9.4 appeared in [40]. Results of Section 9.6 in a more general setting were obtained in [61]. The contents of Section 9.7 is taken from [38].

Chapter 10

Definite Invariant Subspaces

As we have seen in Section 5.12, every H-selfadjoint matrix has an invariant subspace which is maximal H-nonnegative (or maximal H-nonpositive). The proof of this property was based on the canonical form. This problem will be approached in a different way in this chapter. A general scheme will be developed concerning extension of invariant H-nonnegative (or H-nonpositive) subspaces of various classes of linear transformations with respect to an indefinite inner product. The scheme is based on a classical fixed point theorem which is given in Section A.8. To avoid trivialities, we assume throughout the chapter that the indefinite inner product is truly indefinite, i.e., there exist vectors x and y such that $[x, x] < 0 < [y, y]$.

10.1 Semidefinite and Neutral Subspaces: A Particular H

We start with descriptions of nonnegative and neutral subspaces with respect to the indefinite inner product given by a hermitian matrix H in a particular form. As everywhere in the book, for $Y \in C^{m \times n}$, we use the *operator matrix norm*

$$\|Y\| := \max\{\|Yx\| \ : \ \|x\| = 1, \ x \in C^n\}. \tag{10.1.1}$$

For simplicity, it is assumed in this section that the indefinite inner product in C^n is given by

$$[x, y] = (Hx, y), \qquad x, y \in C^n,$$

where the invertible hermitian matrix H has the form

$$H = \begin{bmatrix} I_p & 0 \\ 0 & -I_q \end{bmatrix}. \tag{10.1.2}$$

Clearly, by applying a congruence $H \to S^* H S$ for a suitable invertible matrix S, the form (10.1.2) can always be achieved. The subspaces which are H-nonnegative with respect to the form (10.1.2) can be conveniently described:

Lemma 10.1.1. *Let H be given by (10.1.2). Then a nonzero subspace $\mathcal{M} \subseteq \mathbb{C}^n$ of dimension d is H-nonnegative if and only if $\mathcal{M}$ has the form*

$$\mathcal{M} = \text{Range} \begin{bmatrix} P \\ K \end{bmatrix}, \tag{10.1.3}$$

where P is a $p \times d$ matrix with orthonormal columns, and $K \in \mathbb{C}^{q \times d}$ satisfies $\|K\| \leq 1$.

Matrices K with the property that $\|K\| \leq 1$ are said to be *contractions*. We will use this terminology in the sequel.

Proof. The condition that P has orthonormal columns can be written in the form

$$P^* P = I_d. \tag{10.1.4}$$

It is easy to see that every subspace of the form (10.1.3) is H-nonnegative. Indeed, let

$$\begin{bmatrix} x_1 \\ x_2 \end{bmatrix} = \begin{bmatrix} P \\ K \end{bmatrix} y,$$

for some $y \in \mathbb{C}^d$. Then

$$\left(H \begin{bmatrix} x_1 \\ x_2 \end{bmatrix}, \begin{bmatrix} x_1 \\ x_2 \end{bmatrix} \right) = \left(\begin{bmatrix} I_p & 0 \\ 0 & -I_q \end{bmatrix} \begin{bmatrix} P \\ K \end{bmatrix} y, \begin{bmatrix} P \\ K \end{bmatrix} y \right)$$

$$= (Py, Py) - (Ky, Ky) = (P^* Py, y) - (Ky, Ky)$$

$$= \|y\|^2 - \|Ky\|^2 \geq 0, \tag{10.1.5}$$

where the last inequality follows since $\|K\| \leq 1$.

Conversely, let $\mathcal{M}$ be an H-nonnegative subspace, and dim $\mathcal{M} = d$. Let $f_1, \ldots, f_d$ be a basis of $\mathcal{M}$ and, for $j = 1, \ldots, j$, form the partitions:

$$f_j = \begin{bmatrix} f_{1j} \\ f_{2j} \end{bmatrix}, \quad f_{1j} \in \mathbb{C}^p, \ f_{2j} \in \mathbb{C}^q.$$

We claim that the vectors $f_{11}, \ldots, f_{1d}$ are linearly independent. Otherwise, $\mathcal{M}$ would contain a nonzero vector of the form $\begin{bmatrix} 0_{p \times 1} \\ f \end{bmatrix}$ where $f \in \mathbb{C}^q$, and this contradicts the H-nonnegative property of $\mathcal{M}$. Denote by Q_1 the $p \times d$ matrix with the columns $f_{11}, \ldots, f_{1d}$ (in that order from left to right). Now let

$$P = Q_1 T \in \mathbb{C}^{p \times d},$$

where the invertible matrix $T \in \mathbb{C}^{d \times d}$ is chosen so that $P^*P = I_d$. This choice is always possible because, as $f_{11}, \ldots, f_{1d}$ are linearly independent, the matrix $Q_1^*Q_1$ is invertible, as well as positive definite, and so one can take

$$T = \left(\sqrt{Q_1^*Q_1} \right)^{-1}.$$

Here, $\sqrt{\cdot}$ stands for the positive definite square root of a positive definite matrix. Finally, letting Q_2 be the $q \times d$ matrix with columns $f_{21}, f_{22}, \cdots, f_{2d}$ in that order from left to right, consider

$$K = Q_2 T \in \mathbb{C}^{q \times d}.$$

With these definitions of P and K, the verification of the formula (10.1.3) is immediate, because

$$\mathcal{M} = \text{Range } [f_1 \cdots f_d] = \text{Range } \begin{bmatrix} Q_1 \\ Q_2 \end{bmatrix} = \text{Range } \begin{bmatrix} Q_1 \\ Q_2 \end{bmatrix} T = \text{Range } \begin{bmatrix} P \\ K \end{bmatrix}.$$

Since $P^*P = I$, we verify as in (10.1.5) that

$$\left(H \begin{bmatrix} P \\ K \end{bmatrix} y, \begin{bmatrix} P \\ K \end{bmatrix} y \right) = \|y\|^2 - \|Ky\|^2 \geq 0,$$

for every $y \in \mathbb{C}^d$, and hence $\|K\| \leq 1$. $\qquad\square$

Essentially the same proof provides a description of H-positive subspaces:

Lemma 10.1.2. *Let H be given by* (10.1.2). *Then a nonzero subspace $\mathcal{M} \subseteq \mathbb{C}^n$ of dimension d is H-positive if and only if $\mathcal{M}$ has the form*

$$\mathcal{M} = \text{Range } \begin{bmatrix} P \\ K \end{bmatrix},$$

where P is a $p \times d$ matrix with orthonormal columns, and $K \in \mathbb{C}^{q \times d}$ is a strict contraction, i.e., $\|K\| < 1$.

Note that, for a given $\mathcal{M}$, the representation (10.1.3), where $P^*P = I$ and K is a contraction, is not unique. For example, if $\alpha \in \mathbb{C}$, $|\alpha| = 1$, then replacing P and K with αP and αK, respectively, we obtain a different representation for the same $\mathcal{M}$.

To study this nonuniqueness, first of all observe that if

$$\text{Range } \begin{bmatrix} P \\ K_1 \end{bmatrix} = \text{Range } \begin{bmatrix} P \\ K_2 \end{bmatrix}, \qquad (10.1.6)$$

where $P^*P = I_d$, then $K_1 = K_2$; in other words, the matrix K in (10.1.3) is uniquely determined by $\mathcal{M}$ and P. The verification is easy: if (10.1.6) holds and

$P^*P = I_d$ then, in particular, the columns of $\begin{bmatrix} P \\ K_1 \end{bmatrix}$ and of $\begin{bmatrix} P \\ K_2 \end{bmatrix}$ are linearly independent, and we have

$$\begin{bmatrix} P \\ K_1 \end{bmatrix} = \begin{bmatrix} P \\ K_2 \end{bmatrix} W,$$

for some invertible matrix W. Consequently, $P = PW$. Multiplying this equality on the left by P^*, it follows that $W = I$. But then also $K_1 = K_2$.

As for the nonuniqueness of P in (10.1.3), assume that

$$\text{Range} \begin{bmatrix} P_1 \\ K_1 \end{bmatrix} = \text{Range} \begin{bmatrix} P_2 \\ K_2 \end{bmatrix},$$

where $P_1^* P_1 = P_2^* P_2 = I_d$. Then, as in the preceding paragraph, we obtain

$$\begin{bmatrix} P_1 \\ K_1 \end{bmatrix} = \begin{bmatrix} P_2 \\ K_2 \end{bmatrix} W, \tag{10.1.7}$$

for a unique invertible matrix W. In fact, $W = P_2^* P_1$. This can be verified by multiplying $P_1 = P_2 W$ on the left by P_2^*. Moreover, (10.1.7) implies $P_1 W^{-1} = P_2$, and therefore $W^{-1} = P_1^* P_2$. We see that $W^{-1} = W^*$, in other words, W is unitary. The following result is obtained:

Lemma 10.1.3. *If*

$$\text{Range} \begin{bmatrix} P_1 \\ K_1 \end{bmatrix} = \text{Range} \begin{bmatrix} P_2 \\ K_2 \end{bmatrix}, \tag{10.1.8}$$

where P_1 and P_2 are $p \times d$ matrices such that $P_1^* P_1 = P_2^* P_2 = I$, then $P_1 = P_2 W$, $K_1 = K_2 W$ for some unitary $d \times d$ matrix W, and this unitary matrix is unique.

In particular Lemma 10.1.3 applies to an H-nonnegative subspace of the form (10.1.8). However, for the validity of the lemma, it is not necessary that K_1 and K_2 be contractions.

Corollary 10.1.4. *Every maximal H-nonnegative subspace $\mathcal{M}$ can be written uniquely in the form*

$$\mathcal{M} = \text{Range} \begin{bmatrix} I_p \\ K \end{bmatrix}, \tag{10.1.9}$$

where $K \in \mathbb{C}^{q \times p}$ is a contraction. Conversely, every subspace of the form (10.1.9) is maximal H-nonnegative.

In particular, Corollary 10.1.4 describes a one-to-one correspondence between the set of maximal H-nonnegative subspaces and the set of $q \times p$ contractions.

Proof. The uniqueness of the representation (10.1.9) follows from Lemma 10.1.3. The existence of (10.1.9) follows from Lemma 10.1.1 in which P is unitary because, in view of Theorem 2.3.2, the dimension of $\mathcal{M}$ is equal to p. □

Next, taking advantage of the representation (10.1.3), we consider H-nonnegative subspaces which are contained in a maximal H-nonnegative subspace.

Lemma 10.1.5. *Let*

$$\mathcal{M}_0 = \text{Range} \begin{bmatrix} P_0 \\ K_0 \end{bmatrix} \quad and \quad \mathcal{M} = \text{Range} \begin{bmatrix} I_p \\ K \end{bmatrix}$$

be given, where $P_0 \in \mathbb{C}^{p \times d}$ has orthonormal columns, $K_0 \in \mathbb{C}^{q \times d}$ and $K \in \mathbb{C}^{q \times p}$ are contractions. (In particular, $\mathcal{M}$ is maximal H-nonnegative). Then $\mathcal{M}_0 \subseteq \mathcal{M}$ if and only if $K_0 = K P_0$.

Proof. If $K_0 = K P_0$ then, obviously,

$$\begin{bmatrix} P_0 \\ K_0 \end{bmatrix} = \begin{bmatrix} I \\ K \end{bmatrix} P_0$$

and therefore $\mathcal{M}_0 \subseteq \mathcal{M}$. Conversely, if $\mathcal{M}_0 \subseteq \mathcal{M}$, then there exists a matrix B such that

$$\begin{bmatrix} P_0 \\ K_0 \end{bmatrix} = \begin{bmatrix} I \\ K \end{bmatrix} B.$$

It follows immediately that $B = P_0$. □

Now we consider H-neutral subspaces. Since every H-neutral subspace is, in particular, H-nonnegative, Lemma 10.1.1 applies. However, K now has a more restrictive property:

Lemma 10.1.6. *Let H be given by (10.1.2). Then a nonzero subspace $\mathcal{M} \subseteq \mathbb{C}^n$ of dimension d is H-neutral if and only if $\mathcal{M}$ has the form*

$$\mathcal{M} = \text{Range} \begin{bmatrix} P \\ K \end{bmatrix}, \tag{10.1.10}$$

where P and K are matrices with orthonormal columns and sizes $p \times d$, $q \times d$, respectively.

As we know (Theorem 2.3.4), an H-neutral subspace cannot have dimension larger than $\min\{p, q\}$. This is reflected in Lemma 10.1.6 because a matrix of size $m \times d$ can have orthonormal columns only if $m \geq d$.

Proof. By Lemma 10.1.1 we may assume that $\mathcal{M}$ has the form (10.1.10), with $P^*P = I$ and K a contraction. Now $\mathcal{M}$ is H-neutral if and only if $(Hy, y) = 0$ for every $y \in \mathcal{M}$. But this means that

$$0 = (Hy, y) = (Px, Px) - (Kx, Kx) = ((I - K^*K)x, x),$$

for every $x \in \mathbb{C}^d$. In fact, such an x satisfies

$$y = \begin{bmatrix} P \\ K \end{bmatrix} x.$$

Thus, $K^*K = I$. In other words, the columns of K are orthonormal.   □

By Lemma 10.1.3, for a given H-neutral subspace $\mathcal{M}$, the matrices P and K of the representation (10.1.10) are determined by $\mathcal{M}$ up to a multiplication on the right by the same unitary matrix.

A completely parallel line of statements exists concerning H-nonpositive subspaces. They can be obtained by applying the H-nonnegative subspaces results to the matrix $-H$:

Lemma 10.1.7. *Let H be given by (10.1.2). Then:*

(a) *A nonzero subspace $\mathcal{M} \subseteq \mathbb{C}^n$ of dimension d is H-nonpositive, resp., H-negative, if and only if $\mathcal{M}$ has the form*

$$\mathcal{M} = \text{Range} \begin{bmatrix} K \\ P \end{bmatrix},$$

where P is a $q \times d$ matrix with orthonormal columns, and $K \in \mathbb{C}^{p \times d}$ is a contraction, resp., strict contraction.

(b) *A subspace $\mathcal{M} \subseteq \mathbb{C}^n$ is maximal H-nonpositive subspace if and only if it has the form*

$$\mathcal{M} = \text{Range} \begin{bmatrix} K \\ I_q \end{bmatrix}, \tag{10.1.11}$$

where $K \in \mathbb{C}^{p \times q}$ is a contraction. Moreover, the form (10.1.11) is unique for a given $\mathcal{M}$.

(c) *Let*

$$\mathcal{M}_0 = \text{Range} \begin{bmatrix} K_0 \\ P_0 \end{bmatrix} \quad \text{and} \quad \mathcal{M} = \text{Range} \begin{bmatrix} K \\ I_q \end{bmatrix}$$

be given, where $P_0 \in \mathbb{C}^{q \times d}$ has orthonormal columns, and $K_0 \in \mathbb{C}^{p \times d}$ and $K \in \mathbb{C}^{p \times q}$ are contractions. Then $\mathcal{M}_0 \subseteq \mathcal{M}$ if and only if $K_0 = KP_0$.

10.2 Plus Matrices and Invariant Nonnegative Subspaces

Throughout this section, we fix the indefinite inner product $[\cdot, \cdot]$ determined by an invertible hermitian indefinite $n \times n$ matrix H: $[x, y] = (Hx, y)$.

A matrix $A \in \mathbb{C}^{n \times n}$ is called a *plus matrix*, (or, if the dependence on H is to be emphasized, an *H-plus matrix*) if $[Ax, Ax] \geq 0$ for every $x \in \mathbb{C}^n$ such that $[x, x] \geq 0$.

A key property of plus matrices is given in the following proposition:

Proposition 10.2.1. *A matrix $A \in \mathbb{C}^{n \times n}$ is a plus matrix if and only if there exists a $\mu \geq 0$ such that*

$$[Ax, Ax] \geq \mu[x, x] \quad \text{for every} \quad x \in \mathbb{C}^n. \tag{10.2.12}$$

Proof. The "if" part is clear from the definition of plus matrices. Assume A is a plus matrix. Consider $X := H + iA^*HA$, and its numerical range (see Section A.7):

$$W(X) = \{(Xy, y) \in \mathbb{C} : y \in \mathbb{C}^n, \|y\| = 1\}.$$

Since both H and A^*HA are hermitian, we clearly have

$$
\begin{aligned}
W(X) &= \{((Hy, y), (A^*HAy, y)) \in \mathbb{R}^2 : y \in \mathbb{C}^n, \|y\| = 1\} \\
&= \{([y, y], [Ay, Ay]) \in \mathbb{R}^2 : y \in \mathbb{C}^n, \|y\| = 1\}.
\end{aligned}
$$

By definition of a plus matrix, the set $W(X)$ does not intersect the half closed quadrant $\{(a, b) \in \mathbb{R}^2 : a \geq 0, \ b < 0\}$. Since $W(X)$ is convex by Theorem A.7.1, it follows that

$$W(X) \subseteq \{(a, b) \in \mathbb{R}^2 : b \geq \mu a\}$$

for some fixed $\mu \geq 0$. Thus, (10.2.12) follows for all $x \in \mathbb{C}^n$ with $\|x\| = 1$, and then it follows for all $x \in \mathbb{C}^n$ by homogeneity. $\qquad \square$

Note that (10.2.12) can be re-written in an equivalent form

$$A^*HA - \mu H \geq 0. \tag{10.2.13}$$

The set of all nonnegative numbers μ for which (10.2.13) holds (which is nonempty if A is a plus matrix), is a closed bounded interval $[a, b]$, $0 \leq a \leq b < \infty$. The maximal value of $\mu \geq 0$ for which (10.2.13) holds true will be called the *plus-index* of the plus matrix A.

The set of plus matrices respects unitary similarity transformations in the following sense:

Proposition 10.2.2. *If $A \in \mathbb{C}^{n \times n}$ is an H-plus matrix, then $T^{-1}AT$ is a T^*HT-plus matrix, for every invertible $T \in \mathbb{C}^{n \times n}$.*

The proof is immediate upon noticing that $Tx \in \mathbb{C}^n$ is H-nonnegative if and only if x is T^*HT-nonnegative.

Example 10.2.3. *We describe all plus matrices for $H = \begin{bmatrix} 0 & 1 \\ 1 & 0 \end{bmatrix}$. Let*

$$A = \begin{bmatrix} a & b \\ c & d \end{bmatrix}, \quad a, b, c, d \in \mathbb{C}.$$

For $\mu \in \mathbb{R}$, we have

$$A^*HA - \mu H = \begin{bmatrix} \bar{a}c + \bar{c}a & \bar{a}d + \bar{c}b - \mu \\ a\bar{d} + c\bar{b} - \mu & \bar{b}d + \bar{d}b \end{bmatrix}.$$

Thus, (10.2.13) holds if and only if

$$\bar{a}c + \bar{c}a \geq 0, \quad \bar{b}d + \bar{d}b \geq 0, \tag{10.2.14}$$

and

$$|a\bar{d} + c\bar{b} - \mu|^2 \leq (\bar{a}c + \bar{c}a)(\bar{b}d + \bar{d}b).$$

We obtain the following criterion: The matrix A is a plus matrix if and only if the inequalities (10.2.14) hold and

$$\delta^2 \leq (\bar{a}c + \bar{c}a)(\bar{b}d + \bar{d}b),$$

where δ is the distance from the complex number $a\bar{d} + c\bar{b}$ to the nonnegative half-axis $\{\mu \in \mathbb{C} : \mu \geq 0\}$. The plus-index of A is equal to the largest (out of possibly two solutions) real solution μ_0 of the equation

$$|a\bar{d} + c\bar{b} - \mu|^2 = (\bar{a}c + \bar{c}a)(\bar{b}d + \bar{d}b). \qquad \square$$

We now formulate and prove the main result of this section concerning invariant nonnegative subspaces of invertible plus matrices: They can be extended to invariant maximal nonnegative subspaces.

Theorem 10.2.4. *Let A be an invertible plus matrix, and let $\mathcal{M}_0 \subseteq \mathbb{C}^n$ be an A-invariant subspace which is H-nonnegative. Then there exists an H-nonnegative A-invariant subspace $\widetilde{\mathcal{M}} \supseteq \mathcal{M}_0$ such that $\dim \widetilde{\mathcal{M}} = i_+(H)$.*

Thus, by Theorem 2.3.2, the subspace $\widetilde{\mathcal{M}}$ is maximal H-nonnegative. The hypothesis that A is invertible is essential in Theorem 10.2.4, and the next example illustrates this.

Example 10.2.5. *Let*

$$A = \begin{bmatrix} 0 & 1 & 0 & p \\ -1 & 0 & 1 & q \\ 0 & 0 & 0 & 0 \\ 0 & 0 & 0 & 0 \end{bmatrix}, \qquad H = \begin{bmatrix} 0 & 0 & 1 & 0 \\ 0 & 0 & 0 & 1 \\ 1 & 0 & 0 & 0 \\ 0 & 1 & 0 & 0 \end{bmatrix},$$

where p and q are real numbers such that $4p + q^2 > 0$. Clearly, A is a plus matrix; in fact, $[Ax, Ax] = 0$ for every $x \in \mathbb{C}^4$. Let

$$\mathcal{M}_0 = \mathrm{Span}\, \langle 1, 0, 1, 0 \rangle \in \mathbb{C}^4.$$

Then $A\mathcal{M}_0 = \{0\}$, *in particular,* $\mathcal{M}_0$ *is A-invariant. Since*

$$\langle 1,0,1,0 \rangle^T H \langle 1,0,1,0 \rangle = 2,$$

$\mathcal{M}_0$ *is H-positive. On the other hand,*

$$\text{Ker } A = \text{Span}\{\langle 1,0,1,0 \rangle, \langle 0,-p,-q,1 \rangle\}$$

is not H-nonnegative (this is where the condition $4p + q^2 > 0$ *is needed). Besides* Ker A, *there are two two-dimensional A-invariant subspaces that contain* $\mathcal{M}_0$, *namely,*

$$\text{Span}\{\langle 1,0,1,0 \rangle, \langle 1,\pm i,0,0 \rangle\}.$$

But

$$\begin{bmatrix} 1 & \mp i & 0 & 0 \\ 1 & 0 & 1 & 0 \end{bmatrix} H \begin{bmatrix} 1 & 1 \\ \pm i & 0 \\ 0 & 1 \\ 0 & 0 \end{bmatrix} = \begin{bmatrix} 0 & 1 \\ 1 & 2 \end{bmatrix},$$

so neither of these two subspaces is H-nonnegative. □

Nevertheless, it will be seen later that the hypothesis on the invertibility of A in Theorem 10.2.4 can be relaxed.

Proof of Theorem 10.2.4. By Proposition 10.2.2 it may be assumed that

$$H = \begin{bmatrix} I_p & 0 \\ 0 & -I_q \end{bmatrix},$$

for some positive integers p and q (we do not consider the trivial cases when H is either positive definite or negative definite).

Let $\mathcal{M}_0$ be as in Theorem 10.2.4. Represent $\mathcal{M}_0$ as in Lemma 10.1.1:

$$\mathcal{M}_0 = \text{Range} \begin{bmatrix} P_0 \\ K_0 \end{bmatrix},$$

where $P_0^* P_0 = I$, $\|K_0\| \leq 1$. On the other hand, consider the set S of all maximal H-nonnegative subspaces $\mathcal{M}$ that contain $\mathcal{M}_0$. Writing

$$\mathcal{M} = \text{Range} \begin{bmatrix} I \\ K \end{bmatrix} \in S,$$

then, in view of Lemma 10.1.5, we may identify S with the set, S_0, of all $q \times p$ contractions K such that $K_0 = K P_0$.

At this point we need more basic notions of matrix analysis concerning convergence, limits, etc. Let there be given a sequence

$$[X^{(j)}]_{j=1}^{\infty} \tag{10.2.15}$$

of $m \times n$ complex matrices, written in terms of their entries $X^{(j)} = [x_{u,v}^{(j)}]_{u=1,v=1}^{u=m,v=n}$. The sequence (10.2.15) is said to *converge* to a matrix $X = [x_{u,v}]_{u=1,v=1}^{u=m,v=n} \in \mathbb{C}^{m \times n}$ if

$$\lim_{j \longrightarrow \infty} x_{u,v}^{(j)} = x_{u,v}$$

for every pair of indices (u, v), where $u = 1, \ldots, m$, $v = 1, \ldots, n$. In this case X is the *limit* of (10.2.15). A set of matrices is called *closed* if it contains the limit of every converging sequence $X^{(j)}$, $j = 1, 2, \ldots$, with the $X^{(j)}$'s in the set in question. A set of matrices is called *bounded* if there exists a positive M such that $\|X\| \leq M$ for every matrix X in the set. Finally, we say that a set $T \subseteq \mathbb{C}^{m \times n}$ is *convex* if

$$X, Y \in T \quad \Longrightarrow \quad \alpha X + (1 - \alpha)Y \in T$$

for every α such that $0 \leq \alpha \leq 1$. If one identifies the complex vector space of $m \times n$ matrices with $\mathbb{C}^{pq}$, then the notions of convergence, limits, closedness, boundedness, and convexity of sets of matrices translate into precisely the same notions concerning vectors in $\mathbb{C}^{pq}$.

We now return to the set S_0 introduced above. Note that S_0 is a convex, closed, and bounded subset of $\mathbb{C}^{p \times q}$. Indeed, assume that $K^{(j)}$, $j = 1, 2, \ldots$, is a sequence of $q \times p$ matrices such that

$$\|K^{(j)}\| \leq 1, \quad K_0 = K^{(j)} P_0 \quad \text{for } j = 1, 2, \ldots, \tag{10.2.16}$$

and

$$K = \lim_{j \longrightarrow \infty} K^{(j)}. \tag{10.2.17}$$

Then, letting $x \in \mathbb{C}^p$ be such that $\|K\| = \|Kx\|$ and $\|x\| = 1$, we have

$$\|K\| = \|Kx\| = \lim_{j \longrightarrow \infty} \|K^{(j)}x\| \leq 1$$

because, in view of (10.2.16), $\|K^{(j)}x\| \leq 1$ for every $j = 1, 2, \ldots$. Also, $K_0 = KP_0$, again by (10.2.16). This shows that $K \in S_0$, and hence S_0 is closed. The boundedness of S_0 is obvious from its definition. Finally, the convexity of S_0 is easy to check: If $K_1, K_2 \in \mathbb{C}^{q \times p}$ are contractions such that $K_0 = K_j P_0$ for $j = 1, 2$, then for every α, $0 \leq \alpha \leq 1$, the matrix $K := \alpha K_1 + (1 - \alpha)K_2$ is also a contraction (because of the triangle inequality for the norm $\| \cdot \|$), and obviously satisfies $K_0 = KP$.

Next, we observe that

$$A(\mathcal{M}) := \{Ax | x \in \mathcal{M}\} \in S$$

for every $\mathcal{M} \in S$. Indeed, the hypothesis that A is a plus matrix implies that $A(\mathcal{M})$ is H-nonnegative for every H-nonnegative $\mathcal{M}$. Since A is invertible,

$$\dim (A(\mathcal{M})) = \dim \mathcal{M} = i_+(H)$$

for every maximal H-nonnegative $\mathcal{M}$ and this implies, using Theorem 2.3.2, that $A(\mathcal{M})$ is maximal H-nonnegative as well. Finally, since $A(\mathcal{M}_0) = \mathcal{M}_0$, we obviously have $A(\mathcal{M}) \supseteq \mathcal{M}_0$ for every subspace $\mathcal{M} \supseteq \mathcal{M}_0$.

In other words, A maps S into itself. Again, identify S with the set

$$S_0 := \{K \in C^{q \times p} \,|\, \|K\| \leq 1 \text{ and } K_0 = KP_0\}, \qquad (10.2.18)$$

and let $\widetilde{A}$ be the map induced by A on S_0. Thus,

$$A \left(\text{Range} \begin{bmatrix} I \\ K \end{bmatrix} \right) = \text{Range} \begin{bmatrix} I \\ \widetilde{A}(K) \end{bmatrix}, \qquad K \in S_0. \qquad (10.2.19)$$

Observe that the map $\widetilde{A}$ is a continuous function of $K \in S_0$ (more precisely, a continuous function of the entries of K, considered as pq complex variables subject to the restriction $K \in S_0$ but otherwise independent). To verify this, write

$$A = \begin{bmatrix} A_{11} & A_{12} \\ A_{21} & A_{22} \end{bmatrix},$$

where A_{11} is $p \times p$ and A_{22} is $q \times q$, and note that (10.2.19) can be rewritten in the form

$$\begin{bmatrix} A_{11} & A_{12} \\ A_{21} & A_{22} \end{bmatrix} \begin{bmatrix} I \\ K \end{bmatrix} X = \begin{bmatrix} I \\ \widetilde{A}(K) \end{bmatrix},$$

for some invertible matrix X. It follows that $X = (A_{11} + A_{12}K)^{-1}$ (in particular, $A_{11} + A_{12}K$ is invertible for every $K \in S_0$), and

$$\widetilde{A}(K) = (A_{21} + A_{22}K)(A_{11} + A_{12}K)^{-1}.$$

This is obviously a continuous function of $K \in S_0$. Now the fixed point Theorem A.8.1 guarantees existence of a matrix $K' \in C^{q \times p}$ such that

$$\|K'\| \leq 1, \quad K_0 = K'P_0, \quad \text{and} \quad \widetilde{A}(K') = K'.$$

Then

$$\widetilde{\mathcal{M}} = \text{Range} \begin{bmatrix} I \\ K' \end{bmatrix}$$

satisfies all the requirements of Theorem 10.2.4. $\qquad\qquad\qquad\qquad\qquad\square$

10.3 Deductions from Theorem 10.2.4

Theorem 10.2.4 has several important corollaries. To start with, every H-unitary matrix is obviously invertible and a plus matrix. Thus:

Theorem 10.3.1. *Let A be H-unitary, and let $\mathcal{M}_0 \subseteq C^n$ be an A-invariant H-nonnegative (resp. H-nonpositive) subspace. Then there exists an A-invariant H-nonnegative (resp. H-nonpositive) subspace $\mathcal{M}$ such that $\mathcal{M} \supseteq \mathcal{M}_0$ and dim $\mathcal{M} = i_+(H)$ (resp. dim $\mathcal{M} = i_-(H)$).*

The part of Theorem 10.3.1 concerning H-nonpositive subspaces follows by noticing that A is also a plus matrix with respect to $-H$, and applying Theorem 10.2.4 with H replaced by $-H$.

A similar result holds for H-selfadjoint matrices:

Theorem 10.3.2. *Let A be H-selfadjoint, and let $\mathcal{M}_0 \subseteq C^n$ be an A-invariant H-nonnegative (resp. H-nonpositive) subspace. Then there exists an A-invariant H-nonnegative (resp. H-nonpositive) subspace $\mathcal{M}$ such that $\mathcal{M} \supseteq \mathcal{M}_0$ and dim $\mathcal{M} = i_+(H)$ (resp. dim $\mathcal{M} = i_-(H)$).*

Theorem 10.3.2 follows immediately from Theorem 10.3.1 by using Proposition 4.3.4. Indeed, if A is H-selfadjoint, then U given by (4.3.14) is H-unitary, and since A and U are functions of each other, they have exactly the same set of invariant subspaces.

A matrix $A \in C^{n \times n}$ is called an H-*expansion*, or H-*expansive*, if $[Ax, Ax] \geq [x, x]$ for every $x \in C^n$.

Proposition 10.3.3. *A matrix A is H-expansive if and only if the hermitian matrix $A^*HA - H$ is positive semidefinite.*

The proof is easy:

$$[Ax, Ax] - [x, x] = (HAx, Ax) - (Hx, x) = ((A^*HA - H)x, x), \quad x \in C^n,$$

and therefore $[Ax, Ax] \geq [x, x]$ for every $x \in C^n$ if and only if $((A^*HA - H)x, x) \geq 0$ for every $x \in C^n$, which amounts to the positive semidefiniteness of $A^*HA - H$.

A matrix $B \in C^{n \times n}$ is called H-*dissipative* if the real part of $[Bx, x]$ is nonpositive for every $x \in C^n$. This condition can easily be interpreted in terms of positive semidefiniteness: A matrix B is dissipative if and only if

$$B^*H + HB \leq 0. \tag{10.3.20}$$

Lemma 10.3.4. (a) *Let A be H-expansive, and let $w, \eta \in C$ be such that w has positive imaginary part, $|\eta| = 1$ and η is not an eigenvalue of A. Then the matrix*

$$B = i(wA - \overline{w}\eta I)(A - \eta I)^{-1} \tag{10.3.21}$$

is H-dissipative.

(b) *Let B be H-dissipative, and let $w, \eta \in \mathbb{C}$ be such that $|\eta| = 1$, w has positive imaginary part and is not an eigenvalue of $-iB$. Then the matrix*

$$A = \eta(-iB - \overline{w}I)(-iB - wI)^{-1} \tag{10.3.22}$$

is H-expansive.

Proof. (a) Letting B be defined by (10.3.21), we have to prove that

$$HB + B^*H = iH(wA - \overline{w}\eta I)(A - \eta I)^{-1} - i(A^* - \overline{\eta}I)^{-1}(\overline{w}A^* - w\overline{\eta}I)H \tag{10.3.23}$$

is negative semidefinite. Multiplying (10.3.23) on the right by $A - \eta I$ and on the left by $A^* - \overline{\eta}I$, we obtain (using $\overline{\eta}\eta = 1$)

$$i(A^* - \overline{\eta}I)H(wA - \overline{w}\eta I) - i(\overline{w}A^* - w\overline{\eta}I)H(A - \eta I)$$

$$= (iw - i\overline{w})(A^*HA - H) - i\overline{\eta}wHA - i\overline{w}\eta A^*H + iw\overline{\eta}HA + i\overline{w}\eta A^*H$$

$$= (iw - i\overline{w})(A^*HA - H). \tag{10.3.24}$$

By the hypothesis that A is H-expansive, we have

$$((A^*HA - H)x, x) = (A^*HAx, x) - (Hx, x) = [Ax, Ax] - [x, x] \geq 0,$$

for every $x \in \mathbb{C}^n$. Hence $A^*HA - H$ is positive semidefinite. Since w has positive imaginary part, the real number $iw - i\overline{w}$ is negative, and hence (10.3.24) is negative semidefinite. Finally, (10.3.23) is congruent to (10.3.24), therefore (10.3.23) is negative semidefinite as well.

(b) Letting A be defined by (10.3.22), consider

$$A^*HA - H = (iB^* - \overline{w}I)^{-1}(iB^* - wI)H(-iB - \overline{w}I)(-iB - wI)^{-1} - H$$

$$= (iB^* - \overline{w}I)^{-1}[(iB^* - wI)H(-iB - \overline{w}I)$$

$$\quad - (iB^* - \overline{w}I)H(-iB - wI)](-iB - wI)^{-1}$$

$$= (iB^* - \overline{w}I)^{-1}[iwHB - i\overline{w}B^*H - i\overline{w}HB + iwB^*H](-iB - wI)^{-1}$$

$$= (iB^* - \overline{w}I)^{-1}(-i\overline{w} + iw)(HB + B^*H)(-iB - wI)^{-1}. \tag{10.3.25}$$

Since B is H-dissipative and $-i\overline{w} + iw$ is a negative real number, the matrix $(-i\overline{w} + iw)(HB + B^*H)$ is positive semidefinite, and hence so is also $A^*HA - H$, by (10.3.25). $\qquad\square$

The transformations in (10.3.21) and (10.3.22) are, in fact, the inverses of each other. Thus, denote

$$g(z) = (wz - \overline{w}\eta)(z - \eta)^{-1}, \quad z \in \mathbb{C},$$

and

$$h(z) = \eta(z - \overline{w})(z - w)^{-1}, \quad z \in \mathsf{C},$$

where $|\eta| = 1$ and w has positive imaginary part. Then for every matrix A not having eigenvalue η, the number w is not an eigenvalue of the matrix $B = g(A)$, and

$$A = h(B). \tag{10.3.26}$$

Here the matrices $g(A)$ and $h(B)$ are understood in the sense of functions of matrices; they are also given by formulas (10.3.21) and (10.3.22), respectively. The verification of (10.3.26) is easy:

$$
\begin{aligned}
h(g(z)) &= \eta(g(z) - \overline{w})(g(z) - w)^{-1} \\
&= \left(\frac{wz - \overline{w}\eta}{z - \eta} - \overline{w} \right) \left(\frac{wz - \overline{w}\eta}{z - \eta} - w \right)^{-1} = \eta(wz - \overline{w}z)(w\eta - \overline{w}\eta)^{-1} \\
&= z,
\end{aligned}
$$

and therefore, by the well-known properties of functions of matrices, $h(g(A)) = A$.

As for the invertibility of $B - wI$, note that

$$
\begin{aligned}
B - wI &= (wA - \overline{w}\eta I)(A - \eta I)^{-1} - wI \\
&= [(wA - \overline{w}\eta I) - w(A - \eta I)](A - \eta I)^{-1} = (w\eta - \overline{w}\eta)(A - \eta I)^{-1}.
\end{aligned}
$$

Theorem 10.3.5. *Let B be H-dissipative or H-expansive, and let $\mathcal{M}_0 \subseteq \mathsf{C}^n$ be a B-invariant H-nonnegative (resp. H-nonpositive) subspace. Then there exists a B-invariant maximal H-nonnegative (resp. maximal H-nonpositive) subspace $\mathcal{M}$ such that $\mathcal{M} \supseteq \mathcal{M}_0$.*

Proof. First, suppose that B is H-dissipative. Assume $\mathcal{M}_0$ is H-nonnegative. Let A be given by (10.3.22). Then A is H-expansive and, in particular, A is a plus matrix. Note also that $\mathcal{M}_0$ is A-invariant, because A is a function of B. Selecting w so that $\overline{w} \notin \sigma(-iB)$, we guarantee that A is invertible. By Theorem 10.2.4, there exists an A-invariant subspace $\mathcal{M}$ which is maximal H-nonnegative and contains $\mathcal{M}_0$. Since B is a function of A (given by formula (10.3.21)) $\mathcal{M}$ is also B-invariant. This proves the part of the theorem for H-dissipative B and H-nonnegative subspaces. If $\mathcal{M}_0$ is H-nonpositive, apply the already proved part to the $(-H)$-dissipative matrix $-B$.

If B is H-expansive, then we use Lemma 10.3.4, and the property that the matrices given by formulas (10.3.21) and (10.3.22) have the same set of invariant subspaces, to reduce the proof to the already considered case of H-dissipative B. $\qquad\square$

Finally, we are in a position to prove a stronger form of Theorem 10.2.4.

Theorem 10.3.6. *Let $A \in \mathbb{C}^{n \times n}$ be a plus matrix such that* Range A *is not H-nonnegative. Let $\mathcal{M}_0 \subseteq \mathbb{C}^n$ be an A-invariant subspace which is H-nonnegative (resp. H-nonpositive). Then there exists H-nonnegative (resp. H-nonpositive) A-invariant subspace $\widetilde{\mathcal{M}} \supseteq \mathcal{M}_0$ such that* dim $\widetilde{\mathcal{M}} = i_+(H)$ *(resp.* dim $\widetilde{\mathcal{M}} = i_-(H)$*).*

Proof. By Proposition 10.2.1 there exists a $\mu > 0$ such that $[Ax, Ax] \geq \mu[x, x]$ for all $x \in \mathbb{C}^n$. (The possibility of $\mu = 0$ is excluded by the hypothesis that the range of A is not H-nonnegative.) Scaling A, if necessary, we can assume that $\mu = 1$. Then A is H-expansive, and the result follows from Theorem 10.3.5. $\qquad \square$

10.4 Expansive, Contractive Matrices and Spectral Properties of Invariant Maximal Semidefinite Subspaces

In the previous section we have derived extension results on invariant semidefinite subspaces for various classes of matrices with respect to an indefinite inner product. In particular, the extension results yield existence of invariant maximal semidefinite subspaces. Here, we shall see that, for expansive and contractive matrices, these invariant subspaces have additional spectral properties.

Recall that a matrix $A \in \mathbb{C}^{n \times n}$ is called H-*expansive* if $[Ax, Ax] \geq [x, x]$ for every $x \in \mathbb{C}^n$. A matrix $A \in \mathbb{C}^{n \times n}$ is called H-*strictly expansive* if

$$[Ax, Ax] > [x, x] \quad \text{for every nonzero } x \in \mathbb{C}^n. \tag{10.4.27}$$

Equivalently, A is H-strictly expansive if and only if there is $\varepsilon > 0$ such that

$$[Ax, Ax] \geq [x, x] + \varepsilon \|x\|^2 \quad \text{for every } x \in \mathbb{C}^n. \tag{10.4.28}$$

Indeed, (10.4.28) clearly implies (10.4.27). Conversely, if (10.4.27) holds, then the continuous function $f(x) := [Ax, Ax] - [x, x]$ takes only positive values on the unit sphere $\mathcal{S} = \{x \in \mathbb{C}^n : \|x\| = 1\}$. Since the unit sphere is bounded and closed in $\mathbb{C}^n$, the function $f(x)$ attains its minimum value, call this value ε, on $\mathcal{S}$. Clearly, $\varepsilon > 0$, and by definition of ε, (10.4.28) is satisfied for $x \in \mathcal{S}$. By homogeneity, (10.4.28) is satisfied for every nonzero x, and for $x = 0$ the inequality (10.4.28) is obvious.

We record several useful properties of H-expansive matrices:

Proposition 10.4.1. (a) *If A and B are H-expansive, then so is AB.*

(b) *If A is H-strictly expansive, and B is H-expansive, then BA is H-strictly expansive.*

(c) *Assume that H is a signature matrix: $H = H^* = H^{-1}$. Then A is H-strictly expansive if and only if A^* is H-strictly expansive.*

Proof. The proofs of (a) and (b) follow readily from the definitions.

We now prove (c) for H-expansive matrices. The case of H-strictly expansive matrices is completely analogous. Given $A \in \mathbb{C}^{n \times n}$, let $\eta \in \mathbb{C}$, $|\eta| = 1$, be such that $\eta I + A$ is invertible, and let

$$C = (\eta I - A)(\eta I + A)^{-1} H.$$

Then

$$H + C = [I + (\eta I - A)(\eta I + A)^{-1}] H = 2\eta (\eta I + A)^{-1} H, \qquad (10.4.29)$$

and, in particular, $H + C$ is invertible. Furthermore,

$$H - A^* H A = 2((H + C)^{-1})^*(C^* + C)(H + C)^{-1}, \qquad (10.4.30)$$

$$H - A H A^* = 2(I + CH)^{-1}(C^* + C)((I + CH)^{-1})^*. \qquad (10.4.31)$$

To verify (10.4.30), rewrite the right-hand side of (10.4.30), using (10.4.29), in the form

$$\frac{1}{2}(\overline{\eta} I + A^*) H (C^* + C) H (\eta I + A)$$

$$= \frac{1}{2}(\overline{\eta} I + A^*) H [H(\overline{\eta} I + A^*)^{-1}(\overline{\eta} I - A^*) + (\eta I - A)(\eta I + A)^{-1} H] H (\eta I + A)$$

$$= \frac{1}{2}[(\overline{\eta} I - A^*) H (\eta I + A) + (\overline{\eta} I + A^*) H (\eta I - A)]$$

$$= \frac{1}{2}[(H - A^* H A) + (H - A^* H A)] = H - A^* H A.$$

Equation (10.4.31) can be verified in a similar way. In view of Proposition 10.3.3, it is clear from (10.4.30) and (10.4.31) that A is H-expansive if and only if $C^* + C$ is negative semidefinite, and the same condition is equivalent to A^* being H-expansive. This proves (c). $\square$

For H-strictly expansive matrices A, the A-invariant maximal H-nonnegative and A-invariant maximal H-nonpositive subspaces (which exist by Theorem 10.3.5) have important spectral properties:

Theorem 10.4.2. *Let $A \in \mathbb{C}^{n \times n}$ be an H-strictly expansive matrix. Then A has no eigenvalues of modulus 1, and the following properties hold true for any A-invariant H-nonnegative subspace $\mathcal{M}_+$ of dimension $i_+(H)$, and any A-invariant H-nonpositive subspace $\mathcal{M}_-$ of dimension $i_-(H)$:*

(1) *The subspace $\mathcal{M}_+$ is actually H-positive.*

(2) *$|\lambda| > 1$ for every eigenvalue λ of $A|_{\mathcal{M}_+}$.*

(3) *The subspace $\mathcal{M}_+$ contains all root subspaces of A corresponding to its eigenvalues λ with $|\lambda| > 1$.*

(4) *The subspace $\mathcal{M}_-$ is actually H-negative.*

(5) *$|\lambda| < 1$ for every eigenvalue λ of $A|_{\mathcal{M}_-}$.*

(6) *The subspace $\mathcal{M}_-$ contains all root subspaces of A corresponding to its eigenvalues λ with $|\lambda| < 1$.*

Proof. We prove only parts (1), (2), and (3), leaving the proofs of (4), (5), and (6) for the reader.

First note that by (10.4.28) the matrix $A^*HA - H$ is positive definite. Therefore, by Theorem A.1.5 A has no eigenvalues on the unit circle.

Next, let P be an invertible matrix whose first $i_+(H)$ columns form a basis for $\mathcal{M}_+$. A transformation $A \mapsto P^{-1}AP$, $H \mapsto P^*HP$ will transform A into a block triangular form. So (cf. Exercise 5 in Section 10.6) we may assume that

$$A = \begin{bmatrix} A_0 & A_1 \\ 0 & A_2 \end{bmatrix}, \quad H = \begin{bmatrix} H_0 & H_1 \\ H_1^* & H_2 \end{bmatrix},$$

where A_0 is $i_+(H) \times i_+(H)$, and where $\mathcal{M}_+ = \text{Span}\{e_1, \ldots, e_{i_+(H)}\}$. Since $\mathcal{M}_+$ is H-nonnegative, it follows that H_0 is positive semidefinite. Now the positive definiteness of $A^*HA - H$ implies the positive definiteness of $A_0^*H_0A_0 - H_0$. Now Theorem A.1.5 shows that H_0 is invertible. Thus, in fact, H_0 is positive definite, so (1) holds and, also, the matrix A_0 has no eigenvalues of modulus less than 1. This proves (2).

For (3), observe that by the same Theorem A.1.5 the dimension of the sum of the root subspaces of A corresponding to the eigenvalues of A outside the unit circle, is equal to $i_+(H)$. Thus, in view of the already proven property (2), if (3) were false, the dimension of $\mathcal{M}_+$ would have been less than $i_+(H)$, a contradiction with $\mathcal{M}_+$ being a maximal H-nonnegative subspace. $\square$

Corollary 10.4.3. *Let A be H-strictly expansive. Then the sum of the root subspaces of A corresponding to its eigenvalues λ with $|\lambda| > 1$ (resp., with $|\lambda| < 1$) is the unique A-invariant maximal H-nonnegative (resp., H-nonpositive) subspace.*

For matrices that are H-expansive, but not H-strictly expansive, there is a weaker result:

Theorem 10.4.4. *Let $A \in \mathbb{C}^{n \times n}$ be H-expansive. Then:*

(1) *There exist A-invariant H-nonnegative subspaces $\mathcal{M}_+$ of dimension $i_+(H)$, such that $|\lambda| \geq 1$ for every eigenvalue λ of $A|_{\mathcal{M}_+}$, and containing all root subspaces of A corresponding to its eigenvalues λ with $|\lambda| > 1$.*

(2) *There exist A-invariant H-nonpositive subspaces $\mathcal{M}_-$ of dimension $i_-(H)$, such that $|\lambda| \leq 1$ for every eigenvalue λ of $A|_{\mathcal{M}_-}$, and containing all root subspaces of A corresponding to its eigenvalues λ with $|\lambda| < 1$.*

Proof. The proof uses a perturbation argument. Let P be an invertible matrix such that

$$H = P^* \begin{bmatrix} I_p & 0 \\ 0 & -I_q \end{bmatrix} P, \quad p = i_+(H), \quad q = i_-(H).$$

Then for every ε, $0 < \varepsilon < 2$, the matrix

$$S_\varepsilon := P^{-1} \begin{bmatrix} (1+\varepsilon)I_p & 0 \\ 0 & (1-\varepsilon)I_q \end{bmatrix} P$$

is H-strictly expansive. Indeed,

$$S_\varepsilon^* H S_\varepsilon - H$$

$$= P^* \left(\begin{bmatrix} (1+\varepsilon)I_p & 0 \\ 0 & (1-\varepsilon)I_q \end{bmatrix} \begin{bmatrix} I_p & 0 \\ 0 & -I_q \end{bmatrix} \begin{bmatrix} (1+\varepsilon)I_p & 0 \\ 0 & (1-\varepsilon)I_q \end{bmatrix} \right.$$

$$\left. - \begin{bmatrix} I_p & 0 \\ 0 & -I_q \end{bmatrix} \right) P$$

$$= \varepsilon P^* \begin{bmatrix} (2+\varepsilon)I_p & 0 \\ 0 & (2-\varepsilon)I_q \end{bmatrix} P,$$

which is positive definite.

Consider the matrix $B_\varepsilon := A S_\varepsilon$ where $0 < \varepsilon < 2$. Obviously, $\lim_{\varepsilon \to 0} S_\varepsilon = I$, so we have $\lim_{\varepsilon \to 0} B_\varepsilon = A$. On the other hand, B_ε is H-strictly expansive by Proposition 10.4.1. Thus, Corollary 10.4.3 can be applied to B_ε. Let $\mathcal{M}_{+,\varepsilon}$ and $\mathcal{M}_{-,\varepsilon}$ be the B_ε-invariant maximal H-nonnegative and B_ε-invariant maximal H-nonpositive subspaces, respectively.

At this point we need basic topological properties of the set of subspaces in $\mathbf{C}^n$ (see, for example, [41, Chapter 13] for details, also Section A.5). The set of subspaces in $\mathbf{C}^n$ is a compact complete metric space in the metric defined by the gap (A.5.13):

$$\theta(\mathcal{M}, \mathcal{N}) := \|P_\mathcal{M} - P_\mathcal{N}\|, \quad \mathcal{M}, \mathcal{N} \text{ subspaces in } \mathbf{C}^n.$$

Therefore, there exists the limit subspaces

$$\mathcal{M}_\pm := \lim_{m \to \infty} \mathcal{M}_{\pm, \varepsilon_m}$$

for some sequence $\{\varepsilon_m\}_{m=1}^\infty$ such that $\lim_{m \to \infty} \varepsilon_m = 0$. One verifies that the subspaces $\mathcal{M}_\pm$ have the properties required in (1) and (2). We provide details for $\mathcal{M}_+$ only; the consideration of $\mathcal{M}_-$ is completely analogous.

Since $\mathcal{M}_{+,\varepsilon_m}$ is H-nonnegative, the subspace $\mathcal{M}_+$ is H-nonnegative as well, as one can easily verify arguing by contradiction. Now

$$i_+(H) = \dim \mathcal{M}_{+,\varepsilon_m} = \dim \mathcal{M}_+,$$

for sufficiently large m, where the second equality is guaranteed by Proposition A.5.4. Hence $\mathcal{M}_+$ is maximal H-nonnegative. Next, we check that $\mathcal{M}_+$ is A-invariant. Let $x_0 \in \mathcal{M}_+$. We have

$$Ax_0 = (A - B_{\varepsilon_m})x_0 + B_{\varepsilon_m}x_0 \tag{10.4.32}$$
$$= (A - B_{\varepsilon_m})x_0 + B_{\varepsilon_m}P_{\mathcal{M}_{+,\varepsilon_m}}x_0 + B_{\varepsilon_m}(P_{\mathcal{M}_+} - P_{\mathcal{M}_{+,\varepsilon_m}})x_0.$$

Apply $P_{\mathcal{M}_+}$:

$$\begin{aligned}
P_{\mathcal{M}_+}Ax_0 &= P_{\mathcal{M}_+}\left[(A - B_{\varepsilon_m})x_0 + B_{\varepsilon_m}P_{\mathcal{M}_{+,\varepsilon_m}}x_0 + B_{\varepsilon_m}(P_{\mathcal{M}_+} - P_{\mathcal{M}_{+,\varepsilon_m}})x_0\right] \\
&= P_{\mathcal{M}_+}\left[(A - B_{\varepsilon_m})x_0 + B_{\varepsilon_m}(P_{\mathcal{M}_+} - P_{\mathcal{M}_{+,\varepsilon_m}})x_0\right] \\
&\quad + (P_{\mathcal{M}_+} - P_{\mathcal{M}_{+,\varepsilon_m}})B_{\varepsilon_m}P_{\mathcal{M}_{+,\varepsilon_m}}x_0 + B_{\varepsilon_m}P_{\mathcal{M}_{+,\varepsilon_m}}x_0. \tag{10.4.33}
\end{aligned}$$

In the last equality we have used the property that the range of $P_{\mathcal{M}_{+,\varepsilon_m}}$ is B_{ε_m}-invariant. Subtracting (10.4.32) from (10.4.33) we see that $\|P_{\mathcal{M}_+}Ax_0 - Ax_0\|$ tends to zero as $m \longrightarrow \infty$. Thus, $P_{\mathcal{M}_+}Ax_0 = Ax_0$, and $\mathcal{M}_+$ is A-invariant, as claimed.

Next, let $x \in \mathcal{R}_{\lambda_0}(A)$, where $|\lambda_0| > 1$. Denote by Γ a circle of sufficiently small radius centered at λ_0, and let $P_\Gamma(A)$ be the Riesz projection of A corresponding to Γ (cf. Section A.3). Then, for sufficiently large m, we have

$$x = P_\Gamma(A)x = (P_\Gamma(A) - P_\Gamma(B_{\varepsilon_m}))x + x_m, \tag{10.4.34}$$

where $x_m = P_\Gamma(B_{\varepsilon_m})x$. By Theorem A.3.2, $\|P_\Gamma(A) - P_\Gamma(B_{\varepsilon_m})\|$ tends to zero as $m \to \infty$, and by Corollary 10.4.3, $x_m \in \mathcal{M}_{+,\varepsilon_m}$. Thus, (10.4.34) yields $\lim_{m\to\infty} x_m = x$. Using Theorem A.5.1, we obtain $x \in \mathcal{M}_+$. This proves that $\mathcal{M}_+$ contains all root subspaces of A corresponding to eigenvalues with modulus larger than 1.

Finally, we prove that $\mathcal{M}_+$ does not intersect the root subspaces of A corresponding to eigenvalues with modulus smaller than 1. Let Γ_0 be a simple closed rectifiable contour such that all eigenvalues λ of A with $|\lambda| \geq 1$ are inside Γ_0, and all eigenvalues λ of A with $|\lambda| < 1$ are outside Γ_0. Take $x \in \mathcal{M}_+$. By Theorem A.5.1, $x = \lim_{m\to\infty} x_m$ for some sequence of vectors x_m, $m = 1, 2, \ldots$, such that $x_m \in \mathcal{M}_{+,\varepsilon_m}$. Now for sufficiently large m, the range of the Riesz projection $P_{\Gamma_0}(B_{\varepsilon_m})$ contains all root subspaces of B_{ε_m} corresponding to the eigenvalues of B_{ε_m} with modulus larger than 1, as it is easily seen using the continuity of eigenvalues of B_{ε_m} with $m \to \infty$. In particular, using also Corollary 10.4.3, we have

$$x_m = P_{\Gamma_0}(B_{\varepsilon_m})x_m.$$

Now write, for sufficiently large m:

$$\begin{aligned}
x - P_{\Gamma_0}(A)x &= (x - x_m) + (x_m - P_{\Gamma_0}(A)x) \\
&= (x - x_m) + (P_{\Gamma_0}(B_{\varepsilon_m})x_m - P_{\Gamma_0}(A)x) \\
&= (x - x_m) + (P_{\Gamma_0}(B_{\varepsilon_m})x_m - P_{\Gamma_0}(A)x_m) + P_{\Gamma_0}(A)(x_m - x). \tag{10.4.35}
\end{aligned}$$

In view of Theorem A.3.2, the right-hand side of (10.4.35) tends to zero as $m \to \infty$. Thus, $x = P_{\Gamma_0}(A)x$, and the result follows. $\qquad\square$

We now consider contractions with respect to the indefinite inner product. A matrix $A \in C^{n \times n}$ is said to be *H-contractive*, or an *H-contraction*, if $[Ax, Ax] \leq [x, x]$ for every $x \in C^n$. Many properties of H-contractive matrices are analogous, but with the inequality signs reversed, to the properties of H-expansive matrices. For illustration, we state one such analogue here (without proof); that of Theorem 10.4.4.

Theorem 10.4.5. *Let $A \in C^{n \times n}$ be H-contractive. Then:*

(1) *There exist A-invariant H-nonnegative subspaces $\mathcal{M}_+$ of dimension $i_+(H)$, such that $|\lambda| \leq 1$ for every eigenvalue λ of $A \mid_{\mathcal{M}_+}$, and containing all root subspaces of A corresponding to its eigenvalues λ with $|\lambda| < 1$.*

(2) *There exist A-invariant H-nonpositive subspaces $\mathcal{M}_-$ of dimension $i_-(H)$, such that $|\lambda| \geq 1$ for every eigenvalue λ of $A \mid_{\mathcal{M}_-}$, and containing all root subspaces of A corresponding to its eigenvalues λ with $|\lambda| > 1$.*

10.5   The Real Case

In this section we consider indefinite inner products in a real space R^n. Thus, the inner product is defined by $[x, y] = (Hx, y)$, for $x, y \in R^n$, where $H \in R^{n \times n}$ is a real symmetric invertible matrix, and there exist $x_0, y_0 \in R^n$ such that $[x_0, x_0] < 0 < [y_0, y_0]$.

In the particular case when H has the form

$$H = \begin{bmatrix} I_p & 0 \\ 0 & -I_q \end{bmatrix},$$

all the results and constructions of Section 10.1 are valid in the real case as well, with the subspaces, vectors, and matrices being real rather than complex.

A matrix $A \in R^{n \times n}$ is called an *H-plus matrix* if

$$[x, x] \geq 0, \ x \in R^n \implies [Ax, Ax] \geq 0.$$

As in the complex case (using Theorem A.7.2 in place of Theorem A.7.1) one shows that A is H-plus if and only if there exists $\mu \geq 0$ such that $[Ax, Ax] \geq \mu[x, x]$ for every $x \in R^n$ or, equivalently, if and only if $A^T H A - \mu H$ is positive semidefinite.

Theorem 10.2.4 and its proof hold in the real case. As a consequence, an interesting observation results: *If A is an $n \times n$ real matrix with no real eigenvalues (thus, n is necessarily even), then A cannot be an H-plus matrix, for any real symmetric invertible matrix H that has an odd number of positive eigenvalues, counted with multiplicities.* Indeed, under the hypothesis of the observation, A has no odd dimensional invariant subspaces, and therefore cannot have an invariant maximal H-nonnegative subspace.

Many constructions and results of Section 10.3 involve complex numbers and therefore are not applicable in the real case. However, Theorem 10.3.1 and Proposition 10.3.3 also hold in the real case.

Finally, note that the results of Section 10.4 hold in the real case, with the same proof (except for the proof of Proposition 10.4.1), because the sum of the root subspaces of a real matrix that correspond to its eigenvalues with modulus greater (resp., smaller) than 1, is real. Indeed, as the real Jordan form (Theorem A.2.6) shows, there are bases in these sums of root subspaces that consist of real vectors. As for the proof of Proposition 10.4.1, although it depends on complex numbers, the result itself is valid in the real case, and indeed the real case follows easily from the complex case by considering the linear transformations A and H as acting on C^n rather than on R^n.

10.6 Exercises

1. Prove that the norm (10.1.1) coincides with the largest singular value of Y. Recall that the *singular values* $\alpha_1 \geq \alpha_2 \geq \cdots \geq \alpha_{\min\{n,m\}} \geq 0$ of a matrix $Y \in C^{m \times n}$ are defined by the *singular value decomposition* $Y = UDV$, where U and V are unitary matrices of sizes $m \times m$ and $n \times n$, respectively, and D is an $m \times n$ diagonal matrix with the nonnegative numbers $\alpha_1, \ldots, \alpha_{\min\{n,m\}}$ on the diagonal starting with the upper left corner of D.

2. Let
$$H_0 = \begin{bmatrix} 0 & 1 \\ 1 & 0 \end{bmatrix}. \tag{10.6.36}$$
Find all matrices $A = \begin{bmatrix} a & b \\ c & d \end{bmatrix}$ in the following classes:
(a) A is H_0-expansive, (b) A is H_0-strictly expansive, (c) A is H_0-dissipative, (d) A is singular (=noninvertible) real H_0-expansive, (e) A is singular real H_0-contractive.

3. Let $H = S_n$, the $n \times n$ sip matrix. Answer the questions of the preceding exercise for diagonal matrices $A = \text{diag}(a_1, \ldots, a_n)$, $a_j \in C$.

4. Let H be as in the preceding exercise. Describe all diagonal H-plus matrices.

5. Show that if $A \in C^{n \times n}$ is H-strictly expansive, then $T^{-1}AT$ is also T^*HT-strictly expansive, for every invertible matrix $T \in C^{n \times n}$.

6. Establish an analogue of the preceding exercise for H-dissipative matrices.

7. Provide detailed proof of parts (4), (5), and (6) of Theorem 10.4.2.

8. Give an example of an H-strictly expansive $A \in C^{n \times n}$, and an H-expansive $B \in C^{n \times n}$, such that AB is not H-strictly expansive.

9. Provide details in the proof of Theorem 10.4.5.

10. Prove that if A is H-expansive and invertible, then A^{-1} is H-contractive and, conversely, if A is H-contractive and invertible, then A^{-1} is H-expansive.

11. (a) Verify by a direct computation that the matrix $A = \begin{bmatrix} 0 & 1 \\ -1 & 0 \end{bmatrix}$ is not
     H-plus for any real symmetric invertible indefinite 2×2 matrix H.

     (b) For A given in part (a), find all complex hermitian invertible indefinite
     matrices H such that A is H-plus.

     (c) What are the plus-indices of the H-plus matrix A in (b)?

12. Let H be an invertible indefinite hermitian $n \times n$ matrix and let $A \in \mathsf{C}^{n \times n}$.

     (a) Show that the set of all real numbers μ for which the matrix $A^*HA - \mu H$
     is positive semidefinite is either empty or a bounded closed interval $[a, b]$.

     (b) For every bounded closed interval $[a, b]$ find hermitian H and $A \in \mathsf{C}^{n \times n}$
     such that $[a, b]$ consists of exactly those real numbers μ for which the
     matrix $A^*HA - \mu H$ is positive semidefinite.

13. Let $H = \begin{bmatrix} \alpha & 0 \\ 0 & \beta \end{bmatrix}$, and

$$A = \begin{bmatrix} a & b \\ c & 0 \end{bmatrix}, \quad \text{where} \quad a, b, c \in \mathsf{C}. \tag{10.6.37}$$

Assume that $\alpha > 0$ and $\beta < 0$. Verify that A is H-strictly expansive if and
only if the following inequality holds:

$$-\beta\alpha \left(|a|^2 - 1 - |c|^2 |b|^2 \right) - \beta^2 |c|^2 - \alpha^2 |b|^2 > 0.$$

Conclude that if $\alpha^2 \neq \beta^2$, then there exists a matrix A of the form (10.6.37)
such that A is H-strictly expansive, but A^* is not.

10.7   Notes

The approach to invariant subspaces via fixed point theorems originated with M.G.
Krein [59]. Since then, it was developed in many directions, including classes of
operators in infinite dimensional spaces, see [6] and references there.

The description of H-nonnegative and H-neutral subspaces for $H = I_p \oplus -I_q \oplus 0_r$ which is similar to the one given in Section 10.2 was developed and used
in [83].

The proof of Proposition 10.2.1 using the Toeplitz–Hausdorff theorem goes
back to Ando [1].

Example 10.2.5 is taken from [83]. The material of Sections 10.3 and 10.4 is
based on [83].

Results related to those of Section 10.2 were obtained in [10]. Theorem 10.4.4
(in a slightly stronger version, and in the context of infinite dimensional spaces)
is proved in [57]; see Theorem 11.2 there.

The proof of Proposition 10.4.1 is adapted from [17].

Chapter 11

Differential Equations of First Order

This chapter contains a brief introduction to first order time-invariant (i.e., constant coefficient) systems of differential equations. The objective is to discuss those systems with symmetries in which an indefinite inner product plays a role, so that these applications serve to fix some of the theory already developed. Also, the scene will be set for a more substantial treatment of higher order systems in Chapter 13. The reader is referred to beginning differential equations texts for the details of proofs of basic results that are needed here.

11.1 Boundedness of solutions

Let K and H be $n \times n$ hermitian matrices with H invertible, and consider the differential equation

$$iH\frac{dx}{dt} = Kx; \quad x = x(t) \in \mathsf{C}^n, \quad t \in \mathsf{R}. \tag{11.1.1}$$

The symmetries of the coefficients make this a fundamental "Hamiltonian" system of differential equations; "fundamental" because the coefficients do not depend on t.

The general solutions of (11.1.1) are easily obtained and can be found in many textbooks on differential equations:

Proposition 11.1.1. *The formula*

$$x(t) = e^{-itH^{-1}K}x_0, \quad x_0 \in \mathsf{C}^n \tag{11.1.2}$$

defines the general solution of (11.1.1).

In applications it is important to know when every solution of (11.1.1) is bounded on the real line. Criteria of this kind are well-known for the *general* system of first order linear differential equations:

$$\frac{dx}{dt} = iAx; \quad x = x(t) \in \mathbb{C}^n, \quad t \in \mathbb{R} \tag{11.1.3}$$

where A is an $n \times n$ matrix. Thus:

Proposition 11.1.2. *The following statements are equivalent:*

(a) *Every solution of* (11.1.3) *is bounded on the real line;*

(b) *There exists a positive constant M such that*

$$\|e^{itA}\| \le M \quad \text{for} \quad \text{every} \quad t \in \mathbb{R};$$

(c) *A is diagonalizable with all eigenvalues real.*

The proof can be obtained without difficulty using the Jordan form of A or, more precisely, using the description of a fundamental set of solutions in terms of eigenvalues, eigenvectors, and generalized eigenvectors of the matrix iA. Again, this can be found in many differential equations textbooks.

Returning to the Hamiltonian system (11.1.1), it is clear that the matrix $A := -H^{-1}K$ is H-selfadjoint, and it follows from Proposition 11.1.2 that all solutions of (11.1.1) are bounded on the real line if and only if A is r-diagonalizable. This observation, together with Theorem 9.2.2, leads to the equivalence of (i), (iii), and (iv) in the following result.

We say that the solutions of (11.1.1), where H and K are hermitian matrices with invertible H, are *stably bounded* if every solution of (11.1.1) is bounded on the real line and this property holds for all equations

$$i\tilde{H}\frac{dx}{dt} = \tilde{K}x, \quad \tilde{H} = \tilde{H}^*, \quad \tilde{K} = \tilde{K}^*,$$

with $\tilde{H}$ and $\tilde{K}$ sufficiently close to H and K, respectively. If in the above definition H is kept fixed, i.e.,, we always take $\tilde{H} = H$, then we say that the solutions of (11.1.1) are *H-stably bounded*; if the range of the independent variable t is restricted to the half line $[0, \infty)$, then stable boundedness on the half line is obtained.

Theorem 11.1.3. *Let H and K be $n \times n$ hermitian matrices, with H invertible. Then the following statements are equivalent:*

(i) *the solutions of* (11.1.1) *are stably bounded;*

(ii) *the solutions of* (11.1.1) *are stably bounded on the half line $[0, \infty)$;*

(iii) *the solutions of* (11.1.1) *are H-stably bounded;*

(iv) *$\sigma(-H^{-1}K) \in \mathbb{R}$, and the quadratic form (Hx, x) is definite on the subspace $\mathrm{Ker}(\lambda_i I + H^{-1}K)$ for every $\lambda_i \in \sigma(-H^{-1}K)$.*

Proof. We verify that (i) and (ii) are equivalent. By Proposition 11.1.1, the solutions of (11.1.1) are bounded on the half line $[0, \infty)$ if and only if the matrix $iA := -iH^{-1}K$ has all eigenvalues with nonnegative real parts and the eigenvalues of iA with zero real parts (if any) have only partial multiplicities equal to 1. But since A is H-selfadjoint, the eigenvalues of iA are symmetric with respect to the imaginary axis. Thus, the criterion for boundedness of (11.1.1) on the half line $[0, \infty)$ boils down to r-diagonalizability of A; this is just the criterion for boundedness of solutions of (11.1.1) on the whole real line. $\square$

Now consider the matrix version of equation (11.1.1):

$$iH\frac{dX}{dt} = KX, \tag{11.1.4}$$

together with the initial condition $X(0) = I_n$. The solution $X(t)$ of (11.1.4) satisfying this initial condition is uniquely defined, and the matrix $X(1)$ is called the *monodromy matrix* of the matrix equation (11.1.4) or of the vector equation (11.1.1). Clearly, the monodromy matrix is equal to $e^{-iH^{-1}K}$. Thus, statement (iv) of Theorem 11.1.3 can be re-cast in terms of the monodromy matrix:

Corollary 11.1.4. *Under the hypotheses of Theorem 11.1.3, each of the statements* (i) *and* (ii) *is equivalent to*

(iv) *All eigenvalues of the monodromy matrix X have absolute value 1, and the quadratic form (Hx, x) is definite on the subspace $\mathrm{Ker}(\lambda_i I - X)$ for every $\lambda_i \in \sigma(X)$.*

In view of the formula $X = e^{-iH^{-1}K}$, the proof follows immediately from Theorem 11.1.3 .

This section concludes with a description of the connected components of the set of stably bounded Hamiltonian systems of the form (11.1.1). This is summarized in the next theorem, where the terminology and notation of Section 9.7 are used.

Theorem 11.1.5. (a) *Let an invertible $H = H^* \in \mathbb{C}^{n \times n}$ be fixed. If $n_1, \ldots, n_p$ are integers with the properties* (9.7.11), (9.7.12), *and* (9.7.13), *then the stably bounded first order systems*

$$iH\frac{dx}{dt} = Kx \tag{11.1.5}$$

for which

$$\mathrm{ind}_r (-H^{-1}K, H) = \{n_1, \ldots, n_p\}$$

belong to the same connected component in the set of all stably bounded systems (11.1.5) *with fixed H.*

Conversely, if two stably bounded systems

$$iH\frac{dx}{dt} = K_1 x \quad \text{and} \quad iH\frac{dx}{dt} = K_2 x, \qquad K_1 \text{ and } K_2 \text{ hermitian,}$$

belong to the same connected component in the set of all stably bounded systems (11.1.5) with fixed H, then

$$\operatorname{ind}_r \left(-H^{-1}K_1, H\right) = \operatorname{ind}_r \left(-H^{-1}K_2, H\right).$$

(b) If $n_1, \ldots, n_p$ are integers with the properties (9.7.11) and (9.7.12), then the stably bounded first order systems

$$iH_1 \frac{dx}{dt} = Kx \tag{11.1.6}$$

for which

$$\operatorname{ind}_r \left(-H^{-1}K, H\right) = \{n_1, \ldots, n_p\}$$

belong to the same connected component in the set of all stably bounded systems (11.1.5) with arbitrary $K = K^* \in \mathbb{C}^{n \times n}$ and invertible $H = H^* \in \mathbb{C}^{n \times n}$.

Conversely, if

$$iH_1 \frac{dx}{dt} = K_1 x \quad \text{and} \quad iH_2 \frac{dx}{dt} = K_2 x,$$

where

$$K_1 = K_1^*, K_2 = K_2^* \in \mathbb{C}^{n \times n}, \quad \text{and} \quad H_1 = H_1^*, H_2 = H_2^* \in \mathbb{C}^{n \times n} \quad \text{are invertible},$$

belong to the same connected component in the set of all stably bounded systems (11.1.5) with arbitrary $K = K^* \in \mathbb{C}^{n \times n}$ and invertible $H = H^* \in \mathbb{C}^{n \times n}$, then

$$\operatorname{ind}_r \left(-H_1^{-1}K_1, H_1\right) = \operatorname{ind}_r \left(-H_2^{-1}K_2, H_2\right).$$

The proof is obtained by combining Theorems 11.1.3 and 9.7.1, 9.7.2.

11.2   Hamiltonian Systems of Positive Type with Constant Coefficients

Consider the Hamiltonian system of differential equations with constant coefficients:

$$iH \frac{dx}{dt} = (K + K_0)x, \tag{11.2.7}$$

where $H = H^*$ is invertible, and $K = K^*$, $K_0 = K_0^*$. Here, the matrix K_0 is viewed as a small perturbation of K; so $\|K_0\|$ is small in some sense. We are to study the behavior of the eigenvalues of the monodromy matrix of (11.2.7) (called the *multiplicators* of the system (11.2.7)) as K_0 changes. The case when K_0 is positive definite is of special interest, and the system is then said to be of *positive type*.

Assume now that all solutions of the system

$$iH\frac{dx}{dt} = Kx, \quad t \in \mathsf{R} \tag{11.2.8}$$

are bounded. In this case, it follows from Proposition 11.1.2 that the multiplicators of (11.2.8) are unimodular and, for every multiplicator λ_0, the quadratic form (Hx, x) is nondegenerate on the subspace $\mathrm{Ker}(\lambda_0 I - X)$, where $X = \exp(-iH^{-1}K)$ is the monodromy matrix of (11.2.8).

We say that the multiplicator λ_0 has *positive multiplicity* $r_+(\lambda_0)$ and *negative multiplicity* $r_-(\lambda_0)$ if the quadratic form (Hx, x) defined on $\mathrm{Ker}(\lambda_0 I - X)$ has $r_+(\lambda_0)$ positive squares and $r_-(\lambda_0)$ negative squares in its canonical form. In particular, $r_+(\lambda_0) + r_-(\lambda_0)$ coincides with the dimension of $\mathrm{Ker}(\lambda_0 I - X)$, which in turn is equal to the multiplicity of λ_0 as a zero of $\det(\lambda I - X)$. In particular, the multiplicator λ_0 is said to be of *positive* (resp. *negative*) type if the quadratic form (Hx, x) is positive (resp. negative) definite on $\mathrm{Ker}(\lambda_0 I - X)$.

Theorem 11.2.1. *Let (11.2.7) be a constant coefficient Hamiltonian system of positive type, and assume that all solutions of the unperturbed system (11.2.8) are bounded. Then for $\|K_0\|$ small enough all solutions of (11.2.7) are also bounded.*

Moreover, if λ_0 is a multiplicator of (11.2.8) with positive multiplicity $r_+(\lambda_0)$ and negative multiplicity $r_-(\lambda_0)$, then $r_+(\lambda_0)$ (resp., $r_-(\lambda_0)$) multiplicators of (11.2.7) in a neighborhood of λ_0 are of positive (resp. negative) type and situated on the unit circle in the negative (resp. positive) direction from λ_0.

By convention, the counterclockwise direction is positive, and the clockwise direction is negative.

Proof. As $X = \exp(-iH^{-1}K)$ is the monodromy matrix of (11.2.8) then, for any multiplicator λ_0 of (11.2.8) there is an eigenvalue μ_0 of the matrix $A = -H^{-1}K$ for which $\lambda_0 = e^{i\mu_0}$. Observe that A is H-selfadjoint and $-H^{-1}K_0$ is H-negative for K_0 positive definite; moreover, $\mathrm{Ker}(\lambda_0 I - X) = \mathrm{Ker}(\mu_0 I - A)$.

Now apply Theorem 9.6.1, suitably adapted, to the H-negative perturbation $-H^{-1}K_0$ of $-H^{-1}K$ and, after mapping the perturbed real eigenvalues μ on the unit circle with the map $\lambda = e^{i\mu}$, the theorem is obtained. $\square$

Note that positive definiteness of K_0 is essential in Theorem 11.2.1. Indeed, we know from Theorem 11.1.3 that if λ_0 is not of positive or negative type, then there exists a perturbation $K_0 = K_0^*$ with the norm as small as we wish such that the perturbed system

$$iH\frac{dx}{dt} = (K + K_0)x$$

has an unbounded solution. Theorem 11.2.1 shows, in particular, that this situation is impossible for positive definite perturbations.

Let λ_0 be a multiplicator of the system (11.2.8) with bounded solutions, and let $r_+(\lambda_0)$ (resp. $r_-(\lambda_0)$) be the positive (resp. negative) multiplicity of λ_0.

It is possible to regard λ_0 as $r_+(\lambda_0) + r_-(\lambda_0)$ equal multiplicators; $r_+(\lambda_0)$ of them being of positive type and $r_-(\lambda_0)$ of them being of negative type. With this convention, one can reformulate Theorem 11.2.1 in more informal terms as follows: a multiplicator of positive (resp. negative) type of the system (11.2.8) with bounded solutions moves clockwise (resp. counterclockwise) on the unit circle as the matrix K is perturbed by a positive definite matrix.

11.3   Exercises

1. Show that a matrix A is H-selfadjoint if and only if e^{iA} is H-unitary.

2. Let A be an H-selfadjoint matrix. Then A is stably r-diagonalizable if and only if e^{iA} is stably u-diagonalizable.

3. Let

$$i\frac{dx}{dt} = Ax(t), \quad t \in \mathsf{R} \qquad\qquad (11.3.9)$$

   be a differential equation with an H-selfadjoint $n \times n$ matrix A.

   (a) Under what conditions is every solution of (11.3.9) bounded (on the real line)?

   (b) Under what conditions is every nonzero solution of (11.3.9) unbounded?

   (c) Assume that every nonzero solution of (11.3.9) is unbounded. Show that, similarly, there exists an $\varepsilon > 0$ such that every nonzero solution is unbounded for all differential equations of the form

   $$i\frac{dx}{dt} = Bx(t), \quad t \in \mathsf{R}$$

   where B is an H-selfadjoint matrix and $\|B - A\| < \varepsilon$.

   (d) Is the statement in (c) valid if "unbounded" is replaced by "bounded"?

4. (Floquet's theorem) Let $G \in \mathsf{C}^{n \times n}$ be an invertible hermitian matrix, and let $H(t)$, $t \in \mathsf{R}$, be an hermitian piecewise continuous $n \times n$ matrix function. Let $Z(t)$ be the unique solution of the initial value problem

   $$G\frac{dZ}{dt} = iH(t)Z(t), \quad Z(0) = I_n.$$

   Furthermore, fix a G-skew-adjoint matrix V, in other words, $V = -G^{-1}V^*G$. Then a matrix function $X(t)$ is a solution of

   $$G\frac{dX}{dt} = iH(t)X(t)$$

   if and only if the matrix function

   $$Y(t) := e^{tV}Z(t)^{-1}X(t)$$

satisfies the differential equation with constant coefficients

$$\frac{dY}{dt} = VY.$$

5. Solve the differential equation

$$iH_j \frac{dx(t)}{dt} = G_j x(t), \quad t \in \mathsf{R}$$

for the following pairs of hermitian $n \times n$ matrices. In each case determine whether or not the solutions are stably bounded.

(a)

$$H_1 = \begin{bmatrix} 0 & 0 & 0 & \cdots & 0 & 1 \\ 0 & 0 & 0 & \cdots & 1 & 0 \\ & & \cdots\cdots\cdots & & \\ 0 & 1 & 0 & \cdots & 0 & 0 \\ 1 & 0 & 0 & \cdots & 0 & 0 \end{bmatrix}, \quad G_1 = \begin{bmatrix} 0 & 1 & 0 & \cdots & 0 & 0 \\ 1 & 0 & 0 & \cdots & 0 & 0 \\ 0 & 0 & 1 & \cdots & 0 & 0 \\ & & \cdots\cdots\cdots & & \\ 0 & 0 & 0 & \cdots & 0 & 1 \end{bmatrix}.$$

(b)

$$H_2 = H_1, \quad G_2 = \begin{bmatrix} 1 & 0 & 0 & \cdots & 0 & 0 \\ 0 & 1 & 0 & \cdots & 0 & 0 \\ & & \cdots\cdots\cdots & & \\ 0 & 0 & 0 & \cdots & 1 & 0 \\ 0 & 0 & 0 & \cdots & 0 & -1 \end{bmatrix}.$$

(c)

$$H_3 = H_1, \quad G_3 = \begin{bmatrix} I_k & 0 \\ 0 & -I_k \end{bmatrix}, \quad n = 2k \quad \text{is even.}$$

(d)

$$H_4 = H_1, \quad G_4 = \operatorname{diag}(\lambda_1, \ldots, \lambda_n), \quad \lambda_1, \ldots, \lambda_n \in \mathsf{R}.$$

(e)

$$H_5 = G_1, \quad G_5 = \begin{bmatrix} 0 & 0 & 0 & \cdots & 0 & \lambda_0 \\ 0 & 0 & 0 & \cdots & \lambda_0 & 1 \\ \vdots & \vdots & \vdots & \cdot{}^{\cdot{}^{\cdot}} & \vdots & \vdots \\ 0 & \lambda_0 & 1 & \cdots & 0 & 0 \\ \lambda_0 & 1 & 0 & \cdots & 0 & 0 \end{bmatrix}, \quad \lambda_0 \in \mathsf{R}.$$

6. Let H and K be invertible hermitian $n \times n$ matrices. Prove that the stable boundedness of solutions of

$$iH\frac{dx}{dt} = Kx$$

is equivalent to the stable boundedness of solutions of each of the following three equations:

$$iH\frac{dx}{dt} = -Kx, \quad iK\frac{dx}{dt} = Hx, \quad iK\frac{dx}{dt} = -Hx.$$

7. Provide a detailed proof for Theorem 11.1.5.

11.4   Notes

The main contents of this chapter are applications of the results of Chapter 9 to differential equations of first order with hermitian matrix coefficients. The stability results were in fact obtained by M.G. Krein [61] for a more general case and a more general formulation. The connected components in a more general case were studied in [27], see also [40]. The latter reference is the source of the entire chapter.

Chapter 12

Matrix Polynomials

This chapter concerns the study of matrix polynomials of arbitrary degree with hermitian coefficients. Let

$$L(\lambda) = \sum_{j=0}^{\ell} A_j \lambda^j, \qquad A_j = A_j^* \in \mathbb{C}^{n \times n} \quad \text{for} \quad j = 0, \dots, \ell$$

be such a polynomial. When A_ℓ is invertible the *companion matrix*

$$C_L = \begin{bmatrix} 0 & I & 0 & \cdots & 0 \\ 0 & 0 & I & & 0 \\ & & \ddots & & \vdots \\ & & & & I \\ -\widetilde{A}_0 & -\widetilde{A}_1 & \cdots & & -\widetilde{A}_{\ell-1} \end{bmatrix}, \tag{12.0.1}$$

where $\widetilde{A}_j = A_\ell^{-1} A_j$, $j = 0, \dots, \ell - 1$ plays an important role. It is well-known that the spectral structure of $L(\lambda)$ (i.e., the eigenvalues, eigenvectors and generalized eigenvectors) and that of C_L are intimately related. For example, if x_0 is an eigenvector of $L(\lambda)$ corresponding to λ_0 (so that $L(\lambda_0)x_0 = 0$, $x_0 \neq 0$), then the vector

$$\langle x_0, \lambda_0 x_0, \dots, \lambda_0^{\ell-1} x_0 \rangle := \begin{bmatrix} x_0 \\ \lambda_0 x_0 \\ \vdots \\ \lambda_0^{\ell-1} x_0 \end{bmatrix}$$

is an eigenvector of C_L corresponding to its eigenvalue λ_0. This observation applies generally to matrix polynomials with invertible leading coefficient (not only with hermitian coefficients).

A property enjoyed by matrix polynomials with hermitian coefficients is of particular interest in this work: namely, that the companion matrix is selfadjoint with respect to an indefinite inner product. Thus,

$$B_L C_L = C_L^* B_L,$$

where

$$B_L = \begin{bmatrix} A_1 & A_2 & \cdots & A_\ell \\ A_2 & & \cdot^{\displaystyle\cdot^{\displaystyle\cdot}} & \\ \vdots & A_\ell & & \\ A_\ell & & & 0 \end{bmatrix}$$

is hermitian and invertible (note that B_L is never positive (or negative) definite except for the trivial case when $\ell = 1$ and A_ℓ is positive (or negative) definite). Hence the properties of selfadjoint matrices in indefinite inner products play a key role in the investigation of hermitian matrix polynomials.

The results presented in this chapter are also based on the general theory of matrix polynomials, where a considerable amount of knowledge is available now (see the authors' monograph [39], for example, or Chapter 14 of [70]). Before going on to the situations in which indefinite inner products play a role, we give an account of some basic facts from this theory.

12.1   Standard Pairs and Triples

In this section attention is confined to matrix polynomials $L(\lambda) = \sum_{j=0}^\ell A_j \lambda^j$ ($\lambda \in \mathsf{C}$) where the coefficients $A_0, A_1, \ldots, A_\ell$ are $n \times n$ complex matrices and the leading coefficient A_ℓ is invertible. The *degree* of such a polynomial is ℓ. The degree of the scalar polynomial $\det L(\lambda)$ will obviously be $n\ell$, so the *spectrum* $\sigma(L)$ of L, defined by

$$\sigma(L) = \{\lambda \in \mathsf{C} \mid \det L(\lambda) = 0\},$$

is finite and consists of not more than $n\ell$ different complex numbers; these are called the *eigenvalues* of L.

We now define an $n\ell \times n\ell$ matrix whose spectral properties are intimately related to those of L. Let $\tilde{A}_j = A_\ell^{-1} A_j, j = 0, 1, \ldots, \ell-1$ and define the *companion matrix* of L by (12.0.1). The relationship referred to can be concentrated in the statement that $I\lambda - C_L$ and $L(\lambda) \oplus I_{n(\ell-1)}$ are *equivalent*, where $I_{n(\ell-1)}$ is the identity matrix of size $n(\ell - 1)$. That is, there exist $n\ell \times n\ell$ matrix polynomials $E(\lambda)$ and $F(\lambda)$, whose inverses are also matrix polynomials, for which

$$L(\lambda) \oplus I_{n(\ell-1)} = E(\lambda)(I\lambda - C_L)F(\lambda). \tag{12.1.2}$$

This fact can be demonstrated by writing out $E(\lambda)$ and $F(\lambda)$ explicitly. In fact, we may take

$$F(\lambda = \begin{bmatrix} I & 0 & \cdots & 0 \\ I\lambda & I & & \vdots \\ \vdots & & \ddots & \\ I\lambda^{\ell-1} & I\lambda^{\ell-2} & \cdots & I \end{bmatrix},$$

$$E(\lambda) = \begin{bmatrix} K_{\ell-1}(\lambda) & K_{\ell-2}(\lambda) & \cdots & K_1(\lambda) & K_0(\lambda) \\ I & 0 & & & \\ 0 & I & & & \vdots \\ \vdots & & \ddots & & \\ 0 & 0 & \cdots & I & 0 \end{bmatrix}$$

where $K_0(\lambda) = A_\ell$ and $K_{r+1}(\lambda) = \lambda K_r(\lambda) + A_{\ell-r-1}$ for $r = 0, 1, \ldots, \ell - 2$. Clearly, $\det F(\lambda) \equiv 1$, $\det E(\lambda) \equiv \pm \det A_\ell \neq 0$, and hence $E(\lambda)^{-1}$ and $F(\lambda)^{-1}$ are polynomials. A direct multiplication shows that (12.1.2) is satisfied.

It follows from this equivalence that the elementary divisors of $L(\lambda)$ and $I\lambda - C_L$ coincide; see Section A.6. In particular, the eigenvalues of L coincide with those of C_L and, furthermore, their partial multiplicities agree.

More generally, any $\ell n \times \ell n$ matrix T for which $L(\lambda) \oplus I_{n(\ell-1)}$ and $I\lambda - T$ are equivalent is called a *linearization* of L. It follows that *all linearizations of L are similar* to one another and, in particular, to the companion matrix C_L.

To study the matrix polynomial $L(\lambda)$ it is convenient to introduce pairs of matrices (X, T) as follows: X is an $n \times n\ell$ matrix and T is an $n\ell \times n\ell$ matrix such that

$$X = [I \ 0 \ \cdots \ 0] \, S, \quad T = S^{-1} C_L S \qquad (12.1.3)$$

for some $n\ell \times n\ell$ invertible matrix S, where C_L is the companion matrix of $L(\lambda)$. In particular, T is also a linearization of $L(\lambda)$. Such a pair (X, T) will be called a (right) *standard pair* of $L(\lambda)$.

If the standard pair (X, T) of $L(\lambda)$ is such that T is a matrix in Jordan canonical form (which is necessarily the Jordan form of C_L), then we say that (X, T) is a *Jordan pair*.

The next proposition shows that the definition of a standard pair used here is consistent with that of [39].

Proposition 12.1.1. *The matrices $X \in \mathbb{C}^{n \times 2n}$ and $T \in \mathbb{C}^{2n \times 2n}$ form a standard pair if and only if the $nl \times nl$ matrix*

$$\begin{bmatrix} X \\ XT \\ \vdots \\ XT^{\ell-1} \end{bmatrix} \qquad (12.1.4)$$

is nonsingular and
$$A_\ell X T^\ell + \cdots + A_1 X T + A_0 X = 0. \tag{12.1.5}$$

Proof. Let $P_1 = \begin{bmatrix} I_n & 0 & \cdots & 0 \end{bmatrix}$, and it is easily verified that, for $r = 0, 1, \ldots,$
$\ell - 1$,
$$P_1 C_L^r = \begin{bmatrix} 0 & \cdots & 0 & I & 0 & \cdots & 0 \end{bmatrix}, \tag{12.1.6}$$

where I is in the position $r + 1$, and that

$$P_1 C_L^\ell = \begin{bmatrix} -\tilde{A}_0 & -\tilde{A}_1 & \cdots & -\tilde{A}_{\ell-1} \end{bmatrix}. \tag{12.1.7}$$

Consequently, using (12.1.3),

$$\begin{bmatrix} X \\ XT \\ \vdots \\ XT^{\ell-1} \end{bmatrix} = \begin{bmatrix} P_1 \\ P_1 C_L \\ \vdots \\ P_1 C_L^{\ell-1} \end{bmatrix} S = I_{\ell n} S = S. \tag{12.1.8}$$

But it also follows from (12.1.6) and (12.1.7) that

$$\sum_{r=0}^{\ell-1} \tilde{A}_r P_1 C_L^r = -P_1 C_L^\ell$$

from which (12.1.5) follows.

Conversely, given (12.1.4) and (12.1.5), define S by (12.1.8) and (12.1.3)
follows. $\square$

When the matrix polynomial $L(\lambda)$ is monic (i.e., with leading coefficient
$A_\ell = I$), the following important *representation theorem* holds (see [39]), i.e., the
coefficients are represented in terms of the two matrices of a standard pair:

Theorem 12.1.2. *Let (X, T) be a standard pair of a monic matrix polynomial $L(\lambda)$.*
Then
$$L(\lambda) = \lambda^\ell I - XT^\ell(V_1 + \cdots + V_\ell \lambda^{\ell-1}), \tag{12.1.9}$$

where V_i, $i = 1, \ldots, \ell$ are the $n\ell \times n$ matrices for which

$$[V_1 \cdots V_\ell] = \begin{bmatrix} X \\ XT \\ \vdots \\ XT^{\ell-1} \end{bmatrix}^{-1}.$$

Proof. Let $L(\lambda) = \lambda^\ell I + \sum_{j=0}^{\ell-1} \lambda^j A_j$. It is sufficient to prove Theorem 12.1.2 for
$X = [I \ 0 \ \cdots \ 0]$, $T = C_L$. In this case we have

$$[I \ 0 \ \cdots \ 0] C_L^\ell = [-A_0, -A_1, \ldots, -A_{\ell-1}]$$

and

$$
\begin{bmatrix}
[I\ 0\ \cdots\ 0] \\
[I\ 0\ \cdots\ 0]\, C_L \\
\vdots \\
[I\ 0\ \cdots\ 0]\, C_L^{\ell-1}
\end{bmatrix}
= I,
$$

so the formula (12.1.9) follows. □

Let $L(\lambda) = \sum_{j=0}^{\ell} A_j \lambda^j$ be a matrix polynomial with invertible leading coefficient A_ℓ, and with a standard pair (X,T). Then a third matrix Y can be defined by

$$
Y =
\begin{bmatrix}
X \\
XT \\
\vdots \\
XT^{\ell-1}
\end{bmatrix}^{-1}
\begin{bmatrix}
0 \\
\vdots \\
0 \\
A_\ell^{-1}
\end{bmatrix}.
$$

The (ordered) triple of matrices (X,T,Y) is called a *standard triple* for $L(\lambda)$. If T is a matrix in Jordan form, then the triple (X,T,Y) is said to be a *Jordan triple*. The following proposition follows from the definitions:

Proposition 12.1.3. *If (X,T,Y) is a standard triple for $L(\lambda)$, and if a triple of matrices (X_1,T_1,Y_1) is similar to (X,T,Y), i.e.,,*

$$
X_1 = XS, \quad T_1 = S^{-1}TS, \quad Y_1 = S^{-1}Y
$$

for some invertible matrix S, then (X_1,T_1,Y_1) is also a standard triple for $L(\lambda)$. Conversely, any two standard triples for $L(\lambda)$ are similar.

The following *resolvent form* for the inverse of a matrix polynomial is often useful:

Theorem 12.1.4. *If (X,T,Y) is a standard triple of an $n \times n$ matrix polynomial $L(\lambda)$ of degree ℓ with invertible leading coefficient, then*

$$
L(\lambda)^{-1} = X(I_{n\ell}\lambda - T)^{-1}Y, \quad \lambda \in \mathbb{C} \setminus \sigma(L). \tag{12.1.10}
$$

Conversely, if (12.1.10) holds for a triple of matrices (X,T,Y) of sizes $n \times n\ell$, $n\ell \times n\ell$, $n\ell \times n$, respectively, then (X,T,Y) is a standard triple for $L(\lambda)$.

For a proof see [39, Theorem 2.4] or [70, Section 14.2].

If (X,T,Y) is a standard triple for a matrix polynomial $L(\lambda)$, then the ordered pair (T,Y) is said to be a *left standard pair* of $L(\lambda)$ and if, in addition, T is in a Jordan form, then (T,Y) is a *left Jordan pair*. A dual result for Theorem 12.1.2 holds for left standard pairs; it can be obtained easily from Theorem 12.1.2 by using the observation that (T,Y) is a left standard pair for the matrix polynomial $L(\lambda)$ if and only if the pair of transposed matrices (Y^T,T^T) is a right standard pair for the polynomial $L(\lambda)^T$ with transposed coefficients.

To illustrate the constructions and results of this section, we borrow an example from [39]:

Example 12.1.5. *Let*

$$L(\lambda) = \begin{bmatrix} \lambda^3 & \sqrt{2}\lambda^2 - \lambda \\ \sqrt{2}\lambda^2 + \lambda & \lambda^3 \end{bmatrix}.$$

A Jordan pair (X, T) of $L(\lambda)$ is given by the following two matrices:

$$X = \begin{bmatrix} 1 & 0 & -\sqrt{2}+1 & -\sqrt{2}-2 & \sqrt{2}+1 & \sqrt{2}+2 \\ 0 & 1 & 1 & 0 & 1 & 0 \end{bmatrix},$$

$$T = 0 \oplus 0 \oplus \begin{bmatrix} 1 & 1 \\ 0 & 1 \end{bmatrix} \oplus \begin{bmatrix} -1 & 1 \\ 0 & -1 \end{bmatrix}.$$

Note that T is in Jordan form, as required by the definition of a Jordan pair.
Let

$$Y = \begin{bmatrix} 0 & 1 \\ -1 & 0 \\ (\sqrt{2}+2)/4 & 0 \\ (-\sqrt{2}-1)/4 & 1/4 \\ (-\sqrt{2}+2)/4 & 0 \\ (-\sqrt{2}+1)/4 & -1/4 \end{bmatrix}.$$

Then (X, T, Y) is a Jordan triple for $L(\lambda)$. In particular,

$$\begin{bmatrix} X \\ XT \\ XT^2 \end{bmatrix} Y = \begin{bmatrix} 0 \\ 0 \\ I_2 \end{bmatrix}.$$

This equality is easily verified by a straightforward computation.   □

See [39] for more details concerning standard pairs and triples of matrix polynomials.

12.2   Matrix Polynomials with Hermitian Coefficients

Consider the matrix polynomial $L(\lambda) = \sum_{j=0}^{\ell} \lambda^j A_j$, where A_ℓ is invertible, and assume that all the coefficients A_j are hermitian: $A_j = A_j^*$. In this case the polynomial $L(\lambda)$ is said to be *hermitian*. The special case of *quadratic* hermitian polynomials (with $l = 2$) is particularly important in the analysis of vibrating systems and already requires the full machinery to be developed here.

In this case important roles are played by the companion matrix C_L, by standard pairs (X, T) as discussed above, and by the $\ell n \times \ell n$ matrix B_L defined by

$$B_L = \begin{bmatrix} A_1 & A_2 & \cdots & A_{\ell-1} & A_\ell \\ A_2 & & & A_\ell & 0 \\ \vdots & & \cdot^{\cdot^{\cdot}} & & \vdots \\ A_{\ell-1} & A_\ell & & & \\ A_\ell & 0 & \cdots & & 0 \end{bmatrix}. \tag{12.2.11}$$

It is clear that B_L is hermitian and invertible and so it may be used to form an indefinite inner product in $\mathbb{C}^{n\ell}$. Furthermore, it is easily seen that $B_L C_L = C_L^* B_L$; so C_L is B_L-selfadjoint. In particular, $\sigma(C_L)$ (and therefore also $\sigma(L)$) is symmetric relative to the real axis, i.e., if $\lambda_0 \in \sigma(C_L)$ then $\overline{\lambda_0} \in \sigma(C_L)$. Moreover, the degrees of elementary divisors of $L(\lambda)$ corresponding to λ_0 and to $\overline{\lambda_0}$ are the same[1].

Observe that the signature of B_L is given by:

$$\text{sig } B_L = \begin{cases} 0 & \text{if } \ell \text{ is even} \\ \text{sig } A_\ell & \text{if } \ell \text{ is odd.} \end{cases} \tag{12.2.12}$$

To see this, consider the continuous family of hermitian matrices:

$$B(\varepsilon) = \begin{bmatrix} \varepsilon A_1 & \varepsilon A_2 & \cdots & \varepsilon A_{\ell-1} & A_\ell \\ \varepsilon A_2 & & & A_\ell & 0 \\ \vdots & & & \vdots & \\ \varepsilon A_{\ell-1} & A_\ell & & & \\ A_\ell & 0 & & \cdots & 0 \end{bmatrix}, \quad \varepsilon \in [0,1].$$

Clearly, $B(1) = B_L$ and $B(\varepsilon)$ is invertible for all $\varepsilon \in [0,1]$. Hence sig $B(\varepsilon)$ is independent of ε on this interval; so sig $B_L = $ sig $B(0)$, and the latter is easily calculated to yield (12.2.12).

Real eigenvalues having only linear elementary divisors (i.e., semi-simple eigenvalues) are of special interest. The next proposition gives a geometric characterization of a real eigenvalue with this property.

A nonzero vector $x_0 \in \mathbb{C}^n$ is an *eigenvector* of $L(\lambda)$ corresponding to $\lambda_0 \in \sigma(L)$ if $L(\lambda_0)x_0 = 0$. When this is the case it is easily seen that the vector

$$\widehat{x}_0 = \langle x_0, \lambda_0 x_0, \ldots, \lambda_0^{\ell-1} x_0 \rangle \tag{12.2.13}$$

is an eigenvector of C_L. In fact, $C_L \widehat{x}_0 = \lambda_0 \widehat{x}_0$. Conversely, the structure of C_L implies that every eigenvector $\widehat{x}_0$ of C_L corresponding to the eigenvalue λ_0 of C_L is of the form (12.2.13) where x_0 is an eigenvector of $L(\lambda)$.

It is not difficult to see that an eigenvalue λ_0 has only linear elementary divisors if and only if the dimension of $\text{Ker} L(\lambda_0)$ coincides with the multiplicity of λ_0 as a zero of $\det L(\lambda)$. Or, equivalently, if and only if the dimension of $\text{Ker}(I\lambda_0 - C_L)$ coincides with the multiplicity of λ_0 as a zero of $\det(I\lambda - C_L)$.

Proposition 12.2.1. *Let $L(\lambda)$ be an hermitian matrix polynomial with invertible leading coefficient, and let $\lambda_0 \in \sigma(L)$ be real. Assume that $L(\lambda)$ has only linear elementary divisors corresponding to λ_0, with p_+ (resp. p_-) associated positive (resp. negative) signs in the sign characteristic of (C_L, B_L). Then the quadratic form defined on the $(p_+ + p_-)$-dimensional subspace $\text{Ker} L(\lambda_0)$ by $(x, L^{(1)}(\lambda_0)x)$ is*

---

[1]It is interesting to note that this symmetry of the spectrum also occurs if the coefficients of $L(\lambda)$ are real matrices but not necessarily hermitian (i.e., real symmetric).

nonsingular and has p_+ (resp. p_-) positive (resp. negative) squares in its canonical form. In other words, the hermitian matrix representation of the form in any basis of $\mathrm{Ker} L(\lambda_0)$ has p_+ positive eigenvalues and p_- negative eigenvalues, counted with multiplicities.

Conversely, if the quadratic form $(x, L^{(1)}(\lambda_0)x)$, $x \in \mathrm{Ker} L(\lambda_0)$ is nonsingular, then $L(\lambda)$ has only linear elementary divisors corresponding to λ_0, and the sign characteristic of (C_L, B_L) associated with λ_0 coincides with the number of positive (resp. negative) squares in the canonical form of $(x, L^{(1)}(\lambda_0)x)$, $x \in \mathrm{Ker} L(\lambda_0)$.

In this statement $L^{(1)}(\lambda)$ is the derivative of $L(\lambda)$ with respect to λ. Also, the quadratic form is said to be nonsingular if for a fixed $x_0 \in \mathrm{Ker} L(\lambda_0)$, $(x_0, L^{(1)}(\lambda_0)y) = 0$ for all $y \in \mathrm{Ker} L(\lambda_0)$ implies $x_0 = 0$.

Proof. Let $x, y \in \mathrm{Ker} L(\lambda_0)$. A direct computation shows that

$$(x, L^{(1)}(\lambda_0)y) = (\widehat{x}, B_L \widehat{y}) \tag{12.2.14}$$

where

$$\widehat{x} = \langle x, \lambda_0 x, \ldots, \lambda_0^{\ell-1} x \rangle, \quad \widehat{y} = \langle y, \lambda_0 y, \ldots, \lambda_0^{\ell-1} y \rangle,$$

and therefore $\widehat{x}, \widehat{y} \in \mathrm{Ker}(I\lambda_0 - C_L)$. Now the proposition follows from Theorem 5.8.1. □

In fact, Theorem 5.8.1 can be used to formulate a generalization of this proposition to include any real eigenvalue, i.e., with no hypotheses on the elementary divisors (see [39, Theorem 10.14]).

The fact that a matrix polynomial has hermitian coefficients is reflected in certain symmetries of its standard triples. The following definition captures these symmetries. A triple of matrices (X, T, Y), where $X \in \mathbb{C}^{n \times n\ell}$, $T \in \mathbb{C}^{n\ell \times n\ell}$, $Y \in \mathbb{C}^{n\ell \times n}$ is said to be a *selfadjoint triple* if there exists an invertible hermitian matrix $M \in \mathbb{C}^{n\ell \times n\ell}$ such that

$$Y^* = XM^{-1}, \quad T^* = MTM^{-1}, \quad X^* = MY. \tag{12.2.15}$$

The first and third equations here are actually equivalent, and one of them could equally well have been omitted.

Theorem 12.2.2. *Let $L(\lambda)$ be an $n \times n$ matrix polynomial of degree ℓ with invertible leading coefficient. Then $L(\lambda)$ is hermitian if and only if it has a standard triple which is selfadjoint, and in this case every standard triple of $L(\lambda)$ is selfadjoint.*

Proof. If (X, T, Y) is a selfadjoint standard triple for $L(\lambda)$, then using the resolvent form (12.1.10) we obtain

$$\begin{aligned} (L(\lambda)^{-1})^* &= \left(X(I_{n\ell}\lambda - T)^{-1}Y \right)^* = Y^*(I_{n\ell}\overline{\lambda} - T^*)^{-1}X^* \\ &= XM^{-1}(I_{n\ell}\overline{\lambda} - MTM^{-1})^{-1}MY = L(\overline{\lambda})^{-1}, \end{aligned}$$

and it follows that $L(\lambda)$ is hermitian.

Conversely, if $L(\lambda)$ is hermitian, then the standard triple

$$(X, T, Y) = \left([I \ 0 \ \cdots \ 0], C_L, \begin{bmatrix} 0 \\ \vdots \\ 0 \\ A_\ell^{-1} \end{bmatrix} \right)$$

of $L(\lambda)$ is selfadjoint in the sense of (12.2.15), and B_L of (12.2.11) plays the role of M. Finally, observe that if (X, T, Y) is selfadjoint with the hermitian matrix M satisfying (12.2.15), and if $S \in \mathbb{C}^{n\ell \times n\ell}$ is invertible, then the similar triple $(XS, S^{-1}TS, S^{-1}Y)$ is also selfadjoint, with the corresponding invertible hermitian matrix S^*MS. □

A more general class of matrix polynomials consists of those with H-self-adjoint coefficients, i.e., $L(\lambda) = \sum_{j=0}^{\ell} \lambda^j A_j$, where all coefficients A_j are H-selfadjoint with a fixed H $(HA_j = A_j^*H, \ j = 0, \ldots, \ell)$ and A_ℓ is invertible. However, such polynomials do not exhibit essentially new properties compared with polynomials having hermitian coefficients. The reason is that the companion matrix C_L is selfadjoint in the indefinite inner product determined by the matrix

$$\begin{bmatrix} H & 0 & \cdots & 0 \\ 0 & H & \cdots & 0 \\ \vdots & \vdots & \ddots & \vdots \\ 0 & 0 & \cdots & H \end{bmatrix} \begin{bmatrix} A_1 & A_2 & \cdots & A_{\ell-1} & A_\ell \\ A_2 & & & A_\ell & 0 \\ \vdots & & \cdot^{\cdot^{\cdot}} & & \vdots \\ A_{\ell-1} & A_\ell & & & \\ A_\ell & 0 & \cdots & & 0 \end{bmatrix}.$$

12.3 Factorization of Hermitian Matrix Polynomials

We start with a characterization of divisibility of matrix polynomials (not necessary hermitian) in terms of their standard pairs (see [39] for the proof and more details). A matrix polynomial $L_1(\lambda)$ is said to be a *right divisor* of a matrix polynomial $L(\lambda)$ if $L(\lambda) = L_2(\lambda) L_1(\lambda)$ for some matrix polynomial $L_2(\lambda)$.

Theorem 12.3.1. *Let $L(\lambda)$ be a matrix polynomial of degree ℓ with invertible leading coefficient and with standard pair (X, T). Let $\mathcal{L}$ be a T-invariant subspace such that the linear map*

$$Q_k(\mathcal{L}) := \begin{bmatrix} X \\ XT \\ \vdots \\ XT^{k-1} \end{bmatrix} \Bigg|_{\mathcal{L}} : \mathcal{L} \to \mathbb{C}^{nk}$$

is invertible (in particular, $\dim \mathcal{L} = nk$). Then the matrix polynomial of degree k

$$L_1(\lambda) := \lambda^k I - XT^k \big|_{\mathcal{L}} (V_1 + V_2\lambda + \cdots + V_k \lambda^{k-1}), \tag{12.3.16}$$

where

$$[V_1 \, V_2 \ldots V_k] = Q_k(\mathcal{L})^{-1}, \quad V_i : \mathbf{C}^n \to \mathcal{L},$$

is a right divisor of $L(\lambda)$. Conversely, if a monic matrix polynomial $L_1(\lambda)$ of degree k is a right divisor of $L(\lambda)$, then there exists a unique T-invariant subspace $\mathcal{L}$ such that $Q_k(\mathcal{L})$ is invertible and formula (12.3.16) holds.

This theorem shows that there is a one-to-one correspondence between the set of T-invariant subspaces $\mathcal{L}$ for which $Q_k(\mathcal{L})$ is invertible, and the set of monic right divisors of degree k of $L(\lambda)$. We say that $\mathcal{L}$ is the *supporting subspace (with respect to the standard pair (X,T))* of the right monic divisor $L_1(\lambda)$ of $L(\lambda)$.

Note that, in the notation of Theorem 12.3.1, $(X|_{\mathcal{L}}, T|_{\mathcal{L}})$ is a standard pair for the right divisor $L_1(\lambda)$ (a basis is chosen in $\mathcal{L}$ so that $(X|_{\mathcal{L}}, T|_{\mathcal{L}})$ are represented by matrices).

Assume now that the matrix polynomial $L(\lambda)$ is hermitian. In this case we know that C_L is B_L-selfadjoint. So there are C_L-invariant maximal B_L-nonnegative (or B_L-nonpositive) subspaces (Theorem 5.12.1). It turns out that such subspaces are supporting subspaces for certain monic divisors of hermitian matrix polynomials, provided the leading coefficient is positive definite:

Theorem 12.3.2. *Let $L(\lambda)$ be an hermitian matrix polynomial of degree ℓ with positive definite leading coefficient A_ℓ. Let $\mathcal{L}_+$ (resp. $\mathcal{L}_-$) be a C_L-invariant maximal B_L-nonnegative (resp. maximal B_L-nonpositive) subspace. Then:*

(i) $\dim \mathcal{L}_\pm = \dfrac{\ell n}{2}$ *if ℓ is even;* $\dim \mathcal{L}_\pm = \dfrac{\ell \pm 1}{2} n$ *if ℓ is odd;*

(ii) *the linear transformation*

$$Q_k(\mathcal{L}_\pm) = \left.\begin{bmatrix} P \\ PC_L \\ \vdots \\ PC_L^{k-1} \end{bmatrix}\right|_{\mathcal{L}_\pm} \quad : \mathcal{L}_\pm \to \mathbf{C}^{nk}$$

is invertible, where $P = [I\, 0 \ldots 0]$, and $nk = \dim \mathcal{L}_\pm$;

(iii) *the monic matrix polynomial*

$$L_1(\lambda) = \lambda^k I - P \left(C_L|_{\mathcal{L}_\pm}\right)^k \left(V_1 + V_2\lambda + \cdots + V_{k-1}\lambda^{k-1}\right),$$

where $[V_1 V_2 \ldots V_{k-1}] = \left(Q_k(\mathcal{L}_\pm)\right)^{-1}$, is a right divisor of $L(\lambda)$, with the supporting subspace $\mathcal{L}_\pm$.

Proof. The statement (i) follows from (12.2.12). The statement (iii) follows from (ii) in view of Theorem 12.3.1. So it remains to prove (ii). It is sufficient to check that $Q_k(\mathcal{L}_\pm)$ is one-to-one. Assume first that ℓ is even, and let

$$x = \langle x_1, \ldots, x_\ell \rangle \in \operatorname{Ker} Q_k(\mathcal{L}_\pm),$$

where $x_i \in \mathbb{C}^n$. Since

$$\begin{bmatrix} P \\ PC_L \\ \vdots \\ PC_L^{k-1} \end{bmatrix} = [I_{nk} \quad 0],$$

we have $x_1 = \cdots = x_k = 0$, and this implies that $(B_L x, x) = 0$. However, since $\mathcal{L}_\pm$ is B-nonnegative (or B-nonpositive) and $C_L x \in \mathcal{L}_\pm$, Schwarz' inequality applies and gives,

$$|(B_L C_L x, x)|^2 \le (B_L C_L x, C_L x)(B_L x, x). \tag{12.3.17}$$

(Note that Schwarz' inequality holds if restricted to a nonpositive (or nonnegative) subspace, see (2.3.9).) Now (12.3.17) yields $(B_L C_L x, x) = 0$. But

$$(B_L C_L x, x) = (B_L \langle x_2, \dots, x_\ell, y \rangle, \langle x_1, \dots, x_\ell \rangle)$$

for some $y \in \mathbb{C}^n$. Using the fact that $x_1 = \cdots = x_k = 0$ and the definition of B_L, it follows that $(A_\ell x_{k+1}, x_{k+1}) = 0$. But A_ℓ is positive definite, and so $x_{k+1} = 0$.
Now

$$(B_L C_L x, C_L x) = 0,$$

and using Schwarz' inequality again, it is found that $(B_L C_L^2 x, C_L x) = 0$. This implies that $x_{k+2} = 0$, and the process can be continued.

Assume now that ℓ is odd, and consider $\mathcal{L}_+$ (so that $k = \frac{\ell+1}{2}$). Let $x = \langle x_1, \dots, x_\ell \rangle \in \operatorname{Ker} Q_k(\mathcal{L}_+)$. Then, in particular, $x_1 = \cdots = x_k = 0$. As in the case of even ℓ, we have $(B_L x, x) = 0$ and by Schwarz' inequality,

$$(B_L C_L x, C_L x) = (B_L C_L^2 x, x) = 0.$$

But

$$(B_L C_L x, C_L x) = (A_\ell x_{k+1}, x_{k+1}),$$

so $x_{k+1} = 0$. Applying a similar argument using $(B_L C_L x, C_L x) = 0$ we obtain $x_{k+2} = 0$, and so on.

For $\mathcal{L}_-$, a similar argument is applied starting as follows: Given

$$x = \langle x_1, \dots, x_\ell \rangle \in \operatorname{Ker} Q_{\frac{\ell-1}{2}}(\mathcal{L}_-),$$

we have

$$0 \ge (B_L x, x) = \left(A_\ell x_{\frac{\ell+1}{2}}, x_{\frac{\ell+1}{2}} \right),$$

and this implies that $x_{\frac{\ell+1}{2}} = 0$ and $(B_L x, x) = 0$. $\qquad \square$

For more information about factorizations of hermitian matrix polynomials see [39]. Here, we shall only state the following result (without proof), which describes a factorization of nonnegative matrix polynomials. An $n \times n$ matrix polynomial $L(\lambda)$ is said to be *nonnegative* if $(L(\lambda)x, x) \geq 0$ for all $x \in C^n$ and all $\lambda \in R$. Clearly, a nonnegative matrix polynomial is hermitian (because for every real λ the matrix $L(\lambda)$ is hermitian). Also, the degree of a nonnegative matrix polynomial is necessarily even. To verify this property, assume by contradiction that

$$L(\lambda) = \sum_{j=0}^{\ell} \lambda^j A_j, \qquad A_\ell \neq 0, \quad \ell \quad \text{odd},$$

is a nonnegative matrix polynomial, and let x_0 be an eigenvector of A_ℓ corresponding to a nonzero eigenvalue λ_0. Then

$$(L(\lambda)x_0, x_0) = \lambda^\ell \lambda_0(x_0, x_0) + \text{ lower order terms},$$

which obviously has opposite signs for $\lambda < 0$ and $\lambda > 0$ if $|\lambda|$ is sufficiently large.

Theorem 12.3.3. *The following statements are equivalent for an $n \times n$ matrix polynomial $L(\lambda)$ with hermitian coefficients and invertible leading coefficient:*

(i) *$L(\lambda)$ is nonnegative;*

(ii) *the leading coefficient of $L(\lambda)$ is positive definite, and all elementary divisors of $L(\lambda)$ corresponding to the real eigenvalues (if any) have even degrees;*

(iii) *the leading coefficient of $L(\lambda)$ is positive definite, and all signs in the sign characteristic of (C_L, B_L) are $+1$'s;*

(iv) *$L(\lambda)$ admits the factorization*

$$L(\lambda) = \big(M(\bar{\lambda})\big)^* M(\lambda) \tag{12.3.18}$$

for some $n \times n$ matrix polynomial $M(\lambda)$ of degree k.

In fact, there is a one-to-one correspondence between C_L-invariant nk-dimensional B_L-neutral subspaces in C^{nl} and factorizations (12.3.18). Indeed, we know from Theorem 12.3.2 that each such subspace $\mathcal{L}$ is a supporting subspace for a monic divisor $L_1(\lambda)$ of $L(\lambda)$. It turns out that the B_L-neutrality of $\mathcal{L}$ implies that the quotient $L(\lambda)L_1(\lambda)^{-1}$ is just $\big(L_1(\bar{\lambda})\big)^*$. Conversely, by Theorem 12.3.1, every factorization (12.3.18) is generated by a C_L-invariant supporting subspace, and the special form of the factorization (12.3.18) (i.e., the quotient $L(\lambda)M(\lambda)^{-1}$ is equal to $\big(M(\bar{\lambda})\big)^*$) implies that this subspace is B_L-neutral.

We remark also that if the conditions (i)–(iv) of Theorem 12.3.3 hold, then the matrix polynomial $M(\lambda)$ in (12.3.18) can be chosen in such a way that $\sigma(M)$ lies in the closed upper halfplane (or $\sigma(M)$ lies in the closed lower halfplane). More generally, let S be a set of nonreal eigenvalues of $L(\lambda)$ such that $\lambda \in S$ implies $\bar{\lambda} \notin S$, and S is maximal with respect to this property. Then there exists a monic matrix polynomial $M(\lambda)$ satisfying (12.3.18) for which $\sigma(M) \setminus R = S$.

12.4 The Sign Characteristic of Hermitian Matrix Polynomials

Let $L(\lambda)$ be a hermitian matrix polynomial with invertible leading coefficient and degree ℓ. As we know, $L(\lambda)$ has a selfadjoint standard triple (X, T, Y), i.e., (X, T, Y) is a standard triple for $L(\lambda)$, and the relations

$$Y^* = XM^{-1}, \quad T^* = MTM^{-1}, \quad X^* = MY$$

hold for some invertible hermitian matrix M. The matrix M is actually unique. This follows from the equality

$$\begin{bmatrix} X \\ XT \\ \vdots \\ XT^{\ell-1} \end{bmatrix} M = \begin{bmatrix} Y^* \\ Y^*T^* \\ \vdots \\ Y^*(T^*)^{\ell-1} \end{bmatrix}$$

and the invertibility of the matrix

$$\begin{bmatrix} X \\ XT \\ \vdots \\ XT^{\ell-1} \end{bmatrix}.$$

In particular, T is M-selfadjoint. Thus, we can speak of the sign characteristic of the pair (T, M).

Then the *sign characteristic* of $L(\lambda)$ is defined as the sign characteristic of (T, M). Since any two standard triples of $L(\lambda)$ are similar (Proposition 12.1.3), the definition of the sign characteristic of $L(\lambda)$ is independent of the choice of selfadjoint standard triple (X, T, Y). Thus, the sign characteristic attaches a sign ± 1 to every elementary divisor $(\lambda - \lambda_0)^\alpha$, $\lambda_0 \in \mathbb{R}$ of $L(\lambda)$.

An *eigenvalue* of an $n \times n$ matrix polynomial $M(\lambda)$ (not necessarily hermitian or with invertible leading coefficient) is a number λ_0 for which $M(\lambda_0)$ is not invertible, and a nonzero vector x_1 for which $M(\lambda_0) x_1 = 0$ is called an *eigenvector* of $M(\lambda)$ corresponding to λ_0.

A sequence of vectors $\{x_1, \ldots, x_k\}$, $x_j \in \mathbb{C}^n$, is called a *Jordan chain* of $M(\lambda)$ at λ_0 of length k if $x_1 \neq 0$ and

$$\begin{aligned} M(\lambda_0) x_1 &= 0, \\ M'(\lambda_0) x_1 + M(\lambda_0) x_2 &= 0, \\ &\vdots \end{aligned}$$

$$\frac{1}{(k-1)!} M^{(k-1)}(\lambda_0) x_1 + \frac{1}{(k-2)!} M^{(k-2)}(\lambda_0) x_2 + \cdots + M(\lambda_0) x_k = 0.$$

$$(12.4.19)$$

Here, $M^{(j)}(\lambda_0)$ is the matrix value of the j-th derivative of $M(\lambda)$ at λ_0. It follows easily from the diagonal Smith form of $M(\lambda)$ (Section A.6) that if the determinant of $M(\lambda)$ is not identically zero, then the lengths of Jordan chains of $M(\lambda)$ are uniformly bounded (i. e. with a bound independent of λ_0).

Proposition 12.4.1. *Let $M(\lambda)$ be an $n \times n$ matrix polynomial with invertible leading coefficient, and let C_M be the companion matrix of $M(\lambda)$. Then the columns of an $n \times r$ matrix X_0 form a Jordan chain for $M(\lambda)$ at λ_0 if and only if the columns of the $\ell n \times r$ matrix*

$$
\begin{bmatrix}
X_0 \\
X_0 J_0 \\
\vdots \\
X_0 J_0^{\ell-1}
\end{bmatrix}
\tag{12.4.20}
$$

form a Jordan chain for C_M corresponding to the same λ_0, where J_0 is the $r \times r$ Jordan block with eigenvalue λ_0.

Proof. Write

$$
M(\lambda) = \sum_{j=0}^{\ell} A_j \lambda^j, \qquad A_j \in \mathbb{C}^{n \times n} \quad \text{for} \quad j = 0, \ldots, \ell.
$$

By assumption, A_ℓ is invertible. Denote $\tilde{A}_j = A_\ell^{-1} A_j$, for $j = 0, 1, \ldots, \ell - 1$, and write $X_0 = [x^{(1)} \ x^{(2)} \ \ldots \ x^{(r)}]$, where $x^{(j)} \in \mathbb{C}^n$. Then the kth column of (12.4.20) has the form

$$
\begin{bmatrix}
x^{(k)} \\
\lambda_0 x^{(k)} + x^{(k-1)} \\
\vdots \\
\sum_{i=0}^{j} \binom{j}{i} \lambda_0^{j-i} x^{(k-i)} \\
\vdots \\
\sum_{i=0}^{\ell-1} \binom{\ell-1}{i} \lambda_0^{\ell-1-i} x^{(k-i)}
\end{bmatrix}, \qquad k = 1, 2, \ldots, r,
\tag{12.4.21}
$$

where

$$
\binom{j}{i} = \begin{cases} \dfrac{j!}{i!(j-i)!} & \text{if } 0 \le i \le j \\ 0 & \text{if } i < 0 \text{ or if } i > j \end{cases}
$$

are the binomial coefficients, and where it is assumed that $x^{(i)} = 0$ if $i < 1$. The verification of (12.4.21) is straightforward using induction on k (starting with $k = 1$), and taking advantage of the binomial formula $\binom{j}{i} + \binom{j}{i-1} = \binom{j+1}{i}$.

Now the kth column of

$$
C_M \begin{bmatrix} x^{(k)} \\ \lambda_0 x^{(k)} + x^{(k-1)} \\ \vdots \\ \sum_{i=0}^{j} \binom{j}{i} \lambda_0^{j-i} x^{(k-i)} \\ \vdots \\ \sum_{i=0}^{\ell-1} \binom{\ell-1}{i} \lambda_0^{\ell-1-i} x^{(k-i)} \end{bmatrix} - \lambda_0 \begin{bmatrix} x^{(k)} \\ \lambda_0 x^{(k)} + x^{(k-1)} \\ \vdots \\ \sum_{i=0}^{j} \binom{j}{i} \lambda_0^{j-i} x^{(k-i)} \\ \vdots \\ \sum_{i=0}^{\ell-1} \binom{\ell-1}{i} \lambda_0^{\ell-1-i} x^{(k-i)} \end{bmatrix}
$$

takes the form

$$
\begin{bmatrix} x^{(k-1)} \\ \lambda_0 x^{(k-1)} + x^{(k-2)} \\ \vdots \\ \sum_{i=0}^{j} \binom{j}{i} \lambda_0^{j-i} x^{(k-1-i)} \\ \vdots \\ \sum_{i=0}^{\ell-2} \binom{\ell-2}{i} \lambda_0^{\ell-2-i} x^{(k-1-i)} \\ Z_k \end{bmatrix},
$$

where

$$
Z_k = -\tilde{A}_0 x^{(k)} - \tilde{A}_1 \left(\lambda_0 x^{(k)} + x^{(k-1)} \right) - \cdots - \tilde{A}_{\ell-1} \left(\sum_{i=0}^{\ell-1} \binom{\ell-1}{i} \lambda_0^{\ell-1-i} x^{(k-i)} \right)
$$
$$
- \left(\sum_{i=0}^{\ell-1} \binom{\ell-1}{i} \lambda_0^{\ell-i} x^{(k-i)} \right).
$$

Thus, the columns of (12.4.20) form a Jordan chain of C_M corresponding to the eigenvalue λ_0 if and only if

$$
Z_1 = 0, \quad Z_k = \sum_{i=0}^{\ell-1} \binom{\ell-1}{i} \lambda_0^{\ell-1-i} x^{(k-1-i)}, \quad \text{for} \quad k = 2, 3, \ldots, r. \quad (12.4.22)
$$

It is easy to see that $Z_1 = 0$ is equivalent to $M(\lambda_0) x^{(1)} = 0$, and after some straightforward algebra, the kth equation in (12.4.22) boils down to:

$$
\frac{1}{(k-1)!} M^{(k-1)}(\lambda_0)\, x^{(1)} + \frac{1}{(k-2)!} M^{(k-2)}(\lambda_0)\, x^{(2)} + \cdots + M(\lambda_0)\, x^{(k)} = 0,
$$

and we are done. □

Now let $L(\lambda)$ be a hermitian matrix polynomial with invertible leading coefficient, and let λ_0 be a real eigenvalue of $L(\lambda)$. For an eigenvector $x \in \operatorname{Ker} L(\lambda_0)\backslash\{0\}$ let $\nu(x)$ be the maximal length of a Jordan chain of $L(\lambda)$ beginning with the eigenvector x of λ_0. Let

$$\gamma = \max \{\nu(x) \mid x \in \operatorname{Ker} L(\lambda_0) \setminus \{0\}\},$$

and then define the subspaces $\Psi_i \subseteq \operatorname{Ker} L(\lambda_0)$, $i = 1, \ldots, \gamma$ by:

$$\Psi_i = \operatorname{Span} \{x \in \operatorname{Ker} L(\lambda_0) \setminus \{0\} : \nu(x) \geq i\}.$$

Theorem 12.4.2. *Let λ_0 be a real eigenvalue of L and, for $i = 1, \ldots, \gamma$,*

$$f_i(x, y) = \left(x, \sum_{j=1}^{i} \frac{1}{j!} L^{(j)}(\lambda_0)\, y^{(i+1-j)}\right), \qquad x, y \in \Psi_i,$$

where $y = y^{(1)}, y^{(2)}, \ldots, y^{(i)}$ is a Jordan chain of $L(\lambda)$ corresponding to λ_0 with eigenvector y; if $y = 0$, let $f_i(x, y) = 0$. Then:

(i) *$f_i(x, y)$ does not depend on the choice of $y^{(2)}, \ldots, y^{(i)}$;*

(ii) *there exists a selfadjoint linear transformation $G_i : \Psi_i \to \Psi_i$ such that*

$$f_i(x, y) = (x, G_i y), \qquad x, y \in \Psi_i;$$

(iii) *$\Psi_{i+1} = \operatorname{Ker} G_i$ (by definition, $\Psi_{\gamma+1} = \{0\}$);*

(iv) *the number of positive (negative) eigenvalues of G_i, counted with their multiplicities, coincides with the number of positive (negative) signs in the sign characteristic of $L(\lambda)$ corresponding to the elementary divisors $(\lambda - \lambda_0)^i$.*

Proof. We will use the sign characteristic of the pair (C_L, B_L) for the sign characteristic of $L(\lambda)$. We know by Proposition 12.4.1 that the columns of an $n \times r$ matrix X_0 form a Jordan chain for $L(\lambda)$ at λ_0 if and only if the columns of the $\ell n \times r$ matrix form a Jordan chain for C_L corresponding to the same λ_0. So, in view of Theorem 5.8.1, it remains to check that

$$\left(x, \sum_{j=1}^{r} \frac{1}{j!} L^{(j)}(\lambda_0)\, y^{(r+1-j)}\right) = \left(\widehat{x}, B_L \widehat{y}^{(r)}\right) \qquad (12.4.23)$$

for every eigenvector x and every Jordan chain $y^{(1)}, y^{(2)}, \ldots, y^{(r)}$ of $L(\lambda)$ corresponding to λ_0, where

$$\widehat{x} = \begin{bmatrix} x \\ \lambda_0 x \\ \vdots \\ \lambda_0^{\ell-1} x \end{bmatrix} \quad \text{and} \quad [\widehat{y}^{(1)} \ldots \widehat{y}^{(r)}] = \begin{bmatrix} [y^{(1)} \ldots y^{(r)}] \\ [y^{(1)} \ldots y^{(r)}] J_0 \\ \vdots \\ [y^{(1)} \ldots y^{(r)}] J_0^{\ell-1} \end{bmatrix}, \quad J_0 = J_r(\lambda_0).$$

The proof of (12.4.23) is combinatorial. Let $L(\lambda) = \sum_{j=0}^{\ell} \lambda^j A_j$. Then, substituting

$$\widehat{x} = \begin{bmatrix} x \\ \lambda_0 x \\ \vdots \\ \lambda_0^{\ell-1} x \end{bmatrix} \quad \text{and} \quad \widehat{y}^{(r)} = \begin{bmatrix} y^{(r)} \\ \lambda_0 y^{(r)} + y^{(r-1)} \\ \vdots \\ \sum_{i=0}^{j} \binom{j}{i} \lambda_0^{j-i} y^{(r-i)} \\ \vdots \\ \sum_{i=0}^{\ell-1} \binom{\ell-1}{i} \lambda_0^{\ell-1-i} y^{(r-i)} \end{bmatrix}$$

(by definition $y^{(p)} = 0$ for $p \le 0$) in the expression $(\widehat{x}, B_L \widehat{y}_r)$ we deduce that

$$(\widehat{x}, B_L \widehat{y}^{(r)}) = \left(x, \sum_{i=0}^{\ell-1} \left[\sum_{k=1}^{\ell} \sum_{p=i+k}^{\ell} \binom{p-k}{i} \lambda_0^{p-i-1} A_p \right] y^{(r-i)} \right).$$

It is easy to see that

$$\sum_{k=1}^{\ell} \sum_{p=i+k}^{\ell} \binom{p-k}{i} \lambda_0^{p-i-1} A_p = \frac{1}{(i+1)!} L^{(i+1)}(\lambda_0),$$

and (12.4.23) follows. □

Now let us compute the first two linear transformations G_1 and G_2. Since

$$f_1(x, y) = (x, L'(\lambda_0)y), \quad x, y \in \operatorname{Ker} L(\lambda_0),$$

it is easy to see that

$$G_1 = P_1 L'(\lambda_0) P_1 |_{\Psi_1},$$

where P_1 is the orthogonal projection onto Ψ_1. For brevity, write

$$L_0 = L(\lambda_0), \quad L_0' = L^{(1)}(\lambda_0), \quad \text{and} \quad L_0'' = L^{(2)}(\lambda_0).$$

Then, for $x, y \in \operatorname{Ker} G_1$,

$$f_2(x, y) = \left(x, \frac{1}{2} L_0'' y + L_0' y' \right),$$

where $\{y, y'\}$ is a Jordan chain of $L(\lambda)$ corresponding to λ_0. Thus

$$L_0' y + L_0 y' = L_0' y + L_0 (I - P_1) y' = 0, \tag{12.4.24}$$

(and the last equality follows since $L_0 P_1 = 0$).

Denote by L_0^+ : $\mathbb{C}^n \to \mathbb{C}^n$ the linear transformation which is equal to L_0^{-1} on $\Psi_1^\perp$ and zero on $\Psi_1 = \operatorname{Ker} L_0$. Then $L_0^+ L_0 (I - P_1) = I - P_1$ and (12.4.24) gives

$$(I - P_1)y' = -L_0^+ L_0' y.$$

Now, for $x, y \in \Psi_2$,

$$
\begin{aligned}
(x, L_0' y') &= \left(x, L_0'(P_1 + (I - P_1))y'\right) = (x, L_0' P_1 y') + \left(x, L_0'(I - P_1)y'\right) \\
&= (x, G_1 P_1 y') + \left(x, L_0'(I - P_1)y'\right) = \left(x, L_0'(I - P_1)y'\right) \\
&= (x, -L_0' L_0^+ L_0' y),
\end{aligned}
$$

where the last but one equality follows from the fact that $x \in \operatorname{Ker} G_1$ and $G_1 = G_1^*$. Thus

$$f_2(x, y) = \left(x, \frac{1}{2} L_0'' y - L_0' L_0^+ L_0' y\right),$$

and, finally,

$$G_2 = P_2 \left[\frac{1}{2} L_0'' - L_0' L_0^+ L_0'\right] P_2|_{\Psi_2},$$

where P_2 is the orthogonal projection of $\mathbb{C}^n$ on Ψ_2.

To illustrate this construction consider an example:

Example 12.4.3. *Let*

$$
L(\lambda) = \begin{bmatrix} \lambda^2 & 0 & 0 \\ 0 & \lambda^2 + \lambda & \lambda \\ 0 & \lambda & \lambda^2 + \lambda \end{bmatrix}.
$$

Choose the eigenvalue $\lambda_0 = 0$ of $L(\lambda)$. Then $\operatorname{Ker} L(0) = \mathbb{C}^3$, so $\Psi_1 = \mathbb{C}^3$. Furthermore,

$$
L'(0) = \begin{bmatrix} 0 & 0 & 0 \\ 0 & 1 & 1 \\ 0 & 1 & 1 \end{bmatrix}
$$

and

$$f_1(x, y) = (x, L'(0)y) = x_2(\bar{y}_2 + \bar{y}_3) + x_3(\bar{y}_2 + \bar{y}_3),$$

where $x = \langle x_1, x_2, x_3 \rangle$, $y = \langle y_1, y_2, y_3 \rangle$.

The matrix $L'(0)$ has one nonzero eigenvalue, namely 2, and

$$\operatorname{Ker} L'(0) = \Psi_2 = \operatorname{Span}\{\langle 1, 0, 0 \rangle, \ \langle 0, -1, 1 \rangle\}.$$

Thus there is exactly one partial multiplicity of $L(\lambda)$ corresponding to $\lambda_0 = 0$ which is equal to 1, and its sign is $+1$. It is easily seen that $\{y, 0\}$ is a Jordan chain for any eigenvector $y \in \Psi_2$. Thus

$$f_2(x, y) = \left(x, \frac{1}{2} L''(0)y + L'(0)y'\right) = (x, y) \quad \text{for } x, y \in \Psi_2.$$

Therefore there are exactly two partial multiplicities of $\lambda_0 = 0$ which are equal to 2, and their signs are $+1$. □

For future reference, let us determine the sign characteristic of scalar polynomials with real coefficients.

Example 12.4.4. *Let*

$$L(\lambda) = a_\ell \lambda^\ell + a_{\ell-1} \lambda^{\ell-1} + \cdots + a_1 \lambda + a_0, \quad a_0, \ldots, a_\ell \in \mathsf{R},$$

be a scalar hermitian polynomial. If λ_0 is a zero of $L(\lambda)$, then there is only one partial multiplicity of $L(\lambda)$ associated with λ_0. If moreover λ_0 is real, then the sign in the sign characteristic of $L(\lambda)$ associated with λ_0 coincides with $\operatorname{sgn} L^{(q)}(\lambda_0)$, where q is the multiplicity of λ_0 as a zero of $L(\lambda)$ (and hence $L^{(q)}(\lambda_0) \neq 0$.) This follows immediately from Theorem 12.4.2 using the observation that an eigenvector $x \in \mathsf{C} \setminus \{0\}$ of $L(\lambda)$ at λ_0 has a Jordan chain of the form $\{x, 0, \ldots, 0\}$ with $q - 1$ zeros. □

The next result shows that the sign characteristic of hermitian matrix polynomials is in fact a local notion.

Theorem 12.4.5. *Let $L_1(\lambda)$ and $L_2(\lambda)$ be two hermitian matrix polynomials with invertible leading coefficients. If $\lambda_0 \in \sigma(L_1)$ is real and*

$$L_1^{(i)}(\lambda_0) = L_2^{(i)}(\lambda_0), \quad i = 0, 1, \ldots, \gamma,$$

where the integer γ is greater than or equal to the maximal length of Jordan chains of $L_1(\lambda)$ and of $L_2(\lambda)$ corresponding to λ_0, then the sign characteristics of $L_1(\lambda)$ and $L_2(\lambda)$ at λ_0 are the same.

Naturally, the sign characteristic of a hermitian matrix polynomial *at an eigenvalue* is taken to mean the subset of the sign characteristic, corresponding to the elementary divisors of this eigenvalue. It is clear that Theorem 12.4.5 defines the sign characteristic at λ_0 as a local property of a hermitian matrix polynomial. This result is an immediate corollary of Theorem 12.4.2.

The next theorem concerns the stability of the sign characteristic under hermitian perturbations of the coefficients.

Theorem 12.4.6. *Let $L(\lambda) = \sum_{j=0}^{\ell} A_j \lambda^j$ be a hermitian matrix polynomial with invertible A_ℓ and let $\lambda_0 \in \sigma(L)$ be real. Then there exists a $\delta > 0$ with the following property:*
For every hermitian matrix polynomial

$$\widetilde{L}(\lambda) = \sum_{j=0}^{\ell} \widetilde{A}_j \lambda^j$$

such that

$$\|\widetilde{A}_j - A_j\| < \delta, \quad j = 1, \ldots, \ell, \tag{12.4.25}$$

and for which there exists a unique eigenvalue λ_1 *of* $\widetilde{L}(\lambda)$ *in the disc* $\{\lambda \in \mathbb{C} :$ $|\lambda_0 - \lambda| < \delta\}$ *with the same partial multiplicities as those of* $L(\lambda)$ *at* λ_0, *the sign characteristic of* $\widetilde{L}(\lambda)$ *at* λ_1 *coincides with the sign characteristic of* $L(\lambda)$ *at* λ_0.

Proof. The result follows easily from Theorem 5.9.1. Indeed, let B_L (resp., $\widetilde{B}_L$) and C_L (resp., $\widetilde{C}_L$) be defined as in (12.0.1) and (12.2.11) for L (resp., $\widetilde{L}$). Then for $\delta > 0$ small enough, $\widetilde{B}_L$ is as close as we wish to B_L and $\widetilde{C}_L$ is as close as we wish to C_L. Recall that C_L (resp., $\widetilde{C}_L$) is B_L (resp., $\widetilde{B}_L$)-selfadjoint and the sign characteristic of L (resp., $\widetilde{L}$) coincides with the C_L (resp., $\widetilde{C}_L$)-sign characteristic of B_L (resp., $\widetilde{B}_L$). It remains to apply Theorem 5.9.1.   □

Now let $L(\lambda)$ and $\widetilde{L}(\lambda)$ be as in Theorem 12.4.6 and suppose, in addition, that the real Jordan structure of $\widetilde{L}(\lambda)$ is the same as that of $L(\lambda)$. More precisely, this means that the number r of different real eigenvalues $\lambda_1 < \cdots < \lambda_r$ and $\widetilde{\lambda}_1 < \cdots < \widetilde{\lambda}_r$ of L and $\widetilde{L}$ (respectively) is the same and, furthermore, for every j $(1 \leq j \leq r)$, the partial multiplicities of $L(\lambda)$ at λ_j and those of $\widetilde{L}(\lambda)$ at $\widetilde{\lambda}_j$ are the same.

The following result (which is a particular case of Theorem 12.4.6) shows that stability of the real Jordan structure implies stability of the sign characteristic.

Theorem 12.4.7. *Let* $L(\lambda)$ *and* $\widetilde{L}(\lambda)$ *be as in Theorem 12.4.6. Then there exists a* $\delta > 0$ *such that the sign characteristics of* L *and* $\widetilde{L}$ *are the same for every hermitian matrix polynomial* $\widetilde{L}(\lambda)$ *that satisfies (12.4.25) and such that the real Jordan structures of* $L(\lambda)$ *and* $\widetilde{L}(\lambda)$ *coincide.*

The statement on the agreement of sign characteristics means that, if $\lambda_1 < \cdots < \lambda_r$ and $\widetilde{\lambda}_1 < \cdots < \widetilde{\lambda}_r$ are the different real eigenvalues of of L and $\widetilde{L}$, respectively, then the sign characteristics of $L(\lambda)$ at λ_j and $\widetilde{L}(\lambda)$ at $\widetilde{\lambda}_j$ are the same, for $j = 1, \ldots, r$.

12.5   The Sign Characteristic of Hermitian Analytic Matrix Functions

Let $L(\lambda)$ be a hermitian matrix polynomial with invertible leading coefficient. Then Theorem A.6.7 states that, for real λ the matrix $L(\lambda)$ has a diagonal decomposition

$$L(\lambda) = U(\lambda) \cdot \text{diag}\left[\mu_1(\lambda), \ldots, \mu_n(\lambda)\right] \cdot V(\lambda), \qquad (12.5.26)$$

where $U(\lambda)$ is unitary (for real λ) and $V(\lambda) = (U(\lambda))^*$. Moreover, the functions $\mu_i(\lambda)$ and $U(\lambda)$ can be chosen to be analytic functions of the real parameter λ (but in general $\mu_i(\lambda)$ and $U(\lambda)$ are not polynomials). This result suggests that the notion of a sign characteristic will also apply to analytic matrix functions. This extension of the theory is made here. However, it will also serve as preparation for

the important Theorem 12.5.2 which gives another (frequently useful) definition
of the sign characteristic in the polynomial case.

Let Ω be a connected domain in the complex plane which is symmetric with
respect to the real axis. An analytic $n \times n$ matrix function $A(\lambda)$ in Ω is said to be
hermitian if $A(\lambda) = \left(A(\lambda)\right)^*$ for all real $\lambda \in \Omega$.

In what follows, we only consider those hermitian analytic functions $A(\lambda)$
for which $\det A(\lambda)$ is not identically zero (and this condition will not be repeated
explicitly). For such matrix functions $A(\lambda)$ the spectrum

$$\sigma(\lambda) = \{\lambda \in \Omega \mid \det A(\lambda) = 0\}$$

is a set of isolated points, because they are the zeros of the analytic function
$\det A(\lambda)$. Then, for every $\lambda_0 \in \sigma(A)$ the Jordan chains of $A(\lambda)$ at λ_0 are defined
as follows (cf. equations (12.4.19)). As for matrix polynomials, we say that a chain
of vectors $\{x_1, \ldots, x_k\}$, $x_j \in \mathbb{C}^n$ is a *Jordan chain* of $A(\lambda)$ at λ_0 if $x_1 \neq 0$ and

$$
\begin{aligned}
A(\lambda_0)\, x_1 &= 0, \\
A'(\lambda_0)\, x_1 + A(\lambda_0)\, x_2 &= 0, \\
&\vdots \qquad \vdots
\end{aligned}
$$

$$\frac{1}{(k-1)!}\, A^{(k-1)}(\lambda_0)\, x_1 + \frac{1}{(k-2)!}\, A^{(k-2)}(\lambda_0)\, x_2 + \cdots + A(\lambda_0)\, x_k = 0.$$

$$(12.5.27)$$

Here $A^{(j)}(\lambda_0)$ is the matrix value of the j-th derivative of $A(\lambda)$ at λ_0.

This set of equations can be written conveniently in matrix notation:

$$
\begin{bmatrix}
A(\lambda_0) & 0 & \cdots & 0 \\
A'(\lambda_0) & A(\lambda_0) & \cdots & 0 \\
\vdots & \vdots & \cdots & \vdots \\
\frac{1}{(k-1)!} A^{(k-1)}(\lambda_0) & \frac{1}{(k-2)!} A^{(k-2)}(\lambda_0) & \cdots & A(\lambda_0)
\end{bmatrix}
\begin{bmatrix}
x_1 \\ x_2 \\ \vdots \\ x_k
\end{bmatrix}
= 0.
$$

The integer k is called the *length* of the Jordan chain $\{x_1, \ldots, x_k\}$ and the
condition $\det A(\lambda) \not\equiv 0$ ensures that the Jordan chains of $A(\lambda)$ at λ_0 cannot be
continued indefinitely. More precisely, there is a positive integer γ (which may
depend on λ_0) such that every Jordan chain of $A(\lambda)$ at λ_0 has length at most γ.
This follows easily from the diagonal form of $A(\lambda)$ in a neighborhood of λ_0 (see
Section A.6).

Now let $\lambda_0 \in \sigma(A)$ be real. Then there exists a hermitian matrix polynomial
$L(\lambda)$ with invertible leading coefficient such that

$$L^{(j)}(\lambda_0) = A^{(j)}(\lambda_0), \quad j = 0, \ldots, \gamma, \qquad (12.5.28)$$

where γ is the maximal length of Jordan chains of $A(\lambda)$ corresponding to λ_0.
Bearing in mind that $x_1, \ldots, x_k$ is a Jordan chain of $L(\lambda)$ corresponding to λ_0 if

and only if $x_1 \neq 0$ and

$$
\begin{bmatrix}
L(\lambda_0) & 0 & \cdots & 0 \\
L'(\lambda_0) & L(\lambda_0) & \cdots & 0 \\
\vdots & \vdots & \cdots & \vdots \\
\dfrac{1}{(k-1)!}L^{(k-1)}(\lambda_0) & \dfrac{1}{(k-2)!}L^{(k-2)}(\lambda_0) & \cdots & L(\lambda_0)
\end{bmatrix}
\begin{bmatrix}
x_1 \\ x_2 \\ \vdots \\ x_k
\end{bmatrix}
= 0,
$$

and similarly for $A(\lambda)$, it follows from (12.5.28) that $L(\lambda)$ and $A(\lambda)$ have the same Jordan chains corresponding to λ_0.

In particular, the structure of Jordan chains of $A(\lambda)$ and $L(\lambda)$ at λ_0 agree and we may define the *sign characteristic* of $A(\lambda)$ at λ_0 as the sign characteristic of the pair (C_L, B_L) at λ_0, where C_L is the companion matrix of $L(\lambda)$ and B_L is given by (12.2.11). (In view of Theorem 12.4.5, this definition does not depend on the choice of $L(\lambda)$).

The next theorem will be useful later, but is clearly of independent interest. Given an analytic matrix function $R(\lambda)$ in Ω, let $R^*(\lambda)$ denote the analytic (in Ω) matrix function $\left(R(\bar{\lambda})\right)^*$.

Theorem 12.5.1. *Let A be a hermitian analytic matrix function in Ω, and let $\lambda_0 \in \sigma(A)$ be real. Let $R(\lambda)$ be an analytic matrix function in Ω such that $\det R(\lambda_0) \neq 0$. Then the sign characteristics of A and R^*AR at λ_0 are the same.*

Proof. Consider first the case that $A(\lambda) = L(\lambda)$ is a hermitian matrix polynomial with invertible leading coefficient.

Let γ be the maximal length of Jordan chains of $L(\lambda)$ corresponding to λ_0. Let m be an integer which is large enough that there exist matrix polynomials $S(\lambda)$ and $T(\lambda)$, both of degree m, with invertible leading coefficients, and which are solutions of the following interpolation problems:

$$
S^{(j)}(\lambda_0) = R^{(j)}(\lambda_0) \quad \text{for } j = 0, 1, \ldots, \gamma;
$$

$$
T^{(j)}(\lambda_0) = I \quad \text{for } j = 0, \quad \text{and} \quad T^{(j)}(\lambda_0) = 0 \quad \text{for } j = 1, \ldots, \gamma.
$$

(The proof of existence of $S(\lambda)$ and $T(\lambda)$ is reduced to the scalar case, and their leading coefficients can be chosen diagonal with nonzero diagonal entries.) Then

$$
(S^*LS)^{(j)}(\lambda_0) = (R^*LR)^{(j)}(\lambda_0) \quad \text{for } j = 0, \ldots, \gamma,
$$

thus in view of the definition, the sign characteristics of S^*LS and R^*LR at λ_0 are the same.

On the other hand, let $F(\lambda, t)$, $t \in [0, 1]$ be a continuous family of matrix polynomials of degree m with invertible leading coefficients such that

$$
F(\lambda, 0) = S(\lambda), \quad F(\lambda, 1) = T(\lambda)
$$

and $F(\lambda_0, t)$ is invertible for every $t \in [0, 1]$. For example,

$$F(\lambda, t) = (\lambda - \lambda_0)^m F_m(t) + \sum_{j=1}^{m-1} (\lambda - \lambda_0)^j \left[tT_j + (1-t)S_j \right] + F_0(t),$$

where

$$S(\lambda) = \sum_{j=0}^{m} (\lambda - \lambda_0)^j S_j, \quad T(\lambda) = \sum_{j=0}^{m} (\lambda - \lambda_0)^j T_j$$

and $F_i(t)$ $(i = 0, m)$ is a continuous invertible matrix function such that $F_i(0) = S_i$ and $F_i(1) = T_i$.

Consider the family of hermitian matrix polynomials with invertible leading coefficients:

$$M(\lambda, t) = F^*(\lambda, t) L(\lambda) F(\lambda, t), \qquad t \in [0, 1].$$

Since $\det F(\lambda_0, t) \neq 0$ it follows that $\lambda_0 \in \sigma(M(\lambda, t))$ and the partial multiplicities of $M(\lambda, t)$ at λ_0 do not depend on $t \in [0, 1]$. Applying Theorem 12.4.6 we see that, also, the sign characteristic of $M(\lambda, t)$ at λ_0 does not depend on $t \in [0, 1]$. In particular, the sign characteristics of S^*LS and T^*LT at λ_0 are the same. But the sign characteristic of T^*LT at λ_0 is the same as that of L in view of Theorem 12.4.5.

So Theorem 12.5.1 is proved for the case that $A(\lambda)$ is a polynomial with invertible leading coefficient. The general case can be easily reduced to this. Namely, let $L(\lambda)$ be a hermitian matrix polynomial satisfying (12.5.28). Let $M(\lambda)$ be a matrix polynomial with invertible leading coefficient such that

$$M^{(j)}(\lambda_0) = R^{(j)}(\lambda_0), \quad j = 0, \dots, \gamma.$$

By definition, the sign characteristic of R^*AR at λ_0 is defined by M^*LM, and that of A is defined by L. But in view of the already proved case, the sign characteristics of L and M^*LM at λ_0 are the same.   □

Theorem 12.5.1 will be used to describe the sign characteristic of a hermitian matrix polynomial from the viewpoint of perturbation theory. This description (given in Theorem 12.5.2 below) is one of the main results of this chapter.

As indicated in (12.5.26), let $\mu_1(\lambda), \dots, \mu_n(\lambda)$ be the eigenvalues of the hermitian matrix polynomial $L(\lambda)$ (when considered as a matrix depending on the real parameter λ). Then the real analytic functions $\mu_j(\lambda)$ are the roots of the equation

$$\det (\mu I - L(\lambda)) = 0. \tag{12.5.29}$$

It is easy to see that $\lambda_0 \in \sigma(L)$ if and only if at least one of the $\mu_j(\lambda_0)$ is zero. Moreover, $\dim \operatorname{Ker} L(\lambda_0)$ is exactly the number of indices j $(1 \leq j \leq n)$ such that

$$\mu_j(\lambda_0) = 0.$$

On the other hand, it is possible to consider the solutions $\mu_1(\lambda), \dots, \mu_n(\lambda)$ of equation (12.5.29) (for every fixed λ) as analytic functions of λ. As the next theorem shows, the partial multiplicities of $L(\lambda)$ at λ_0 coincide with the multiplicities of λ_0 as a zero of the analytic functions $\mu_1(\lambda), \ldots, \mu_n(\lambda)$. Moreover, the sign characteristic of $L(\lambda)$ at λ_0 can be described in terms of the functions $\mu_j(\lambda)$.

Theorem 12.5.2. *Let $L = L^*$ be a hermitian matrix polynomial having an invertible leading coefficient, and let $\mu_1(\lambda), \ldots, \mu_n(\lambda)$ be real analytic functions of real λ such that*

$$\det\left(\mu_j I - L(\lambda)\right) = 0, \quad j = 1, \ldots, n.$$

Let $\lambda_1 < \cdots < \lambda_r$ be the distinct real eigenvalues of $L(\lambda)$. For every $i = 1, \ldots, r$ write

$$\mu_j(\lambda) = (\lambda - \lambda_0)^{m_{ij}} \nu_{ij}(\lambda), \quad m_{ij} \geq 0,$$

where $\nu_{ij}(\lambda_i) \neq 0$ is real. Then the nonzero numbers among $m_{i1}, \ldots, m_{in}$ are the partial multiplicities of $L(\lambda)$ associated with λ_i, and $\operatorname{sign} \nu_{ij}(\lambda_i)$ (for $m_{ij} \neq 0$) is the sign attached to the partial multiplicity m_{ij} of $L(\lambda)$ at λ_i in its (possibly nonnormalized) sign characteristic.

Proof. The decomposition (12.5.26) which, a priori, holds only for real λ in a neighborhood of λ_i can be extended to those complex λ which are close enough to λ_i. Then $U(\lambda)$, $\mu_j(\lambda)$ and $V(\lambda)$ can be regarded as analytic functions in some complex neighborhood of λ_i in $\mathbb{C}$. This is possible since $U(\lambda)$, $\mu_j(\lambda)$ and $V(\lambda)$ can be expressed as convergent series in a real neighborhood of λ_i. Consequently, these series also converge in some complex neighborhood of λ_i. (But then of course it is no longer true that $U(\lambda)$ is unitary and $V(\lambda) = \left(U(\lambda)\right)^*$.)

Now the first assertion of Theorem 12.5.2 follows from (12.5.26) and Theorem A.6.6.

Further, in view of Theorem 12.5.1, the sign characteristics of L and $\operatorname{diag}\left[\mu_j(\lambda)\right]_{j=1}^n$ at λ_i are the same. Let us compute the latter. Choose scalar polynomials $\widetilde{\mu}_1(\lambda), \ldots, \widetilde{\mu}_n(\lambda)$ of the same degree with real coefficients and with the properties:

$$\widetilde{\mu}_j^{(k)}(\lambda_i) = \mu_j^{(k)}(\lambda_i)$$

for $k = 0, \ldots, m_{ij}$; $i = 1, \ldots, r$; $j = 1, \ldots, n$.

By definition, the sign characteristics of $\operatorname{diag}\left[\mu_j(\lambda)\right]_{j=1}^n$ and $\operatorname{diag}\left[\widetilde{\mu}_j(\lambda)\right]_{j=1}^n$ are the same. Using the description of the sign characteristic of $\operatorname{diag}\left[\widetilde{\mu}_j(\lambda)\right]_{j=1}^n$ given in Example 12.4.4, we see that the first nonzero derivative $\widetilde{\mu}_j^{(k)}(\lambda_i)$ (for fixed i and j) is positive or negative depending on whether the sign of the Jordan block corresponding to the Jordan chain

$$\langle 0, \ldots, 0, 1, 0, \ldots, 0 \rangle, 0, \ldots, 0$$

(with "1" in the j-th place) of λ_i is $+1$ or -1. Thus, the second assertion of Theorem 12.5.2 follows. $\qquad \square$

As for the proof of Theorem 5.11.2, simply observe that it is a particular case of Theorem 12.5.2, when the hermitian matrix polynomial $L(\lambda)$ is of the first degree.

12.6 Hermitian Matrix Polynomials on the Unit Circle

Let $L(\lambda) = \sum_{j=0}^{\ell} \lambda^j A_j$ be a matrix polynomial, where $\ell = 2k$ is even and the leading coefficient A_ℓ is invertible. Assume also that the rational function $\widehat{L}(\lambda) = \lambda^{-k} L(\lambda)$ is hermitian on the unit circle, i.e.,

$$\left(\widehat{L}(\lambda)\right)^* = \widehat{L}(\bar{\lambda}^{-1}), \quad \lambda \in \mathsf{C}.$$

This class of matrix polynomials plays a role in the study of systems of difference equations (see Section 13.7).

Matrix polynomials of this kind satisfy the condition $\left(\widehat{L}(\lambda)\right)^* = \widehat{L}(\lambda)$ for $|\lambda| = 1$ and so the matrix polynomial $L(\lambda)$ is said to be *hermitian with respect to the unit circle* . By analogy with the case of hermitian matrix polynomials we might expect that the companion matrix C_L will be unitary in a suitable indefinite scalar product. This is the case. In fact,

$$C_L^* \widehat{B}_L C_L = \widehat{B}_L, \tag{12.6.30}$$

where

$$\widehat{B}_L = i \left[\begin{array}{ccc|ccc} & & & A_\ell & \cdots & 0 \\ & 0 & & \vdots & \ddots & \vdots \\ & & & A_{k+1} & \cdots & A_\ell \\ \hline -A_0 & \cdots & -A_{k-1} & & & \\ \vdots & \ddots & \vdots & & 0 & \\ 0 & \cdots & -A_0 & & & \end{array} \right]$$

is hermitian and invertible. To check (12.6.30), observe that

$$C_L^* = \left[\begin{array}{cccc} 0 & \cdots & 0 & -A_\ell A_0^{-1} \\ I & \cdots & 0 & -A_{\ell-1} A_0^{-1} \\ \vdots & \ddots & \vdots & \vdots \\ 0 & \cdots & I & -A_1 A_0^{-1} \end{array} \right]$$

and

$$(C_L^*)^{-1} = \left[\begin{array}{ccccc} -A_{\ell-1} A_\ell^{-1} & I & 0 & \cdots & 0 \\ -A_{\ell-2} A_\ell^{-1} & 0 & I & \cdots & 0 \\ \vdots & \vdots & \vdots & \ddots & \vdots \\ -A_1 A_\ell^{-1} & 0 & 0 & \cdots & I \\ -A_0 A_\ell^{-1} & 0 & 0 & \cdots & 0 \end{array} \right].$$

Now the condition $\widehat{B}_L C_L = C_L^{*-1} \widehat{B}_L$ can be verified by a direct computation.

Let λ_0 be a unimodular eigenvalue of $L(\lambda)$ (i. e. $|\lambda_0| = 1$). For each $x \in$ $\mathrm{Ker}L(\lambda_0)$ write $\widehat{x} = \langle x, \lambda_0 x, \ldots, \lambda_0^{\ell-1}x \rangle$. Then for any $x, y \in \mathrm{Ker}L(\lambda_0)$ it follows that

$$(x, i\lambda_0 \widehat{L}^{(1)}(\lambda_0)\, y) = (\widehat{x}, \widehat{B}_L \widehat{y}), \qquad\qquad (12.6.31)$$

where $\widehat{L}^{(1)}(\lambda)$ is the derivative of $\widehat{L}(\lambda)$ with respect to λ. This relation can be checked by a direct computation of $(\widehat{x}, \widehat{B}_L \widehat{y})$ using the property $\overline{\lambda}_0 = \lambda_0^{-1}$.

Using the definition of the sign characteristic of a unitary matrix in an indefinite scalar product (Section 5.15) the following statement (which is an analogue of Proposition 12.2.1) is obtained.

Proposition 12.6.1. *Let $L(\lambda)$ be a matrix polynomial of even degree ℓ, which is hermitian on the unit circle, and with invertible leading coefficient. Let λ_0 be a unimodular eigenvalue of $L(\lambda)$. Then the quadratic form*

$$\left(x, i\lambda_0 \big(\lambda^{-\frac{\ell}{2}} L(\lambda)\big)^{(1)}(\lambda_0)x\right), \quad x \in \mathrm{Ker}L(\lambda_0)$$

is nonsingular if and only if all the elementary divisors of $L(\lambda)$ corresponding to λ_0 are linear. In this case the number of positive (resp. negative) squares in the canonical form of

$$\left(x, i\lambda_0 \big(\lambda^{-\frac{\ell}{2}} L(\lambda)\big)^{(1)}(\lambda_0)x\right), \quad x \in \mathrm{Ker}L(\lambda_0)$$

coincides with the number of signs $+1$ (resp. -1) in the sign characteristic of $(C_L, \widehat{B}_L)$ (here the $\widehat{B}_L$-unitary matrix C_L is the companion matrix of $L(\lambda)$).

In concluding this section, note that one can obtain factorization results for hermitian matrix polynomials on the unit circle from corresponding results for hermitian matrix polynomials on the real line. To illustrate this approach, let us prove the following result, which is an analogue of Theorem 12.3.3.

Theorem 12.6.2. *Let*

$$R(\lambda) = \sum_{j=-k}^{k} \lambda^j R_j$$

be a rational $n \times n$ matrix function such that $\big(R(\lambda)x, x\big) \geq 0$ for all $|\lambda| = 1$ and $x \in \mathbb{C}^n$, and $\det R(\lambda)$ is not identically zero. Then $R(\lambda)$ admits a factorization

$$R(\lambda) = \big(A(\overline{\lambda}^{-1})\big)^* A(\lambda), \quad \lambda \in \mathbb{C} \qquad\qquad (12.6.32)$$

where $A(\lambda)$ is a matrix polynomial.

Proof. Let

$$L(\lambda) = \sum_{j=0}^{2k} (1+i\lambda)^j (1-i\lambda)^{2k-j} R_{j-k}.$$

It is easily checked that $L(\lambda)$ is a matrix polynomial which is nonnegative on the real line, and $\det L(\lambda) \not\equiv 0$. Let $a \in \mathbb{R}$ be such that $L(a)$ is invertible, hence $L(a)$ is positive definite. Then the matrix polynomial

$$M(\lambda) = \lambda^{2k} L(a)^{-\frac{1}{2}} L(\lambda^{-1} + a) L(a)^{-\frac{1}{2}}$$

is monic and nonnegative on the real line. By Theorem 12.3.3, $M(\lambda)$ admits a factorization

$$M(\lambda) = \left(M_1(\bar{\lambda})\right)^* M_1(\lambda)$$

for some monic matrix polynomial $M_1(\lambda)$ of degree k. This factorization leads to the factorization

$$L(\lambda) = \left(L_1(\bar{\lambda})\right)^* L_1(\lambda)$$

with matrix polynomial

$$L_1(\lambda) = (\lambda - a)^k M_1\left((\lambda - a)^{-1}\right) L(a)^{\frac{1}{2}}.$$

Now

$$R\left(\frac{1+i\lambda}{1-i\lambda}\right) = \frac{L(\lambda)}{(1+\lambda^2)^k} = \left[\frac{L_1(\bar{\lambda})}{(1-i\bar{\lambda})^k}\right]^* \frac{L_1(\lambda)}{(1-i\lambda)^k}$$

and denoting

$$\mu = \frac{1+i\lambda}{1-i\lambda},$$

we obtain

$$R(\mu) = \left[\frac{(1+\bar{\mu})^k}{2^k} L_1\left(\frac{-i(1-\bar{\mu})}{1+\bar{\mu}}\right)\right]^* \frac{(1+\mu)^k}{2^k} L_1\left(\frac{i(1-\mu)}{1+\mu}\right),$$

which is a factorization of type (12.6.32) with

$$A(\lambda) = \frac{(1+\lambda)^k}{2^k} L_1\left(\frac{i(1-\lambda)}{1+\lambda}\right). \qquad \square$$

As with Theorem 12.3.3, the matrix polynomial $A(\lambda)$ in (12.6.32) can be chosen with an additional spectral property. Thus, if S is a set of nonunimodular eigenvalues of $R(\lambda)$ such that $\lambda \in S$ implies $\bar{\lambda}^{-1} \notin S$, and S is maximal with respect to this property, then there exists a matrix polynomial $A(\lambda)$ satisfying (12.6.32) for which the nonunimodular spectrum of $A(\lambda)$ coincides with S.

12.7 Exercises

1. Prove that a pair of matrices (T, Y) of sizes $n\ell \times n\ell$ and $n\ell \times n$, respectively, is a left standard pair for

$$L(\lambda) = I\lambda^\ell + \sum_{j=0}^{\ell-1} A_j \lambda^j$$

if and only if $[Y, TY, \dots, T^{\ell-1}Y]$ is invertible and

$$Y A_0 + T Y A_1 + \dots + T^{\ell-1} Y A_{\ell-1} + T^\ell Y = 0.$$

2. Show that given a right standard pair (X, T) of a monic matrix polynomial $L(\lambda)$, there exists a unique Y such that (X, T, Y) is a standard triple for $L(\lambda)$, and in fact Y is given by formula

$$Y = \left(\begin{bmatrix} X \\ XT \\ \vdots \\ XT^{\ell-1} \end{bmatrix} \right)^{-1} \begin{bmatrix} 0 \\ \vdots \\ 0 \\ I \end{bmatrix}.$$

3. Show that, given a left standard pair (T, Y) of a monic matrix polynomial $L(\lambda)$, there exists a unique X such that (X, T, Y) is a standard triple of $L(\lambda)$ and, in fact,

$$X = [0 \dots 0\, I][Y, TY, \dots, T^{\ell-1}Y]^{-1}.$$

4. Prove that (T, Y) is a left standard pair for $L(\lambda) = I\lambda^\ell + \sum_{j=0}^{\ell-1} A_j \lambda^j$ if and only if (Y^*, T^*) is a right standard pair for the monic matrix polynomial

$$I\lambda^\ell + \sum_{j=0}^{\ell-1} A_j^* \lambda^j.$$

5. Let $L(\lambda) = \lambda^\ell + \sum_{i=0}^{\ell-1} a_i \lambda^i$ be a scalar polynomial with ℓ distinct zeros $\lambda_1, \dots, \lambda_\ell$.

   (a) Show that
   $$(X, T) = \left([1 \ 1 \ \dots 1], \mathrm{diag}\,[\lambda_1, \dots, \lambda_\ell]\right)$$
   is a right standard pair for $L(\lambda)$. Find a matrix Y such that (X, T, Y) is a standard triple for $L(\lambda)$.

   (b) Show that
   $$(T, Y) = \left(\mathrm{diag}\,[\lambda_1, \dots, \lambda_\ell], \begin{bmatrix} 1 \\ 1 \\ \vdots \\ 1 \end{bmatrix} \right)$$
   is a left standard pair for $L(\lambda)$, and find X such that (X, T, Y) is a standard triple for $L(\lambda)$.

6. Let $L(\lambda) = (\lambda - \lambda_1)^{\ell_1} \cdots (\lambda - \lambda_k)^{\ell_k}$ be a scalar polynomial, where $\lambda_1, \ldots, \lambda_k$ are distinct complex numbers. Show that

$$\left([X_1, \ldots, X_k], J_{\ell_1}(\lambda_1) \oplus \cdots \oplus J_{\ell_k}(\lambda_k)\right)$$

and

$$\left(J_{\ell_1}(\lambda_1) \oplus \cdots \oplus J_{\ell_k}(\lambda_k), \begin{bmatrix} Y_1 \\ Y_2 \\ \vdots \\ Y_k \end{bmatrix}\right)$$

are right and left standard pairs, respectively, of $L(\lambda)$, where $X_i = [1\,0 \ldots 0]$ is a $1 \times \ell_i$ matrix and

$$Y_i = \begin{bmatrix} 0 \\ 0 \\ \vdots \\ 1 \end{bmatrix}$$

is an $\ell_i \times 1$ matrix.

7. Let

$$L(\lambda) = \begin{bmatrix} L_1(\lambda) & 0 \\ 0 & L_2(\lambda) \end{bmatrix}$$

be a monic matrix polynomial, and let (X_1, T_1, Y_1) and (X_2, T_2, Y_2) be standard triples for the polynomials $L_1(\lambda)$ and $L_2(\lambda)$, respectively. Find a standard triple for the polynomial $L(\lambda)$.

8. Given a standard triple for the matrix polynomial $L(\lambda)$, find a standard triple for the polynomial $S^{-1}L(\lambda + \alpha)S$ where S is an invertible matrix, and α is a complex number.

9. Let (X, T, Y) be a standard triple for $L(\lambda)$. Show that

$$\left([X\ \ 0], \begin{bmatrix} 0 & I \\ T & 0 \end{bmatrix}, \begin{bmatrix} 0 \\ Y \end{bmatrix}\right)$$

is a standard triple for the matrix polynomial $L(\lambda^2)$.

10. Given a standard triple for the matrix polynomial $L(\lambda)$, find a standard triple for the polynomial $L(p(\lambda))$, where $p(\lambda) = \sum_{j=0}^{m-1} \lambda^j \alpha_j$ is a scalar polynomial.

11. Let $L(\lambda) = I\lambda^\ell + \sum_{j=0}^{\ell-1} A_j \lambda^j$ be a 3×3 matrix polynomial whose coefficients are circulants:

$$A_k = \begin{bmatrix} a_k & b_k & c_k \\ c_k & a_k & b_k \\ b_k & c_k & a_k \end{bmatrix}, \quad k = 0, 1, \ldots, \ell - 1$$

(a_k, b_k and c_k are complex numbers). Describe right and left standard pairs of $L(\lambda)$.

12. Let
$$L_2(\lambda) = \begin{bmatrix} 2 & 1 \\ 1 & 1 \end{bmatrix} \lambda^2 + \begin{bmatrix} 4 & 2 \\ 2 & 0 \end{bmatrix} \lambda + \begin{bmatrix} 2 & 1 \\ 1 & 1 \end{bmatrix}.$$

(a) Find a left standard (Jordan) pair.

(b) Find a right standard (Jordan) pair.

(c) Find the sign characteristic of $L(\lambda)$.

(d) Find a selfadjoint triple.

(e) Reconstruct the polynomial from its left and right Jordan pairs.

(f) Prove that $L_2(\lambda)$ is positive semidefinite on the real line.

(g) Factorize $L_2(\lambda)$ in the form $L_2(\lambda) = L_1^*(\lambda)L_1(\lambda)$.

13. Solve the preceding exercise for the following matrix polynomials:

(α) $L_3(\lambda) = L_2(\lambda^2)$;

(β) $L_4(\lambda) = \begin{bmatrix} 2 & 1 \\ 1 & 1 \end{bmatrix} \lambda^2 + \begin{bmatrix} 0 & -2 \\ -2 & -8 \end{bmatrix} \lambda + \begin{bmatrix} 0 & 0 \\ 0 & 16 \end{bmatrix}$;

(γ) $L_5(\lambda) = L_4(\lambda^3)$.

14. Let $L_6(\lambda)$ be the hermitian matrix polynomial

$$L_6(\lambda) = \begin{bmatrix} 0 & 1 \\ 1 & -1 \end{bmatrix} \begin{bmatrix} 2(\lambda^2+1)^2 & (\lambda^2+1)^2 \\ (\lambda^2+1)^2 & \lambda^4+1 \end{bmatrix} \begin{bmatrix} 0 & 1 \\ 1 & -1 \end{bmatrix}.$$

(a) Show that $L_6(\lambda_0$ is positive semidefinite for every real λ_0.

(b) Find one factorization

$$L_6(\lambda) = (M(\bar{\lambda}))^* M(\lambda), \qquad (12.7.33)$$

where $M(\lambda)$ is a monic matrix polynomial of degree 2.

(c) Find another factorization of the type (12.7.33), for the same $L(\lambda)$.

(d) How many factorizations of the type (12.7.33) are there?

12.8   Notes

The main results of this chapter related to perturbation and stability were obtained by the authors in their papers [35, 36, 37, 38]. The first five sections and the part of the sixth section describing the spectral theory of matrix polynomials in general, and selfadjoint polynomials in particular, are taken from the authors' papers [34, 35, 36, 37]. In addition, consult [39, 40] where more details and further results can be found. See also [70, Chapter 14].

Chapter 13

Differential and Difference Equations of Higher Order

The notions and results developed in Chapter 12 for matrix polynomials are used in this chapter to study systems of differential and difference equations of higher order with constant coefficients.

13.1 General Solution of a System of Differential Equations

Consider the system of differential equations with constant coefficients

$$\sum_{j=0}^{\ell} i^j A_j \frac{d^j x}{dt^j} = 0; \quad t \in \mathsf{R}, \quad (i = \sqrt{-1}) \tag{13.1.1}$$

where A_j are $n \times n$ (complex) matrices and A_ℓ is invertible, and $x = x(t) \in \mathsf{C}^n$ is an unknown vector-valued function. It turns out that the general solution $x(t)$ of (13.1.1) can be conveniently expressed via a standard pair (X, T) of the corresponding matrix polynomial, as defined in Section 12.1:

Theorem 13.1.1. *Let* $L(\lambda) = \sum_{j=0}^{\ell} i^j \lambda^j A_j$, *be an* $n \times n$ *matrix polynomial with invertible leading coefficient* A_ℓ, *and let* (X, T) *be a standard pair for* $L(\lambda)$. *Then the general solution of the system of differential equations* (13.1.1) *is given by*

$$x(t) = X e^{tT} x_0, \quad t \in \mathsf{R}, \tag{13.1.2}$$

where $x_0 \in \mathsf{C}^{n\ell}$ *is an arbitrary constant vector.*

Proof. Since all standard pairs for $L(\lambda)$ are similar then, by definition, it is sufficient to prove (13.1.2) for the pair $(X, T) = ([I \ 0 \ \cdots \ 0], C_L)$. Let us first verify that any $x(t)$ of the form (13.1.2) is, indeed, a solution of (13.1.1). It is easily seen that

$$x^{(j)}(t) = [I\ 0\ \cdots\ 0]C_L^j e^{tC_L} x_0, \quad j = 0, 1, 2, \ldots,$$

and therefore

$$\sum_{j=0}^{\ell} i^j A_j \frac{d^j x}{dt^j} = \left(\sum_{j=0}^{\ell} i^j A_j [I\ 0\ \cdots\ 0] C_L^j \right) e^{tC_L} x_0. \tag{13.1.3}$$

But $[I\ 0\ \cdots\ 0]\,C_L^j = [0\ \cdots\ 0\ I\ 0\ \cdots\ 0]$ with I_n in position $(j+1)$, for $j = 0, \ldots, \ell - 1$; and

$$[I\ 0\ \cdots\ 0]\,C_L^\ell = \left[-(i^\ell A_\ell)^{-1} A_0,\ \ -(i^\ell A_\ell)^{-1}(iA_1),\ \ \ldots,\ \ -(i^\ell A_\ell)^{-1}(i^{\ell-1} A_{\ell-1}) \right].$$

Substitution in (13.1.3) yields

$$\sum_{j=1}^{\ell} i^j A_j \frac{d^j x}{dt^j} = 0.$$

To see that formula (13.1.2) gives *all* solutions of (13.1.1), observe first that, since A_l is invertible, the space of all solutions of (13.1.1) is $n\ell$-dimensional (see Theorem S1.6 of [39], for example). So it is sufficient to check that the formula (13.1.2) gives the zero solution only for $x_0 = 0$. Assume that $Xe^{tT}x_0 \equiv 0$. Then repeated differentiation gives $XT^i e^{tT} x_0 \equiv 0$ and hence

$$\begin{bmatrix} X \\ XT \\ \vdots \\ XT^{\ell-1} \end{bmatrix} e^{tT} x_0 \equiv 0.$$

But, by Proposition 12.1.1, the first matrix in the left is invertible, and so is e^{tT}. Consequently, $x_0 = 0$. $\qquad\square$

13.2   Boundedness for a System of Differential Equations

Now consider a system of differential equations with constant coefficients of the form (13.1.1), where $A_0, A_1, \ldots, A_\ell$ are $n \times n$ hermitian matrices, and A_ℓ is invertible. Such systems also arise in applications, and form a natural generalization of the Hamiltonian equations studied in Chapter 11 (i.e., the case $\ell = 1$ of (13.1.1)). Obviously, the properties of solutions of (13.1.1) are closely related to the properties of the matrix polynomial

$$L(\lambda) = \sum_{j=0}^{\ell} \lambda^j A_j, \tag{13.2.4}$$

and note that this polynomial is hermitian: $A_j = A_j^*$ for $j = 0, 1, \ldots, \ell$.

Again, the cases for which all solutions of (13.1.1) are bounded are of primary interest. To describe these cases in terms of $L(\lambda)$, it is convenient to introduce the following definition: A hermitian matrix polynomial has *simple structure* if all its eigenvalues are real and all elementary divisors are linear (i.e., real and semi-simple).

Theorem 13.2.1. *The solutions of* (13.1.1) *are bounded on the whole real line, correspondingly on the half line* $[0, \infty)$, *if and only if* $L(\lambda)$ *has simple structure.*

Proof. By Theorem 13.1.1, the general solution of the system $L\left(\frac{d}{dt}\right)x(t) = 0$ is given by the formula $x(t) = Pe^{tC_L}x_0$, where $P = [I\ 0\ \cdots\ 0]$ and $x \in \mathbb{C}^{n\ell}$ is arbitrary. Then it is easily seen that the general solution of (13.1.1) is given by the formula $x(t) = Pe^{-itC_L}x_0$. So if all eigenvalues of C_L (or, what is the same, the eigenvalues of $L(\lambda)$) are real with all partial multiplicities equal to 1, then $\sup_{-\infty<t<\infty}||x(t)|| < \infty$ for all x_0.

Conversely, assume that the solution of (13.1.1) are bounded ($t \in \mathbb{R}$). Since $x^{(i)}(t)$, $k = 1, 2, \ldots$ are also solutions of (13.1.1) and therefore are bounded, it follows that, for all x_0, the vector

$$\begin{bmatrix} x(t) \\ x'(t) \\ \vdots \\ x^{(\ell-1)}(t) \end{bmatrix} = \begin{bmatrix} P \\ -iPC_L \\ \vdots \\ \pm i^{\ell-1}PC_L^{\ell-1} \end{bmatrix} e^{-itC_L}x_0$$

is bounded on the real line. But $PC_L^{k-1} = [0 \cdots I\ 0 \cdots 0]$ with I in the k-th place so, in fact,

$$\sup_{-\infty<t<\infty} ||e^{-itC_L}x_0|| < \infty \quad \text{for all } x_0.$$

This happens only if all eigenvalues of C_L are real with all partial multiplicities equal to 1.

If the solutions of (13.1.1) are bounded on the half line $[0, \infty)$ then, similarly, we obtain

$$\sup_{0\leq t<\infty} ||e^{-itC_L}x_0|| < \infty \quad \text{for all } x_0.$$

Thus, iC_L has all its eigenvalues with nonnegative real parts, and the eigenvalues of iC_L with zero real parts (if any) have only partial multiplicities equal to 1. But since C_L is B_L-selfadjoint, the eigenvalues of iC_L are symmetric with respect to the imaginary axis. It follows again that all eigenvalues of C_L are real and semi-simple. □

Finally, we remark that when (13.1.1) has only bounded solutions, Proposition 12.2.1 can be applied to each eigenvalue of L and will, in fact, play an important part in the characterization of those systems (13.1.1) for which *all neighboring* systems of the same kind have only bounded solutions.

13.3   Stable Boundedness for Differential Equations

Consider the system of differential equations with constant coefficients of (13.1.1), where A_j, $j = 0, \dots, \ell$ are $n \times n$ hermitian matrices. It is assumed that the leading coefficient A_ℓ is invertible, and we let $L(\lambda)$ be the associated hermitian matrix polynomial given by (13.2.4). The equation (11.1.1) studied in Section 11.1 is the particular case of (13.1.1) with $\ell = 1$.

Our main objective will be the characterization of those cases in which (13.1.1) has *stably bounded solutions*, i.e., the solutions of (13.1.1) are bounded and remain bounded under any sufficiently small hermitian perturbations of the coefficients $A_0, \dots, A_\ell$. The next theorem provides a criterion for this stable boundedness property of solutions of (13.1.1) in terms of the underlying polynomial $L(\lambda)$.

Theorem 13.3.1. *Equation (13.1.1) has stably bounded solutions if and only if $\sigma(L)$ is real and, for every $\lambda_0 \in \sigma(L)$, the quadratic form $(L'(\lambda_0)x, x)$ is definite on the subspace $\mathrm{Ker}L(\lambda_0)$.*

Here $L'(\lambda)$ denotes the derivative of $L(\lambda)$ with respect to λ. Theorem 13.3.1 will follow as an easy corollary of certain stability properties of the hermitian matrix polynomial $L(\lambda)$, which are independently interesting.

Recall that the polynomial (13.2.4) is said to have *simple structure* if all its eigenvalues are real and all elementary divisors of $L(\lambda)$ are linear (the latter property means that $\dim \mathrm{Ker}L(\lambda_0)$ coincides with the multiplicity of λ_0 as a zero of $\det L(\lambda)$ for all $\lambda_0 \in \sigma(L)$). If these properties also hold for all polynomials obtained from (13.2.4) by sufficiently small hermitian perturbations of its coefficients $A_0, \dots, A_\ell$, then the polynomial (13.2.4) is said to have *stable simple structure* . For example, every polynomial $L(\lambda)$ of the form (13.2.4) with $n\ell$ *distinct* real eigenvalues, has stable simple structure.

As an illustration, consider the "gyroscopic" systems arising in mechanics. In this case equation (13.2.4) applies with $\ell = 2$, A_2 positive definite, $A_1^* = A_1$ and A_0 is negative definite. Then $L(\lambda)$ has stable simple structure (see [64, Section 7.7] and [16, Ex. 7(a), p. 91] for details).

The following theorem describes polynomials with stable simple structure in terms of the definiteness of certain quadratic forms; an analogue of the description of Theorem 9.2.1.

Theorem 13.3.2. *The polynomial $L(\lambda)$ given by (13.2.4) has stable simple structure if and only if $\sigma(L) \subseteq \mathbb{R}$ and, for every $\lambda_0 \in \sigma(L)$, the quadratic form $(L'(\lambda_0)x, x)$ is definite on the subspace $\mathrm{Ker}L(\lambda_0)$.*

Note that Theorem 9.2.1 (more exactly, the equivalence (i) ⇔ (iii) there) is a particular case of Theorem 13.3.2 which may be obtained by putting $L(\lambda) = \lambda H - HA$. Note also that the definiteness of the quadratic form $(L'(\lambda_0)x, x)$ on $\mathrm{Ker}L(\lambda_0)$ implies that all elementary divisors of L corresponding to λ_0 are linear (see Proposition 12.2.1).

Theorem 13.3.1 follows immediately from Theorem 13.3.2. Indeed, the solutions of (13.1.1) are bounded if and only if $L(\lambda)$ has simple structure (Theorem 13.2.1).

Proof of Theorem 13.3.2. Together with the polynomial $L(\lambda)$ consider the block matrices:

$$C_L = \begin{bmatrix} 0 & I & 0 & \cdots & 0 \\ 0 & 0 & I & \cdots & 0 \\ \vdots & & & \ddots & \vdots \\ & & & & I \\ -A_\ell^{-1}A_0 & -A_\ell^{-1}A_1 & & \cdots & -A_\ell^{-1}A_{\ell-1} \end{bmatrix};$$

$$B_L = \begin{bmatrix} A_1 & \cdots & A_{\ell-1} & A_\ell \\ \vdots & & & \vdots \\ A_{\ell-1} & A_\ell & 0 \cdots & 0 \\ A_\ell & 0 & \cdots & 0 \end{bmatrix}$$

$(13.3.5)$

Then C_L is B_L-selfadjoint (see Section 12.2). Also, the eigenvalues of $L(\lambda)$ coincide with those of C_L, and the elementary divisors of λ_0 as an eigenvalue of $L(\lambda)$, and the partial multiplicities of λ_0 as an eigenvalue of $\lambda I - C_L$, are the same (see Chapter 12).

For the converse, suppose that the quadratic form $(L'(\lambda_0)x, x)$ is definite on the subspace $\mathrm{Ker}L(\lambda_0)$ for every $\lambda_0 \in \sigma(L)$ so that, in particular, L has simple structure. Then C_L is r-diagonalizable and, moreover, the quadratic form $(B_L y, y)$ is definite on the subspace $\mathrm{Ker}(\lambda_0 I - C_L)$, for every $\lambda_0 \in \sigma(L)$ (see Proposition 12.2.1). Hence, by Theorem 9.2.2, $L(\lambda)$ has stable simple structure.

Assume now that $L(\lambda)$ has simple structure, but for some $\lambda_0 \in \sigma(L)$, the quadratic form $(L'(\lambda_0)x, x)$ is not definite on $\mathrm{Ker}L(\lambda_0)$. We shall prove that by small hermitian perturbations of the coefficients of L one can make some eigenvalues of the perturbed polynomial nonreal (and this will show that L does not have stable simple structure). Let

$$\mathbf{C}^n = \mathrm{Ker}L(\lambda_0) \oplus (\mathrm{Ker}L(\lambda_0))^\perp;$$

and write (with respect to this decomposition):

$$L(\lambda) = \begin{bmatrix} D_1(\lambda - \lambda_0) + E_1(\lambda)(\lambda - \lambda_0)^2 & D_2^*(\lambda - \lambda_0) + (E_2(\bar{\lambda}))^*(\lambda - \lambda_0)^2 \\ D_2(\lambda - \lambda_0) + E_2(\lambda)(\lambda - \lambda_0)^2 & E_3(\lambda) \end{bmatrix}.$$

Here $D_1 = D_1^*$ is a constant matrix, and $E_1(\lambda)$, $E_2(\lambda)$, $E_3(\lambda)$ are matrix polynomials. Since all elementary divisors of L corresponding to λ_0 are linear, it follows that D_1 is invertible, and since $(L'(\lambda_0)x, x)$ is not definite on $\mathrm{Ker}(L(\lambda_0))$, D_1 has both negative and positive eigenvalues.

Replacing $L(\lambda)$ by $M(\lambda) := WL(\lambda)W^*$, where $W = \begin{bmatrix} I & 0 \\ -D_2 D_1^{-1} & I \end{bmatrix}$, we can assume that $D_2 = 0$. Let x, y be orthonormal eigenvectors of D_1 corresponding to a positive eigenvalue α and a negative eigenvalue β, respectively. With respect to the orthogonal decomposition

$$C^n = \text{Span } \{x\} \oplus \text{Span } \{y\} \oplus (\text{Span } \{x, y\})^\perp$$

write:

$$M(\lambda) = \begin{bmatrix} \alpha & 0 & 0 \\ 0 & \beta & 0 \\ 0 & 0 & 0 \end{bmatrix} (\lambda - \lambda_0) + F(\lambda)(\lambda - \lambda_0)^2 + \begin{bmatrix} 0 & 0 & 0 \\ 0 & 0 & 0 \\ 0 & 0 & G(\lambda) \end{bmatrix},$$

where $F(\lambda)$, $G(\lambda)$ are selfadjoint matrix polynomials and the lower right block corner of $F(\lambda)$ is zero. Now let ζ be a small positive number and put

$$M_\zeta(\lambda) = M(\lambda) + \begin{bmatrix} 0 & \alpha\beta & 0 \\ \alpha\beta & 0 & 0 \\ 0 & 0 & 0 \end{bmatrix} \zeta - F(\lambda)\beta\zeta^2.$$

It is easily verified that $\lambda' := \lambda_0 + i\zeta(\alpha|\beta|)^{1/2}$ is a zero of $(\lambda - \lambda_0)^2 - \alpha\beta\zeta^2$ and a nonreal eigenvalue of $M_\zeta(\lambda)$. Indeed,

$$M(\lambda') = \begin{bmatrix} \alpha(\lambda' - \lambda_0) & \alpha\beta\zeta \\ \alpha\beta\zeta & \beta(\lambda' - \lambda_0) \end{bmatrix} \oplus G(\lambda'),$$

and therefore

$$\det M(\lambda') = \alpha\beta \left[(\lambda' - \lambda_0)^2 - \alpha\beta\zeta^2 \right] \det G(\lambda').$$

This concludes the proof of Theorem 13.3.2.   □

Now consider an hermitian matrix polynomial $L(\lambda)$ (as in (13.2.4)) which has simple structure but not *stable* simple structure. The proof of Theorem 13.3.2 shows that there is a perturbation of $L(\lambda)$ of type $L(\lambda, \zeta) = L(\lambda) + \zeta M_1 + \zeta^2 M_2(\lambda)$, where $\zeta > 0$ is small, with the following properties (we assume $n > 1$):

(i) M_1 is a hermitian matrix of rank 2.

(ii) $M_2(\lambda)$ is a hermitian matrix polynomial whose degree does not exceed $\ell - 2$;

(iii) $L(\lambda, \zeta)$ has a nonreal eigenvalue $\lambda_0(\zeta)$ of the form $\lambda_0(\zeta) = \lambda_0 + i\zeta\gamma$, for all sufficiently small positive ζ, and where γ is a fixed real number.

Thus, the unperturbed multiple eigenvalue "splits" under such a perturbation in such a way that *nonreal* eigenvalues appear.

13.4 The Strongly Hyperbolic Case

In this section we describe an important class of differential equations which necessarily have stably bounded solutions.

An $n \times n$ matrix polynomial $L(\lambda) = \sum_{j=0}^{\ell} \lambda^j A_j$ will be called *strongly hyperbolic* if the leading coefficient A_ℓ is positive definite and for every $x \in \mathsf{C}^n \setminus \{0\}$, the scalar equation

$$(L(\lambda)x, x) = \sum_{j=0}^{\ell} (A_j x, x) = 0 \qquad (13.4.6)$$

has ℓ distinct real zeros. All coefficients of a strongly hyperbolic polynomial $L(\lambda) = \sum_{j=0}^{\ell} \lambda^j A_j$ are hermitian (indeed, all the numbers $(A_j x, x)$, $x \in \mathsf{C}^n \setminus \{0\}$, $j = 1, \ldots, \ell$ are real, which implies the hermitian property of each A_j; see [70], for example).

Theorem 13.4.1. *The differential equation*

$$\sum_{j=0}^{\ell} i^j A_j \frac{d^j x}{dt^j} = 0, \quad t \in \mathsf{R}$$

has stably bounded solutions provided the matrix polynomial $L(\lambda) = \sum_{j=0}^{\ell} \lambda^j A_j$ *is strongly hyperbolic.*

The proof will be based on properties of strongly hyperbolic polynomials which must first be developed. Let $L(\lambda)$ be a strongly hyperbolic $n \times n$ matrix polynomial of degree ℓ, and let

$$\lambda_1(x) > \cdots > \lambda_\ell(x)$$

be the zeros of $(L(\lambda)x, x) = 0$, $x \in \mathsf{C}^n \setminus \{0\}$. As all zeros of the scalar polynomial $(L(\lambda)x, x) = 0$, $x \neq 0$ are real and distinct, and its leading coefficient is positive, it follows that

$$(-1)^{j-1}(L'(\lambda_j(x))x, x) > 0, \quad j = 1, \ldots, \ell; \quad x \neq 0. \qquad (13.4.7)$$

For $j = 1, 2, \ldots, \ell$ define

$$\triangle_j = \{\lambda_j(x) \mid x \in \mathsf{C}^n \setminus \{0\}\} = \{\lambda_j(x) \mid x \in \mathsf{C}^n, \ ||x|| = 1\}.$$

The set $\triangle_j$ is called the j-th *spectral zone* of $L(\lambda)$. The spectral zones are compact and connected (because $\triangle_j$ is the image of the unit sphere under the continuous map λ_j, and the unit sphere is compact and connected); so in fact $\triangle_j = [\alpha_j, \beta_j]$ for some real numbers $\alpha_j \leq \beta_j$. Further, the spectrum $\sigma(L)$ of L is real and

$$\sigma(L) \subseteq \sum_{j=1}^{\ell} \triangle_j. \qquad (13.4.8)$$

Indeed, for $\lambda \in C \setminus \bigcup_{j=1}^{\ell} \triangle_j$ and x on the unit sphere we have

$$\|L(\lambda)x\| \geq |(L(\lambda)x, x)| = (A_n x, x) \prod_{j=1}^{\ell} |\lambda - \lambda_j(x)| > 0,$$

so $L(\lambda)$ is invertible which means that $\lambda \notin \sigma(L)$.

The following property of the spectral zones is deeper.

Lemma 13.4.2. *Different spectral zones of a strongly hyperbolic matrix polynomial do not intersect.*

Proof. We have to prove that $\triangle_j \cap \triangle_k = \emptyset$ for $j \neq k$. It will suffice to show that $\triangle_j \cap \triangle_{j+1} = \emptyset$ for $j = 1, \ldots, \ell - 1$. Assume the contrary, so $\triangle_j \cap \triangle_{j+1} \neq \emptyset$ for some j. So there exist vectors x and y of norm 1 and a real number $\alpha = \lambda_j(x) = \lambda_{j+1}(y)$ such that $(L(\alpha)x, x) = (L(\alpha)y, y) = 0$. By (13.4.7) we have $(L'(\alpha)x, x)(L'(\alpha)y, y) < 0$. Define the $n \times n$ matrix $C = L'(\alpha) + iL(\alpha)$. Then the nonzero numbers (Cx, x) and (Cy, y) are real and have opposite signs. Now use the Toeplitz–Hausdorff theorem (Theorem A.7.1) according to which the numerical range of C is convex. So there exists a vector z of norm 1 such that $(Cz, z) = 0$. This means that $(L(\alpha)z, z) = (L'(\alpha)z, z) = 0$; so α is at least a double zero of $(L(\lambda)z, z) = 0$, and this contradicts the strongly hyperbolic property of $L(\lambda)$. $\square$

Proof of Theorem 13.4.1. By Theorem 13.3.1, and in view of the inclusion (13.4.8), we have only to check that for every $\lambda_0 \in \sigma(L)$ the quadratic form $(L'(\lambda_0)x, x)$ is definite on $\mathrm{Ker}L(\lambda_0)$. Evidently, $\lambda_0 \in \triangle_j$ for some j. Moreover, by Lemma 13.4.2, $\lambda_0 = \lambda_j(x)$ for every $x \in \mathrm{Ker}L(\lambda_0) \setminus \{0\}$. Now (13.4.7) shows that the quadratic form $(L'(\lambda_0)x, x)$ is positive definite on $\mathrm{Ker}L(\lambda_0)$ if j is odd, and negative definite if j is even. $\square$

13.5 Connected Components of Differential Equations with Hermitian Coefficients and Stably Bounded Solutions

Consider the system of differential equations

$$L\left[i\frac{d}{dt}\right]x = \sum_{j=0}^{\ell} i^j A_j \frac{d^j x}{dt^j} = 0; \quad t \in R \qquad (13.5.9)$$

where $L(\lambda) = \sum_{i=0}^{\ell} A_i \lambda^i$ is a polynomial with $n \times n$ hermitian coefficients ($A_i = A_i^*$) and invertible leading coefficient A_ℓ. We shall assume that (13.5.9) has stably bounded solutions. Then Theorem 13.3.1 implies that $\sigma(L)$ is real and the quadratic form $(L'(\lambda_0)x, x)$ is definite on the subspace $\mathrm{Ker}L(\lambda_0)$ for every $\lambda_0 \in \sigma(L)$. So we can assign a sign $\varepsilon(\lambda_0)$ to every eigenvalue of L; namely,

$\varepsilon(\lambda_0) = +1$ (resp. $\varepsilon(\lambda_0) = -1$) if $(L'(\lambda_0)x, x)$ is positive (resp. negative) definite on the subspace $\mathrm{Ker} L(\lambda_0)$.

Now construct the index

$$\mathrm{Ind}\,(L) := \{r; n_1, \ldots, n_p\}$$

associated with the polynomial L as follows (cf. the definition of the index of an H-selfadjoint stably r-diagonalizable matrix). Put $r = \mathrm{sig}\, A_\ell$. Further, define $\alpha_0 = -\infty$, $\alpha_p = \infty$ and let

$$(\alpha_0, \alpha_1), (\alpha_1, \alpha_2), \ldots, (\alpha_{p-1}, \alpha_p)$$

be consecutive intervals on the real line such that every interval (α_i, α_{i+1}) contains the maximal number of eigenvalues of $L(\lambda)$ of the same sign. Let $n_1, \ldots, n_p$ be the sum of multiplicities of the eigenvalues of $L(\lambda)$ lying in (α_{i-1}, α_i), multiplied by (-1) if the sign of these eigenvalues in the sign characteristic is negative.

For brevity, a matrix polynomial L with hermitian coefficients and invertible leading coefficient, and such that the differential equation (13.5.9) has stably bounded solutions, will be called an *SB polynomial*.

Theorem 13.5.1. *Let L_1 and L_2 be SB polynomials. If $\mathrm{Ind}\, L_1 \neq \mathrm{Ind}\, L_2$, then L_1 and L_2 belong to different connected components in the set of all SB polynomials.*

The topology in the set Ω of all SB polynomials is introduced naturally as follows: Ω is a disconnected union of the sets Ω_ℓ, $\ell = 1, 2, \ldots$, where Ω_ℓ is the set of all SB polynomials of fixed degree ℓ. In turn, Ω_ℓ is given the topology as a subset of $\mathbf{C}^{n \times n} \times \cdots \times \mathbf{C}^{n \times n}$ ($\ell + 1$ times), where $\mathbf{C}^{n \times n}$ is the set of all $n \times n$ complex matrices with the standard topology, by identifying the SB polynomial $\sum_{i=0}^{\ell} A_i \lambda^i$ with the ordered set $(A_0, \ldots, A_\ell) \in \mathbf{C}^{n \times n} \times \cdots \times \mathbf{C}^{n \times n}$.

Proof. Write $L_i(\lambda) = \sum_{j=0}^{\ell} \lambda^j A_{ij}$, $i = 1, 2$, and let

$$\mathrm{Ind}\,(L_i) = \left\{ r^{(i)}; n_1^{(i)}, \ldots, n_{p_i}^{(i)} \right\}.$$

We can assume that L_1 and L_2 have the same degree ℓ (otherwise Theorem 13.5.1 is trivial). Clearly, if the leading coefficients of L_1 and L_2 have different signatures (i.e., if $r^{(1)} \neq r^{(2)}$), then L_1 and L_2, belong to different connected components in the set Ω_ℓ.

Let C_{L_1} and C_{L_2} be the companion matrices of L_1 and L_2, respectively, and introduce the matrices

$$B_{L_i} = \begin{bmatrix} A_{i0} & A_{i1} & \cdots & A_{i\ell} \\ A_{i1} & & & 0 \\ \vdots & A_{i\ell} & & \vdots \\ A_{i\ell} & 0 & \cdots & 0 \end{bmatrix}, \quad i = 1, 2.$$

As observed in Section 12.2 (in particular, Proposition 12.2.1), for $i = 1, 2$ the companion matrix C_{L_i} is B_{L_i}-selfadjoint and also B_{L_i}-stably r-diagonalizable. Also for $i = 1, 2$,

$$\operatorname{ind}_r(C_{L_i}, B_{L_i}) = \left\{ n_1^{(i)}, \ldots, n_{p_i}^{(i)} \right\}.$$

By Theorem 9.7.2, if $\left\{ n_1^{(1)}, \ldots, n_p^{(1)} \right\} \neq \left\{ n_1^{(2)}, \ldots, n_p^{(2)} \right\}$, then (C_{L_1}, B_{L_1}) and (C_{L_2}, B_{L_2}) belong to different connected components in the set S_r. So in this case the polynomials L_1 and L_2 also belong to different connected components in Ω_ℓ. $\square$

13.6   A Special Case

In this section we study in detail the class $\mathcal{A}$ of SB matrix polynomials of second degree and size 2×2 with positive definite leading coefficient. In this case every $L(\lambda) \in \mathcal{A}$ can be written in the form

$$L(\lambda) = \lambda^2 A_2 + \lambda A_1 + A_0, \tag{13.6.10}$$

where A_i are 2×2 hermitian matrices $(i = 0, 1, 2)$, and A_2 is positive definite.

A preliminary result is required:

Lemma 13.6.1. *Let $\underset{\sim}{\lambda}$ (resp. $\widetilde{\lambda}$) be the smallest (resp. the greatest) eigenvalue of $L(\lambda) \in \mathcal{A}$. Then the form $(L'(\underset{\sim}{\lambda})x, x)$ is negative definite on the space $\operatorname{Ker}L(\underset{\sim}{\lambda})$, and the form $(L'(\widetilde{\lambda})x, x)$ is positive definite on the space $\operatorname{Ker}L(\widetilde{\lambda})$.*

Proof. Let $L(\lambda)$ be given by (13.6.10). Let us show first that the hermitian matrix $L(\lambda)$ is positive definite for $\lambda < \underset{\sim}{\lambda}$. Indeed, since A_2 is positive definite, $L(\lambda) = \lambda^2 \left(A_2 + \frac{1}{\lambda} A_1 + \frac{1}{\lambda^2} A_0 \right)$ is certainly positive definite for λ real and $|\lambda|$ large enough. Let $\mu = \inf \{ \lambda \in \mathbb{R} \mid L(\lambda) \text{ is not positive definite} \}$. Then $L(\lambda)$ is positive definite for $\lambda < \mu$ and, by continuity, $L(\mu)$ is positive semidefinite. But $L(\mu)$ cannot be positive definite (otherwise $L(\lambda)$ would be positive definite also for $\mu < \lambda < \mu + \varepsilon$, for small $\varepsilon > 0$; a contradiction with the definition of μ). This means that $L(\mu)x = 0$ for some $x \neq 0$, i.e., μ is an eigenvalue of $L(\lambda)$, and therefore $\mu \geq \underset{\sim}{\lambda}$. So $L(\lambda)$ is positive definite for $\lambda < \underset{\sim}{\lambda}$.

Now pick $x \in \operatorname{Ker}L(\underset{\sim}{\lambda}) \setminus \{0\}$ and put $f(\lambda) = (L(\lambda)x, x)$. Then $f(\lambda) < 0$ for $\lambda < \underset{\sim}{\lambda}$ and $f(\underset{\sim}{\lambda}) = 0$. Consequently, $f'(\underset{\sim}{\lambda}) = (L'(\underset{\sim}{\lambda})x, x) \leq 0$. But the equality $(L'(\underset{\sim}{\lambda})x, x) = 0$ is impossible because $L(\lambda)$ is SB.

Similarly, one checks Lemma 13.6.1 for the largest eigenvalue $\widetilde{\lambda}$. $\square$

This proof shows that the result of Lemma 13.6.1 remains valid for every SB matrix polynomial with positive definite leading coefficient and even degree.

Lemma 13.6.1 shows that the index for $L \in A$ can be only one of the following two:

$$\operatorname{Ind}_1 = \{2; -1, 1, -1, 1\}, \qquad \operatorname{Ind}_2 = \{2; -2, 2\}.$$

Let $\mathcal{A}_1$ (resp. $\mathcal{A}_2$) be the set of all polynomials $L \in \mathcal{A}$ whose index is Ind_1 (resp. Ind_2). It is easily seen that both sets $\mathcal{A}_1$ and $\mathcal{A}_2$ are nonempty; for instance,

$$\begin{bmatrix} \lambda(\lambda - 1) & 0 \\ 0 & (\lambda - 2)(\lambda - 3) \end{bmatrix} \in \mathcal{A}_1, \quad \begin{bmatrix} \lambda(\lambda - 2) & 0 \\ 0 & (\lambda - 1)(\lambda - 3) \end{bmatrix} \in \mathcal{A}_2.$$

Theorem 13.5.1 shows that if $L_1 \in \mathcal{A}_1$, $L_2 \in \mathcal{A}_2$, then L_1 and L_2 belong to different connected components in $\mathcal{A}$. It will be proved that the sets $\mathcal{A}_1$ and $\mathcal{A}_2$ are connected, thereby verifying the conjecture for SB polynomials of size 2×2, degree 2 and with positive definite leading coefficients.

To this end consider the set $\mathcal{A}_0$ of all matrix polynomials $L \in \mathcal{A}$ which have four distinct eigenvalues (necessarily real).

Lemma 13.6.2. *The set $\mathcal{A}_0$ consists of exactly two connected components.*

Proof. Pick

$$L(\lambda) = \lambda^2 A_2 + \lambda A_1 + A_0 \in \mathcal{A}_0. \tag{13.6.11}$$

Since A_2 is positive definite, write $A_2 = S^*S$ for some invertible 2×2 matrix S. Let $S(t)$, $t \in [0, 1]$ be a continuous path of invertible 2×2 matrices such that $S(0) = S$; $S(1) = I$. Then putting

$$L(\lambda, t) = (S(t))^*(\lambda^2 I + (S^*)^{-1}A_1 S + (S^*)^{-1}A_0 S)S(t), \quad t \in [0, 1]$$

we connect L with a polynomial $L_1(\lambda) = L(\lambda, 1)$ with leading coefficient I, and the connecting path lies entirely in $\mathcal{A}_0$. Further, let $\underset{\sim}{\lambda}$ be the smallest eigenvalue of L_1. Putting

$$L_1(\lambda, t) = L_1(\lambda + t\underset{\sim}{\lambda}), \quad t \in [0, 1]$$

we obtain a continuous path in $\mathcal{A}_0$ connecting $L_1(\lambda) = L_1(\lambda, 0)$ with $L_2(\lambda) \overset{\text{def}}{=} L_1(\lambda + \underset{\sim}{\lambda})$, and the latter polynomial has leading coefficient I and its smallest eigenvalue is zero. Finally, by putting

$$L_2(\lambda, t) = U(t)^* L_2(\lambda) U(t), \quad t \in [0, 1]$$

where $U(t)$ is a suitable continuous path in the connected group of 2×2 unitary matrices, we connect $L_2(\lambda)$ in $\mathcal{A}_0$ with a polynomial $L_3(\lambda) \in \mathcal{A}_0$ of the following form:

$$L_3(\lambda) = \begin{bmatrix} \lambda^2 + \lambda a & \lambda(b + ic) \\ \lambda(b - ic) & \lambda^2 + \lambda d + x \end{bmatrix}, \tag{13.6.12}$$

where a, b, c, d, x are real numbers with $x \neq 0$, and $\det L_3(\lambda)$ has three different positive zeros.

Let $\widetilde{\mathcal{A}}_0$ be the set of all polynomials of type (13.6.12). Let $0 < p < q < r$ be the eigenvalues of $L_3(\lambda) \in \widetilde{\mathcal{A}}_0$, which is given by (13.6.12). Then the following relations hold:

$$ax = -pqr; \quad d + a = -p - q - r; \quad x + ad - b^2 - c^2 = pq + pr + qr, \tag{13.6.13}$$

which, together with inequalities $0 < p < q < r$, are necessary and sufficient in order that $L_3 \in \tilde{\mathcal{A}}_0$. It follows from (13.6.13) that, given $x \neq 0$ and p, q, r such that $0 < p < q < r$, one can solve (13.6.13) uniquely for a, d and $b^2 + c^2$ if and only if the number

$$x + ad - pq - pr - qr = \frac{1}{x^2} \left[x^3 - x^2(pq + pr + qr) + x(p + q + r)pqr - (pqr)^2 \right]$$

is nonnegative. As the polynomial $x^3 - x^2(pq + pr + qr) + x(p + q + r)pqr - (pqr)^2$ has three different positive real zeros $pq < pr < qr$, we find that $\tilde{\mathcal{A}}_0$ has exactly two connected components, which can be identified in terms of the numbers p, q, r, x, b, c introduced above in the following way. One component is given by

$$\{(p, q, r, x, b, c) \in \mathbb{R}^6 \mid 0 < p < q < r;\ pq \leq x \leq pr;$$
$$b^2 + c^2 = x^3 - x^2(pq + pq + qr) + x(p + q + r)pqr - (pqr)^2\};$$

the second is

$$\{(p, q, r, x, b, c) \in \mathbb{R}^6 \mid 0 < p < q < r;\ x \geq qr;$$
$$b^2 + c^2 = x^3 - x^2(pq + pr + qr) + x(p + q + r)pqr - (pqr)^2\}.$$

Hence the set $\mathcal{A}_0$ also has exactly two connected components.   □

Now one can easily show that the sets $\mathcal{A}_1$ and $\mathcal{A}_2$ are connected. Indeed, since $\mathcal{A}_0$ is dense in $\mathcal{A}$, the set $\mathcal{A}$ consists of at most two connected components, and because of Theorem 13.5.1, $\mathcal{A}$ has exactly two connected components. Now $\mathcal{A} = \mathcal{A}_1 \cup \mathcal{A}_2$ and $\mathcal{A}_1 \cap \mathcal{A}_2 = \emptyset$; so in view of the same Theorem 13.5.1, it is clear that $\mathcal{A}_1$ and $\mathcal{A}_2$ are the connected components of $\mathcal{A}$.

13.7   Difference Equations

Consider the following difference equation:

$$A_0 x_i + A_1 x_{i+1} + \cdots + A_\ell x_{i+\ell} = 0, \quad i = 0, 1, \ldots, \tag{13.7.14}$$

where $\{x_i\}_{i=0}^\infty$ is a sequence of n-dimensional complex vectors to be found, and $A_0, \ldots, A_\ell$ are given complex $n \times n$ matrices. It is assumed throughout this section that A_ℓ is invertible.

As in the case of the system of differential equations (13.1.1), a general solution of (13.7.14) can be found in terms of a standard pair (X, T) of the associated matrix polynomial

$$L(\lambda) = \sum_{j=0}^\ell \lambda^j A_j.$$

Theorem 13.7.1. *The general solution of* (13.7.14) *is given by the formula*

$$x_i = XT^i z, \quad i = 0, 1, \ldots, \tag{13.7.15}$$

where z is an arbitrary $n\ell$-dimensional vector.

If T is invertible, then the general solution of the bilateral difference equation

$$A_0 x_i + A_1 x_{i+1} + \cdots + A_\ell x_{i+\ell} = 0, \quad i = 0, \pm 1, \pm 2, \ldots, \tag{13.7.16}$$

is given by the formula

$$x_i = XT^i z, \quad i = 0, \pm 1, \pm 2, \ldots, \quad z \in \mathbb{C}^{n\ell}. \tag{13.7.17}$$

Proof. We will prove formula (13.7.15) only; the proof of (13.7.17) is completely analogous.

Since any two standard pairs of $L(\lambda)$ are similar, we can assume $X = [I \ 0 \ \ldots \ 0]$ and $T = C_L$. Then

$$\sum_{j=0}^{\ell} A_j X T^j = 0$$

(cf. the proof of (13.1.3)), so the formula (13.7.15) indeed gives solutions for (13.7.14). Further, invertibility of A_ℓ implies that

$$x_{\ell+i} = -A_\ell^{-1}(A_0 x_i + \cdots + A_{\ell-1} x_{i+\ell-1}), \quad i = 0, 1, \ldots,$$

and therefore the solution of (13.7.14) is determined by the values of $x_0, \ldots, x_{\ell-1}$. In other words, the dimension of the space of solutions of (13.7.14) is $n\ell$. So in order to prove that (13.7.15) gives all solutions of (13.7.14) we have to check that $x_i = 0$, $i = 0, 1, 2, \ldots, \ell - 1$, implies $z = 0$. But this follows from the invertibility of the matrix

$$\begin{bmatrix} X \\ XT \\ \vdots \\ XT^{\ell-1} \end{bmatrix}$$

which was established in Proposition 12.1.1. □

We will be interested in description of systems of difference equations (13.7.14) with the property that every solution sequence is bounded:

Proposition 13.7.2. *The system* (13.7.14) *has all solutions bounded if and only if any linearization T of the associated matrix polynomial $L(\lambda) = \sum_{j=0}^{\ell} \lambda^j A_j$ has all its eigenvalues in the closed unit disc, and the eigenvalues of T on the unit circle (if any) are all semi-simple.*

Assuming that the matrix A_0 is invertible, the bilateral system (13.7.16) *has all solutions bounded if and only if any linearization T of the associated matrix polynomial $L(\lambda)$ is diagonalizable and has all eigenvalues on the unit circle.*

Recall that an $n\ell \times n\ell$ matrix T, a linearization of $L(\lambda)$, appears in a standard pair (X, T) of $L(\lambda)$.

Proof. Formula (13.7.15) shows that all solutions of (13.7.14) are bounded if and only if the set $\{T^q\}_{q=0}^{\infty}$ of powers T with nonnegative integer exponents is bounded. This property in turn is well known (and easily observed using the Jordan form of T) to be equivalent to the properties of T stated in the proposition.

Similarly, formula (13.7.17) shows that all solutions of (13.7.16) are bounded if and only if the set $\{T^q\}_{q=-\infty}^{\infty}$ is bounded (the invertibility of T follows from the hypothesis that A_0 is invertible). The boundedness of $\{T^q\}_{q=-\infty}^{\infty}$ is easily seen to be equivalent to the properties of T as stated in the second part of the proposition. $\qquad\square$

Now consider the system (13.7.14) with the additional assumptions that ℓ is even, say $\ell = 2k$, and

$$A_i^* = A_{\ell-i}, \quad i = 0, \dots, \ell. \tag{13.7.18}$$

Note, in particular, that $A_0^* = A_\ell$ and is therefore invertible. A matrix polynomial with coefficients satisfying this condition is sometimes said to be *palindromic*. For example, palindromic polynomials arise if the differential equation

$$\sum_{j=0}^{l} i^j \tilde{A}_j \frac{d^j x}{dt^j} = 0$$

with hermitian coefficients $\tilde{A}_j$ is approximated using a symmetric finite difference technique (i. e. replacing the derivative $\dfrac{dx}{dt}$ at the point kh ($k = 0, \pm 1, \pm 2, \dots$) by $\frac{1}{2h}\big[x\big((k+1)h\big) - x\big((k-1)h\big)\big]$; here h is a fixed positive number).

If (13.7.18) holds, then C_L is $\widehat{B}_L$-unitary and therefore the spectrum of C_L is symmetric relative to the unit circle. We obtain the following theorem. We say that a matrix polynomial $L(\lambda)$ has *simple structure* with respect to the unit circle if the spectrum $\sigma(L)$ lies on the unit circle and all eigenvalues are semi-simple.

Theorem 13.7.3. *Let there be given $n \times n$ matrices $A_0, \dots, A_\ell$ such that $A_j^* = A_{\ell-j}$ for $j = 0, \dots, \ell$, the integer ℓ is even, and A_ℓ is invertible. Then the following statements are equivalent:*

(a) *All solutions of the system*

$$A_0 x_i + A_1 x_{i+1} + \dots + A_\ell x_{i+\ell} = 0, \quad i = 0, 1, \dots, \tag{13.7.19}$$

are bounded.

(b) *All solutions of the bilateral system*

$$A_0 x_i + A_1 x_{i+1} + \dots + A_\ell x_{i+\ell} = 0, \quad i = 0, \pm 1, \pm 2, \dots, \tag{13.7.20}$$

are bounded.

(c) *The matrix polynomial $L(\lambda) = \sum_{j=0}^{\ell} \lambda^j A_j$ has simple structure with respect to the unit circle.*

Proof. By Proposition 13.7.2 we know that all solutions of (13.7.20) are bounded if and only if the linearization of $L(\lambda)$ is diagonalizable and has all eigenvalues on the unit circle. In view of formula (12.1.2) (cf. Theorems A.6.2 and A.6.3) this condition is equivalent to (c).

By the same Proposition 13.7.2, all solutions of (13.7.19) are bounded if and only if C_L has all eigenvalues in the closed unit disc, with the eigenvalues on the unit circle (if any) have all partial multiplicities equal to 1. But since the spectrum of C_L is symmetric relative to the unit circle, we arrive at statement (c), as before. $\qquad\square$

In fact, Theorem 13.7.3 holds for systems (13.7.19) also when ℓ is odd. Indeed, the condition $A_j^* = A_{\ell-j}$ for $j = 0, \ldots, \ell$ implies that

$$L(\lambda) = \lambda^\ell \left(L(\overline{\lambda}^{-1}) \right)^*, \quad \lambda \in \mathbb{C} \setminus \{0\}.$$

Thus, $\sigma(L)$ is symmetric relative to the unit circle. Therefore, if the solutions of (13.7.19) are bounded, then $L(\lambda)$ cannot have spectrum outside of the unit circle, and the same holds for any linearization of $L(\lambda)$. Now use Proposition 13.7.2.

13.8 Stable Boundedness for Difference Equations

We continue with investigations of the difference equation (13.7.14) under the assumptions of the preceding section, namely, that A_ℓ is invertible, ℓ is even, and the palindromic conditions (13.7.18) hold. We are interested in the case when the solutions of (13.7.14) are *stably bounded*, i.e.,, all solutions of (13.7.14) are bounded, and all solutions of every system

$$\tilde{A}_0 y_i + \tilde{A}_1 y_{i+1} + \cdots + \tilde{A}_\ell y_{i+\ell} = 0, \quad i = 0, 1, \ldots,$$

with

$$\tilde{A}_j^* = \tilde{A}_{\ell-j}, \quad j = 0, 1, \ldots, \ell,$$

are bounded provided $\|\tilde{A}_j - A_j\|$ is small enough (for $j = 0, 1, \ldots, \ell$).

Theorem 13.8.1. *The solutions of (13.7.14) are stably bounded if and only if the spectrum of the associated matrix polynomial $L(\lambda) = \sum_{j=0}^{\ell} \lambda^j A_j$ lies on the unit circle, and the quadratic form*

$$\left(x, i\lambda_0 \left(\lambda^{-\ell/2} L(\lambda) \right)^{(1)} (\lambda_0) x \right), \quad x \in \operatorname{Ker} L(\lambda_0) \tag{13.8.21}$$

is either positive definite or negative definite, for every $\lambda_0 \in \sigma(L)$.

In this statement, the superscript $^{(1)}$ denotes the derivative.

Proof. We use the companion matrix for the polynomial $L(\lambda)$:

$$
C_L = \begin{bmatrix} 0 & I & 0 & \cdots & 0 \\ 0 & 0 & I & & 0 \\ & & & \ddots & \vdots \\ & & & & I \\ \vdots & & & & \\ -\tilde{A}_0 & -\tilde{A}_1 & \cdots & & -\tilde{A}_{\ell-1} \end{bmatrix}, \quad \tilde{A}_j = A_\ell^{-1} A_j, \quad j = 0, 1, \ldots, \ell-1,
$$

and the following matrix introduced in Section 12.6:

$$
\widehat{B}_L = i \left[\begin{array}{ccc|ccc} & & & A_\ell & \cdots & 0 \\ & 0 & & \vdots & \ddots & \vdots \\ & & & A_{k+1} & \cdots & A_\ell \\ \hline -A_0 & \cdots & -A_{k-1} & & & \\ \vdots & \ddots & \vdots & & 0 & \\ 0 & \cdots & -A_0 & & & \end{array} \right], \quad k = \frac{\ell}{2}.
$$

Assume that the spectrum of $L(\lambda)$ is unimodular, and that the quadratic form (13.8.21) is definite for every $\lambda_0 \in \sigma(C_L)$. This means that, for every eigenvalue λ_0 of C_L, the quadratic form $(x, \widehat{B}_L x)$, $x \in \operatorname{Ker}(\lambda_0 I - C_L)$ is either positive definite or negative definite (see formula (12.6.31) and Proposition 12.6.1). By Theorem 9.8.3 there exists an $\varepsilon > 0$ such that every H-unitary matrix U with

$$
\|U - C_L\| + \|H - \widehat{B}_L\| < \varepsilon
$$

is similar to a diagonal matrix with unimodular spectrum. In particular, this is true for every matrix

$$
U = \begin{bmatrix} 0 & I & 0 & \cdots & 0 \\ 0 & 0 & I & & 0 \\ & & & \ddots & \vdots \\ & & & & I \\ \vdots & & & & \\ -(A_\ell')^{-1} A_0' & -(A_\ell')^{-1} A_1' & \cdots & & -(A_\ell')^{-1} A_{\ell-1}' \end{bmatrix},
$$

under the palindromic conditions (13.7.18), and provided that $\sum_{j=0}^{\ell} \|A_j' - A_j\|$ is sufficiently small. By Theorem 13.7.3 the solutions of (13.7.14) are stably bounded.

For the converse, assume now that the solutions of (13.7.14) are stably bounded. Then, using Theorem 13.7.3 again, the polynomial $L(\lambda)$ must have only unimodular semi-simple eigenvalues, and every matrix polynomial

$$
\widehat{L}(\lambda) = \sum_{j=0}^{\ell} \lambda^j \widehat{A}_j \tag{13.8.22}
$$

with

$$\widehat{A}_j = \widehat{A}^*_{\ell-j} \quad \text{for } j = 0, 1, \ldots, \ell \quad \text{and} \quad \sum_{j=0}^{\ell} \|\widehat{A}_j - A_j\| \quad \text{sufficiently small}$$

(13.8.23)

has this property. Choose $w \in \mathbb{C}$, $|w| = 1$ such that $L(w)$ is invertible, and put

$$R(\lambda) = \sum_{j-0}^{\ell} (1 + i\lambda)^j (1 - i\lambda)^{\ell-j} B_j, \qquad (13.8.24)$$

where $B_0, \ldots, B_\ell$ are the coefficients of the polynomial

$$L_w(\lambda) := (-w)^{-\frac{\ell}{2}} L(-w\lambda) = \sum_{j=0}^{\ell} \lambda^j B_j.$$

Note that $L_w(\lambda)$ is hermitian with respect to the unit circle, i.e., the rational matrix function $(\lambda^{-\ell/2}) L_w(\lambda)$ has the property that

$$(\lambda^{-\ell/2}) L_w(\lambda) = ((\overline{\lambda})^{-\ell/2}) L_w(\overline{\lambda}) \quad \text{for } \lambda \in \mathsf{T}.$$

Therefore, $R(\lambda)$ is a hermitian (relative to the real line) matrix polynomial of degree ℓ. Moreover, the leading coefficient of $R(\lambda)$ is easily seen to be equal to $i^\ell L_w(-1)$, and therefore is invertible.

Clearly, all eigenvalues of $R(\lambda)$ are real and semi-simple. Moreover, all eigenvalues of hermitian matrix polynomials of degree ℓ which are sufficiently close to $R(\lambda)$ (in the sense that their coefficients are close to the corresponding coefficients of $R(\lambda)$) are also semi-simple. Indeed, this follows from the property that every matrix polynomial (13.8.22) satisfying (13.8.23) has only unimodular semi-simple eigenvalues, in view of the inverse formula for (13.8.24):

$$L(\mu) = \left[\frac{1 - w^{-1}\mu}{2}\right]^\ell (-w)^k R\left(\frac{i(1 + w^{-1}\mu)}{1 - w^{-1}\mu}\right), \qquad k = \frac{\ell}{2}.$$

By Theorem 13.3.2 the quadratic form $(x, R^{(1)}(\lambda_0)x)$ is either positive definite or negative definite on $\operatorname{Ker} R(\lambda_0)$ for every eigenvalue λ_0 of $R(\lambda)$. A computation shows that, putting

$$\widehat{L}(\lambda) = \lambda^{-k} L(\lambda), \qquad k = \frac{\ell}{2},$$

we have

$$R^{(1)}(\lambda) = 2\lambda k (1 + \lambda^2)^{k-1} \widehat{L}\left(\frac{-w(1 + i\lambda)}{1 - i\lambda}\right)$$

$$+ (-w)(1 + \lambda^2)^k \left[\frac{2i}{(1 - i\lambda)^2}\right] \widehat{L}^{(1)}\left(\frac{-w(1 + i\lambda)}{1 - i\lambda}\right).$$

Let λ_0 be a (necessarily real) eigenvalue of $R(\lambda)$. Then

$$\mu_0 := \frac{-w(1+i\lambda_0)}{1-i\lambda_0}$$

is a (necessarily unimodular) eigenvalue of $L(\lambda)$, and

$$\operatorname{Ker} R(\lambda_0) = \operatorname{Ker} L(\mu_0) = \operatorname{Ker} \widehat{L}(\mu_0).$$

Hence, for every $x \in \operatorname{Ker} L(\mu_0)$ we have

$$i\mu_0 \widehat{L}^{(1)}(\mu_0)x = \frac{1}{2}(1+\lambda_0^2)^{-k+1} R^{(1)}(\lambda_0)x,$$

so the form

$$(x, i\mu_0 \widehat{L}^{(1)}(\mu_0)x), \qquad x \in \operatorname{Ker} \widehat{L}(\mu_0)$$

is positive definite or negative definite together with the form $(x, R^{(1)}(\lambda_0)x)$, $x \in \operatorname{Ker} R(\lambda_0)$. $\qquad\square$

13.9   Connected Components of Difference Equations

Consider again the class of difference equations studied in Section 13.7. To repeat, a typical equation has the form

$$A_0 x_i + A_1 x_{i+1} + \cdots + A_\ell x_{i+l} = 0, \quad i = 0, 1, \ldots, \tag{13.9.25}$$

where $\{x_i\}_{i=0}^{\infty}$ is a sequence of n-dimensional complex vectors to be found, and $A_0, \ldots, A_\ell$ are given complex $n \times n$ matrices. As before, it is assumed that ℓ is even, the coefficient symmetries (13.7.18) hold, and A_ℓ is invertible. In particular, $A_{\ell/2}$ is hermitian and A_0 is invertible as well. We say that a difference equation (13.9.25) has *stably bounded* solutions if there exists an $\varepsilon > 0$ such that all solutions of every equation

$$\sum_{m=0}^{\ell} A'_m x_{i+m} = 0, \quad i = 0, 1, \ldots,$$

with $A'^*_j = A'_{\ell-j}$, $j = 0, \ldots, \ell$, and with

$$\sum_{m=0}^{\ell} \| A'_m - A_m \| < \varepsilon$$

are bounded. A description of the difference equations with stably bounded solutions was given in Theorem 13.8.1.

There exists a natural topology on the set Ω_u of all equations (13.9.25) with ℓ even, $A^*_{\ell-j} = A_j$ for $j = 0, \ldots, \ell$, A_ℓ invertible, and with stably bounded solutions. Observe that Ω_u is a disconnected union of the sets $\Omega_{u,\ell}$, $\ell = 2, 4, \ldots$, where $\Omega_{u,\ell}$

is the set of equations (13.9.25) with stably bounded solutions of degree ℓ. In turn, $\Omega_{u,\ell}$ is given a topology as a subset of $C^{n\times n} \times \cdots \times C^{n\times n}$ ($\ell+1$ times), where we identify the difference equation (13.9.25) with the ordered set of its coefficients

$$(A_0, \ldots, A_\ell) \in C^{n\times n} \times \cdots \times C^{n\times n}.$$

So it makes sense to speak about the connected components of Ω_u.

In what follows we shall identify any system (13.9.25) with the matrix polynomial $L(\lambda)$ formed by its coefficients. To study the connected components of Ω_u, introduce the notion of an index for $L(\lambda)$:

$$\mathrm{Ind}_u(L) = \{n_1, \ldots, n_p\}. \tag{13.9.26}$$

As we noticed above, every $\lambda_0 \in \sigma(L)$ is unimodular: we say that the sign of λ_0 is negative (resp. positive) if the quadratic form

$$\left(x, i\lambda_0 \left(\lambda^{-\ell/2} L(\lambda) \right)^{(1)} (\lambda_0) x \right), \quad x \in \mathrm{Ker}\, L(\lambda_0)$$

is negative (resp. positive) definite. Let

$$[\alpha_0, \alpha_1), (\alpha_1, \alpha_2), \ldots, (\alpha_{p-1}, \alpha_p), \quad \alpha_p - \alpha_0 = 2\pi,$$

be consecutive intervals such that every arc

$$\left[e^{i\alpha_0}, e^{i\alpha_1} \right), \ldots, \left(e^{i\alpha_{p-1}}, e^{i\alpha_p} \right)$$

contains the largest possible number of eigenvalues of $L(\lambda)$ having the same sign. We let n_i ($i = 1, \ldots, p$) in (13.9.26) be the sum of the multiplicities of eigenvalues of $L(\lambda)$ lying on the i-th arc and multiplied by (-1) if the sign of these eigenvalues is negative.

Theorem 13.9.1. *Let*

$$A_{01} x_i + A_{11} x_{i+1} + \cdots + A_{\ell,1} x_{i+\ell} = 0, \quad i = 0, 1, \ldots, \tag{13.9.27}$$

and

$$A_{02} x_i + A_{12} x_{i+1} + \cdots + A_{\ell,2} x_{i+\ell} = 0, \quad i = 0, 1, \ldots, \tag{13.9.28}$$

be two difference equations of type (13.9.25), i.e., $A_{ji}^* = A_{\ell-j,i}$ for $j = 0, \ldots, \ell$ and for $i = 1, 2$; ℓ is even; and $A_{\ell,1}$, $A_{\ell,2}$ are invertible. Suppose that both equations (13.9.27) and (13.9.28) have stably bounded solutions, and the indices $\mathrm{Ind}_u(L_1)$ and $\mathrm{Ind}_u(L_2)$ of

$$L_1(\lambda) = \sum_{j=0}^{\ell} \lambda^j A_{j1}$$

and

$$L_2(\lambda) = \sum_{j=0}^{\ell} \lambda^j A_{j2},$$

respectively, cannot be obtained from each other by a cyclic permutation. Then the equations (13.9.27) and (13.9.28) belong to different connected components in the set of all equations of type (13.9.25) with stably bounded solutions.

Proof. Let

$$
C_j = \begin{bmatrix}
0 & I & 0 & \cdots & 0 \\
0 & 0 & I & \cdots & 0 \\
\vdots & \vdots & \vdots & \ddots & \vdots \\
0 & 0 & 0 & \cdots & I \\
-A_{\ell,j}^{-1}A_{0j} & -A_{\ell,j}^{-1}A_{1j} & -A_{\ell,j}^{-1}A_{2j} & \cdots & -A_{\ell,j}^{-1}A_{\ell-1,j}
\end{bmatrix}, \quad j = 1,2,
$$

$$
\widehat{B}_j = i \left[\begin{array}{ccc|ccc}
 & & & A_{\ell,j} & \cdots & 0 \\
 & 0 & & \vdots & \ddots & \vdots \\
 & & & A_{k+1,j} & \cdots & A_{\ell,j} \\ \hline
-A_{0j} & \cdots & -A_{k-1,j} & & & \\
\vdots & \ddots & \vdots & & 0 & \\
0 & \cdots & -A_{0j} & & &
\end{array} \right], \quad j = 1,2,
$$

where $k = \ell/2$. As we know from Section 12.6, C_j is $\widehat{B}_j$-unitary and, moreover, C_j is stably u-diagonalizable. Further, $\mathrm{Ind}_u(L_j)$ coincides with the index $\mathrm{ind}_u(C_j, \widehat{B}_j)$ (see equation (9.9.16)). Now Theorem 9.9.1 shows that $(C_1, \widehat{B}_1)$ and $(C_2, \widehat{B}_2)$ belongs to different connected components in the set of all pairs (U, H) where U is H-unitary and stably u-diagonalizable, and the conclusion of Theorem 13.9.1 follows. □

13.10   Exercises

1. Are the solutions of the differential equation

$$
L_j \left[i\frac{d}{dt} \right] x = 0, \qquad t \in \mathrm{R},
$$

stably bounded, in each of the following cases?

(a) $j = 1$ and $L_1(\lambda)$ is a scalar polynomial with real coefficients.

(b) $j = 2$ and $L_2(\lambda) = I_2 \lambda^2 + \begin{bmatrix} 0 & 2 \\ 2 & 0 \end{bmatrix} \lambda + I_2$.

(c) $j = 3$ and $L_3(\lambda) = L_2(\lambda^2)$.

2. Factorize the matrix polynomials $L_1(\lambda)$, $L_2(\lambda)$, $L_3(\lambda)$ of Exercise 1, when possible.

3. For those polynomials $L_j(\lambda)$, $j=1,2,3$, of Exercise 1 that are stably bounded, compute their indices.

4. Consider the system

$$A_0 x_k + A_1 x_{k+1} + A_2 x_{k+2} = 0, \qquad k = 0,1,\ldots, \qquad (13.10.29)$$

where $A_2 = A_0^* \in \mathbb{C}^{n\times n}$ is invertible and $A_1 \in \mathbb{C}^{n\times n}$ is hermitian. Under what conditions has the difference equation (13.10.29) stably bounded solutions? Answer the question for each of the following situations:

(a) $n = 1$, (i.e., A_0 and A_1 are scalars);

(b) A_0 and A_1 are diagonal matrices;

(c) A_0 and A_1 are circulant matrices: (An $n \times n$ matrix $B = [b_{jk}]_{j,k=1}^n$ is called a *circulant* if the equality $b_{jk} = b_{rs}$ holds for all ordered pairs of indices (j,k) and (r,s) such that $j - k = r - s$ modulo n.)

5. Under what conditions does the difference equation

$$A_0 x_k + A_1 x_{k+1} + \cdots + A_\ell x_{k+\ell} = 0, \quad k = 0,1,\ldots, \qquad (13.10.30)$$

have stably bounded solutions? Assume that $\ell = 2m$ is even and $A_r = A_{\ell-r}^*$ for $r = 1,\ldots,m$ with invertible A_0. Answer the question for the following cases:

(a) The matrices A_j, $j = 0,1,\ldots,\ell$ are scalar matrices;

(b) The matrices A_j, $j = 0,1,\ldots,\ell$ are diagonal;

(c) The matrices A_j, $j = 0,1,\ldots,\ell$ are circulants.

6. Assuming that (13.10.29) has stably bounded solutions, compute the index under each of the hypotheses (a), (b), (c) of the preceding exercise.

7. Repeat the preceding exercise for the difference equation (13.10.30).

8. Given that $A, H \in \mathbb{C}^{n\times n}$ are hermitian and H is invertible, under what conditions are the solutions of the differential equation $iH\frac{dx}{dt} = Ax$ stably bounded on the half line $[0,\infty)$?

9. Let $L(\lambda)$ be a hermitian matrix polynomial with invertible leading coefficient. Under what conditions are the solutions of the differential equation $L\left(i\frac{d}{dt}\right)x = 0$ stably bounded on the half line $[0,\infty)$?

10. Under what conditions are the solutions of the difference equation

$$A_0 x_k + A_1 x_{k+1} + \cdots + A_{2m} x_{k+2m} = 0, \quad k - 0, \pm 1, \pm 2,\ldots,$$

stably bounded? Here, A_j are $n \times n$ matrices such that $A_j = A_{2m-j}^*$ for $j = 0,1,\ldots,m$, and A_0 is invertible.

13.11   Notes

This chapter includes results about stable boundedness of the solutions of linear
differential and difference equations of higher order with selfadjoint coefficients.
These results are based on theorems from Chapter 12 and were obtained by the
authors in the papers [37], [38]; see also [39], [40]. For more on hyperbolic systems
see the monographs [81] and [92].

Chapter 14

Algebraic Riccati Equations

The subject of this chapter is the description of the solution set of a certain nonlinear matrix equation which arises in several different fields. Fortunately, the nonlinear equation is amenable to solution by linear techniques and, in particular, some of the techniques developed in this book are fundamental. It will be found that the solution of certain quadratic equations in matrices depends heavily on the structures introduced in earlier chapters. Briefly, the equation for $X \in C^{n \times n}$ has the form

$$XBX - XA - A^*X - C = 0,$$

where the coefficients A, $B = B^*$, $C = C^* \in C^{n \times n}$ are known. In particular, this problem can be resolved using our knowledge of matrices which are selfadjoint in an indefinite inner product.

The origin of this equation in systems theory will be summarized in Section 14.2, and will be seen to involve *differential systems*. There is a parallel problem area concerning *discrete systems* and this leads to nonlinear equations of a different kind. However, it is interesting that the complete solution of this problem requires knowledge of matrices which are *unitary* in an indefinite inner product. Unfortunately, space limitations prevent the development of this theory here, and the reader is referred to [67] for a detailed investigation of this problem.

It is necessary to begin with some concepts from systems theory. This is the subject of Section 14.1. After discussing the origins of the problem in Section 14.2, the systematic theory begins in Section 14.3. A fundamental existence theorem appears in Section 14.6 and the existence of extremal solutions is the topic of Section 14.9. The theory is presented first in the context of (possibly) complex matrix solutions of equations with complex coefficients. Investigation of real symmetric solutions for real systems is treated in Section 14.10 as a special case of the more general theory over the complex field.

14.1   Matrix Pairs in Systems Theory and Control

The following definitions play an important role in systems theory and, consequently, in the theory to be developed in this chapter.

Definition 14.1.1. *The matrix pair, $A \in \mathbb{C}^{n \times n}$ and $B \in \mathbb{C}^{n \times m}$ is said to be controllable if*

$$\text{rank} \begin{bmatrix} B & AB & A^2B & \cdots & A^{n-1}B \end{bmatrix} = n. \qquad (14.1.1)$$

More generally, the *controllable subspace* of such a pair (A, B) is the subspace of $\mathbb{C}^n$:

$$\mathcal{C}_{A,B} := \text{Range} \begin{bmatrix} B & AB & \cdots & A^{n-1}B \end{bmatrix} = \sum_{r=0}^{n-1} \text{Range}(A^r B). \qquad (14.1.2)$$

Clearly, when the pair is controllable, $\mathcal{C}_{A,B} = \mathbb{C}^n$, the whole space (see also Exercise 1 of this chapter).

The following property of the controllable subspace is often helpful:

Proposition 14.1.2. *The subspace $\mathcal{C}_{A,B}$ coincides with $\sum_{r=0}^{s-1} \text{Range}(A^r B)$ for every $s \geq n$.*

We relegate the proof to the exercises. There is also a nice geometric description of the controllable subspace:

Proposition 14.1.3. *Given a pair (A, B) as above, the controllable subspace $\mathcal{C}_{A,B}$ is the smallest A-invariant subspace containing $\text{Range } B$.*

Proof. For $r = 0, 1, \ldots$, define

$$\mathcal{C}_r = \text{Range} \begin{bmatrix} B & AB & \cdots & A^{r-1}B \end{bmatrix},$$

and let k be the smallest integer for which $\mathcal{C}_{k+1} = \mathcal{C}_k$. Clearly, if $x \in \mathcal{C}_r$ then $Ax \in \mathcal{C}_{r+1}$. Thus, if $x \in \mathcal{C}_k$ then $Ax \in \mathcal{C}_{k+1}$ and so $\mathcal{C}_k \subseteq \mathcal{C}_{A,B}$ is A-invariant. Also, it is clear that $\text{Range } B \subseteq \mathcal{C}_{A,B}$.

Finally, let $\mathcal{S}$ be any subspace of $\mathbb{C}^n$ for which $A\mathcal{S} \subseteq \mathcal{S}$ and $\text{Range } B \subseteq \mathcal{S}$. It is to be shown that $\mathcal{C}_{A,B} \subseteq \mathcal{S}$. But for $r = 1, 2, \ldots$ we have

$$\text{Range} \, (A^{r-1}B) = A^{r-1}\text{Range } B \subseteq A^{r-1}\mathcal{S} \subseteq \mathcal{S}.$$

It follows from equation(14.1.2) that $\mathcal{C}_{A,B} \subseteq \mathcal{S}$. $\square$

A condition which is weaker than controllability is also important for the theory:

Definition 14.1.4. *A matrix pair (A, B) (as in the preceding definition) is said to be* stabilizable *if there is a matrix $K \in \mathbb{C}^{m \times n}$ such that $A + BK$ has all of its eigenvalues in the open left half of the complex plane.*

Note that a square matrix with the property that all of its eigenvalues lie in the open left half of the complex plane is often said to be *stable*. Thus, the pair (A, B) (as above) is stabilizable if and only if there is a K such that $A + BK$ is stable. It is not hard to see that controllable pairs are necessarily stabilizable, but not conversely (see the Exercises).

The notion of "observability" is dual to that of controllability. Consider a pair $C \in \mathbb{C}^{m \times n}$, $A \in \mathbb{C}^{n \times n}$:

Definition 14.1.5. *The matrix pair* (C, A) *is said to be* observable *if*

$$\text{rank} \begin{bmatrix} C \\ CA \\ \vdots \\ CA^{n-1} \end{bmatrix} = n.$$

More generally, the *unobservable subspace* of a pair (C, A), as above, is the subspace of $\mathbb{C}^n$:

$$\mathcal{U}_{C,A} = \text{Ker} \begin{bmatrix} C \\ CA \\ \vdots \\ CA^{n-1} \end{bmatrix} = \bigcap_{r=0}^{n-1} \text{Ker}(CA^r).$$

Clearly, the pair (C, A) is observable if and only if the unobservable subspace is the trivial subspace, $\{0\}$.

The duality of the notions of controllability and observability is made apparent in:

Proposition 14.1.6. *The pair* (C, A) *is observable if and only if the pair* (A^*, C^*) *is controllable.*

The following lemma concerning controllable pairs will be useful.

Lemma 14.1.7. *Let the pair* $A, B \in \mathbb{C}^{n \times n}$ *be controllable and* B *be positive semidefinite. Then the matrix*

$$\Omega(t) := -\int_0^t e^{-\tau A} B e^{-\tau A^*} d\tau$$

is negative (resp. positive) definite for all $t > 0$ *(resp. $t < 0$) and the matrix* $\widehat{A} := A + B\Omega(t)^{-1}$ *is stable for any $t > 0$.*

Proof. Clearly, $\Omega(0) = 0$ and, when $t_1 \geq t_2$, $\Omega(t_1) \leq \Omega(t_2)$. In particular, $\Omega(t)$ is negative (positive) semidefinite for $t > 0$ ($t < 0$, respectively). To prove that $\Omega(t) < 0$ (resp. $\Omega(t) > 0$) for $t > 0$ (resp. $t < 0$) it is enough to check that $(\Omega(t)x, x) = 0$ only if $t = 0$ or $x = 0$.

Assume the contrary so that $(\Omega(t)x, x) = 0$ for some $t \neq 0$, say $t > 0$. Since

$$(\Omega(t)x, x) = -\int_0^t \|B^{\frac{1}{2}} e^{-\tau A^*} x\|^2 \, d\tau,$$

it follows that $B^{\frac{1}{2}} e^{-\tau A^*} x = 0$ for all $\tau \in (0, t)$. Differentiating repeatedly with respect to τ it follows that

$$\begin{bmatrix} B^{\frac{1}{2}} \\ B^{\frac{1}{2}} A^* \\ \vdots \\ B^{\frac{1}{2}} (A^*)^{n-1} \end{bmatrix} e^{-\tau A^*} x = 0.$$

Consequently, since

$$\begin{bmatrix} B & AB & \cdots & A^{n-1}B \end{bmatrix} = \begin{bmatrix} B^{\frac{1}{2}} & AB^{\frac{1}{2}} & \cdots & A^{n-1}B^{\frac{1}{2}} \end{bmatrix} B^{\frac{1}{2}},$$

we have

$$\operatorname{rank} \begin{bmatrix} B & AB & \cdots & A^{n-1}B \end{bmatrix} \leq \operatorname{rank} \begin{bmatrix} B^{\frac{1}{2}} & AB^{\frac{1}{2}} & \cdots & A^{n-1}B^{\frac{1}{2}} \end{bmatrix} < n,$$

which contradicts the controllability of (A, B).

To prove the last statement observe that, if $t > 0$, then

$$A\Omega(t) + \Omega(t)A^* = \int_0^t \left\{\frac{d}{d\tau} e^{-\tau A} B e^{-\tau A^*}\right\} d\tau = e^{-tA} B e^{-tA^*} - B, \qquad (14.1.3)$$

and hence

$$\widehat{A}\Omega(t) + \Omega(t)\widehat{A}^* = e^{-tA} B e^{-tA^*} + B.$$

Now let λ_0 be an eigenvalue of $\widehat{A}^*$ with corresponding eigenvector x. Then for $t > 0$,

$$\begin{aligned}
(\lambda_0 + \overline{\lambda_0}) x^* \Omega(t) x &= x^* (\widehat{A}\Omega(t) + \Omega(t)\widehat{A}^*) x \\
&= x^* (e^{-tA} B e^{-tA^*} + B) x \geq 0.
\end{aligned}$$

But $\Omega(t)$ is negative definite, so this inequality must imply that $\lambda_0 + \overline{\lambda_0} \leq 0$. Furthermore, if $\lambda_0 + \overline{\lambda_0} = 0$, then $Bx = 0$ and the definition of $\widehat{A}^*$ implies that $A^* x = \lambda_0 x$. Hence

$$\Omega(t)x = -\int_0^t e^{-\tau A} B x e^{-\lambda_0 \tau} \, d\tau = 0, \qquad t > 0,$$

which contradicts the negative definite property of $\Omega(t)$. Thus $\lambda_0 + \overline{\lambda_0} < 0$ for every eigenvalue λ_0 of $\widehat{A}^*$, and hence for every eigenvalue of $\widehat{A}$. $\square$

Another proposition will be useful in the sequel. The proof is deferred to the exercises. In this proposition matrix K is sometimes referred to as a "feedback" matrix.

Proposition 14.1.8. *Let A, B, K be complex matrices of sizes $n \times n$, $n \times m$ and $m \times n$, respectively and write $\widehat{A} = A + BK$. Then:*

1. *For $r = 0, 1, 2, \ldots$,*

$$\text{Range} \begin{bmatrix} B & \widehat{A}B & \ldots \widehat{A}^r B \end{bmatrix} = \text{Range} \begin{bmatrix} B & AB & \ldots A^r B \end{bmatrix}.$$

2. $C_{\widehat{A},B} = C_{A,B}$.

3. $(\widehat{A}, B)$ *is controllable if and only if (A, B) is controllable.*

14.2 Origins in Systems Theory

In this section the main objective is an account of a classical optimization problem with many applications. It is known as the "LQR" problem where the letters LQR stand for "linear quadratic regulator". There are no formal proofs and those who can accept these nonlinear matrix equations at face value may wish to omit this section.

Description of the optimization problem begins with an apparently simple differential system which is, however, of central importance. An "input" or "control" vector function $u(t)$ and an "output" vector function $y(t)$ are connected as follows:

$$\dot{x}(t) = Ax(t) + Bu(t), \quad x(0) = x_0$$

$$y(t) = Cx(t) + Du(t).$$

The intermediate function $x(t)$ (generally living in a space of relatively high dimension) is known as the "state" of the system. The *cost* of applying a control $u(t)$ with initial vector x_0 is defined in terms of the input and output functions as follows:

$$J^u(x_0) = \|u\|_{R_1}^2 + \|y\|_{R_2}^2,$$

where $R_1 > 0$, $R_2 > 0$, and

$$\|u\|_{R_1}^2 = \int_0^\infty u(t)^* R_1 u(t) dt, \quad \|y\|_{R_2}^2 = \int_0^\infty y(t)^* R_2 y(t) dt.$$

Then, substituting for y,

$$\|y\|_{R_2}^2 = \int_0^\infty [x^* \; u^*] \begin{bmatrix} C^* \\ D^* \end{bmatrix} R_2 [C \; D] \begin{bmatrix} x \\ u \end{bmatrix} dt,$$

and

$$J^u(x_0) = \int_0^\infty [x^* \ u^*]\widehat{R}\begin{bmatrix} x \\ u \end{bmatrix} dt,$$

where

$$\widehat{R} = \begin{bmatrix} 0 & 0 \\ 0 & R_1 \end{bmatrix} + \begin{bmatrix} C^*R_2C & C^*R_2D \\ D^*R_2C & D^*R_2D \end{bmatrix} =: \begin{bmatrix} Q & S \\ S^* & R \end{bmatrix}.$$

Notice that this formulation expresses the cost in terms of the input and the state. The output no longer appears explicitly. It follows from the hypotheses made above that $\widehat{R} \geq 0$, $Q = C^*R_2C \geq 0$, and $R = R_1 + D^*R_2D > 0$.

The LQR problem is now formulated as follows: *Minimize the cost $J^u(x_0)$ by choice of $u(t)$ subject to the linear constraint*

$$\dot{x}(t) = Ax(t) + Bu(t), \quad x(0) = x_0.$$

An *optimal control* is a function $\widehat{u}(t)$ for which the optimal cost $\widehat{J}(x_0) = J^{\widehat{u}}(x_0) = \inf_u J^u(x_0)$ is attained.

In order to describe a solution to the problem it turns out to be convenient (and physically significant) to restrict attention to matrix coefficients which are linked in particular ways.

Theorem 14.2.1. *Given the hypotheses:*

(1) *the pair (A, B) is stabilizable,*

(2) *rank $\widehat{R}$ = rank Q + rank R,*

(3) *the pair (Q, A) is observable,*

then there exists an optimal control of the form

$$\widehat{u}(t) = Ke^{(A+BK)t}x_0,$$

where $K = -R^{-1}(S^* + B^*X)$ and X is the maximal[1] hermitian solution of the algebraic Riccati equation (ARE)

$$X(BR^{-1}B^*)X - X\widehat{A} - \widehat{A}^*X - \widehat{Q} = 0,$$

where

$$\widehat{A} = A - BR^{-1}S^*, \quad \widehat{Q} = Q - SR^{-1}S^*.$$

Furthermore, $X > 0$, $\widehat{J}(x_0) = x_0^*Xx_0$, and $A + BK$ is stable.

---

[1]The sense in which the solution is "maximal" will be clarified in Section 14.9 below.

14.3 Preliminaries on the Riccati Equation

We start with a matrix quadratic equation which includes that arising in the LQR problem of the preceding section. Thus, consider the equation

$$XDX + XA + BX - C = 0, \qquad (14.3.4)$$

where D, A, B, C are matrices of sizes $n \times m$, $n \times n$, $m \times m$, and $m \times n$ respectively and may be real or complex. Note that the coefficients fit into a square matrix of size $(n + m) \times (n + m)$:

$$T := \begin{bmatrix} A & D \\ C & -B \end{bmatrix}. \qquad (14.3.5)$$

It has been seen above, however, that our main interest lies in equations with $D^* = D$, $C^* = C$, and $B = A^*$. Thus,

$$XDX + XA + A^*X - C = 0, \qquad (14.3.6)$$

and it is natural to anticipate hermitian solutions, $X^* = X$, as required for the optimization problem. This equation is the object of interest for most of this chapter and, for easy reference, it is given a name "the CARE" (abbreviating continuous algebraic Riccati equation).

The solution set of such an equation can take a variety of forms, and it will be useful to illustrate this in some primitive examples.

Example 14.3.1. *Scalar equations of the form $dx^2 + 2ax - c = 0$, where $a, d, c \in \mathbb{R}$ and $d > 0$, $a \geq 0$. The hermitian property of solutions means just real solutions here, and they may or may not exist. The existence of a nonnegative solution requires $c \geq 0$.* □

Example 14.3.2. *The equation (14.3.6) with*

$$D = \begin{bmatrix} 0 & 1 \\ 1 & 0 \end{bmatrix}, \quad A = 0, \quad C = \begin{bmatrix} 1 & 0 \\ 0 & 0 \end{bmatrix}$$

has no solution (either real or complex). □

Example 14.3.3. *The equation (14.3.6) with*

$$D = \begin{bmatrix} 1 & 0 \\ 0 & 0 \end{bmatrix}, \quad A = \begin{bmatrix} 0 & 0 \\ 1 & 0 \end{bmatrix}, \quad C = \begin{bmatrix} -2 & 0 \\ 0 & 1 \end{bmatrix}$$

has the one hermitian solution $X = \begin{bmatrix} 0 & -1 \\ -1 & 0 \end{bmatrix}$ and two nonhermitian solutions. □

Example 14.3.4. *The equation* (14.3.6) *with*

$$D = \begin{bmatrix} 0 & 0 \\ 0 & 1/3 \end{bmatrix}, \quad A = \begin{bmatrix} 0 & -1 \\ 2/3 & 0 \end{bmatrix}, \quad C = \begin{bmatrix} 5/3 & 0 \\ 0 & 1 \end{bmatrix}$$

has four hermitian solutions including the maximal solution $X = \begin{bmatrix} 3 & 1 \\ 1 & 3 \end{bmatrix}$, *and two nonhermitian solutions.*   □

Example 14.3.5. *The equation* (14.3.6) *with*

$$D = \begin{bmatrix} 1 & 0 \\ 0 & 0 \end{bmatrix}, \quad A = \begin{bmatrix} 0 & 0 \\ 0 & 1 \end{bmatrix}, \quad C = \begin{bmatrix} 1 & 0 \\ 0 & 0 \end{bmatrix}$$

has an isolated solution $X = \begin{bmatrix} 1 & 0 \\ 0 & 0 \end{bmatrix}$, *and a continuum of hermitian solutions depending on a parameter* $a \in \mathbb{C}$:

$$X_a = \begin{bmatrix} -1 & a \\ \bar{a} & -\frac{1}{2}|a|^2 \end{bmatrix},$$

and also a continuum of nonhermitian solutions.   □

14.4   Solutions and Invariant Subspaces

Now let us begin a systematic analysis of solutions of the general equation (14.3.4) via the properties of the matrix T of (14.3.5). First, for any $m \times n$ matrix X, the n-dimensional subspace

$$\mathcal{G}(X) := \mathrm{Range} \begin{bmatrix} I_n \\ X \end{bmatrix} \subseteq \mathbb{C}^{m+n}$$

is called the *graph* of X, or a *graph subspace* of $\mathbb{C}^{m+n}$. The first proposition is a candidate for most important fact in the geometric study of Riccati equations. Recall the definition of T in equation (14.3.5).

Proposition 14.4.1. *Let X be $m \times n$. Then X is a solution of* (14.3.4) *if and only if the graph of X is T-invariant.*

Proof. If $\mathcal{G}(X)$ is T-invariant, then

$$\begin{bmatrix} A & D \\ C & -B \end{bmatrix} \begin{bmatrix} I \\ X \end{bmatrix} = \begin{bmatrix} I \\ X \end{bmatrix} Z \qquad (14.4.7)$$

for a suitable matrix Z. The first block row of this equation gives $Z = A + DX$, and then the second row gives $C - BX = X(A + DX)$. In other words, X solves (14.3.4).

Conversely, if X solves (14.3.4), then (14.4.7) holds with $Z = A + DX$.   □

Representing the T-invariant subspace $\mathcal{G}(X)$ as the linear span of a set of Jordan chains (or eigenvectors and generalized eigenvectors) of T the following constructive result is obtained. The preceding statement makes equal sense over R and over C. Now it is more convenient to work over C.

Theorem 14.4.2. *Equation (14.3.4) has a solution $X \in \mathbb{C}^{m \times n}$ if and only if there is a set of vectors $v_1, \ldots, v_n \in \mathbb{C}^{m+n}$ forming a set of Jordan chains for T and, if*

$$v_j = \begin{bmatrix} y_j \\ z_j \end{bmatrix}, \quad j = 1, 2, \ldots,$$

where $y_j \in \mathbb{C}^n$, then $y_1, y_2, \ldots, y_n$ form a basis for $\mathbb{C}^n$.
Furthermore, if

$$Y = [y_1 \ y_2 \ \cdots \ y_n], \quad Z = [z_1 \ z_2 \ \cdots \ z_n],$$

every solution of (14.3.4) has the form $X = ZY^{-1}$ for some set of Jordan chains $v_1, v_2, \ldots, v_n$ for T such that Y is nonsingular.

These results focus our attention on n-dimensional T-invariant subspaces. It is not difficult to see that, generically, the number of solutions of (14.3.4) is finite and bounded above by the number of ways of choosing n vectors from a basis of $m + n$ eigenvectors and generalized eigenvectors of T. This applies to Example 14.3.4. By design, the other examples of that collection are not generic.

14.5 Symmetric Equations

Now focus on equation (14.3.6): the CARE. Instead of T, define

$$M = i \begin{bmatrix} A & D \\ C & -A^* \end{bmatrix} \tag{14.5.8}$$

and associated matrices

$$H = \begin{bmatrix} -C & A^* \\ A & D \end{bmatrix}, \quad \widehat{H} = i \begin{bmatrix} 0 & I \\ -I & 0 \end{bmatrix}. \tag{14.5.9}$$

Notice the following connections:

$$H^* = H, \quad \widehat{H}^* = \widehat{H}, \quad HM = M^*H, \quad \widehat{H}M = M^*\widehat{H}, \quad H = \widehat{H}M.$$

In the language of Chapter 4, these relations mean that M is self-adjoint in each of the two indefinite inner products generated on $\mathbb{C}^{2n}$ by H and by $\widehat{H}$ (given that they are nonsingular). In particular, the spectrum of M is symmetric with respect to the real axis. Since solutions of the CARE are now linked with certain invariant subspaces of M, we see that the canonical form (over C) discussed in Section 5.1 has a vital role to play.

Lemma 14.5.1. *It can be assumed that H is nonsingular with inertia $(n, n, 0)$.*

Proof. Notice first that the Riccati equation, CARE, is invariant if the coefficient A is replaced by $A + i\alpha I$ where $\alpha \in \mathbb{R}$. Thus, if H does not already have the desired inertia, then A is replaced by $A + i\alpha I$. Then, for α sufficiently large, the inertia of H will be equal to that of

$$\begin{bmatrix} 0 & -i\alpha I \\ i\alpha I & 0 \end{bmatrix},$$

and this inertia is easily seen to be $(n, n, 0)$.   □

This lemma means that, without loss of generality, the *existence* of n-dimensional neutral subspaces of H can be assumed.

Returning to the graph subspace of a solution X of (14.3.6), another very nice property arises from the symmetry of the CARE. It is easily seen that

$$\begin{bmatrix} I \\ X \end{bmatrix}^* H \begin{bmatrix} I \\ X \end{bmatrix} = (X^* - X)(A + DX) + XDX + XA + A^*X - C.$$

As in Section 2.3, a subspace S is said to be $\widehat{H}$-neutral if $(\widehat{H}x, y) = 0$ for all $x, y \in S$. (In particular, the generalized eigenspace corresponding to a nonreal eigenvalue of M is necessarily both H-neutral and $\widehat{H}$-neutral.) Now it is seen that:

Proposition 14.5.2. *If X is a solution of the CARE with graph subspace $\mathcal{G}(X)$, then X is hermitian if and only if $\mathcal{G}(X)$ is $\widehat{H}$-neutral. Also, X is hermitian if and only if $\mathcal{G}(X)$ is H-neutral.*

Note that the property in Proposition 14.5.2 that $\mathcal{G}(X)$ is $\widehat{H}$-neutral if and only if $\mathcal{G}(X)$ is H-neutral follows from Theorem 7.5.3.

An important special case is, of course, that in which the coefficients of the CARE are all real matrices. Then we define

$$M_r = \begin{bmatrix} A & D \\ C & -A^T \end{bmatrix}, \quad \widehat{H}_r = \begin{bmatrix} 0 & I \\ -I & 0 \end{bmatrix}, \quad H_r = \begin{bmatrix} -C & A^T \\ A & D \end{bmatrix}.$$

Now the spectrum of M is symmetric with respect to *both* axes in the complex plane, and M_r is said to have Hamiltonian symmetry. We have,

$$\widehat{H}_r M_r = -M_r^T \widehat{H}_r, \quad H_r M_r = -M_r^T H_r, \quad H_r = -\widehat{H}_r M_r,$$

and now M_r can be analyzed as an iH_r-self-adjoint matrix.

14.6   An Existence Theorem

For simplicity, this presentation is first focussed on the complex case. The objective is to examine hermitian solutions of the CARE in terms of the M-invariant subspaces which are both graph subspaces (and so necessarily n-dimensional) and neutral with respect to $\widehat{H}$ or H. If the nonsingular matrix H_0 defining an indefinite inner product has inertia $(\pi, \nu, 0)$ then the dimension of a maximal H_0-neutral subspace is $\min(\pi, \nu)$ (see Theorem 2.3.4). Thus, using either H or $\widehat{H}$, the dimension of a maximal neutral subspace matches that of a graph subspace.

The problem of finding a formal proof of existence of hermitian solutions of the Riccati equation has now been transformed to that of proving the existence of n-dimensional subspaces which are neutral with respect to $\widehat{H}$ or H, and this can be done using the theory of Chapters 4 and 5. However, an essential feature is still missing. According to Proposition 14.4.1, such a subspace must also be a *graph* subspace. This property holds under a further condition which links the coefficients of the quadratic and linear terms of the Riccati equation (14.3.6):

Lemma 14.6.1. *Assume that $D \geq 0$ and the pair (A, D) is controllable. Let $\mathcal{L}$ be an n-dimensional M-invariant H-nonpositive subspace of $\mathbb{C}^{2n}$. Then $\mathcal{L}$ is a graph subspace.*

Proof. For a subspace $\mathcal{L}$ defined in the statement, write

$$\mathcal{L} = \mathrm{Range} \begin{bmatrix} X_1 \\ X_2 \end{bmatrix}$$

for some $n \times n$ matrices X_1 and X_2. We are going to prove that X_1 is invertible.
First, observe that M-invariance of $\mathcal{L}$ means that

$$\begin{bmatrix} A & D \\ C & -A^* \end{bmatrix} \begin{bmatrix} X_1 \\ X_2 \end{bmatrix} = \begin{bmatrix} X_1 \\ X_2 \end{bmatrix} T$$

for some $n \times n$ matrix T. In other words,

$$AX_1 + DX_2 = X_1 T; \tag{14.6.10}$$

$$CX_1 - A^* X_2 = X_2 T. \tag{14.6.11}$$

Then H-nonpositivity of $\mathcal{L}$ means that the matrix

$$[X_1^* \ X_2^*] \begin{bmatrix} -C & A^* \\ A & D \end{bmatrix} \begin{bmatrix} X_1 \\ X_2 \end{bmatrix} = X_2^* DX_2 + X_1^* A^* X_2 + X_2^* AX_1 - X_1^* CX_1 \tag{14.6.12}$$

is negative semidefinite.
Let $\mathcal{K} = \mathrm{Ker} X_1$. Since (14.6.12) is negative semidefinite, we have for every $x \in \mathcal{K}$:

$$0 \geq x^* X_2^* DX_2 x + x^* X_1^* A^* X_2 x + x^* X_2^* AX_1 x - x^* X_1^* CX_1 x = x^* X_2^* DX_2 x,$$

and since $D \geq 0$, $X_2 x \in \mathrm{Ker} D$, i.e.,

$$X_2 \mathcal{K} \subseteq \mathrm{Ker} D. \tag{14.6.13}$$

Further, equation (14.6.10) implies that

$$T\mathcal{K} \subseteq \mathcal{K}. \tag{14.6.14}$$

Indeed, for $x \in \mathcal{K}$ we have in view of (14.6.10) and (14.6.13):

$$X_1 T x = A X_1 x + D X_2 x = 0.$$

Now equation (14.6.11) gives for every $x \in \mathcal{K}$:

$$A^* X_2 x = -C X_1 x + A^* X_2 x = -X_2 T x \in X_2 \mathcal{K}$$

and so

$$A^* X_2 \mathcal{K} \subseteq X_2 \mathcal{K}. \tag{14.6.15}$$

We see from (14.6.13) that $A^* X_2 \mathcal{K} \subseteq \mathrm{Ker} D$ and we now claim, more generally, that

$$A^{*r} X_2 \mathcal{K} \subseteq \mathrm{Ker} D, \quad r = 0, 1, 2, \ldots. \tag{14.6.16}$$

We have already proved this inclusion for $r = 0$ and $r = 1$. Assuming, inductively, that (14.6.16) holds for $r - 1$, and using (14.6.15) it is found that

$$A^{*r}(X_2 \mathcal{K}) = A^{*r-1}(A^* X_2 \mathcal{K}) \subseteq A^{*r-1} X_2 \mathcal{K} \subseteq \mathrm{Ker} D;$$

so (14.6.16) holds. Now for every $x \in \mathcal{K}$:

$$\begin{bmatrix} D \\ DA^* \\ \vdots \\ DA^{*n-1} \end{bmatrix} (X_2 x) = 0,$$

or $(X_2 x)^* [D, AD, \ldots, A^{n-1} D] = 0$. But $\mathrm{rank}[D, AD, \ldots, A^{n-1} D] = n$, so $X_2 x = 0$. But the only n-dimensional vector x for which $X_1 x = X_2 x = 0$ is the zero vector; otherwise $\dim \mathcal{L} < n$, which contradicts our assumptions. So $\mathcal{K} = \{0\}$ and X_1 is invertible. Then we can write

$$\mathcal{L} = \mathrm{Range} \begin{bmatrix} I \\ X \end{bmatrix},$$

where $X = X_2 X_1^{-1}$, and so $\mathcal{L}$ is indeed a graph subspace. $\square$

Recall that M is H-self-adjoint and that H is assumed to have n positive eigenvalues and n negative eigenvalues (Lemma 14.5.1). Thus, the existence of n-dimensional M-invariant H-nonpositive subspaces (or, what is the same, M-invariant maximal H-nonpositive subspaces) is ensured — as discussed in Chapter 5. The next result associates subspaces of this kind (including H-neutral subspaces) with certain solutions of (14.3.6) (including the hermitian solutions).

Theorem 14.6.2. *Assume that $D \geq 0$ and the pair (A, D) is controllable, and let $\mathcal{L}$ be an n-dimensional M-invariant, H-nonpositive subspace. Then*

$$\mathcal{L} = \mathrm{Range} \begin{bmatrix} I \\ X \end{bmatrix}, \tag{14.6.17}$$

where X is a solution of the CARE such that

$$(X^* - X)(A + DX) \tag{14.6.18}$$

is negative semidefinite.

Conversely, if X is a solution of (14.3.6) such that $(X^* - X)(A + DX)$ is negative semidefinite, then the subspace

$$\mathrm{Range} \begin{bmatrix} I \\ X \end{bmatrix}$$

is n-dimensional, M-invariant and maximal H-nonpositive.

Proof. As in Proposition 14.4.1, if X is a solution of (14.3.6) then $\mathrm{Range} \begin{bmatrix} I \\ X \end{bmatrix}$ is M-invariant. Indeed,

$$i \begin{bmatrix} A & D \\ C & -A^* \end{bmatrix} \begin{bmatrix} I \\ X \end{bmatrix} = \begin{bmatrix} I \\ X \end{bmatrix} i(A + DX). \tag{14.6.19}$$

Furthermore,

$$\begin{bmatrix} I & X^* \end{bmatrix} \begin{bmatrix} -C & A^* \\ A & D \end{bmatrix} \begin{bmatrix} I \\ X \end{bmatrix} = -C + X^*A + A^*X + X^*DX \tag{14.6.20}$$

and, since X is a solution of (14.3.6), this is equal to $(X^* - X)(A + DX)$. So if this matrix is negative semidefinite, so is

$$\begin{bmatrix} I & X^* \end{bmatrix} \begin{bmatrix} -C & A^* \\ A & D \end{bmatrix} \begin{bmatrix} I \\ X \end{bmatrix}$$

and, in turn, this means that $\mathrm{Range} \begin{bmatrix} I \\ X \end{bmatrix}$ is an H-nonpositive subspace. As $\dim(\mathrm{Range} \begin{bmatrix} I \\ X \end{bmatrix}) = n$ and the signature of H is zero, the subspace $\mathrm{Range} \begin{bmatrix} I \\ X \end{bmatrix}$ is maximal H-nonpositive (see Theorem 2.3.2).

Conversely, if $\mathcal{L}$ is an n-dimensional M-invariant H-nonpositive subspace then, by the preceding lemma, it may be assumed to be a graph subspace, say $\mathcal{L} = \mathrm{Range} \begin{bmatrix} I \\ X \end{bmatrix}$. The M-invariance of $\mathcal{L}$ implies that

$$\begin{bmatrix} A & D \\ C & -A^* \end{bmatrix} \begin{bmatrix} I \\ X \end{bmatrix} = \begin{bmatrix} I \\ X \end{bmatrix} T_0,$$

for some $T_0 \in \mathbb{C}^{n \times n}$. But this gives $T_0 = A + DX$ and then

$$C - A^* X = X T_0 = X(A + DX),$$

i.e., X is a solution of (14.3.6). Further, using (14.6.20), H-nonpositivity of $\mathcal{L}$ implies that

$$X^* DX + A^* X + X^* A - C = (X^* - X)(A + DX)$$

is a negative semidefinite matrix. $\square$

Theorem 14.6.2 shows, in particular, that there is a one-to-one correspondence between the n-dimensional, M-invariant, maximal H-nonpositive subspaces $\mathcal{L}$ and solutions of (14.3.6) with the property that $(X^* - X)(A + DX)$ is negative semidefinite. This correspondence is given by the formula (14.6.17).

Now take advantage of the constructive development of M-invariant, H-nonpositive subspaces of Section 5.12. Let $\mathcal{C}$ be a maximal set of nonreal eigenvalues of M with the property that, if $\lambda \in \mathcal{C}$ then $\overline{\lambda} \notin \mathcal{C}$. Then Theorem 5.12.1 ensures the existence of an n-dimensional M-invariant, H-nonpositive subspace $\mathcal{L}_{\mathcal{C}}$ such that the nonreal spectrum of the restriction of M to this subspace is $\mathcal{C}$. By Theorem 14.6.2,

$$\mathcal{L}_{\mathcal{C}} = \text{Range} \begin{bmatrix} I \\ X \end{bmatrix},$$

where X is a solution of the CARE such that $(X^* - X)(A + DX)$ is negative semidefinite. Furthermore, (14.6.19) shows that $\sigma(M|_{\mathcal{L}_{\mathcal{C}}}) = \sigma(i(A + DX))$. The following result is obtained:

Theorem 14.6.3. *Assume that $D \geq 0$ and that the pair (A, D) is controllable. Then for every maximal set $\mathcal{C}'$ of nonreal eigenvalues of M with the property that $\lambda \in \mathcal{C}'$ implies $\overline{\lambda} \notin \mathcal{C}'$, there exists a solution of (14.3.6) such that $(X^* - X)(A + DX)$ is negative semidefinite and $\sigma(i(A + DX)) \setminus \mathbb{R} = \mathcal{C}'$. In particular, there exist solutions X_1 and X_2 of (14.3.6) such that $(X_i^* - X_i)(A + DX_i)$ is negative semidefinite for $i = 1, 2$, and*

$$\Re \, \sigma(A + DX_1) \leq 0, \quad \Re \, \sigma(A + DX_2) \geq 0.$$

It is apparent that the construction of hermitian solutions via invariant subspaces of M is considerably simplified if M has no real eigenvalues. The following result suggests that there are some useful conditions under which this happens. (Indeed, this is a simplified version of a more general result. See, for example, [67, Theorem 7.2.8]).

Theorem 14.6.4. *If $D \geq 0$, $C \geq 0$, (A, D) is controllable and (C, A) observable, then M has no real eigenvalues.*

Proof. The argument is by contradiction. Thus, suppose $\lambda \in \mathbf{R}$ and

$$M \begin{bmatrix} x_1 \\ x_2 \end{bmatrix} = \lambda \begin{bmatrix} x_1 \\ x_2 \end{bmatrix}, \quad \begin{bmatrix} x_1 \\ x_2 \end{bmatrix} \neq \begin{bmatrix} 0 \\ 0 \end{bmatrix}.$$

Thus

$$iAx_1 + iDx_2 = \lambda x_1, \quad iCx_1 - iA^*x_2 = \lambda x_2, \tag{14.6.21}$$

or,

$$Ax_1 + Dx_2 = -i\lambda x_1, \quad -Cx_1 + A^*x_2 = i\lambda x_2.$$

The first equation implies

$$x_2^* Dx_2 = -i\lambda x_2^* x_1 - x_2^* Ax_1, \tag{14.6.22}$$

and the second gives

$$-x_1^* Cx_1 = i\lambda x_1^* x_2 - x_1^* Ax_2 = (-i\lambda x_2^* x_1 - x_2^* Ax_1)^* = x_2^* Dx_2, \tag{14.6.23}$$

having used (14.6.21) at the last step. But, as $C \geq 0$ and $D \geq 0$ this implies that, in fact, $Cx_1 = Dx_2 = 0$.

Now equations (14.6.21) reduce to

$$(A + i\lambda I)x_1 = 0,$$

which, together with $Cx_1 = 0$, implies $CA^r x_1 = 0$ for $r = 0, 1, 2, \ldots$, i.e., x_1 is in the unobservable subspace of the pair (C, A) (see Definition 14.1.5). But, as the pair (C, A) is assumed to be observable this means that $x_1 = 0$.

A similar argument using the controllability of (A, D) shows that $x_2 = 0$ as well, and a contradiction is obtained. $\qquad\square$

14.7 Existence when M has Real Eigenvalues

In contrast to Theorem 14.6.4, when M has real eigenvalues they must have a special structure. The first indication of this comes from a simple proposition:

Proposition 14.7.1. *If there is a hermitian solution X of equation (14.3.6), then*

$$M = i \begin{bmatrix} A & D \\ C & -A^* \end{bmatrix} \quad \text{and} \quad i \begin{bmatrix} A + DX & D \\ 0 & -(A + DX)^* \end{bmatrix}$$

are similar.

Proof. Given the existence of the solution $X = X^*$, consider the matrix $T = \begin{bmatrix} I & 0 \\ X & I \end{bmatrix}$ and verify that $T^{-1}MT$ has the required form. $\qquad\square$

It follows from this block similarity that M has real eigenvalues if and only if $A + DX$ (or $-(A + DX)^*$) has a pure imaginary eigenvalue. But such an eigenvalue will re-appear as an eigenvalue of the block $-(A + DX)^*$ (or $A + DX$, resp.) and, therefore, as a *multiple* eigenvalue of M itself. Thus, when the symmetric Riccati equation has a hermitian solution, real eigenvalues of M (if any) cannot be *simple* eigenvalues.

This section is devoted to resolution of the more delicate case in which real eigenvalues for M are admitted. The first result provides conditions under which the multiplicities of the real eigenvalues of M have a relatively simple structure. First recall Definition 14.1.1 and the notion of a "controllable subspace" (equation (14.1.2)).

It will also be convenient to use notation for the "root subspace", or "generalized eigenspace" of an eigenvalue. Thus if $\lambda_0 \in \sigma(A)$ and A is $n \times n$, then

$$\mathcal{R}_{\lambda_0}(A) = \mathrm{Ker}(A - \lambda_0 I)^n.$$

Notice that condition (14.7.24) below is automatically satisfied if the pair (A, D) is controllable, as assumed in Theorems 14.6.2 and 14.6.3.

Theorem 14.7.2. *Assume that $D \geq 0$ and $C^* = C$. If the CARE has a hermitian solution X, and*

$$\mathcal{R}_{\lambda_0}(A + DX) \subseteq \mathcal{C}_{A,D} \tag{14.7.24}$$

for every pure imaginary or zero eigenvalue λ_0 of $A + DX$, then all the partial multiplicities corresponding to the real eigenvalues of M (if any) are even. In fact, the partial multiplicities of a real eigenvalue λ_0 of M are twice the partial multiplicities of $-i\lambda_0$ as an eigenvalue of $A + DX$.

For the proof of this theorem a technical lemma is needed which is, however, of independent interest. (This proof, in turn, depends on Lemma A.2.3.)

Lemma 14.7.3. *Let Y, D be $n \times n$ matrices with $D \geq 0$ and let X_+, X_- be matrices of the form*

$$X_\pm = \begin{bmatrix} Y & D \\ 0 & \pm Y^* \end{bmatrix}.$$

Assume also that, with the minus signs (with the plus signs) in this formula,

$$\mathcal{R}_{\lambda_0}(Y) \subseteq \mathcal{C}_{Y,D} \tag{14.7.25}$$

for every eigenvalue λ_0 of Y on the imaginary axis (on the real axis, respectively).
Then the partial multiplicities of any eigenvalue λ_0 of X_- (or X_+) on the imaginary axis (on the real axis) are all even. Indeed, they are twice the partial multiplicities of λ_0 as an eigenvalue of Y.

Proof. We consider the case of X_-; the argument for X_+ is similar. Let Z be a Jordan form for Y and $Z = SYS^{-1}$. Then

$$\begin{bmatrix} Y & D \\ 0 & -Y^* \end{bmatrix} = \begin{bmatrix} S^{-1} & 0 \\ 0 & S^* \end{bmatrix} \begin{bmatrix} Z & D_0 \\ 0 & -Z^* \end{bmatrix} \begin{bmatrix} S & 0 \\ 0 & (S^{-1})^* \end{bmatrix}$$

where $D_0 = SDS^* \geq 0$, and it is sufficient to prove that all eigenvalues of $\begin{bmatrix} Z & D_0 \\ 0 & -Z^* \end{bmatrix}$ on the imaginary axis (if any) have the required properties.

Let λ_0 be such a pure imaginary or zero eigenvalue. Let $Z_1, \ldots, Z_k$ be the Jordan blocks of Z corresponding to λ_0, let their sizes be $\alpha_1, \ldots, \alpha_k$, and denote $\alpha = \alpha_1 + \cdots + \alpha_k$. Without loss of generality we can suppose that these blocks are in the top left corner of Z. So we can write

$$\begin{bmatrix} Z & D_0 \\ 0 & -Z^* \end{bmatrix} = \begin{bmatrix} Z' & 0 & D_1 & D_2 \\ 0 & Z'' & D_2^* & D_3 \\ 0 & 0 & -Z'^* & 0 \\ 0 & 0 & 0 & -Z''^* \end{bmatrix}$$

where $Z' = Z_1 \oplus \cdots \oplus Z_k$, Z'' is the 'rest' of Z, and

$$D_0 = \begin{bmatrix} D_1 & D_2 \\ D_2^* & D_3 \end{bmatrix}$$

is the corresponding partition of D_0. The condition (14.7.25) now takes the form

$$\text{Range}\left(\begin{bmatrix} D_1 & D_2 \\ D_2^* & D_3 \end{bmatrix}, \begin{bmatrix} Z' & 0 \\ 0 & Z'' \end{bmatrix} \begin{bmatrix} D_1 & D_2 \\ D_2^* & D_3 \end{bmatrix}, \ldots, \right.$$
$$\left. \begin{bmatrix} (Z')^{n-1} & 0 \\ 0 & (Z'')^{n-1} \end{bmatrix} \begin{bmatrix} D_1 & D_2 \\ D_2^* & D_3 \end{bmatrix} \right) \supseteq \begin{bmatrix} \mathbb{C}^\alpha \\ 0 \end{bmatrix} . \quad (14.7.26)$$

Let $D_1 = (D_{1ij})_{i,j=1}^k$ be the partition of D_1 consistent with the partitioning $Z' = Z_1 \oplus \cdots \oplus Z_k$. It is enough to prove that in the Jordan form of the matrix

$$\begin{bmatrix} Z_1 - \lambda_0 I & 0 & \cdot & 0 & D_{111} & D_{112} & \cdot & D_{11k} \\ 0 & Z_2 - \lambda_0 I & \cdot & 0 & D_{121} & D_{122} & \cdot & D_{12k} \\ \cdot & \cdot & \cdot & \cdot & \cdot & \cdot & \cdot & \cdot \\ 0 & 0 & \cdot & Z_k - \lambda_0 I & D_{1k1} & D_{1k2} & \cdot & D_{1kk} \\ 0 & 0 & \cdot & 0 & -Z_1^* - \lambda_0 I & 0 & \cdot & 0 \\ 0 & 0 & \cdot & 0 & 0 & -Z_2^* - \lambda_0 I & \cdot & 0 \\ \cdot & \cdot & \cdot & \cdot & \cdot & \cdot & \cdot & \cdot \\ 0 & 0 & \cdot & 0 & 0 & 0 & \cdot & -Z_k^* - \lambda_0 I \end{bmatrix}$$

the blocks with eigenvalue 0 have sizes $2\alpha_1, \ldots, 2\alpha_k$. Let f_{ij} $(i,j = 1, \ldots, k)$ be the entry in the bottom right corner of D_{1ij}, and consider the matrix $F = [f_{ij}]_{i,j=1}^k$

formed by all these entries. Since F is a principal submatrix of D_1, and hence of D_0, $F \geq 0$.

Let us show that F is invertible. If not, then there exists an invertible matrix $U = [u_{ij}]_{i,j=1}^{k}$ such that UFU^* has a zero in the bottom right corner. Let $G = [g_{ij}]_{i,j=1}^{\alpha}$ be an $\alpha \times \alpha$ invertible matrix of the following structure:

$$
\begin{aligned}
g_{\beta_i \beta_j} &= u_{ij} & \text{where} \quad & \beta_i = \alpha_1 + \cdots + \alpha_i; \quad i, j = 1, \ldots, k, \\
g_{qq} &= 1 & \text{for} \quad & q \notin \{\beta_1, \ldots, \beta_k\}, \\
g_{pq} &= 0 & \text{for} \quad & p \neq q \text{ and } \{p, q\} \not\subseteq \{\beta_1, \ldots, \beta_k\}.
\end{aligned}
$$

Then the matrix GD_1G^* has a zero in the bottom right corner, and since $GD_1G^* \geq 0$, the last column and last row of GD_1G^* are also zeros. On the other hand, from the structure of G it is seen (bearing in mind that the β_1-th,$\ldots$, β_k-th rows of $Z' - \lambda_0 I$ are zeros) that the bottom row of $G(Z' - \lambda_0 I)G^{-1}$ is also zero. Now let

$$
\widetilde{G} = \begin{bmatrix} G & 0 \\ 0 & I \end{bmatrix}
$$

be an $n \times n$ matrix where I is the $(n - \alpha) \times (n - \alpha)$ unit matrix. It is clear that the β_k-th row of $\widetilde{G}D_0\widetilde{G}^*$ is zero, as well as the β_k-th row of $\widetilde{G}(Z - \lambda_0 I)\widetilde{G}^{-1}$. However, this contradicts (14.7.26).

So the matrix F is invertible, and since $F > 0$, every principal submatrix of F is also positive definite, hence invertible. Thus, Lemma A.2.3 applies and the proof is complete. $\qquad \Box$

Proof of Theorem 14.7.2. Observe that (by Proposition 14.1.8) the right-hand side of (14.7.24) can be written in the form

$$
\mathcal{C}_{A,D} = \text{Range}[D, \ (A + DX)D, \ \ldots, \ (A + DX)^{n-1}D]
$$

for any $n \times n$ matrix X. Let $X = X^*$ be a solution of the CARE. Then the similarity

$$
\begin{bmatrix} I & 0 \\ -X & I \end{bmatrix} \begin{bmatrix} A & D \\ C & -A^* \end{bmatrix} \begin{bmatrix} I & 0 \\ X & I \end{bmatrix} = \begin{bmatrix} A + DX & D \\ 0 & -(A^* + XD) \end{bmatrix} \qquad (14.7.27)
$$

is easily verified. So we can consider the matrix

$$
i \begin{bmatrix} A + DX & D \\ 0 & -(A^* + XD) \end{bmatrix}
$$

in place of M. It remains to apply Lemma 14.7.3. $\qquad \Box$

Of particular interest in Theorem 14.7.2 is the case in which the pair (A, D) is controllable. In this case condition (14.7.24) is always satisfied, and we obtain the following corollary.

Corollary 14.7.4. *Assume that $D \geq 0$ and the pair (A, D) is controllable. If the CARE admits a hermitian solution, then all the partial multiplicities of M corresponding to real eigenvalues (if any) are even.*

We now return to Theorem 14.7.2. As observed previously, the matrix M is $\widehat{H}$-self-adjoint, where $\widehat{H} = i \begin{bmatrix} 0 & I \\ -I & 0 \end{bmatrix}$. It turns out that, in this context, the sign characteristic of the pair $(M, \widehat{H})$ is completely determined.

Theorem 14.7.5. *Under the hypotheses of Theorem 14.7.2, and assuming that the CARE has a hermitian solution X satisfying the condition (14.7.24), the sign characteristic of $(M, \widehat{H})$ consists of $+1$'s only.*

The proof of this elegant result is, unfortunately, long and technical, so it is omitted and the reader is referred to [40, Section II.4.4], or [67, Theorem 7.3.5], for the details.

14.8 Description of Hermitian Solutions

Existence theorems for hermitian solutions of the CARE have been formulated in the preceding sections. Here, it is assumed that hermitian solutions exist and they are to be described in terms of the invariant subspaces of matrix M of equation (14.5.8). The first result is really a special case of Theorem 14.6.2.

Theorem 14.8.1. *Assume that, in the CARE, $D \geq 0$ and (A, D) is a controllable pair. Then every n-dimensional, M-invariant, H-neutral subspace $\mathcal{L}$ has the form*

$$\mathcal{L} = \text{Range} \begin{bmatrix} I \\ X \end{bmatrix}, \qquad (14.8.28)$$

where X is a hermitian solution of the CARE and, in this case, $M|_{\mathcal{L}}$ is similar to $i(A + DX)$.

Conversely, for every hermitian solution X of the CARE, the subspace $\mathcal{L}$ given by (14.8.28) is n-dimensional, M-invariant, and H-neutral.

Proof. Since an H-neutral subspace is also H-nonpositive it follows from Theorem 14.6.2 that $\mathcal{L}$ is a graph subspace. Now apply Proposition 14.5.2 to conclude that X is an hermitian solution of the CARE. The converse follows immediately from Proposition 14.5.2. $\qquad \square$

This important theorem establishes a one-to-one correspondence between the hermitian solutions of the CARE and the n-dimensional, M-invariant, H-neutral subspaces by means of equation (14.8.28). In this statement matrix H can be replaced by $\widehat{H}$ of equation (14.5.9), since $\mathcal{L}$ is H-neutral if and only if it is $\widehat{H}$-neutral. The next theorem describes the hermitian solutions of the CARE in terms of all M-invariant subspaces. We adopt the following terminology: If Z is an $n \times n$

matrix, and Ω is a subset of the complex plane, the sum of the root subspaces of Z corresponding to the eigenvalues in Ω will be referred to as the *spectral subspace* of Z corresponding to the eigenvalues in Ω. If Ω does not contain any eigenvalues of Z, the corresponding spectral subspace is taken to be the zero subspace.

Theorem 14.8.2. *Assume that $D \geq 0$ and (A, D) is a controllable pair. Assume that the CARE has a hermitian solution, and let $\mathcal{N}_+$ be the spectral subspace of M corresponding to all eigenvalues in the open upper half-plane. Then for every M-invariant subspace $\mathcal{N} \subseteq \mathcal{N}_+$ there exists a unique hermitian solution X of the CARE such that*

$$\text{Range} \begin{bmatrix} I \\ X \end{bmatrix} \cap \mathcal{N}_+ = \mathcal{N}.$$

Conversely, if X is a hermitian solution of the CARE, then $\text{Range} \begin{bmatrix} I \\ X \end{bmatrix} \cap \mathcal{N}_+$ is M-invariant.

The proof is obtained by combining Theorem 14.7.5 with Theorem 5.12.4.

The subspace $\mathcal{N}_+$ of Theorem 14.8.2 can be replaced by the spectral subspace of M corresponding to any maximal set $\mathcal{C}$ of eigenvalues with the property that $\lambda_0 \in \mathcal{C}$ implies that $\overline{\lambda_0} \notin \mathcal{C}$. Indeed, the following Theorem 14.8.4 is arrived at in this way. It is obtained by taking $\mathcal{N}_+$ to be the spectral subspace of M determined by the set iS and choosing $\mathcal{N} = \mathcal{N}_+$. The assertion about partial multiplicities follows from Theorem 5.12.3.

Theorem 14.8.2 allows us to count the number of hermitian solutions of the CARE by counting the number of invariant subspaces of the restriction $M|_{\mathcal{N}_+}$. First, let J_+ be the Jordan form of $M|_{\mathcal{N}_+}$ and, clearly J_+ and $M|_{\mathcal{N}_+}$ have the same number of invariant subspaces. Now a Jordan block of size k has exactly $k + 1$ invariant subspaces. Hence, *if for each eigenvalue of J_+ there is exactly one associated Jordan block, then J_+ has exactly* $\prod_{i=1}^{\alpha}(k_i + 1)$ *invariant subspaces, where $k_i = \dim \text{Ker}(\lambda_i I - J_+)$, for $i = 1, 2, \ldots, \alpha$ and $\lambda_1, \ldots, \lambda_\alpha$ are all the distinct eigenvalues of J_+.*

If, on the other hand, $\dim \text{Ker}(\lambda I - J_+) \geq 2$ for some λ, then J_+ has a continuum of invariant subspaces. To see this observe that, if x and y are linearly independent eigenvectors of J_+ corresponding to the same eigenvalue λ, then all of the 1-dimensional subspaces spanned by $\{x + cy\}$, $c \in \mathbb{C}$, are different and J_+-invariant. The following corollary has been established:

Corollary 14.8.3. *Assume that $D \geq 0$, (A, D) is a controllable pair, and that the CARE has a hermitian solution. If $\dim \text{Ker}(\lambda I - M) \leq 1$ for all λ in the open upper half-plane, then the number of hermitian solutions of the CARE is exactly $\prod_{i=1}^{\alpha}(k_i + 1)$, where $k_1, \ldots, k_\alpha$ are the algebraic multiplicities of all distinct eigenvalues $\lambda_1, \ldots, \lambda_\alpha$ of M in the open upper half-plane.*

If $\dim \text{Ker}(\lambda I - M) \geq 2$ for some nonreal λ, then the CARE has a continuum of hermitian solutions.

In particular, it follows from this corollary that the CARE has a *unique* solution if and only if all eigenvalues of M are real and all their partial multiplicities are even.

For the following theorem and the next section, it is convenient to work with the spectrum of the matrix

$$T := \begin{bmatrix} A & D \\ C & -A^* \end{bmatrix} = -iM, \qquad (14.8.29)$$

rather than M itself. Of course, this simply involves a rotation of the spectrum in the complex plane through a right angle and has no effect on the invariant subspaces.

Theorem 14.8.4. *Assume that $D \geq 0$, (A, D) is a controllable pair, and that the CARE has a hermitian solution. Let S be a set of nonreal eigenvalues of T with nonzero real parts for which $\lambda \in S$ implies $-\overline{\lambda} \notin S$ and which is maximal with respect to this property. Then there exists a unique hermitian solution X of the CARE such that S is exactly the set of eigenvalues of $A + DX$ having nonzero real parts.*

Furthermore, the partial multiplicities for every pure imaginary or zero eigenvalue λ_0 (if any) of $A + DX$ are equal to $m_1, \ldots, m_k$, where the partial multiplicities of T corresponding to λ_0 are $2m_1, \ldots, 2m_k$.

Notice that, in this statement, the fact that the partial multiplicities of T corresponding to purely imaginary λ_0 are all even follows from Theorem 14.7.2.

14.9 Extremal Hermitian Solutions

In this section we continue our study of the CARE (14.3.6) under the assumptions that $C = C^*$, $D \geq 0$ and (A, D) is controllable. It will also be assumed throughout this section that the CARE has at least one hermitian solution. We shall prove the existence of maximal and minimal solutions under these hypotheses. Recall that, for the optimal control problem discussed in Section 14.2, it is precisely the *maximal* solution which is of interest.

The extreme solutions are defined with respect to a natural order relation in the set of all hermitian matrices. Namely, $X_1 \leq X_2$ for hermitian matrices X_1 and X_2 means that $X_2 - X_1$ is positive semidefinite. A hermitian solution X_+ (resp. X_-) of the CARE is called *maximal* (resp. *minimal*) if $X \leq X_+$ (resp. $X_- \leq X$) for every hermitian solution X. Obviously, if a maximal (resp. minimal) solution exists, it is unique. The following theorem establishes the existence of extremal hermitian solutions and characterizes them in spectral terms.

Theorem 14.9.1. *If $C^* = C$, $D \geq 0$, (A, D) is controllable, and the CARE has a hermitian solution, then there exist a maximal hermitian solution X_+, and a minimal hermitian solution, X_-. The solution X_+ is the unique hermitian solution* *for which $\sigma(A + DX_+)$ lies in the closed right half-plane, and is obtained from Theorem 14.8.4 by taking S to be the set of eigenvalues of T (of equation (14.8.29)) having positive real parts.*

The solution X_- is the unique hermitian solution with $\sigma(A + DX_-)$ in the closed left half-plane, and is obtained from Theorem 14.8.4 by taking S to be the set of eigenvalues of T having negative real parts.

Proof. Let X be a hermitian solution of the CARE and define $\widetilde{A} = A + DX$ and

$$U(t) = -\int_0^t e^{-\tau \widetilde{A}} D e^{-\tau \widetilde{A}^*}\, d\tau, \quad t \in \mathsf{R}.$$

Since (A, D) is controllable, so is $(\widetilde{A}, D)$ (by Proposition 14.1.8). So Lemma 14.1.7 is applicable, and $U(t)$ is positive (resp. negative) definite for $t < 0$ (resp. $t > 0$). Also,

$$U(t_1) \le U(t_2) \quad \text{for} \quad t_1 \ge t_2,$$

and therefore (using the easily verified property that $X \ge Y > 0$ implies $0 < X^{-1} \le Y^{-1}$)

$$U(t_1)^{-1} \ge U(t_2)^{-1} > 0 \quad \text{for} \quad 0 > t_1 \ge t_2. \tag{14.9.30}$$

It follows that the limit $\lim_{t \to -\infty} U(t)^{-1}$ exists; indeed, this limit can be uniquely determined by

$$\left(\lim_{t \to -\infty} U(t)^{-1}x, x\right) = \lim_{t \to -\infty} (U(t)^{-1}x, x)$$

for any $x \in \mathsf{C}^n$, where the limit on the right-hand side exists by (14.9.30). Moreover, $\lim_{t \to -\infty} U(t)^{-1}$ is positive semidefinite.

For any positive real T and $t < T$ define

$$X_T(t) = X + (U(t - T))^{-1}. \tag{14.9.31}$$

Since $X_T(t) = [XU(t - T) + I]U(t - T)^{-1}$, it follows that $X_T(t)$ is invertible for $t \in [T - \delta, T)$, where $\delta > 0$ is small enough, and (because $U(0) = 0$)

$$\lim_{t \to T} X_T(t)^{-1} = 0.$$

By a direct computation, using the facts that $X = X^*$ is a solution of the CARE and that $U(t)$ satisfies the differential equation (cf. (14.1.3)),

$$U(t)' = -\widetilde{A}U - U\widetilde{A}^* - D,$$

and one checks easily that $X_T(t)$ satisfies the differential Riccati equation

$$X_T(t)' + C - X_T(t)A - A^*X_T(t) - X_T(t)DX_T(t) = 0, \quad t < T. \tag{14.9.32}$$

Equation (14.9.32), together with the boundary condition $\lim_{t \to T} X_T(t)^{-1} = 0$, determine $X_T(t)$ uniquely as a quantity independent of the choice of the hermitian solution X of the CARE. Indeed, let $V_T(t) = X_T(t)^{-1}$. Then, by differentiating the equation $X_T(t) \cdot V_T(t) = I$, it is found that

$$V_T(t)' = -V_T(t)X_T(t)'V_T(t),$$

and, using (14.9.32), $V_T(t)$ satisfies the differential equation

$$-V_T(t)' + V_T(t)CV_T(t) - AV_T(t) - V_T(t)A^* - D = 0$$

for $T - \delta < t < T$ (this is the interval where the invertibility of $X_T(t)$ is guaranteed). Also, $V_T(t)$ satisfies the initial condition

$$\lim_{t \to T} V_T(t) = 0.$$

Hence, by the uniqueness theorem for solutions of the initial value problem we find that $V_T(t)$ and, consequently, also $X_T(t)$ are uniquely determined (i.e., independent of the choice of X) for $T - \delta_0 < t < T$. Here $\delta_0 \le \delta$ is a suitable positive number. But then $X_T(t)$ is uniquely determined for all $t < T$, as is easily seen from the formula (14.9.31). (Indeed, it follows from that formula and the definition of $U(t)$ that $X_T(t)$ is a real analytic function of $t < T$; now use the property that if two real analytic functions coincide on some open interval, they coincide on the whole domain of definition.)

As $T \to \infty$, the matrix function $X_T(t)$ converges uniformly on compact intervals to the constant matrix

$$X_+ = X + \lim_{t \to -\infty} U(t)^{-1}. \tag{14.9.33}$$

Of course, X_+ is independent of X. Now (14.9.33) implies that $X_+ = X_+^*$ and $X \le X_+$. Further,

$$
\begin{aligned}
X_T'(t) &= \frac{dU(t-T)^{-1}}{dt} = -U(t-T)^{-1} \cdot \frac{dU(t-T)}{dt} U(t-T)^{-1} \\
&= U(t-T)^{-1}[\tilde{A}U(t-T) + U(t-T)\tilde{A}^* + D]U(t-T)^{-1} \\
&= U(t-T)^{-1}\tilde{A} + \tilde{A}^*U(t-T)^{-1} + U(t-T)^{-1}DU(t-T)^{-1}
\end{aligned}
$$

has a limit when $T \to \infty$ (keeping t fixed). As the function $U(t-T)^{-1}$ itself has a limit when $T \to \infty$ (and t is fixed), it follows that

$$\lim_{T \to \infty} X_T'(t) = 0.$$

For a fixed t, passing to the limit when $T \to \infty$ in (14.9.32), we find that X_+ is a solution of the CARE. Since $X \le X_+$ for any hermitian solution X, X_+ is maximal.

Furthermore,

$$A + DX_+ = \lim_{t \to -\infty} (A + D(X + U(t)^{-1})) = \lim_{T \to \infty} (A + DX_T(0)), \tag{14.9.34}$$

and

$$A + DX_T(0) = \tilde{A} + D(U(-T))^{-1} = \tilde{A} - D\left[\int_0^{-T} e^{-\tau \tilde{A}} De^{-\tau \tilde{A}^*}\, d\tau\right]^{-1}$$

$$= -\left[(-\tilde{A}) - D\left(\int_0^T e^{-\tau(-\tilde{A})} De^{-\tau(-\tilde{A}^*)}\, d\tau\right)^{-1}\right].$$

Since $\mathrm{rank}[D, -\tilde{A}D, \ldots, (-1)^{n-1}\tilde{A}^{n-1}D] = n$, the second part of Lemma 14.1.7 ensures that $\sigma(A + DX_T(0))$ lies in the open right half-plane. Using (14.9.34) and the continuity of eigenvalues of $A + DX_T(0)$ as functions of T, we find that $\Re\lambda_0 \geq 0$ for every $\lambda_0 \in \sigma(A + DX_+)$. Uniqueness of a hermitian solution X of the CARE with the additional property that $\Re\sigma(A + DX) \geq 0$ follows from Theorem 14.8.4.

Applying the results we have obtained concerning maximal solutions of the CARE to the equation

$$XDX - XA - A^*X - C = 0 \tag{14.9.35}$$

and noting that X is a solution of (14.9.35) if and only if $-X$ is a solution of (14.3.6), we obtain the corresponding results for minimal solutions.   □

Although Theorem 14.9.1 refers only to extremal solutions, the idea emerging here is that the choice of S in the spectrum of T (of equation (14.8.29)) determines a hermitian solution X (by Theorem 14.8.4) and the set S reappears as that part of the spectrum of the resulting modified state matrix $A + DX$ which is not on the imaginary axis.

14.10   The CARE with Real Coefficients

Riccati equations frequently arise with *real* coefficient matrices, A, D, C, and it is natural to investigate, not just the set of hermitian solutions, but the subset of real symmetric solutions. Such equations are the topic of this section and, of course, the preceding analysis applies but more structure can be expected in the solution set of a "real" CARE (14.3.6). The definitions of matrices M and H are retained (as in (14.5.8) and (14.5.9)):

$$M = i\begin{bmatrix} A & D \\ C & -A^* \end{bmatrix}, \quad H = \begin{bmatrix} -C & A^* \\ A & D \end{bmatrix}. \tag{14.10.36}$$

Recall that M is H-selfadjoint and, in particular, the spectrum of M is symmetric with respect to the real axis. However, as A, D, C are now real, the spectrum of M is also symmetric with respect to the imaginary axis. Thus, if $\lambda_0 \in \sigma(M)$ then $\overline{\lambda_0}$, $-\lambda_0$, and $-\overline{\lambda_0}$ are also in $\sigma(M)$ and the partial multiplicities of such a quadruple of eigenvalues will all be the same.

A description of the set of real symmetric solutions of the CARE with real coefficients is provided by the next theorem.

Theorem 14.10.1. *Assume that, for the CARE with real coefficient matrices A, D, C, $D \geq 0$, the pair (A, D) is controllable, and there is at least one hermitian (not necessarily real symmetric) solution. Let $\widetilde{\mathcal{N}}_+$ be the spectral subspace of M corresponding to the eigenvalues in the halfopen first quadrant:*

$$\{\lambda \in \mathsf{C} \mid \Im \lambda > 0, \ \Re \lambda \geq 0\}.$$

Then for every M-invariant subspace $\widetilde{\mathcal{L}} \subseteq \widetilde{\mathcal{N}}_+$ there exists a unique real symmetric solution X of the CARE such that

$$\text{Range} \begin{bmatrix} I \\ X \end{bmatrix} \cap \widetilde{\mathcal{N}}_+ = \widetilde{\mathcal{L}}. \tag{14.10.37}$$

Conversely, if X is a real symmetric solution of the real CARE, then the subspace $\text{Range} \begin{bmatrix} I \\ X \end{bmatrix} \cap \widetilde{\mathcal{N}}_+$ is M-invariant.

Proof. Denote by $\mathcal{N}_{++}$ (resp. $\mathcal{N}_+$) the spectral subspace of M in the open first quadrant, $\{\lambda \in \mathsf{C} \mid \Im \lambda > 0, \ \Re \lambda > 0\}$, (resp. in the open upper half-plane). Then $\mathcal{N}_{++} \subseteq \widetilde{\mathcal{N}}_+ \subseteq \mathcal{N}_+$.

Given an M-invariant subspace $\widetilde{\mathcal{L}} \subseteq \widetilde{\mathcal{N}}_+$, let $\mathcal{L}_{++} = \widetilde{\mathcal{L}} \cap \mathcal{N}_{++}$, and let $\mathcal{L}$ be the sum of two subspaces, namely $\widetilde{\mathcal{L}}$ and

$$\overline{\mathcal{L}_{++}} = \{< x_1, \ldots, x_{2n} >\in \mathsf{C}^{2n} \mid < \overline{x_1}, \ldots, \overline{x_{2n}} >\in \mathcal{L}_{++}\}.$$

It is easily seen that, if the vectors $f_0, \ldots, f_k \in \mathsf{C}^{2n}$ form a Jordan chain for M with the eigenvalue λ_0, then the vectors

$$\overline{f_0}, \ -\overline{f_1}, \ \overline{f_2}, \ldots, \pm \overline{f_k}$$

form a Jordan chain of M with eigenvalue $-\overline{\lambda_0}$. Hence the subspace $\overline{\mathcal{L}_{++}}$ is M-invariant and $\overline{\mathcal{L}_{++}} \cap \widetilde{\mathcal{N}}_+ = \{0\}$. Moreover, the subspace $\mathcal{L}$ enjoys the property that

$$< x_1, \ldots, x_{2n} >\in \mathcal{L} \quad \text{implies} \quad < \overline{x_1}, \ldots, \overline{x_{2n}} >\in \mathcal{L}. \tag{14.10.38}$$

In particular, $\mathcal{L}$ is M-invariant and $\mathcal{L} \cap \widetilde{\mathcal{N}}_+ = \widetilde{\mathcal{L}}$. By Theorem 14.8.2 there is a unique hermitian solution X of the CARE such that

$$\text{Range} \begin{bmatrix} I \\ X \end{bmatrix} \cap \mathcal{N}_+ = \mathcal{L},$$

and we claim that this X is real. Indeed, $\overline{X}$ is also a solution of the CARE, and both $\mathcal{L}$ and $\mathcal{N}_+$ enjoy the property (14.10.38). Hence

$$\text{Range} \begin{bmatrix} I \\ \overline{X} \end{bmatrix} \cap \mathcal{N}_+ = \mathcal{L}.$$

However, the solution X is unique and so $\overline{X} = X$. Thus, there is a real symmetric solution X of the real CARE such that (14.10.37) holds.

Now let X and Y be two real symmetric solutions of the real CARE, and assume that

$$\text{Range} \begin{bmatrix} I \\ X \end{bmatrix} \cap \tilde{\mathcal{N}}_+ = \text{Range} \begin{bmatrix} I \\ Y \end{bmatrix} \cap \tilde{\mathcal{N}}_+. \tag{14.10.39}$$

Theorem 14.8.2 ensures that the subspace (14.10.39) is M-invariant. Taking complex conjugates it follows that

$$\text{Range} \begin{bmatrix} I \\ X \end{bmatrix} \cap \tilde{\mathcal{N}}_- = \text{Range} \begin{bmatrix} I \\ Y \end{bmatrix} \cap \tilde{\mathcal{N}}_-, \tag{14.10.40}$$

where

$$\tilde{\mathcal{N}}_- := \left\{ <x_1, \ldots, x_{2n}> \in \mathbb{C}^{2n} \mid <\overline{x_1}, \ldots, \overline{x_{2n}}> \in \tilde{\mathcal{N}}_+ \right\}.$$

But $\tilde{\mathcal{N}}_-$ is the spectral subspace of M corresponding to the eigenvalues in the quadrant

$$\{\lambda \in \mathbb{C} \mid \Re\lambda \leq 0,\ \Im\lambda > 0\}$$

(cf. the property of Jordan chains of M mentioned above). Combining (14.10.39) and (14.10.40) it is found that

$$\text{Range} \begin{bmatrix} I \\ X \end{bmatrix} \cap \mathcal{N}_+ = \text{Range} \begin{bmatrix} I \\ Y \end{bmatrix} \cap \mathcal{N}_+$$

and the uniqueness statement of Theorem 14.8.2 ensures that $X = Y$.   □

Notice the special role played in this proof by the subspace $\tilde{\mathcal{N}}_+$. In fact, instead of this subspace, it is possible to take in its place a spectral subspace of M corresponding to any set $\mathcal{C}_+$ of nonreal eigenvalues of M which is maximal with respect to the two following properties:

(a) If $\lambda_0 \in \mathcal{C}_+$ and is pure imaginary, then $\overline{\lambda_0} \notin \mathcal{C}_+$.

(b) If $\lambda_0 \in \mathcal{C}_+$ and is *not* pure imaginary, then $\overline{\lambda_0}$, $-\lambda_0$, and $-\overline{\lambda_0}$ are not in $\mathcal{C}_+$.

Now consider the possibility of real symmetric extremal solutions for the real CARE. Recall that the extremal solutions are defined in Theorem 14.9.1 in terms of subsets S of eigenvalues of matrix T (defined in (14.8.29)) which contain no pairs $(\lambda_0, -\overline{\lambda_0})$, and are maximal in this respect. It has been shown in Theorem 14.8.4 that, given the existence of one hermitian solution, every such subset S determines a unique hermitian solution X_S of the CARE such that S is the set of eigenvalues of $A + DX$ having nonzero real parts. The next result provides necessary and sufficient conditions under which X_S is real.

Theorem 14.10.2. *Consider the real* CARE. *Assume that $D \geq 0$, (A, D) is a controllable pair and that there exists at least one hermitian solution (not necessarily real). Then a set S of eigenvalues of M (as defined above) consists of conjugate pairs of eigenvalues if and only if the solution X_S is real. In particular, the maximal and minimal solutions of the real* CARE *are real and symmetric.*

The proof of this theorem uses the same line of argument as the proof of Theorem 14.10.1 and will not be reproduced here (see the Exercises to this chapter). The statement concerning the maximal and minimal solutions follows from Theorem 14.9.1.

14.11 The Concerns of Numerical Analysis

First of all, general purpose algorithms for solving the CARE are designed for equations with *real* matrix coefficients and are designed to compute the *maximal* solution, when it exists. The analysis is generally undertaken in terms of the real $2n \times 2n$ matrix

$$T = \begin{bmatrix} A & D \\ C & -A^T \end{bmatrix}$$

of equation (14.8.29). Defining $N = \begin{bmatrix} 0 & I \\ -I & 0 \end{bmatrix}$, it follows that

$$NT = \begin{bmatrix} C & -A^T \\ -A & D \end{bmatrix},$$

a real symmetric matrix, and this (or the equivalent relation, $N^{-1}T^T N = -T$) is the defining property of a *Hamiltonian* matrix. The spectrum of such a matrix is distributed symmetrically with respect to *both* the real and the imaginary axes (cf. the discussion of equations (14.10.36)).

It should also be recognized that, in most applications, *the Hamiltonian has no purely imaginary or zero eigenvalues*. Recall that, when they do exist, they are necessarily multiple eigenvalues (Theorem 14.7.2). As a consequence, many of the analytical difficulties encountered in this chapter do not arise and, generally, algorithms are designed for this less intricate case.

There are iterative methods for finding solutions of the real CARE (some based on Newton's method, for example). However, the methods of choice are usually based on so-called "subspace" methods and the theoretical ideas developed in this chapter. Conceivably, this could be done by direct application of Theorem 14.4.2 via the calculation of eigenvectors and, possibly, generalized eigenvectors. But this strategy presents difficult problems of numerical stability and is generally avoided for this reason. A class of more stable algorithms originates with Laub [72]. They are known as Schur methods because they involve reduction to triangular (or quasi-triangular) form by a *real* orthogonal similarity — following a technique of Schur from 1909. The use of a *real orthogonal* similarity is the basis of more reliable numerical performance.

Thus, a real orthogonal matrix $U = [U_1\ U_2]$ is to be found so that

$$[U_1\ U_2]^T \begin{bmatrix} A & D \\ C & -A^T \end{bmatrix} [U_1\ U_2] = \begin{bmatrix} S_1 & S_{12} \\ 0 & S_2 \end{bmatrix}, \tag{14.11.41}$$

where U_1 and U_2 are $2n \times n$, and S_1, S_2 are real $n \times n$ upper quasi-triangular[2] matrices. Their diagonal blocks of size 1 determine real eigenvalues, and those of size 2 determine conjugate pairs. The presence of the zero block on the right shows immediately that Range U_1 is T-invariant. One must therefore ensure that the eigenvalues of S_1 are those required to generate the stabilizing subspace (i.e., associated with the eigenvalues of the Hamiltonian in the right half-plane, cf. Theorem 14.9.1). When this is the case, we write

$$U_1 = \begin{bmatrix} X_1 \\ X_2 \end{bmatrix} \quad \text{and} \quad \begin{bmatrix} A & D \\ C & -A^T \end{bmatrix} \begin{bmatrix} X_1 \\ X_2 \end{bmatrix} = \begin{bmatrix} X_1 \\ X_2 \end{bmatrix} S_1,$$

so that $X = X_2 X_1^{-1}$ (see Theorem 14.4.2).

Another (more substantial) difficulty is the design of an algorithm which uses a sequence of elementary real orthogonal similarities but also retains the Hamiltonian symmetry at each stage; a property which is also of great importance in accurate numerical computation.

This difficulty can be resolved by further restricting the matrix U to be *symplectic*, i.e., satisfying $U^T N U = N$. The reason for this is the property described in Lemma 14.11.1 below: that if U is both real orthogonal and symplectic and A is Hamiltonian, then $U A U^T$ is Hamiltonian.

There is a remarkable theorem of Paige and van Loan [87] which has played an important role in this connection. Observe first that a matrix A has the Hamiltonian structure of M_r if and only if $NA = (NA)^T$. As noted, if A is Hamiltonian and U is orthogonal and symplectic, then $U A U^T$ is also Hamiltonian. In this way, the Hamiltonian structure is maintained through each step of the recursive algorithm used to generate the matrix U of (14.11.41).

The next lemma is easy preparation for Theorem 14.11.2 to follow. The reader is referred to [87] for the proof, as the technique of proof is quite different from the methods used elsewhere in this book.

Lemma 14.11.1. *If A is Hamiltonian and U is orthogonal and symplectic, then $U A U^T$ is Hamiltonian.*

Indeed, if A is Hamiltonian, in other words, the equation $N^{-1} A^T N = -A$

---

[2]An upper quasi-triangular matrix with entries a_{ij} has the property that $a_{ij} = 0$ whenever $i < j - 1$.

holds, then

$$(UAU^T)^T N = UA^T U^T N = UA^T NU^{-1} = -UNAU^{-1} = -NUAU^{-1}$$
$$= -N(UAU^T),$$

(14.11.42)

where we have used the properties of the orthogonal and symplectic matrix U, namely, $U^T NU = N$ and $U^{-1} = U^T$. Now equality (14.11.42) shows that UAU^T is Hamiltonian as well.

Theorem 14.11.2. *If* $T = \begin{bmatrix} A & D \\ C & -A^T \end{bmatrix}$ *has no real eigenvalues,* $D^T = D$ *and* $C^T = C$, *then there exists a real matrix* U *which is both orthogonal and symplectic for which*

$$U^T \begin{bmatrix} A & D \\ C & -A^T \end{bmatrix} U = \begin{bmatrix} R & G \\ 0 & -R^T \end{bmatrix},$$

where all matrices are real, R is upper quasi-triangular, and $G^T = G$.

14.12 Exercises

1. Let $A \in \mathbb{C}^{n \times n}$ and $B \in \mathbb{C}^{n \times m}$ and $\mathcal{C}_r = \text{Range}[B \ AB \ \ldots \ A^r B]$. Show that for $r = 0, 1, 2, \ldots$, $\mathcal{C}_r \subseteq \mathcal{C}_{r+1}$ and, if $\mathcal{C}_r = \mathcal{C}_{r+1}$ then $\mathcal{C}_s = \mathcal{C}_r$ for all $s \geq r$. (In the latter case $\mathcal{C}_r$ is the *controllable subspace* of the pair (A, B).)

2. Prove Proposition 14.1.2. Hint: Use the previous exercise and the Cayley–Hamilton theorem, which implies that A^n is a linear combination of matrices $I, A, A^2, \ldots, A^{n-1}$.

3. Show that a controllable matrix pair is stabilizable, but that the converse statement is not necessarily true.

4. Establish Propositions 14.1.6 and 14.1.8.

5. Verify the results claimed in the five examples of Section 14.3.

6. Prove that every invertible matrix A has a square root, i.e., there is a matrix X such that $X^2 = A$. Give an example of a 2×2 real matrix with no square root (either real or complex).

7. Consider the CARE (14.3.6) with

$$D = \begin{bmatrix} 1 & 1 \\ 1 & 1 \end{bmatrix}, \quad A = \begin{bmatrix} 0 & -1 \\ 1 & 0 \end{bmatrix}, \quad C = \begin{bmatrix} 1 & 1 \\ 1 & 1 \end{bmatrix}.$$

   (a) Find all hermitian solutions.

   (b) Identify the maximal and minimal solutions.

   (c) Find all solutions X for which $(X^* - X)(A + DX) \leq 0$.

(Notice that the matrix M of (14.5.8) has no real eigenvalues but is not diagonalizable.)

8. Consider the CARE of the preceding exercise once more.

   (a) Find all real symmetric solutions.

   (b) Identify the maximal and minimal real symmetric solutions.

   (c) Find all real solutions satisfying $(X^* - X)(A + DX) \leq 0$.

   (d) Find all real solutions satisfying $(X^* - X)(A + DX) \geq 0$.

9. Consider the CARE (14.3.6) with $D^* = D$, $C \geq 0$, and (C, A) observable. Show that all hermitian solutions (if any) are nonsingular.

10. Complete the "similar argument" in the proof of Theorem 14.6.4.

11. Consider the real CARE, and assume that $D \geq 0$, the pair (A, D) is controllable, and that the CARE has real symmetric solutions. Find conditions in terms of the eigenvalues of T of (14.8.29) which guarantee that every hermitian solution is real.

12. Use the techniques developed in the proof of Theorem 14.10.1 to establish Theorem 14.10.2.

13. Provide a detailed proof of Theorem 14.11.2 for the cases $n = 1$ and $n = 2$.

14.13 Notes

The exposition of this chapter is based on the extensive treatment of algebraic Riccati equations in [67] and earlier papers quoted there, together with expository material from [65].

Appendix A

Topics from Linear Algebra

This Appendix contains a collection of concepts and results from linear algebra and matrix theory which may serve as a refresher for some readers, but also serves as a source of materials needed in the main body of the text.

Some of the basic definitions and results concerning linear spaces with an inner product (*not* indefinite) are reviewed for easy comparison with the theory of linear spaces with an *indefinite inner product* of the main text. (In this appendix an inner product is always definite unless stated otherwise.) The reader is expected to be familiar with most of the material presented here. For more details, including proofs of statements not included here, see, for example, [70].

A.1 Hermitian Matrices

Some useful and well-known properties of hermitian matrices are collected in this section.

Hermitian matrices H_1, $H_2 \in C^{n \times n}$ are said to be *congruent* if $H_1 = T^* H_2 T$ for some invertible $T \in C^{n \times n}$. It is easily seen that congruent matrices form an equivalence class in which the rank is an invariant. Furthermore, congruence has the important property of preserving the "inertia" of a hermitian matrix, i.e., the numbers of positive, negative, and zero eigenvalues as specified more precisely in:

Theorem A.1.1. *Each equivalence class of congruent matrices in* $C^{n \times n}$ *contains exactly one matrix with the partitioned form*

$$D = \begin{bmatrix} I_s & 0 & 0 \\ 0 & -I_{r-s} & 0 \\ 0 & 0 & 0_{n-r} \end{bmatrix},$$

where r is the rank of all matrices in the class and s is the number of positive eigenvalues, each counted as many times as its algebraic multiplicity.

In this partitioned matrix the last row and column simply does not appear if $r = n$. Clearly, the invariant $r - s$ is just the number of negative eigenvalues (counted with algebraic multiplicities). The triple of integers $(s, r - s, n - r)$ is an invariant of the equivalence class of all congruent matrices, and summarizes the numbers of positive, negative, and zero eigenvalues. It is called the *inertia* — of each matrix in the equivalence class.

It will be convenient to use the following notation for a hermitian matrix H: $i_+(H)$, resp., $i_-(H)$, is the number of positive, resp., negative eigenvalues of H, counted with their multiplicities, and $i_0(H)$ is the number of zero eigenvalues of H (multiplicities counted). Thus, H is congruent to the matrix D of Theorem A.1.1 if and only if $i_+(H) = s$, $i_-(H) = r - s$, and $i_0(H) = n - r$.

Continuity of the eigenvalues (properly enumerated) of a matrix, as functions of the entries of the matrix, yields the part (a) of the next theorem (part (b) follows easily from part (a)):

Theorem A.1.2. *Let there be given a hermitian $X \in \mathbb{C}^{n \times n}$. Then, there exists an $\varepsilon > 0$ such that:*

$$\text{(a)} \qquad i_+(Y) \geq i_+(X), \qquad i_-(Y) \geq i_-(X)$$

for every hermitian $Y \in \mathbb{C}^{n \times n}$ satisfying $\|Y - X\| < \varepsilon$, and

$$\text{(b)} \qquad i_+(Y) = i_+(X), \qquad i_-(Y) = i_-(X)$$

for every hermitian $Y \in \mathbb{C}^{n \times n}$ satisfying $\|Y - X\| < \varepsilon$ and $\operatorname{rank} Y = \operatorname{rank} X$.

Given the topic of this book, it is necessary to have constructive methods for the determination of the inertia of an hermitian matrix, or, more specifically, the determination of the eigenvalues themselves. Classical results concerning the former problem can be formulated in terms of the "leading principal minors" of an $n \times n$ matrix. They are, by definition, just the determinants $h_1, h_2, \ldots, h_n$ of the leading principal submatrices of matrix H of sizes $1 \times 1, 2 \times 2, \ldots, n \times n$. Then theorems of Frobenius and Jacobi, respectively, are as follows:

Theorem A.1.3. *An $n \times n$ hermitian matrix H is positive definite if and only if all its leading principal minors, $h_1, h_2, \ldots, h_n$ are positive.*

Theorem A.1.4. *If H is an hermitian matrix of rank r and $h_1, h_2, \ldots, h_r$ are its leading principal minors, then the number of negative (respectively, positive) eigenvalues of H is equal to the number of changes (respectively, constancies) of sign in the sequence $1, h_1, h_2, \ldots, h_r$.*

A result describing the relationship between the inertia of a hermitian matrix and the number of eigenvalues of another (not necessarily hermitian) matrix is known as an *inertia theorem*. A result of this kind (that is needed elsewhere) applies when the two matrices are connected by a linear inequality. Thus:

Theorem A.1.5. *Let $A \in \mathbb{C}^{n \times n}$ and $G \in \mathbb{C}^{n \times n}$ be matrices such that G is hermitian and $A^*GA - G$ is positive definite. Then:*

(a) G is invertible;

(b) A has no eigenvalues of modulus 1;

(c) the number of eigenvalues of A, counted with their algebraic multiplicities, that have modulus larger than 1, is equal to $i_+(G)$;

(d) the number of eigenvalues of A, counted with their algebraic multiplicities, that have modulus smaller than 1, is equal to $i_-(G)$.

This statement is Theorem 13.2.2 of [70]. The result is originally due to Taussky [98], Hill [49], and Wimmer [106]. A proof is available in [70] as well. This theorem differs from Theorem 4.6.1, because here the invertibility of G is a priori not assumed. However, one can easily deduce a proof of Theorem A.1.5 from that of Theorem 4.6.1.

It is useful to have different ways of characterizing eigenvalues and eigenvectors. In the case of hermitian matrices *variational methods* admit a characterization of eigenvalues which is independent of the eigenvectors, and is frequently known as the "mini-max" method. To describe this, let H be an hermitian matrix with eigenvalues $\lambda_1 \geq \lambda_2 \geq \cdots \geq \lambda_n$. It is well-known that the largest and smallest eigenvalues λ_1 and λ_n can be written in the form

$$\lambda_1 = \max_{x \neq 0} \frac{x^* H x}{x^* x}, \qquad \lambda_n = \min_{x \neq 0} \frac{x^* H x}{x^* x},$$

and that these extrema are attained at eigenvectors of H. The mini-max theorem admits generalizations of these formulas to *any* eigenvalue of H.

Let $\mathcal{S}_j$ be any subspace of $\mathbf{C}^n$ of dimension j, for $j = 1, 2, \ldots, n$. Then the theorem can be stated as follows:

Theorem A.1.6. *For $j = 1, 2, \ldots, n$,*

$$\lambda_j = \min_{\mathcal{S}_{n-j+1}} \max_{0 \neq x \in \mathcal{S}_{n-j+1}} \frac{x^* H x}{x^* x},$$

or, in dual form,

$$\lambda_{n-j+1} = \max_{\mathcal{S}_{n-j+1}} \min_{0 \neq x \in \mathcal{S}_{n-j+1}} \frac{x^* H x}{x^* x}.$$

A.2 The Jordan Form

The complex vector space $\mathbf{C}^{n \times n}$ can be subdivided into disjoint equivalence classes of similar matrices. Each equivalence class is determined by a unique matrix in canonical form. Conventions vary to some degree on the specification of this canonical form. It forms a building block for other forms developed in the body of this text.

First define a typical *Jordan block*

$$J_m(\lambda) = \begin{bmatrix} \lambda & 1 & 0 & \cdots & 0 \\ 0 & \lambda & 1 & \cdots & 0 \\ \vdots & \vdots & \ddots & \ddots & 0 \\ & & & \lambda & 1 \\ 0 & 0 & \cdots & 0 & \lambda \end{bmatrix} \in C^{m\times m}.$$

The fundamental theorem specifying the Jordan canonical form is as follows:

Theorem A.2.1. *For any $A \in C^{n\times n}$ there is a block-diagonal matrix $J \in C^{n\times n}$ of the form*

$$J = \begin{bmatrix} J_{m_1}(\lambda_1) & 0 & \cdots & 0 \\ 0 & J_{m_2}(\lambda_2) & \cdots & 0 \\ \vdots & \vdots & \ddots & 0 \\ 0 & 0 & \cdots & J_{m_r}(\lambda_r) \end{bmatrix} \tag{A.2.1}$$

which is similar to A.

Moreover, the matrix J is uniquely determined by A up to permutation of the diagonal blocks, i.e., if A is also similar to a matrix

$$\begin{bmatrix} J_{n_1}(\mu_1) & 0 & \cdots & 0 \\ 0 & J_{n_2}(\mu_2) & \cdots & 0 \\ \vdots & \vdots & \ddots & 0 \\ 0 & 0 & \cdots & J_{n_s}(\mu_s) \end{bmatrix},$$

then $r = s$ and the collection $\{J_{n_1}(\mu_1), J_{n_2}\mu_2, \ldots, J_{n_s}(\mu_s)\}$ (possibly with repeated elements) can be re-arranged so that $J_{n_k}(\mu_k) = J_{m_k}(\lambda_k)$ for $k = 1, 2, \ldots, r$.

Proofs of this theorem can be found in many textbooks on linear algebra and the theory of matrices such as Finkbeiner [23], Gantmacher [26], Gohberg et al. [41], Horn and Johnson [53], Lancaster and Tismenetsky [70], Smith [94].

Here we present a simple and short proof of existence of the Jordan form of a linear transformation on a finite dimensional vector space over the complex numbers. The proof is taken from [32]. It is based on an algorithm that allows one to build the Jordan form of a linear transformation A on an n-dimensional space if the Jordan form of A restricted to an $(n-1)$-dimensional invariant subspace is known.

Let A be a linear transformation on C^n. A subspace of C^n is called *cyclic* if it is of the form

$$\text{Span}\,\{\varphi, (A - \lambda I)\varphi, \ldots, (A - \lambda I)^{m-1}\varphi\}$$

with

$$(A - \lambda I)^{m-1}\varphi \neq 0 \quad \text{and} \quad (A - \lambda I)^m \varphi = 0.$$

Such a subspace is A-invariant and has dimension m. This follows immediately from the fact that if for some r $(r = 0, 1, \ldots, m-1)$,

$$c_r(A - \lambda I)^r \varphi + \cdots + c_{m-1}(A - \lambda I)^{m-1}\varphi = 0 \quad \text{and } c_r \neq 0,$$

then, after multiplying on the left with $(A - \lambda I)^{m-r-1}$, we obtain $c_r(A - \lambda)^{m-1}\varphi = 0$, which is a contradiction.

The argument of the proof can be reduced to two cases. In one case there is a vector g outside of an $(n-1)$-dimensional A-invariant subspace $\mathcal{F}$ of $\mathbf{C}^n$ such that $Ag = 0$. In this case

$$\mathbf{C}^n = \mathcal{F} \dotplus \mathrm{Span}\{g\}$$

and the proof is clear from the induction hypothesis on $\mathcal{F}$. The difficult case is when no such g exists. It turns out that one of the cyclic subspaces of the restriction of A to $\mathcal{F}$ is replaced by a cyclic subspace of A in $\mathbf{C}^n$ which is larger by one dimension while keeping the other cyclic subspaces unchanged.

We make the following observation: Assume

$$\mathcal{W} = \mathcal{H} \dotplus \mathrm{Span}\{\varphi, A\varphi, \ldots, A^{m-1}\varphi\}$$

with

$$A^{m-1}\varphi \neq 0, \quad A^m\varphi = 0,$$

where $\mathcal{H}$ is an A-invariant subspace of $\mathbf{C}^n$ and $A^m\mathcal{H} = \{0\}$. Given $h \in \mathcal{H}$, let $\varphi' = \varphi + h$. Then

$$\mathcal{W} = \mathcal{H} \dotplus \mathrm{Span}\{\varphi', A\varphi', \ldots, A^{m-1}\varphi'\},$$

with

$$A^{m-1}\varphi' \neq 0, \quad A^m\varphi' = 0.$$

This observation follows immediately from the fact that, if a linear combination of the vectors φ', $A\varphi'$, $\ldots$, $A^{m-1}\varphi'$ belongs to $\mathcal{H}$, then the same linear combination of vectors φ, $A\varphi$, $\ldots$, $A^{m-1}\varphi$ also belongs to $\mathcal{H}$.

Proof of Theorem A.2.1 (existence). The proof proceeds by induction on the dimension n of $\mathbf{C}^n$.

The decomposition is trivial if the space is $\mathbf{C}$. Assume that it holds for spaces of dimension $n-1$. Let the space be $\mathbf{C}^n$. First we assume that A is singular. Then Range A has dimension at most $n-1$. Let $\mathcal{F}$ be an $(n-1)$-dimensional space of $\mathbf{C}^n$ which contains Range A. Since $A\mathcal{F} \subseteq \mathrm{Range}\, A \subseteq \mathcal{F}$, the induction hypothesis guarantees that $\mathcal{F}$ is the direct sum of cyclic subspaces

$$\mathcal{M}_j = \mathrm{Span}\left\{\varphi_j, (A - \lambda_j I)\varphi_j, \ldots, (A - \lambda_j I)^{m_j - 1}\varphi_j\right\}, \quad 1 \leq j \leq k.$$

The subscripts are chosen so that $\dim \mathcal{M}_j \leq \dim \mathcal{M}_{j+1}$, $1 \leq j \leq k-1$.

Define $S = \{j \mid \lambda_j = 0\}$. Take $g \notin \mathcal{F}$. We claim that Ag is of the form

$$Ag = \sum_{j \in S} \alpha_j \varphi_j + Ah, \quad h \in \mathcal{F}, \tag{A.2.2}$$

if $S \neq \emptyset$. If $S = \emptyset$, then $Ag = Ah$. To verify (A.2.2), note that $Ag \in \operatorname{Range} A \subseteq \mathcal{F}$. Hence Ag is a linear combination of vectors of the form

$$(A - \lambda_j I)^q \varphi_j, \quad 0 \leq q \leq m_j - 1, \quad 1 \leq j \leq k.$$

For $\lambda_j = 0$, the vectors $A\varphi_j, \ldots, A^{m-1}\varphi_j$ are in $A(\mathcal{F})$. If $\lambda_j \neq 0$, then from $(A - \lambda_j I)^{m_j} \varphi_j = 0$ and the binomial theorem we find that φ_j is of the form

$$\sum_{m=1}^{m_j} b_m A^m \varphi_j.$$

Thus all vectors $(A - \lambda_j I)^q \varphi_j$ belongs to $A\mathcal{F}$ and equation (A.2.2) holds. Let $g_1 = g - h$, where h is given in (A.2.2). Since $g \notin \mathcal{F}$ and $h \in \mathcal{F}$, we have $g_1 \notin \mathcal{F}$ and from equation (A.2.2),

$$Ag_1 = \sum_{j \in S} \alpha_j \varphi_j. \tag{A.2.3}$$

If $Ag_1 = 0$, then $\operatorname{Span}\{g_1\}$ is cyclic and $\mathbb{C}^n = \mathcal{F} \dotplus \operatorname{span}\{g_1\}$.

Suppose $Ag_1 \neq 0$. Let p be the largest of the integers j in (A.2.3) for which $\alpha_j \neq 0$. Then for $\tilde{g} = g_1/\alpha_p$,

$$A\tilde{g} = \varphi_p + \sum_{j \in S, j < p} \frac{\alpha_j}{\alpha_p} \varphi_j. \tag{A.2.4}$$

Define

$$\mathcal{H} = \sum_{j \in S, j < p} \dotplus \mathcal{M}_j.$$

The subspace $\mathcal{H}$ is A-invariant and since $\dim \mathcal{M}_j \leq \dim \mathcal{M}_p$, $j < p$, it follows that $A^{m_p}(\mathcal{H}) = \{0\}$. Thus by the observation (before the proof) applied to $\mathcal{H} \dotplus \mathcal{M}_p$ and equality (A.2.4), we have

$$\mathcal{H} \dotplus \mathcal{M}_p = \mathcal{H} \dotplus \operatorname{Span}\{A\tilde{g}, \ldots, A^m \tilde{g}\}.$$

Hence,

$$\mathcal{F} = \left(\sum_{j \neq p} \dotplus \mathcal{M}_j \right) \dotplus \operatorname{Span}\{A\tilde{g}, \ldots, A^m \tilde{g}\}.$$

Since $\tilde{g} \notin \mathcal{F}$,

$$\mathbb{C}^n = \mathcal{F} \dotplus \operatorname{Span}\{\tilde{g}\} = \left(\sum_{j \neq p} \dotplus \mathcal{M}_j \right) \dotplus \operatorname{Span}\{A\tilde{g}, \ldots, A^m \tilde{g}\}.$$

This completes the proof of the theorem under the assumption that A is singular.

For the general case, let μ be an eigenvalue of A. Then $A - \mu I$ is singular and by the above result applied to $A - \mu I$, it follows that $\mathbb{C}^n$ is the direct sum of cyclic subspaces for A. $\square$

This proof shows how to extend a Jordan form for A on an $(n-1)$-dimensional invariant subspace $\mathcal{F}$ to an n-dimensional A-invariant subspace containing $\mathcal{F}$.

The matrix (A.2.1) is called the *Jordan form* of A (or possibly the *complex* Jordan form of A). Note that a matrix which is represented by a Jordan form in some basis can be transformed by a similarity transformation to the same Jordan form in the standard orthonormal basis (cf. Exercise 6).

Example A.2.2. *Let*

$$A = \begin{bmatrix} 0 & 1 & 0 & 0 & a \\ 0 & 0 & 0 & 0 & b \\ 0 & 0 & 0 & 1 & c \\ 0 & 0 & 0 & 0 & d \\ 0 & 0 & 0 & 0 & 0 \end{bmatrix}, \quad a, b, c, d \in \mathbb{C}.$$

Then

$$Ae_2 = e_1, \quad Ae_1 = 0, \quad Ae_4 = e_3, \quad Ae_3 = 0.$$

We take

$$\mathcal{F} = \mathrm{Span}\{e_1, e_2, e_3, e_4\} = \mathrm{Span}\{e_2, Ae_2\} \dotplus \mathrm{Span}\{e_4, Ae_4\}.$$

Now $e_5 \notin \mathcal{F}$ and

$$Ae_5 = ae_1 + be_2 + ce_3 + de_4 = be_2 + de_4 + A(ae_2 + ce_4).$$

If $d \neq 0$, take $\widetilde{g} = e_5 - ae_2 - ce_4/d$. Then

$$A\widetilde{g} = e_4 + \frac{b}{d} e_2, \quad A^2 \widetilde{g} = e_3 + \frac{b}{d} e_1$$

and

$$\mathbb{C}^5 = \mathrm{Span}\{e_2, Ae_2\} \dotplus \mathrm{Span}\{\widetilde{g}, A\widetilde{g}, A^2\widetilde{g}\}.$$

If $d = 0$ and $b \neq 0$, take $\widetilde{g} = e_5 - ae_2 - ce_4/b$. Then $A\widetilde{g} = e_2$, and $Ae_2 = e_1$. Hence

$$\mathbb{C}^5 = \mathrm{Span}\{\widetilde{g}, A\widetilde{g}, A^2\widetilde{g}\} \dotplus \mathrm{Span}\{e_4, Ae_4\}.$$

Finally, if $d = b = 0$, take $\widetilde{g} = e_5 - ae_2 - ce_4$. Then $A\widetilde{g} = 0$ and

$$\mathbb{C}^5 = \mathrm{Span}\{e_2, Ae_2\} \dotplus \mathrm{Span}\{e_4, Ae_4\} \dotplus \mathrm{Span}\{\widetilde{g}\}.$$ $\square$

This example is taken from [32].

The following terminology is used consistently throughout this book: The numbers $\lambda_1, \lambda_2, \ldots, \lambda_r$ in the Jordan form (A.2.1) of A are the *eigenvalues* of A (not necessarily distinct), and the set of all eigenvalues is the *spectrum* of A, denoted by $\sigma(A)$.

For an eigenvalue λ_0 of A, the *geometric multiplicity*, $\gamma(\lambda_0)$ is the number of Jordan blocks in the Jordan form of A in which λ_0 appears as an eigenvalue (i.e., in which $\lambda_j = \lambda_0$). The geometric multiplicity of an eigenvalue can also be defined as the dimension of the corresponding eigenspace.

The *algebraic multiplicity* , $\alpha(\lambda_0)$, of eigenvalue λ_0 is the sum of the sizes of Jordan blocks in which λ_0 appears as an eigenvalue. It is easily verified that

$$\gamma(\lambda_0) = \dim \operatorname{Ker}(A - \lambda_0 I), \quad \alpha(\lambda_0) = \dim \operatorname{Ker}(A - \lambda_0 I)^n. \qquad \text{(A.2.5)}$$

The *partial multiplicities* of the eigenvalue λ_0 are just the sizes of Jordan blocks with the eigenvalue λ_0 in the Jordan form of A. If several such Jordan blocks have the same size, then the size is repeated as a partial multiplicity as many times as the number of these Jordan blocks indicates. Thus, the number of partial multiplicities is equal to the geometric multiplicity.

A chain of vectors $v_1, \ldots, v_k \in \mathbf{C}^n$ is called a *Jordan chain* of an $n \times n$ matrix A corresponding to its eigenvalue λ_0 if $v_1 \neq 0$ and

$$(A - \lambda I)v_1 = 0, \quad (A - \lambda I)v_j = v_{j-1}, \quad j = 2, 3, \ldots, k.$$

The positive integer k is called the *length* of the Jordan chain $v_1, \ldots, v_k$.

Generally speaking, determination of the Jordan form of a matrix is a difficult problem, both theoretically and numerically, unless a special structure is present. One such result, that determines the Jordan form of a matrix with a special structure, is the next technical lemma, which is however of independent interest, and is used in Chapter 14.

Lemma A.2.3. *Let $J_1, \ldots, J_k$ be the (upper triangular) nilpotent Jordan blocks of sizes $\alpha_1 \geq \alpha_2 \geq \cdots \geq \alpha_k$ respectively. Let*

$$J_0 = \begin{bmatrix} J_1 & 0 & \cdots & 0 \\ 0 & J_2 & \cdots & 0 \\ \vdots & & \cdots & \vdots \\ 0 & 0 & \cdots & J_k \end{bmatrix}$$

and

$$\Phi = \begin{bmatrix} J_0 & \Phi_0 \\ 0 & J_0^T \end{bmatrix}$$

where $\Phi_0 = [\phi_{ij}]_{i,j=1}^{\alpha}$ is a matrix of size $\alpha \times \alpha$ and $\alpha = \alpha_1 + \cdots + \alpha_k$. Let

$$\beta_i = \alpha_1 + \cdots + \alpha_i, \quad i = 1, \ldots, k$$

and suppose that the $k \times k$ submatrix $\Psi = [\phi_{\beta_i \beta_j}]_{i,j=1}^k$ of Φ_0 is invertible and moreover is such that the principal submatrices $[\phi_{\beta_i \beta_j}]_{i,j=1}^\ell$, $\ell = 1, 2, \ldots, k-1$ are invertible as well. Then the sizes of Jordan blocks in the Jordan form of Φ are $2\alpha_1, \ldots, 2\alpha_k$.

Proof. The proof will proceed in several steps. We start with a general fact of independent interest:

Step 1. An $n \times n$ matrix $A = [a_{i,j}]_{i,j=1}^n$, $a_{i,j} \in C$ has the property that the consecutive principal submatrices, including A itself, $[a_{i,j}]_{i,j=1}^k$, $k = 1, 2, \ldots, n$, are invertible if and only if A admits a factorization $A = LU$, where L is an invertible lower triangular matrix, and U is an invertible upper triangular matrix.

Proof of Step 1. The "if" part is clear by inspection, because the consecutive (starting with the top left corner) principal submatrices of L and of U are invertible in view of the triangular property of L and U. For the "only if" part, argue by induction on n, the case $n = 1$ being trivial. Partition

$$A = \begin{bmatrix} A_0 & y \\ x & z \end{bmatrix},$$

where A_0 is $(n-1) \times (n-1)$, x is an $(n-1)$-component row, y is an $(n-1)$-component column, and $z \in C$. Then by the induction hypothesis we have $A_0 = L_0 U_0$ for some invertible lower and upper triangular matrices L_0 and U_0, respectively. Let

$$\ell_1 = x U_0^{-1}, \qquad u_1 = L_0^{-1} y,$$

and ℓ_2, u_2 any pair of nonzero complex numbers such that

$$\ell_2 u_2 = z - x A_0^{-1} y.$$

Note that the formula

$$\begin{bmatrix} I_{n-1} & 0 \\ -x A_0^{-1} & 1 \end{bmatrix} \begin{bmatrix} A_0 & y \\ x & z \end{bmatrix} = \begin{bmatrix} A_0 & y \\ 0 & z - x A_0^{-1} y \end{bmatrix}$$

guarantees that $z - x A_0^{-1} y \neq 0$, and therefore the choice of ℓ_2 and u_2 as above is possible. Now a straightforward verification shows that

$$\begin{bmatrix} A_0 & y \\ x & z \end{bmatrix} = \begin{bmatrix} L_0 & 0 \\ \ell_1 & \ell_2 \end{bmatrix} \begin{bmatrix} U_0 & u_1 \\ 0 & u_2 \end{bmatrix},$$

and we are done. $\qquad\qquad\square$

In the subsequent steps, similarity transformations are to be applied to the matrix Φ, to eventually obtain a matrix similar to Φ for which the sizes of the Jordan blocks are obviously $2\alpha_1, \ldots, 2\alpha_k$.

Step 2. Let $i \notin \{\beta_1, \beta_2, \ldots, \beta_k\}$ be an integer; $1 \leq i \leq \alpha$. Let U_{i1} be the $\alpha \times \alpha$ matrix such that all its rows (except for the $(i+1)$-th row) are zeros, and row number $(i+1)$ is $[\phi_{i1} \, \phi_{i2} \ldots \phi_{i\alpha}]$. Put

$$S_{i1} = \begin{bmatrix} I & U_{i1} \\ 0 & I \end{bmatrix};$$

then

$$S_{i1} \Phi S_{i1}^{-1} = \begin{bmatrix} J_0 & \Phi_{i0} \\ 0 & J_0^T \end{bmatrix},$$

where the i-th row of Φ_{i0} is zero, and all other rows (except for the $(i+1)$-th) of Φ_{i0} are the same as in Φ_0. Note also that the submatrix Ψ is the same in Φ_{i0} and in Φ_0. Applying this transformation sequentially for every $i \in \{1, \ldots, \alpha\} \setminus \{\beta_1, \ldots, \beta_k\}$, starting with the smallest index i in the set $\{1, \ldots, \alpha\} \setminus \{\beta_1, \ldots, \beta_k\}$, we find that Φ is similar to a matrix of the form

$$\Phi_1 = \begin{bmatrix} J_0 & V \\ 0 & J_0^T \end{bmatrix},$$

where the i-th row of V is zero for $i \notin \{\beta_1, \ldots, \beta_k\}$, and $v_{\beta_p \beta_q} = \phi_{\beta_p \beta_q}$, $p, q = 1, \ldots, k$, where $V = (v_{ij})_{i,j=1}^{\alpha}$.

Step 3. We show now that, by applying a similarity transformation to Φ_1, it is possible to make $v_{ij} = 0$ for $i \notin \{\beta_1, \ldots, \beta_k\}$ or $j \notin \{\beta_1, \ldots, \beta_k\}$, without changing the matrix Ψ.

Let $V_j = [v_{ij}]_{i=1}^{\alpha}$ be the j-th column of V. For fixed $j \notin \{\beta_1, \ldots, \beta_k\}$ we have $v_{ij} = 0$ for $i \notin \{\beta_1, \ldots, \beta_k\}$, and, since Ψ is invertible, there exist $\sigma_{j,1}, \ldots, \sigma_{j,k} \in \mathbb{C}$ such that

$$V_j + \sum_{i=1}^{k} \sigma_{j,i} V_{\beta_i} = 0.$$

Let $S_{2j} = \begin{bmatrix} I_\alpha & 0 \\ 0 & I_\alpha + U_{2j} \end{bmatrix}$, where, but for the j-th column, U_{2j} consists of zeros, and the i-th entry in the j-th column of U_{2j} is σ_{j,β_m} if $i = \beta_m$ for some m and zero otherwise. Then

$$S_{2j}^{-1} \Phi_1 S_{2j} = \begin{bmatrix} J_0 & V - Z_1 \\ 0 & J_0^T - Z_2 \end{bmatrix},$$

with the following structure of the matrices Z_1 and Z_2:

$$Z_1 = [0 \, \ldots \, 0 V_j 0 \, \ldots \, 0], \quad \text{where } V_j \text{ is in the } j\text{-th position,}$$

$$Z_2 = [0 \, \ldots \, 0 Z_{2,j-1} 0 \, \ldots \, 0],$$

where

$$Z_{2,j-1} = \langle \underbrace{0, \ldots, 0, -\sigma_{j,1}}_{\alpha_1}, \underbrace{0, \ldots, 0, -\sigma_{j,2}}_{\alpha_2}, \ldots, \underbrace{0, \ldots, 0, -\sigma_{j,k}}_{\alpha_k} \rangle$$

is an $\alpha \times 1$ column in the $(j-1)$-th position in Z_2. If $j = \beta_m + 1$ for some m, then we put $Z_2 = 0$. It is easy to see that $S_{2j}^{-1}\Phi_1 S_{2j}$ can be reduced by a similarity transformation

$$S_{2j}^{-1}\Phi_1 S_{2j} \quad \mapsto \quad \begin{bmatrix} I_\alpha & 0 \\ 0 & U_{3j}^{-1} \end{bmatrix} S_{2j}^{-1}\Phi_1 S_{2j} \begin{bmatrix} I_\alpha & 0 \\ 0 & U_{3j}, \end{bmatrix}$$

with a suitable invertible matrix U_{3j} to the form $\begin{bmatrix} J_0 & W_{1,j} \\ 0 & J_0^T \end{bmatrix}$, where the m-th column of $W_{1,j}$ coincides with the m-th column of $V - Z_1$ for $m \geq j$ and for $m \in \{\beta_1, \ldots, \beta_k\}$. (The matrix U_{3j} affects column operations that transform $J_0^T - Z_2$ into J_0^T.) Applying the similarity

$$\Phi_1 \quad \mapsto \quad \begin{bmatrix} I_\alpha & 0 \\ 0 & U_{3j}^{-1} \end{bmatrix} S_{2j}^{-1}\Phi_1 S_{2j} \begin{bmatrix} I_\alpha & 0 \\ 0 & U_{3j}, \end{bmatrix}$$

sequentially for every $j \notin \{\beta_1, \ldots, \beta_k\}$, starting with the largest index j in the set

$$\{1, 2, \ldots, \alpha\} \setminus \{\beta_1, \ldots, \beta_k\} \tag{A.2.6}$$

and finishing with the smallest index j in the set (A.2.6), we find that Φ_1 is similar to the matrix

$$\Phi_2 := \begin{bmatrix} J_0 & W_2 \\ 0 & J_0^T \end{bmatrix},$$

where the (β_i, β_j)-entries of W_2 $(i, j = 1, \ldots, k)$ form the matrix Ψ, and all other entries of W_2 are zeros.

Step 4. We replace the invertible submatrix Ψ in W_2 by an invertible diagonal matrix.

By Step 1, we have $\Psi_1^{-1}\Psi = D\Psi_2$, where $\Psi_1^{-1} = (b_{ij})_{i,j=1}^k$, $b_{ij} = 0$ for $i < j$, is a lower triangular matrix with 1s on the diagonal: $b_{11} = \cdots = b_{kk} = 1$; $\Psi_2 = (c_{ij})_{i,j=1}^k$, $c_{ij} = 0$ for $i > j$ is an upper triangular matrix with 1s on the diagonal; and D is an invertible diagonal matrix. Define the $\alpha \times \alpha$ invertible matrix $S_3 = (s_{p,q}^{(3)})_{p,q=1}^\alpha$ as follows:

$$s_{p,q}^{(3)} = \begin{cases} b_{ij} & \text{for } p = \beta_i, q = \beta_j, \text{ where } i, j = 1, \ldots, k; \\ 1 & \text{for } p = q \notin \{\beta_1, \ldots, \beta_k\}; \\ 0 & \text{in all other cases.} \end{cases}$$

Then

$$\begin{bmatrix} S_3 & 0 \\ 0 & I_\alpha \end{bmatrix} \Phi_2 \begin{bmatrix} S_3^{-1} & 0 \\ 0 & I_\alpha \end{bmatrix} = \begin{bmatrix} J_0 + Z_3 & W_3 \\ 0 & J_0^T \end{bmatrix},$$

where the (β_i, β_j) entries of W_3 $(i, j = 1, \ldots, k)$ form the upper triangular matrix $D\Psi_2$, and all other entries of W_3 are zeros; the $\alpha \times \alpha$ matrix Z_3 may contain

nonzero entries only in the positions $(\beta_j - 1, \beta_i)$ for $i < j$ and such that $\alpha_j > 1$.
It is easy to see that $\begin{bmatrix} J_0 + Z_3 & W_3 \\ 0 & J_0^T \end{bmatrix}$ is similar to

$$\Phi_3 := \begin{bmatrix} J_0 & W_3 \\ 0 & J_0^T \end{bmatrix};$$

here the hypothesis $\alpha_1 \geq \cdots \geq \alpha_k$ plays a role. Indeed, assuming for a moment for notational simplicity that $k = 2$, let the $(\beta_2 - 1, \beta_1)$ entry of Z_3 be equal to $\gamma \in \mathbb{C}$, and let T be the $\alpha_2 \times \alpha_1$ matrix

$$T = \begin{bmatrix} 0_{\alpha_2 \times (\alpha_1 - \alpha_2)} & \operatorname{diag}(\pm\gamma, \mp\gamma, \ldots, \tau\gamma, 0) \end{bmatrix}, \quad \tau \in \{\mp, \pm\}.$$

Then a computation shows that

$$\begin{bmatrix} I_{\alpha_1} & 0 & 0 \\ T & I_{\alpha_2} & 0 \\ 0 & 0 & I_\alpha \end{bmatrix} \begin{bmatrix} J_0 + Z_3 & W_3 \\ 0 & J_0^T \end{bmatrix} \begin{bmatrix} I_{\alpha_1} & 0 & 0 \\ -T & I_{\alpha_2} & 0 \\ 0 & 0 & I_\alpha \end{bmatrix} = \Phi_3.$$

Define the $\alpha \times \alpha$ invertible matrix $S_4 = (s_{p,q}^{(4)})_{p,q=1}^{\alpha}$ by the following equalities:

$$s_{p,q}^{(4)} = \begin{cases} c_{ij} & \text{for } p = \beta_i, q = \beta_j, \text{ where } i, j = 1, \ldots, k; \\ 1 & \text{for } p = q \notin \{\beta_1, \ldots, \beta_k\}; \\ 0 & \text{in all other cases.} \end{cases}$$

Then

$$\begin{bmatrix} I_\alpha & 0 \\ 0 & S_4 \end{bmatrix} \Phi_3 \begin{bmatrix} I_\alpha & 0 \\ 0 & S_4^{-1} \end{bmatrix} = \begin{bmatrix} J_0 & W_4 \\ 0 & J_0^T + Z_4 \end{bmatrix},$$

where the (β_i, β_j) entries of W_4 form the diagonal matrix D, and all other entries are zeros; the $\alpha \times \alpha$ matrix Z_4 can contain nonzero entries only in the positions $(\beta_i, \beta_j - 1)$ for $i < j$ and such that $\alpha_j > 1$. Again, one verifies that $\begin{bmatrix} J_0 & W_4 \\ 0 & J_0^T + Z_4 \end{bmatrix}$ is similar to

$$\Phi_4 := \begin{bmatrix} J_0 & W_4 \\ 0 & J_0^T \end{bmatrix}.$$

This completes the proof of Step 4.

Now, by inspection, the sizes of the Jordan blocks of Φ_4 are $2\alpha_1, \ldots, 2\alpha_k$. So the same is true for Φ, and the lemma is proved. $\quad\square$

The proof of Lemma A.2.3 is taken from [36] (see also [42]), where the result is stated under the more restrictive hypothesis that Φ_0 is positive definite.

The Jordan canonical form allows one to obtain easily many properties of matrices that relate to eigenvalues and eigenvectors. For $A \in \mathsf{C}^{n \times n}$, the *root subspace* $\mathcal{R}_{\lambda_0}(A)$ corresponding to an eigenvalue λ_0 of A is defined as follows:

$$\mathcal{R}_{\lambda_0}(A) = \{x \in \mathsf{C}^n \mid (A - \lambda_0 I)^s x = 0 \quad \text{for some positive integer } s\}. \quad \text{(A.2.7)}$$

In fact,

$$\mathcal{R}_{\lambda_0}(A) = \mathrm{Ker}\,(A - \lambda_0 I)^n,$$

where n is the size of A. In particular, we obtain that $\mathcal{R}_{\lambda_0}(A)$ is indeed a subspace. We have a direct sum decomposition:

Theorem A.2.4. *If A is an $n \times n$ matrix with the distinct eigenvalues $\lambda_1, \ldots, \lambda_k$, then C^n is a direct sum of the root subspaces $\mathcal{R}_{\lambda_j}(A)$, $j = 1, 2, \ldots, k$:*

$$\mathsf{C}^n = \mathcal{R}_{\lambda_1}(A) \dotplus \cdots \dotplus \mathcal{R}_{\lambda_k}(A).$$

For the proof observe that Theorem A.2.4 is easy if A is a Jordan form (the root subspaces are spanned by appropriate standard unit vectors), and for a general A use the similarity to a Jordan form.

More generally:

Theorem A.2.5. *If A is an $n \times n$ matrix with the distinct eigenvalues $\lambda_1, \ldots, \lambda_k$, and if $\mathcal{M}$ is an A-invariant subspace, i.e., $Ax \in \mathcal{M}$ for every $x \in \mathcal{M}$, then*

$$\mathcal{M} = (\mathcal{M} \cap \mathcal{R}_{\lambda_1}(A)) \dotplus (\mathcal{M} \cap \mathcal{R}_{\lambda_2}(A)) \dotplus \cdots \dotplus (\mathcal{M} \cap \mathcal{R}_{\lambda_k}(A)).$$

Now the same questions concerning Jordan forms arise for equivalence classes of *real* square matrices generated by real similarity transformations. Thus, our concern is now with the real vector space $\mathsf{R}^{n \times n}$. The results are similar to those above but the canonical forms are rather more complicated and, perhaps for this reason, may be less familiar.

Description of an appropriate canonical form requires the introduction of another class of matrices in standard form. For real numbers, λ and $\mu \neq 0$ define the *real Jordan block* of even size, say $2m \times 2m$ by

$$J_{2m}(\lambda \pm i\mu) = \begin{bmatrix} \lambda & \mu & 1 & 0 & \cdots & 0 & 0 \\ -\mu & \lambda & 0 & 1 & \cdots & 0 & 0 \\ 0 & 0 & \lambda & \mu & \cdots & 0 & 0 \\ 0 & 0 & -\mu & \lambda & \cdots & & \\ \vdots & \vdots & \vdots & \vdots & & 1 & 0 \\ \vdots & \vdots & \vdots & \vdots & & 0 & 1 \\ 0 & 0 & 0 & 0 & & \lambda & \mu \\ 0 & 0 & 0 & 0 & & -\mu & \lambda \end{bmatrix}.$$

Thus, there are m real 2×2 blocks on the main diagonal and 2×2 identity matrices making up the super-diagonal blocks. Clearly $J_{2m}(\lambda \pm i\mu)$ is a real matrix, and $\sigma(J_{2m}(\lambda \pm i\mu)) = \{\lambda + i\mu, \lambda - i\mu\}$.

Theorem A.2.6. *For any $A \in \mathbb{R}^{n \times n}$ there is a block-diagonal matrix $J \in \mathbb{R}^{n \times n}$ which is similar to A over the reals (i.e there is an invertible $S \in \mathbb{R}^{n \times n}$ such that $A = S^{-1}JS$) and has the form*

$$
\begin{aligned}
J \;=\; & J_{m_1}(\lambda_1) \oplus J_{m_2}(\lambda_2) \oplus \cdots \oplus J_{m_r}(\lambda_r) \\
& \oplus J_{2m_{r+1}}(\lambda_{r+1} \pm i\mu_{r+1}) \oplus \cdots \oplus J_{2m_q}(\lambda_q \pm i\mu_q), \qquad (A.2.8)
\end{aligned}
$$

where the λ_j are real and the μ_j are real and positive.

Moreover, the matrix J of (A.2.8) is uniquely determined by A up to permutation of the diagonal blocks.

Naturally, the matrix J of (A.2.8) is known as the *real Jordan form* of A. A complete proof of this theorem can be found in Chapter 12 of Gohberg et al. [41]. (Another source is Shilov [93].)

The root subspaces and Theorems A.2.4 and A.2.5 have natural analogues in real spaces. Thus, if $A \in \mathbb{R}^{n \times n}$ and if λ_0 is a real eigenvalue of A, we define the *real root subspace*

$$
\mathcal{R}_{\mathsf{R},\lambda_0}(A) = \{ x \in \mathbb{R}^n \mid (A - \lambda_0 I)^s x = 0 \quad \text{for some positive integer } s \}, \quad (A.2.9)
$$

and if $\mu \pm i\nu$ is a pair of nonreal complex conjugate eigenvalues of A, the corresponding real root subspace is defined by

$$
\begin{aligned}
& \mathcal{R}_{\mathsf{R},\mu \pm i\nu}(A) \\
& = \{ x \in \mathbb{R}^n \mid (A^2 - 2\mu A + (\mu^2 + \nu^2)I)^s x = 0 \quad \text{for some positive integer } s \}.
\end{aligned}
$$
$$(A.2.10)$$

Theorem A.2.7. *If A is a real $n \times n$ matrix with the distinct real eigenvalues $\lambda_1, \ldots, \lambda_k$ and the distinct pairs of nonreal complex conjugate eigenvalues $\mu_1 \pm i\nu_1, \ldots, \mu_\ell \pm i\nu_\ell$, then we have a direct sum decomposition*

$$
\mathbb{R}^n = \mathcal{R}_{\mathsf{R},\lambda_1}(A) \dotplus \cdots \dotplus \mathcal{R}_{\mathsf{R},\lambda_k}(A) \dotplus \mathcal{R}_{\mathsf{R},\mu_1 \pm i\nu_1}(A) \dotplus \cdots \dotplus \mathcal{R}_{\mathsf{R},\mu_\ell \pm i\nu_\ell}(A).
$$

For a proof see, for example, [93, Section 6.34].

Theorem A.2.8. *If A and $\lambda_1, \ldots, \lambda_k$, $\mu_1 \pm i\nu_1, \ldots, \mu_\ell \pm i\nu_\ell$ are as in Theorem A.2.7, and if $\mathcal{M} \subseteq \mathbb{R}^n$ is an A-invariant subspace, then*

$$
\begin{aligned}
\mathcal{M} \;=\; & (\mathcal{M} \cap \mathcal{R}_{\lambda_1}(A)) \dotplus (\mathcal{M} \cap \mathcal{R}_{\lambda_2}(A)) \dotplus \cdots \dotplus (\mathcal{M} \cap \mathcal{R}_{\lambda_k}(A)) \\
& \dotplus (\mathcal{M} \cap \mathcal{R}_{\mathsf{R},\mu_1 \pm i\nu_1}(A)) \dotplus \cdots \dotplus (\mathcal{M} \cap \mathcal{R}_{\mathsf{R},\mu_\ell \pm i\nu_\ell}(A)).
\end{aligned}
$$

A proof of Theorem A.2.8 can be found in [41], Chapters 2 and 12.

A.3 Riesz Projections

In this section we use basic complex analysis, in particular, contour integration and the theorem of residues. All this background material is standard, and is found in many undergraduate texts on complex analysis; see for example, [82], [89], [47].

Let Γ be a simple (without self-intersections), closed, rectifiable contour in the complex plane. The contour integral of a $\mathbb{C}^{n\times n}$-valued function $F(z) = [f_{i,j}(z)]_{i,j=1}^n$ defined on Γ is understood in the entry-wise sense:

$$\int_\Gamma F(z)dz = \left[\int_\Gamma f_{i,j}(z)dz\right]_{i,j=1}^n.$$

We will be particularly interested in the *resolvent functions* $F(z) = (zI - X)^{-1}$, where $X \in \mathbb{C}^{n\times n}$. Clearly, $\int_\Gamma (zI - X)^{-1}dz$ is well defined provided Γ does not intersect the spectrum of X. Furthermore, every entry of $(zI - X)^{-1}$ is a rational function of $z \in \mathbb{C}$, as readily seen from the formula

$$(zI - X)^{-1} = \frac{\text{adj}\,(zI - X)}{\det\,(zI - X)},$$

where $\text{adj}\,(zI - X)$ is the algebraic adjoint of $zI - X$. Thus, by the theorem of residues, the integral

$$P_\Gamma(X) := \frac{1}{2\pi i} \int_\Gamma (zI - X)^{-1}dz \tag{A.3.11}$$

is equal to the sum of the residues of $(zI - X)^{-1}$ inside Γ.

The next proposition and theorem are basic in operator theory, see, for example, [33, Section I.2]. We provide here elementary proofs using matrix analysis.

Proposition A.3.1. *Let $X \in \mathbb{C}^{n\times n}$ and let Λ be a subset of $\sigma(X)$. Assume that Γ is a contour such that Λ is inside Γ and $\sigma(X) \setminus \Lambda$ is outside Γ. Then $P_\Gamma(X)$ is a projection with $\text{Range}\,P_\Gamma(X)$ equal to the sum of the root subspaces of X corresponding to the eigenvalues in Λ, and $\text{Ker}\,P_\Gamma(X)$ equal to the sum of the root subspaces of X corresponding to the eigenvalues in $\sigma(X) \setminus \Lambda$.*

It is understood in this proposition that $P_\Gamma(X) = I$, resp. $P_\Gamma(X) = 0$, if $\Lambda = \sigma(X)$, resp. $\Lambda = \emptyset$. The projection $P_\Gamma(X)$ is called the *Riesz projection* of X corresponding to Γ.

Proof. Using Theorem A.2.1, we may assume without loss of generality that X is in the Jordan form. Then use the formula

$$(zI - J_m(\lambda))^{-1} = \begin{bmatrix} (z-\lambda)^{-1} & (z-\lambda)^{-2} & \cdots & (z-\lambda)^{-m} \\ 0 & (z-\lambda)^{-1} & \cdots & (z-\lambda)^{-m+1} \\ \vdots & \vdots & \ddots & \vdots \\ 0 & 0 & \cdots & (z-\lambda)^{-1} \end{bmatrix},$$

which is straightforward to verify. Now the theorem of residues gives

$$\frac{1}{2\pi i} \int_\Gamma (zI - J_m(\lambda))^{-1}dz = \begin{cases} 0 & \text{if } \lambda \text{ is outside } \Gamma, \\ I_m & \text{if } \lambda \text{ is inside } \Gamma, \end{cases}$$

and we are done. $\qquad\square$

Note that the function

$$f(\mu, X) := \|(\mu I - X)^{-1}\|$$

is continuous on the set

$$\{(\mu, X) : \mu \in \mathbb{C} \setminus \sigma(X), \ X \in \mathbb{C}^{n \times n}\} \qquad (A.3.12)$$

(with n fixed), and therefore attains its maximum on every compact subset of (A.3.12).

Theorem A.3.2. *Let $A \in \mathbb{C}^{n \times n}$, and let Γ be a simple closed rectifiable contour that does not intersect $\sigma(A)$. Then there exist positive constants ε and M which depend on A and Γ only such that for every matrix $B \in \mathbb{C}^{n \times n}$ with $\|B - A\| \leq \varepsilon$ the following properties hold:*

(a) *B has no eigenvalues on Γ;*

(b) *$\|P_\Gamma(A) - P_\Gamma(B)\| \leq M\|A - B\|$.*

Proof. The existence of $\varepsilon > 0$ such that the property (a) holds, is an easy consequence of continuity of eigenvalues of B as functions of the entries of B; in turn, the continuity of eigenvalues follows from the continuity of the roots of the characteristic equation $\det(\lambda I - B)$.

Using the definition of Riesz projections, with $\varepsilon > 0$ such that (a) holds and with B satisfying $\|B - A\| \leq \varepsilon$, we have

$$P_\Gamma(A) - P_\Gamma(B) = \frac{1}{2\pi i} \int_\Gamma (zI - A)^{-1}(A - B)(zI - B)^{-1} dz.$$

Hence, using approximation of the integral by Riemann sums and triangle inequality for the operator matrix norm,

$$\|P_\Gamma(A) - P_\Gamma(B)\|$$
$$\leq \frac{\text{length of } \Gamma}{2\pi} \left(\max_{z \in \Gamma} \|(zI - A)^{-1}\| \right) \left(\max_{z \in \Gamma} \|(zI - B)^{-1}\| \right) \|A - B\|,$$

where the maxima exist (are finite) because the set

$$\{(z, B) : z \in \Gamma, \ B \in \mathbb{C}^{n \times n} \ \text{and} \ \|B - A\| \leq \varepsilon\}$$

is a compact subset of (A.3.12). Thus, we may take

$$M = \frac{\text{length of } \Gamma}{2\pi} \left(\max_{z \in \Gamma, \ \|B - A\| \leq \varepsilon} \|(zI - B)^{-1}\| \right)^2$$

to satisfy the property (b). $\qquad \square$

A.4 Linear Matrix Equations

Theorem A.4.1. *Let A and B be (real or complex) matrices of sizes $m \times m$ and $n \times n$, respectively. Then the matrix equation*

$$AX - XB = 0$$

with the unknown (real or complex) $m \times n$ matrix X has only the trivial solution $X = 0$ if and only if $\sigma(A) \cap \sigma(B) = \emptyset$.

See, e.g, [70, Chapter 12] for a proof.

A.5 Perturbation Theory of Subspaces

Throughout this section, we let $P_{\mathcal{L}}$ be the orthogonal projection on a subspace $\mathcal{L} \subseteq \mathbb{C}^n$: $P_{\mathcal{L}}x = x$ if $x \in \mathcal{L}$ and $P_{\mathcal{L}}x = 0$ if $x \perp \mathcal{L}$.

If $\mathcal{M}, \mathcal{N}$ are subspaces in $\mathbb{C}^n$, we define the *gap* between $\mathcal{M}$ and $\mathcal{N}$ by

$$\theta(\mathcal{M}, \mathcal{N}) := \|P_{\mathcal{M}} - P_{\mathcal{N}}\|. \tag{A.5.13}$$

The set of subspaces in $\mathbb{C}^n$ is a compact complete metric space in the metric defined by the function $\theta(\mathcal{M}, \mathcal{N})$; see, for example, [41, Chapter 13] for details and more information on the gap.

We first collect some elementary and well-known properties of the gap. We begin with convergence in the gap matric, which can be conveniently expressed in terms of convergence of vectors.

Theorem A.5.1. *Let $\mathcal{M}_j$, $j = 1, 2, \ldots,$, be a sequence of subspaces in $\mathbb{C}^n$ such that*

$$\lim_{j \to \infty} \theta(\mathcal{M}_j, \mathcal{N}) = 0$$

for some subspace $\mathcal{N}$. Then $\mathcal{N}$ consists of exactly those vectors $x \in \mathbb{C}^n$ for which there exists a sequence of vectors $x_j \in \mathbb{C}^n$, $j = 1, 2, \ldots,$ such that

$$x_j \in \mathcal{M}_j, \quad j = 1, 2, \ldots, \quad \text{and} \quad \lim_{j \to \infty} x_j = x.$$

A proof is given in [41, Chapter 13], for example.

Proposition A.5.2. *If $Q_{\mathcal{M}}$ and $Q_{\mathcal{N}}$ are projections, not necessarily orthogonal, on the subspaces $\mathcal{M} \subseteq \mathbb{C}^n$ and $\mathcal{N} \subseteq \mathbb{C}^n$ respectively, then*

$$\theta(\mathcal{M}, \mathcal{N}) \leq \|Q_{\mathcal{M}} - Q_{\mathcal{N}}\|.$$

For the proof see [40, Proposition III.A.3] or [41, Theorem 13.1.1].

Proposition A.5.3. *Let $X \in \mathsf{C}^{m \times n}$. Then there exist positive constants K, ε such that*

$$\theta(\mathrm{Ker}\, X, \mathrm{Ker}\, Y) \leq K\|X - Y\|, \qquad \theta(\mathrm{Range}\, X, \mathrm{Range}\, Y) \leq K\|X - Y\|$$

for every $Y \in \mathsf{C}^{m \times n}$ satisfying $\|X - Y\| < \varepsilon$ and $\mathrm{rank}\, X = \mathrm{rank}\, Y$.

For the proof see, for example, [40, Lemma III.5.4] or [41, Theorem 13.5.1].

Proposition A.5.4. *Assume*

$$\theta(\mathcal{M}, \mathcal{N}) < 1. \tag{A.5.14}$$

Then

$$\dim \mathcal{M} = \dim \mathcal{N} \tag{A.5.15}$$

and $P_{\mathcal{M}}(\mathcal{N}) = \mathcal{M}$, $P_{\mathcal{N}}(\mathcal{M}) = \mathcal{N}$.

Proof. If $\dim \mathcal{M} < \dim \mathcal{N}$ then there exists an $x \in \mathcal{N} \cap (\mathcal{M}^{\perp})$ with $\|x\| = 1$. But then

$$\|(P_{\mathcal{M}} - P_{\mathcal{N}})x\| = \|x\| = 1 \quad \text{and} \quad \|P_{\mathcal{M}} - P_{\mathcal{N}}\| \geq 1,$$

which contradicts (A.5.14), and (A.5.15) follows. If $P_{\mathcal{M}}(\mathcal{N}) \neq \mathcal{M}$ then, using the already proved property (A.5.15), we see that there exists an $x \in \mathcal{N}$ with $\|x\| = 1$ such that $P_{\mathcal{M}}x = 0$, i.e., $x \perp \mathcal{M}$. Then we obtain a contradiction with (A.5.14) as above.   □

Two strictly increasing sequences of subspaces of C^n,

$$\mathcal{M}_1 \subseteq \mathcal{M}_2 \subseteq \cdots \subseteq \mathcal{M}_k, \quad \mathcal{M}_j \neq \mathcal{M}_{j+1}, \quad \text{and}$$
$$\mathcal{N}_1 \subseteq \mathcal{N}_2 \subseteq \cdots \subseteq \mathcal{N}_k, \quad \mathcal{N}_j \neq \mathcal{N}_{j+1},$$

are said to be δ-*close* if

$$\|P_{\mathcal{M}_j} - P_{\mathcal{N}_j}\| < \delta < 1 \qquad \text{for } j = 1, 2, \ldots, k.$$

The next result is [40, Lemma III.5.2].

Theorem A.5.5. *Given $\varepsilon > 0$ there exists $\delta > 0$ which depends on ε and on n only such that for every pair of δ-close sequences of subspaces $\{\mathcal{M}_j\}_{j=1}^k$ and $\{\mathcal{N}_j\}_{j=1}^k$ of C^n, there exists an invertible $n \times n$ matrix S (depending on the sequences) for which*

$$S(\mathcal{M}_1) = \mathcal{N}_1, \quad \ldots, \quad S(\mathcal{M}_k) = \mathcal{N}_k, \quad \text{and} \quad \|I - S\| < \varepsilon.$$

Proof. We use induction on k. Put $S_1 = I - (P_{\mathcal{M}_1} - P_{\mathcal{N}_1})$ and observe that $S_1(\mathcal{M}_1) = \mathcal{N}_1$. (Use Proposition A.5.4.) Then $\|P_{\mathcal{M}_1} - P_{\mathcal{N}_1}\| < \delta < 1$ implies $\|S_1 - I\| < \delta$, and we are done in the case $k = 1$. Note also that

$$\|S_1^{-1} - I\| < \frac{\delta}{1 - \delta}.$$

Now consider the general case. Since $S_1 P_{\mathcal{M}_j} S_1^{-1}$ is a (not necessarily orthogonal) projection on $S_1(\mathcal{M}_j)$, we have by Proposition A.5.2

$$\theta(S_1(\mathcal{M}_j), \mathcal{N}_j) \leq \|S_1 P_{\mathcal{M}_j} S_1^{-1} - P_{\mathcal{N}_j}\|,$$

and therefore

$$\theta(S_1(\mathcal{M}_j), \mathcal{N}_j) \leq \|S_1 P_{\mathcal{M}_j}(S^{-1} - I)\| + \|(S_1 - I)P_{\mathcal{M}_j}\| + \|P_{\mathcal{M}_j} - P_{\mathcal{N}_j}\|$$
$$< (1 + \delta)\frac{\delta}{1 - \delta} + \delta + \delta = \delta\left(\frac{3 - \delta}{1 - \delta}\right).$$

$$(A.5.16)$$

By the induction hypothesis there is a $\widetilde{\delta} > 0$ and an invertible linear transformation $\widetilde{S} : \mathcal{N}_k \to \mathcal{N}_k$ with the properties

$$\widetilde{S}(S_1(\mathcal{M}_j)) = \mathcal{N}_j, \quad j = 1, 2, \ldots, k - 1,$$

and $\|I - \widetilde{S}\| < \frac{1}{2}\varepsilon$, provided that

$$\|P_{S_1(\mathcal{M}_j)} - P_{\mathcal{N}_j}\| < \widetilde{\delta}, \quad j = 1, 2, \ldots, k - 1. \quad (A.5.17)$$

Here S_1 is a linear transformation such that $S_1(\mathcal{M}_k) = \mathcal{N}_k$ and $\|S_1 - I\| = \|P_{\mathcal{M}_k} - P_{\mathcal{N}_k}\|$. The existence of S_1 is guaranteed by the case $k = 1$ which has already been considered.

Define the linear transformation $S : \mathbb{C}^n \to \mathbb{C}^n$ by $S\,|_{\mathcal{M}_k} = \widetilde{S}S_1\,|_{\mathcal{M}_k}$ and $S\,|_{(\mathcal{M}_k)^\perp} = I$. Then, given (A.5.17) and

$$\|P_{\mathcal{M}_k} - P_{\mathcal{N}_k}\| < \delta,$$

we have

$$\|I - S\| = \|I - S\,|_{\mathcal{M}_k}\| \leq \|(I - \widetilde{S}S_1\,|_{\mathcal{M}_k}\| + \|I - S_1\| < \frac{1}{2}\varepsilon(1 + \delta) + \delta.$$

Using (A.5.16) we see that it is sufficient to choose $\delta > 0$ so that

$$\delta\left(\frac{3 - \delta}{1 - \delta}\right) < \widetilde{\delta} \quad \text{and} \quad \delta < \min\left\{\frac{\varepsilon}{2 + \varepsilon}, 1\right\}. \qquad \square$$

A closer inspection of the proof of Theorem A.5.5 reveals that, in fact, one can choose $\delta = \varepsilon/(12^{n-1})$ (provided $\varepsilon \leq 1/2$). If the sequences of subspaces are of length not exceeding k, then one can take $\delta = \varepsilon/(12^{k-1})$.

As an immediate corollary from Theorem A.5.5 we obtain:

Corollary A.5.6. *Let $\mathcal{L}_1 \subseteq \mathcal{L}_2$ be two subspaces of $\mathbb{C}^n$, and let $x_1, \ldots, x_k$ be a basis in some direct complement to $\mathcal{L}_1$ in $\mathcal{L}_2$. Then for every $\varepsilon > 0$ there exists $\delta > 0$ which depends on ε, n, and the basis $\{x_j\}_{j=1}^k$, such that if $\widehat{\mathcal{L}}_1 \subseteq \widehat{\mathcal{L}}_2$ are subspaces of $\mathbb{C}^n$ with*

$$\theta(\widehat{\mathcal{L}}_1, \mathcal{L}_1) < \delta, \quad \theta(\widehat{\mathcal{L}}_2, \mathcal{L}_2) < \delta,$$

then for some basis $\{x'_j\}^k_{j=1}$ *in a direct complement to* $\widehat{\mathcal{L}}_1$ *in* $\widehat{\mathcal{L}}_2$ *the inequalities*

$$\|x_1 - x'_1\| < \varepsilon, \ldots, \|x_k - x'_k\| < \varepsilon$$

hold.

In particular, under the hypotheses of Corollary A.5.6, any pair of direct complements of $\mathcal{L}_1$ in $\mathcal{L}_2$, and of $\widehat{\mathcal{L}}_1$ in $\widehat{\mathcal{L}}_2$, have the same dimension.

A.6   Diagonal Forms for Matrix Polynomials and Analytic Matrix Functions

Some necessary background material from the theory of matrix polynomials and analytic matrix functions will be surveyed in this section. The concepts and results presented here are well-known, and proofs of most of the results can be found in [70, Chapter 7] and [39, Chapters S1, S6], for example.

Consider an $n \times n$ matrix $A(\lambda)$ whose elements are polynomials in the independent variable λ with complex coefficients:

$$A(\lambda) = \begin{bmatrix} a_{1,1}(\lambda) & a_{1,2}(\lambda) & \cdots & a_{1,n}(\lambda) \\ a_{2,1}(\lambda) & a_{2,2}(\lambda) & \cdots & a_{2,n}(\lambda) \\ \vdots & \vdots & \ddots & \vdots \\ a_{n,1}(\lambda) & a_{n,2}(\lambda) & \cdots & a_{n,n}(\lambda) \end{bmatrix}. \tag{A.6.18}$$

Such a matrix $A(\lambda)$ is known as a *matrix polynomial*, or a *λ-matrix*. (In the literature, the concept of matrix polynomials includes rectangular matrices, but in this book we work only with square matrix polynomials.) The *degree* of $A(\lambda)$ is defined to be that of the scalar polynomial of largest degree among the entries of $A(\lambda)$. Addition and multiplication of $n \times n$ matrices (for a fixed n) is well-defined in the usual way.

An $n \times n$ matrix polynomial $A(\lambda)$ is said to be *invertible* if $A(\lambda)B(\lambda) = I_n$ for all λ and some matrix polynomial $B(\lambda)$ (the *inverse* of $A(\lambda)$). It is easy to see that $A(\lambda)$ is invertible if and only if $\det A(\lambda)$ is a nonzero constant, i.e., a (scalar) polynomial of degree zero. Indeed, the "if" part is clear in view of the formula

$$A(\lambda_0)^{-1} = \frac{\mathrm{adj}\, A(\lambda_0)}{\det A(\lambda_0)}, \quad \lambda_0 \in \mathbb{C},$$

where $\mathrm{adj}\, A(\lambda_0)$ is the algebraic adjoint of the matrix $A(\lambda_0)$. For the "only if" part, apply the determinant function to both sides of the equality $A(\lambda)B(\lambda) = I_n$ to conclude that $\det A(\lambda)$ is invertible as a scalar polynomial, and therefore must be a nonzero constant.

We say that a matrix polynomial $A(\lambda)$ is *unimodular* if $\det A(\lambda)$ is a nonzero constant. Thus, $A(\lambda)$ is unimodular if and only if $A(\lambda)$ is invertible.

Two $n \times n$ matrix polynomials $A(\lambda)$ and $B(\lambda)$ are said to be *equivalent* if

$$A(\lambda) = P(\lambda)B(\lambda)Q(\lambda) \qquad (A.6.19)$$

for some unimodular matrix polynomials $P(\lambda)$ and $Q(\lambda)$. Clearly, polynomial equivalence is an equivalence relation (it is reflexive, symmetric, and transitive).

Theorem A.6.1. *Every $n \times n$ matrix polynomial $A(\lambda)$ is equivalent to a diagonal matrix polynomial of the form*

$$D(\lambda) = \mathrm{diag}\, [\; i_1(\lambda) \quad i_2(\lambda) \quad \ldots \quad i_r(\lambda) \quad 0 \quad 0 \quad \ldots \quad 0 \;], \qquad (A.6.20)$$

where $i_1(\lambda), \ldots, i_r(\lambda)$ are scalar polynomials with complex coefficients and leading coefficient 1, and with the property that $i_j(\lambda)$ is divisible by $i_{j-1}(\lambda)$, for $j = 2, \ldots, r$. Moreover, the parameter

$$r := \max\{\mathrm{rank}\, A(\lambda_0) \,:\, \lambda_0 \in \mathsf{C}\}$$

and the polynomials $i_1(\lambda), \ldots, i_r(\lambda)$ are uniquely determined by $A(\lambda)$.

The form (A.6.20) is called the *Smith form* of $A(\lambda)$ and the polynomials $i_1(\lambda), \ldots, i_r(\lambda)$ are known as the *invariant polynomials* of $A(\lambda)$. Those complex numbers λ_0 for which $\mathrm{rank}\, A(\lambda_0) < r$, i.e., λ_0 is a root of at least one of the invariant polynomials, are the *eigenvalues* of $A(\lambda)$.

For every index j such that $i_j(\lambda)$ is not identically equal to 1, write the complete factorization

$$i_j(\lambda) = \prod_{k=1}^{\alpha_j} (\lambda - \lambda_{k,j})^{\beta_{k,j}},$$

where for every j, the complex numbers $\lambda_{1,j}, \ldots, \lambda_{\alpha_j,j}$ are distinct, and $\beta_{k,j}$ are positive integers. The factors $(\lambda - \lambda_{k,j})^{\beta_{k,j}}$ are called the *elementary divisors* of $A(\lambda)$. In the collection of all the elementary divisors of $A(\lambda)$, a particular polynomial $(\lambda - \lambda_0)^\beta$ is repeated as many times as the number of invariant polynomials in which it appears as a factor. The elementary divisors $(\lambda - \lambda_{k,j})^{\beta_{k,j}}$ where the numbers $\lambda_{k,j}$ are all equal to a fixed eigenvalue λ_0, are said to correspond to, or to be associated with, the eigenvalue λ_0. In this case, the degrees of the elementary divisors associated with λ_0 are called the *partial multiplicities* of $A(\lambda)$ at λ_0. An elementary divisor $(\lambda - \lambda_{k,j})^{\beta_{k,j}}$ is said to be *linear* if $\beta_{k,j} = 1$.

It follows from Theorem A.6.1 that:

Theorem A.6.2. *Two $n \times n$ matrix polynomials are equivalent if and only if they have the same collection of elementary divisors.*

A comparison with the Jordan form is instructive:

Theorem A.6.3. *Let $X \in \mathsf{C}^{n \times n}$, and consider the $n \times n$ matrix polynomial $A(\lambda) = \lambda I - X$. Then:*

(1) *The eigenvalues of $A(\lambda)$ coincide with the eigenvalues of the matrix X.*

(2) *The degrees $\beta_1, \ldots, \beta_k$ of the elementary divisors $(\lambda - \lambda_0)^{\beta_1}, \ldots, (\lambda - \lambda_0)^{\beta_k}$ corresponding to the eigenvalue λ_0 of $A(\lambda)$ coincide with the sizes of Jordan blocks with the eigenvalue λ_0 in the Jordan form of X.*

(3) *The invariant polynomials of $A(\lambda)$ are:*

$$
\begin{aligned}
i_n(\lambda) &= (\lambda - \lambda_1)^{\gamma_{1,n}}(\lambda - \lambda_2)^{\gamma_{2,n}} \cdots (\lambda - \lambda_p)^{\gamma_{p,n}}, \\
i_{n-1}(\lambda) &= (\lambda - \lambda_1)^{\gamma_{1,n-1}}(\lambda - \lambda_2)^{\gamma_{2,n-1}} \cdots (\lambda - \lambda_p)^{\gamma_{p,n-1}}, \quad \cdots \\
&\;\;\vdots \\
i_1(\lambda) &= (\lambda - \lambda_1)^{\gamma_{1,1}}(\lambda - \lambda_2)^{\gamma_{2,1}} \cdots (\lambda - \lambda_p)^{\gamma_{p,1}}.
\end{aligned}
$$

Here, $\lambda_1, \ldots, \lambda_p$ are the distinct eigenvalues of X and for $k = 1, 2, \ldots, p$,

$$\gamma_{k,n} \geq \gamma_{k,n-1} \geq \cdots \geq \gamma_{k,1}$$

are the sizes of Jordan blocks, in nonincreasing order, corresponding to the eigenvalue λ_k ($k = 1, 2, \ldots, p$). We set $\gamma_{k,m} = 0$ for $m = n-q, n-q-1, \ldots, 1$ if the geometric multiplicity of λ_k is equal to $q < n$.

The proof is easily obtained by considering the Smith form (A.6.20) for $\lambda I - J$, where J is the Jordan form of X. To analyze this Smith form note first that if scalar polynomials $\phi(\lambda)$ and $\psi(\lambda)$ are relatively prime, then the 2×2 matrix polynomial diag $(\phi(\lambda), \psi(\lambda))$ is equivalent to diag $(\phi(\lambda)\psi(\lambda), 1)$:

$$
\begin{bmatrix} \psi(\lambda) & \phi(\lambda) \\ \rho(\lambda) & \tau(\lambda) \end{bmatrix} \begin{bmatrix} \phi(\lambda) & 0 \\ 0 & \psi(\lambda) \end{bmatrix} \left(\begin{bmatrix} 1 & -1 \\ 0 & 1 \end{bmatrix} \begin{bmatrix} 1 & 0 \\ -\rho(\lambda)\phi(\lambda) & 1 \end{bmatrix} \right) = \begin{bmatrix} \phi(\lambda)\psi(\lambda) & 0 \\ 0 & 1 \end{bmatrix},
$$

for some polynomials $\rho(\lambda)$ and $\tau(\lambda)$ such that $\rho(\lambda)\phi(\lambda) - \tau(\lambda)\psi(\lambda) = 1$, and the existence of such $\rho(\lambda)$ and $\tau(\lambda)$ is guaranteed by the relative primeness assumption.

In particular, it follows from Theorem A.6.3 that X is *diagonalizable* if and only if all elementary divisors of $A(\lambda) = \lambda I - X$ are linear (i.e., all eigenvalues are semi-simple). Also, X is *nonderogatory* (i.e., for every eigenvalue the corresponding eigenvectors span a one-dimensional subspace) if and only if exactly one invariant polynomial of $A(\lambda)$ is nonconstant.

A diagonal Smith form is available also for a more general class of matrix functions, namely, for matrices whose entries are *analytic* functions of the complex variable in a fixed open set of the complex plane, or analytic functions of the real variable defined on a fixed interval of the real axis. In short, such matrices will be called *analytic matrix functions*. The next result is a local version of this Smith form valid for analytic matrix functions in a (complex) neighborhood of a fixed $\lambda_0 \in \mathsf{C}$.

Theorem A.6.4. *Let $A(\lambda)$ be an analytic $n \times n$ matrix function of the complex variable $\lambda \in \mathcal{U}(\lambda_0)$, where $\mathcal{U}(\lambda_0)$ is an open neighborhood of $\lambda_0 \in \mathsf{C}$. Then there exist analytic $n \times n$ matrix functions $P(\lambda), Q(\lambda), \lambda \in \mathcal{U}'(\lambda_0)$, where $\mathcal{U}'(\lambda_0) \subseteq \mathcal{U}(\lambda_0)$ is an open neighborhood of λ_0, with the following properties:*

(a) $\det P(\lambda) \neq 0$ and $\det Q(\lambda) \neq 0$ for $\lambda \in \mathcal{U}'(\lambda_0)$;

(b) *for $\lambda \in \mathcal{U}'(\lambda_0)$ the equation*

$$A(\lambda) = P(\lambda) \operatorname{diag} \left[\begin{array}{ccccccc} (\lambda - \lambda_0)^{\gamma_1} & (\lambda - \lambda_0)^{\gamma_2} & \cdots & (\lambda - \lambda_0)^{\gamma_r} & 0 & \cdots & 0 \end{array} \right] Q(\lambda) \tag{A.6.21}$$

holds, where $\gamma_1 \leq \gamma_2 \leq \cdots \leq \gamma_r$ are nonnegative integers.

Moreover, the parameter

$$r := \max\{\operatorname{rank} A(\lambda) \ : \ \lambda \in \mathcal{U}'(\lambda_0)\} \tag{A.6.22}$$

and the integers γ_j, $j = 1, 2, \ldots, r$, are uniquely determined by $A(\lambda)$ and by λ_0.

A proof of Theorem A.6.4 can be developed along the same lines as the proof of Theorem A.6.1 (see, for example, [70, Chapter 7], [39, Chapter S1]). The result of Theorem A.6.4 is also valid (with exactly the same proof) for real analytic matrix functions $A(\lambda)$ of the real variable λ in a neighborhood of $\lambda_0 \in \mathsf{R}$ in the domain of definition of $A(\lambda)$. In this case, $\mathcal{U}'(\lambda_0)$ is a real neighborhood of λ_0, and $P(\lambda)$ and $Q(\lambda)$ are real analytic matrix functions. The nonzero integers among γ_j, $j = 1, 2, \ldots, r$ are called the *partial multiplicities* of $A(\lambda)$ at λ_0. In fact:

Proposition A.6.5. *The sum*

$$\gamma_1 + \cdots + \gamma_s \tag{A.6.23}$$

is equal to the minimal multiplicity of λ_0 as a zero of the determinants of the $s \times s$ submatrices of $A(\lambda)$, for $s = 1, 2, \ldots, r$.

Note that determinants of submatrices of $A(\lambda)$ are scalar analytic functions of λ, so that the multiplicities of their zeros are well-defined. To prove Proposition A.6.5, use (A.6.21) and the Binet–Cauchy formula for subdeterminants of products of matrices ([70, Section 2.5] or [41, Section A.2]).

Theorem A.6.6. *The partial multiplicities of an analytic matrix function at λ_0 are invariant under multiplication on the left and on the right by analytic matrix functions invertible at λ_0.*

Proof. Suppose that the equation

$$A(\lambda) = P(\lambda)B(\lambda)Q(\lambda), \quad \lambda \in \mathcal{U}(\lambda_0) \tag{A.6.24}$$

holds, where $A(\lambda)$, $P(\lambda)$, $B(\lambda)$, and $Q(\lambda)$ are analytic $n \times n$ matrix functions in an open neighborhood $\mathcal{U}(\lambda_0)$ of $\lambda_0 \in \mathsf{C}$, and that $\det P(\lambda_0) \neq 0$, $\det P(\lambda_0) \neq 0$. Then

$$\det P(\lambda) \neq 0, \quad \det P(\lambda) \neq 0, \quad \lambda \in \mathcal{U}'(\lambda_0),$$

where $\mathcal{U}'(\lambda_0) \subseteq \mathcal{U}(\lambda_0)$ is an open neighborhood of λ_0.

The definition (A.6.22) shows that the parameter r is the same for $A(\lambda)$ and for $B(\lambda)$. Let

$$\gamma_1(A) \le \gamma_2(A) \le \cdots \le \gamma_r(A), \quad \gamma_1(B) \le \gamma_2(B) \le \cdots \le \gamma_r(B)$$

be the partial multiplicities of $A(\lambda)$ and of $B(\lambda)$ at λ_0, respectively. Using Proposition A.6.5 and the Binet–Cauchy formula, we obtain

$$\gamma_1(A) + \cdots + \gamma_s(A) \ge \gamma_1(B) + \cdots + \gamma_s(B), \quad s = 1, 2, \ldots, r. \qquad (A.6.25)$$

Write (A.6.24) in the form $B(\lambda) = P(\lambda)^{-1} A(\lambda) Q(\lambda)^{-1}$, note that $P(\lambda)^{-1}$ and $Q(\lambda)^{-1}$ are analytic in an open neighborhood of λ_0, and argue in a similar way to obtain

$$\gamma_1(B) + \cdots + \gamma_s(B) \ge \gamma_1(A) + \cdots + \gamma_s(A), \quad s = 1, 2, \ldots, r. \qquad (A.6.26)$$

Now the inequalities (A.6.25) and (A.6.26) yield $\gamma_s(A) = \gamma_s(B)$, $s = 1, 2, \ldots, r$, i.e., the result of Theorem A.6.6. $\qquad \square$

The final result on diagonal forms concerns analytic diagonalizability of hermitian matrix polynomials:

Theorem A.6.7. *Let $L(\lambda)$ be a hermitian matrix polynomial. Then the matrix $L(\lambda)$ has a diagonal form*

$$L(\lambda) = U(\lambda) \cdot \text{diag} \left[\mu_1(\lambda), \ldots, \mu_n(\lambda) \right] \cdot V(\lambda), \quad \lambda \in \mathsf{R}, \qquad (A.6.27)$$

where $U(\lambda)$ is unitary for real λ and $V(\lambda) = (U(\lambda))^*$.

Moreover, the functions $\mu_i(\lambda)$ and $U(\lambda)$ can be chosen to be analytic functions of the real parameter λ.

This result is a particular case of Rellich's theorem [58]. For a proof see [39, Chapter S6]) for example, where the result is proved in the more general setting of selfadjoint analytic matrix functions.

A.7 Convexity of the Numerical Range

The *numerical range* (also known as the *field of values*), $W(A)$, of a matrix $A \in \mathsf{C}^{n \times n}$ is a subset of the complex plane defined by

$$W(A) = \{ (Ay, y) \in \mathsf{C} \ : \ y \in \mathsf{C}^n, \ \|y\| = 1 \}.$$

The numerical range is compact, i.e., closed and bounded, because it is the range of the continuous complex-valued function $f(y) = (Ay, y)$ defined in the compact set $\{ y \in \mathsf{C}^n \ : \ \|y\| = 1 \}$. Numerical ranges of matrices (and linear transformations) have been extensively studied in the literature. Here, we need the celebrated Toeplitz–Hausdorff theorem:

Theorem A.7.1. *For every $A \in \mathbb{C}^n$, the set $W(A)$ is convex.*

Many proofs of this theorem are available in the literature and, rather than reproduce one of them, note that references [3], [54, Chapter 1], [76], contain the more elementary and detailed proofs.

Letting

$$X_1 = \frac{1}{2}(A + A^*), \quad X_2 = \frac{1}{2i}(A - A^*),$$

and identifying $\mathbb{C}$ with $\mathbb{R}^2$, it can be seen that $W(A)$ coincides with the *joint numerical range* $W(X_1, X_2)$ of the two hermitian matrices X_1 and X_2:

$$W(X_1, X_2) := \{((X_1 y, y), (X_2 y, y)) \in \mathbb{R}^2 \; : \; y \in \mathbb{C}^n, \; \|y\| = 1\}. \tag{A.7.28}$$

Thus, Theorem A.7.1 can be restated as follows:

Theorem A.7.2. *For every pair of hermitian $n \times n$ matrices (X_1, X_2), the set $W(X_1, X_2)$ is convex.*

For real symmetric matrices, the real joint numerical range $W_{\mathsf{R}}(X_1, X_2)$ is defined in a similar way:

$$W_{\mathsf{R}}(X_1, X_2) := \{((X_1 y, y), (X_2 y, y)) \in \mathbb{R}^2 \; : \; y \in \mathbb{R}^n, \; \|y\| = 1\},$$

where X_1 and X_2 are $n \times n$ real symmetric matrices. The real Toeplitz–Hausdorff theorem, proved by Brinkman [12], reads slightly differently from Theorem A.7.2; the real joint numerical range is generally nonconvex if $n = 2$:

Theorem A.7.3. *For every pair of real symmetric $n \times n$ matrices (X_1, X_2), the set $W_{\mathsf{R}}(X_1, X_2)$ is convex, provided $n > 2$.*
For every pair of real symmetric 2×2 matrices (X_1, X_2), the set $W_{\mathsf{R}}(X_1, X_2)$ is either an ellipse, a line segment, or a singleton. In particular, $W_{\mathsf{R}}(X_1, X_2)$ bounds a convex set.

An elementary proof of Theorem A.7.3 for the case $n \geq 3$, which also includes a proof of Theorem A.7.1, is given in [2], [3].

Proof. Here, we only prove the case $n = 2$. Note first that the following transformations do not change the required properties of $W_{\mathsf{R}}(X_1, X_2)$:

- Simultaneous orthogonal transformations

$$(X_1, X_2) \mapsto (U^T X_1 U, U^T X_2 U),$$

  where $U \in \mathbb{R}^{2 \times 2}$ is an orthogonal matrix;

- Shifts $(X_1, X_2) \mapsto (X_1 + aI, X_2 + bI)$ where $a, b \in \mathbb{R}$; scalings $(X_1, X_2) \mapsto (rX_1, sX_2)$, where r, s are nonzero real numbers; and

- Linear combinations: adding to one of the X_1, X_2 a scalar multiple of the other.

Thus, applying suitable transformations of these kinds, we may assume without loss of generality that one of the following cases occurs if at least one of X_1, X_2 is nonzero:

(1) $X_1 = \begin{bmatrix} 1 & 0 \\ 0 & -1 \end{bmatrix}$, $X_2 = \begin{bmatrix} 0 & 1 \\ 1 & 0 \end{bmatrix}$,

(2) $X_1 = \begin{bmatrix} 1 & 0 \\ 0 & -1 \end{bmatrix}$, $X_2 = 0$,

(3) $X_1 = 0$, $X_2 = \begin{bmatrix} 0 & 0 \\ 0 & 1 \end{bmatrix}$.

The cases (2) and (3) are easy: $W_{\mathsf{R}}(X_1, X_2)$ is a line segment, possibly degenerated into a singleton. In case (1), upon writing

$$y = \begin{bmatrix} \cos \alpha \\ \sin \alpha \end{bmatrix} \in \mathsf{R}^2, \quad 0 \le \alpha \le 2\pi,$$

we have

$$W_{\mathsf{R}}(X_1, X_2) = \{(\cos(2\alpha), \sin(2\alpha)) \in \mathsf{R}^2 \ : \ 0 \le \alpha \le 2\pi\},$$

which is the unit circle.   □

A.8   The Fixed Point Theorem

In this section we present a well-known theorem asserting existence of a fixed point of a continuous map. It is known as Brouwer's fixed point theorem.

Theorem A.8.1. *Let S be a nonempty closed bounded convex subset of R^m, and let $f : S \longrightarrow S$ be a continuous function. Then f has a fixed point, i.e, there exists $x \in S$ such that $f(x) = x$.*

An elementary and complete proof of the theorem can be found in [102, Section 7.5], or [31, Chapter 7], for example.

The result of Theorem A.8.1 is clearly valid also for every (nonempty) topological space which is homeomorphic to a closed bounded subset of R^m for some m. On the other hand, each of the three hypotheses: convexity, closedness, and boundedness, is essential for the result. Indeed, the unit circle is closed, bounded, but not convex; a rotation of the unit circle has no fixed point (unless the angle of the rotation is an integer multiple of 2π). The real line is closed and convex, but not bounded; a shift $f(x) = x + a$, $x \in \mathsf{R}$, where $a \in \mathsf{R} \setminus \{0\}$ has no fixed points. The open interval $(0, 1)$ is bounded, convex, but not closed; the function $f : (0, 1) \longrightarrow (0, 1)$ defined by $f(x) = (x + 1)/2$ has no fixed points in $(0, 1)$.

A.9 Exercises

1. Prove that a selfadjoint, unitary, or normal linear transformation is unitarily similar to a diagonal linear transformation in the standard basis.

2. Prove that any square matrix is unitarily similar to an upper (or lower) triangular matrix in the standard basis. (Hint: this is close to a well-known theorem which can be attributed to either Schur or Toeplitz.)

3. Show that any square matrix is similar to a matrix with orthogonal root subspaces.

4. Given any set of linearly independent vectors $x_1, x_2, \ldots, x_n \in \mathbf{C}^n$, show that there is a positive definite inner product on $\mathbf{C}^n$ in which these vectors form an orthogonal basis.

5. Let $\mathcal{M}_1$ and $\mathcal{M}_2$ be two subspaces of $\mathbf{C}^n$ for which $\mathcal{M}_1 \dotplus \mathcal{M}_2 = \mathbf{C}^n$. Show that there is a positive definite inner product in which $\mathcal{M}_2 = \mathcal{M}_1^{[\perp]}$.

6. Let $\phi_j = [\phi_{jk}]_{k=1}^n$, $j = 1, 2, \ldots, n$ be vectors from $\mathbf{C}^n$. Show that:

   (a) These vectors form a basis for $\mathbf{C}^n$ if and only if

   $$\det[\phi_{jk}]_{j,k=1}^n \neq 0. \qquad (A.9.29)$$

   (b) The linear transformation defined by the matrix $\Phi = [\phi_{jk}]_{j,k=1}^n$ transforms the standard basis $e_1, e_2, \ldots, e_n$ into $\phi_1, \phi_2, \ldots, \phi_n$.

   (c) If the condition (A.9.29) holds then the matrix

   $$H := ([\phi_{jk}]_{j,k=1}^n)^{-1} \left[([\phi_{jk}]_{j,k=1}^n)^* \right]^{-1}$$

   is positive definite.

7. Let $a_2, a_3, \ldots, a_n \in \mathbf{C}$ and define

$$H_1 = \begin{bmatrix} 1 & a_2 & a_3 & \cdots & a_n \\ \overline{a_2} & 1 & 0 & \cdots & 0 \\ \overline{a_3} & 0 & 1 & \cdots & 0 \\ \vdots & \vdots & \vdots & & \vdots \\ \overline{a_n} & 0 & \cdots & 0 & 1 \end{bmatrix}.$$

   (a) Under what conditions is the matrix H_1 positive or negative definite?

   (b) When the matrix H_1 is indefinite, how many negative eigenvalues does it have?

8. Let $a \in \mathbf{C}$ and define

$$H_3 = \begin{bmatrix} 1 & a & a^2 & \cdots & a^{n-1} \\ \overline{a} & 1 & a & \cdots & a^{n-2} \\ \vdots & & & \cdots & \vdots \\ \overline{a}^{n-1} & \overline{a}^{n-2} & \overline{a}^{n-3} & \cdots & 1 \end{bmatrix}.$$

(a) When is H_3 positive or negative definite?

(b) How many negative eigenvalues does H_3 have?

(c) When is H_3 invertible? Find the inverse when it exists. When are all principal minors nonzero?

9. Find the Jordan forms for the following matrices:

$$A_1 = \begin{bmatrix} -2 & 2 & -4 \\ 1 & 0 & 1 \\ 4 & -3 & 6 \end{bmatrix}, \quad A_2 = \begin{bmatrix} 0 & 4 & -2 \\ 4 & 10 & 3 \\ -9 & -18 & -4 \end{bmatrix}, \quad A_3 = \begin{bmatrix} 0 & -4 & -2 \\ 1 & 4 & 1 \\ 0 & 0 & 2 \end{bmatrix},$$

$$A_4 = \begin{bmatrix} 0 & 0 & 0 & 1 & 0 & 0 \\ 0 & 0 & 0 & 0 & 1 & 0 \\ 0 & 0 & 0 & 0 & 0 & 0 \\ -1 & 0 & 0 & 0 & 0 & 0 \\ 0 & -1 & 0 & 0 & 0 & 0 \\ 0 & 0 & -1 & 0 & 0 & 0 \end{bmatrix}, \quad A_5 = \begin{bmatrix} a_1 & a_2 & \cdots & a_n \\ a_n & a_1 & \cdots & a_{n-1} \\ \cdots\cdots\cdots\cdots \\ a_2 & a_3 & \cdots & a_1 \end{bmatrix}.$$

The rows of A_5 are cyclic permutations of the first row.

$$A_6 = \begin{bmatrix} 0 & 1 & 0 & \cdots & 0 & 0 \\ -1 & 0 & 1 & \cdots & 0 & 0 \\ 0 & -1 & 0 & \cdots & 0 & 0 \\ \cdots\cdots\cdots\cdots \\ 0 & 0 & 0 & \cdots & 0 & 1 \\ 0 & 0 & 0 & \cdots & -1 & 0 \end{bmatrix}, \quad A_7 = \begin{bmatrix} 1 & \alpha & \alpha & \cdots & \alpha & \alpha \\ \beta & 1 & \alpha & \cdots & \alpha & \alpha \\ \beta & \beta & 1 & \cdots & \alpha & \alpha \\ \cdots\cdots\cdots\cdots \\ \beta & \beta & \beta & \cdots & 1 & \alpha \\ \beta & \beta & \beta & \cdots & \beta & 1 \end{bmatrix},$$

$\alpha, \beta \in \mathsf{C}$.

10. Find the Smith form of the following matrix polynomials:

(a) $L_1(\lambda) = \begin{bmatrix} p_1(\lambda) & p_2(\lambda) \\ 0 & p_3(\lambda) \end{bmatrix}$, where $p_1(\lambda)$, $p_2(\lambda)$, $p_3(\lambda)$ are scalar polynomials.

(b) $L_2(\lambda) = I_2\lambda^2 + \begin{bmatrix} 0 & a \\ a & 0 \end{bmatrix}\lambda + I_2$, $a \in \mathsf{C}$.

(c) $L_3(\lambda) = \begin{bmatrix} 2 & 1 \\ 1 & 1 \end{bmatrix}\lambda^2 + \begin{bmatrix} 4 & 2 \\ 2 & 0 \end{bmatrix}\lambda + \begin{bmatrix} 2 & 1 \\ 1 & 1 \end{bmatrix}$.

(d) $L_4(\lambda) = \begin{bmatrix} 2 & 1 \\ 1 & 0 \end{bmatrix}\lambda^2 + \begin{bmatrix} 4 & 2 \\ 2 & 0 \end{bmatrix}\lambda + \begin{bmatrix} 2 & 1 \\ 1 & 0 \end{bmatrix}$.

(e) $L_5(\lambda) = \begin{bmatrix} p(\lambda)q(\lambda) + r(\lambda) & q(\lambda) \\ p(\lambda) & 1 \end{bmatrix}$, where $p(\lambda)$, $q(\lambda)$, $r(\lambda)$ are scalar polynomials.

(f) $L_6(\lambda) = \begin{bmatrix} \lambda+1 & \lambda & \lambda^3 \\ \lambda+2 & \lambda^2 & \lambda-1 \end{bmatrix}$.

11. If diag $(i_1(\lambda), \ldots, i_r(\lambda), 0, \ldots, 0)$ is the Smith form of an $n \times n$ matrix polynomial $L(\lambda)$, what are the Smith forms of the following matrix polynomials?

   (a) $L(p(\lambda))$, where $p(\lambda)$ is a scalar polynomial;

   (b) $\begin{bmatrix} L(\lambda) & (L(\lambda))^2 + 2L(\lambda) \\ 0 & -L(\lambda) \end{bmatrix}$.

12. For the following collections of elementary divisors of a 6×6 matrix polynomial, find the invariant polynomials:

   (a) λ, $\lambda+1$, $(\lambda+1)^2$, $(\lambda+1)^2$, $\lambda+2$, $\lambda+2$, $(\lambda+2)^2$, $(\lambda+2)^2$, $\lambda+3$, $(\lambda+3)^2$, $(\lambda+3)^3$.

   (b) $(\lambda-i)^3$, $(\lambda-i)^3$, $(\lambda-i)^3$, $(\lambda-i)^3$, $(\lambda-i)^3$, $(\lambda-i)^3$, $(\lambda+i)^2$, $(\lambda+i)^2$, $(\lambda+i)^2$, $(\lambda+i)^2$.

13. Give an example of a pair of 2×2 matrix polynomials $L_1(\lambda)$ and $L_2(\lambda)$ such that $L_1(\lambda)$ and $L_2(\lambda)$ are equivalent, but $(L_1(\lambda))^2$ and $(L_2(\lambda))^2$ are not equivalent.

Bibliography

[1] T. Ando. *Linear Operators on Kreĭn Spaces*. Hokkaido University, Research Institute of Applied Electricity, Division of Applied Mathematics, Sapporo, 1979.

[2] Y. H. Au-Yeung. A theorem on a mapping from a sphere to the circle and the simultaneous diagonalization of two hermitian matrices. *Proc. of Amer. Math. Soc.*, 20:545-548, 1969.

[3] Y. H. Au-Yeung. A simple proof of the convexity of the field of values defined by two hermitian forms. *Aequationes Math.*, 12:82-83, 1975.

[4] Y. H. Au-Yeung, C.-K. Li, and L. Rodman. H-unitary and Lorentz matrices. *SIAM J. of Matrix Analysis*, 25:1140-1162, 2004.

[5] T. Ya. Azizov and I. S. Iokhvidov. Linear operators on spaces with an indefinite metric and their applications. *Itogi Nauki i Techniki*, 17:113-206, 1979. (Russian).

[6] T. Ya. Azizov and I. S. Iokhvidov. *Linear Operators in Spaces with an Indefinite Metric*. John Wiley and Sons, Chichester, 1989. (Translated from Russian.)

[7] H. Baumgärtel. *Analytic Perturbation Theory*. Operator Theory: Advances and Applications, Vol 15, Birkhäuser Verlag, Basel, 1990.

[8] G. R. Belitskii and V. V. Sergeichuk. Complexity of matrix problems. *Linear Algebra Appl.* 361:302-222, 2003.

[9] A. Ben-Artzi and I. Gohberg. Orthogonal polynomials over Hilbert modules. *Operator Theory: Advances and Applications*, 73:96-126, Birkhäuser Verlag, Basel, 1994.

[10] A. Ben-Artzi and I. Gohberg. On contractions in spaces with indefinite metric: G-norms and spectral radii. *Integral Equations and Operator Theory*, 24:422-469, 1996.

[11] J. Bognar. *Indefinite Inner Product Spaces*. Springer-Verlag, New York, 1974.

[12] L. Brinkman. On the field of values of a matrix. *Proc. of Amer. Math. Soc.*, 12:61-66, 1961.

[13] N. Burgoyne and R. Cushman. Normal forms for real linear Hamiltonian systems. *Lie Groups: History Frontiers and Applications*, VII:482-528, Math Sci Press, Brookline, Mass., 1977.

[14] B. E. Cain. Inertia theory. *Linear Algebra and Appl.*, 30:211-240, 1980.

[15] W. A. Coppel and A. Howe. On the stability of linear canonical systems with periodic coefficients. *J. Austral. Math. Soc.*, 5:169-195, 1965.

[16] J. L. Daleckii and M. G. Krein. *Stability of Solutions of Differential Equations in Banach Space.* American Math Society, Providence, 1974. (Transl. Math. Monographs, vol. 43).

[17] V. K. Dubovoj, B. Fritzsche, and B. Kirstein. *Matricial version of the classical Schur problem.* Teubner Texts in Mathematics, vol.129, B. G. Teubner, 1992.

[18] R. L. Ellis and I. Gohberg. *Orthogonal Systems and Convolution Operators.* Operator theory: Advances and Applications, Vol. 140. Birkhäuser Verlag, Basel, 2003.

[19] R. L. Ellis, I. Gohberg, and D. C. Lay. Invertible selfadjoint extensions of band matrices and their entropy. *SIAM J. Alg. Disc. Methods,* 8:483–500, 1987.

[20] R. L. Ellis, I. Gohberg, and D. C. Lay. On two theorems of M. G. Krein concerning polynomials orthogonal on the unit circle. *Integral Equations and Operator Theory,* 11:87–104, 1988.

[21] R. L. Ellis, I. Gohberg, and D. C. Lay. On a class of block Toeplitz matrices. *Linear Algebra Appl.,* 241/243:225–245, 1996.

[22] A. Ferrante and B. C. Levy. Canonical form of symplectic matrix pencils. *Linear Algebra and Appl.,* 274:259-300, 1998.

[23] D. T. Finkbeiner. *Introduction to Matrices and Linear Transformations.* W. H. Freeman and Co., San Francisco, Calif.-London, 1966. Second Edition.

[24] P. A. Furhmann. On symmetric rational transfer functions. *Linear Algebra and Appl.,* 50:167-150, 1983.

[25] P. A. Fuhrmann. On Hamiltonian rational transfer functions. *Linear Algebra and Appl.,* 63:1-93, 1984.

[26] F. R. Gantmacher. *The Theory of Matrices,* Volume 1. Chelsea Publishing Company, New York, N. Y., 1959.

[27] I. M. Gelfand and V. B. Lidskii. On the structure of the regions of stability of linear canonical systems of differential equations with periodic coefficients. *Amer. Math. Soc. Transl.,* 8:143-181, 1958.

[28] I. M. Gelfand and V. A. Ponomarev. Remarks on the classification of a pair of commuting linear transformations in a finite-dimensional space. *Funkcional. Anal. i Priložen.,* 3: 81-82, 1969. (Russian).

[29] I. M. Glazman and Yu. I. Lyubich. *Finite-dimensional Linear Analysis: A Systematic Presentation in Problem Form.* MIT Press, Cambridge, 1974. Translation from Russian: *Finite-dimensional Analysis in Problems,* Izdat. "Nauka", Moscow, 1969.

[30] S. K. Godunov and M. Sadkane. Numerical determination of the canonical form of a symplectic matrix. *Siberian Math. J.,* 42:629-647, 2001. (Translation from Russian).

[31] K. Goebel. *Concise Course on Fixed Point Theorems.* Yokohama Publishers, Yokohama, 2002.

[32] I. Gohberg and S. Goldberg. A simple proof of the Jordan decomposition theorem for matrices. *Amer. Math. Monthly,* 103: 157-159, 1996.

[33] I. Gohberg, S. Goldberg, and M. A. Kaashoek. *Classes of linear operators. Vol. I.* Operator Theory: Advances and Applications, Vol. 49, Birkhäuser Verlag, Basel, 1990.

[34] I. Gohberg, P. Lancaster, and L. Rodman. Spectral analysis of matrix polynomials I: Canonical forms and divisors. *Linear Algebra and Appl.*, 20:1-44, 1978.

[35] I. Gohberg, P. Lancaster, and L. Rodman. Spectral analysis of matrix polynomials II: The resolvent form and spectral divisors. *Linear Algebra and Appl.*, 21:65-88, 1978.

[36] I. Gohberg, P. Lancaster, and L. Rodman. Spectral analysis of selfadjoint matrix polynomials. *Report 419, Dept. of Math. and Stat., University of Calgary*, 1979.

[37] I. Gohberg, P. Lancaster, and L. Rodman. Spectral analysis of selfadjoint matrix polynomials. *Annals of Mathematics*, 112:33-71, 1980.

[38] I. Gohberg, P. Lancaster, and L. Rodman. Perturbations of H-selfadjoint matrices with applications to differential equations. *Integral Equations and Operator Theory*, 5:718-757, 1982.

[39] I. Gohberg, P. Lancaster, and L. Rodman. *Matrix Polynomials*. Academic Press, New York, 1982.

[40] I. Gohberg, P. Lancaster, and L. Rodman. *Matrices and Indefinite Scalar Products*. Operator theory: Advances and Applications, Vol. 8, Birkhäuser Verlag, Basel, 1983.

[41] I. Gohberg, P. Lancaster, and L. Rodman. *Invariant Subspaces of Matrices with Applications*. Wiley - Interscience, New York etc., 1986.

[42] I. Gohberg, P. Lancaster, and L. Rodman. On hermitian solutions of the symmetric algebraic Riccati equations, *SIAM J. Control and Optim.*, 24:1323-1334, 1986.

[43] I. Gohberg and B. Reichstein. On classification of normal matrices in an indefinite scalar product. *Integral Equations and Operator Theory*, 13:364-394, 1990.

[44] I. Gohberg and B. Reichstein. On H-unitary and block-Toeplitz H-normal operators. *Linear and Multilinear Algebra*, 30:17-48, 1991.

[45] I. Gohberg and B. Reichstein. Classification of block-Toeplitz H-normal operators. *Linear and Multilinear Algebra*, 34:213-245, 1993.

[46] I. Gohberg, and E. I. Sigal. On operator generalizations of the logarithmic residue theorem and the theorem of Rouché. *Math. USSR-Sb.*, 13:603-625, 1971. (Translation from Russian).

[47] R. E. Greene, and S. G. Krantz. *Function Theory of one Complex Variable*. 2nd edition, John Wiley and Sons, Inc., New York, 1997.

[48] N. J. Higham, D. S. Mackey, N. Mackey, and F. Tisseur. Functions preserving matrix groups and iterations for the matrix square root. *SIAM J. of Matrix Analysis*, 26:849-877, 2005.

[49] R. D. Hill. Inertia theory for simultaneously triangulable complex matrices. *Linear Algebra and Appl.*, 2:131–142, 1969.

[50] O. V. Holtz. On indecomposable normal matrices in spaces with indefinite scalar product. *Linear Algebra and Appl.*, 259:155-168, 1997.

[51] O. V. Holtz and V. A. Strauss. Classification of normal operators in spaces with indefinite scalar product of rank 2. *Linear Algebra and Appl.*, 241/243:455-517, 1996.

[52] O. V. Holtz and V. A. Strauss. On classification of normal operators in real spaces with indefinite scalar product. *Linear Algebra and Appl.*, 255:113-155, 1997.

[53] R. Horn and C. R. Johnson. *Matrix Analysis.* Cambridge University Press, Cambridge, 1985.

[54] R. Horn and C. R. Johnson. *Topics in Matrix Analysis.* Cambridge University Press, Cambridge, 1991.

[55] L.-K. Hua. On the theory of automorphic functions of a matrix variable. II. The classification of hypercircles under the symplectic group. *Amer. J. of Math.*, 66:531–563, 1944.

[56] C. R. Johnson, and R. DiPrima, The range of $A^{-1}A^*$ in $GL(n, C)$. *Linear Algebra and Appl.*, 9:209-222, 1974.

[57] I. S. Iohvidov, M. G. Krein, and H. Langer. *Introduction to the Spectral Theory of Operators in Spaces with an Indefinite Metric.* Akademie-Verlag, Berlin, 1982.

[58] T. Kato. *Perturbation Theory for Linear Operators.* Springer-Verlag, Berlin and New York, 1966.

[59] M. G. Krein. On an application of the fixed point principle in the theory of linear transformations of spaces with an indefinite metric. *Uspehi Matem. Nauk (N.S.)*, 5:180-190, 1950.

[60] M.G. Krein. Distribution of roots of polynomials which are orthogonal on the unit circle with respect to a sign alternating weight. *Teor. Funkcii, Funkcional. Anal. i Prilozen*, 2:131–137, 1966. (Russian).

[61] M. G. Krein. The basic propositions of the theory of λ-zones of stability of a canonical system of linear differential equations with periodic coefficients. *In memory of Aleksandr Aleksandrovič Andronov*, pp. 413–498. Izdat. Akad. Nauk SSSR, Moscow, 1955; English Transl. in: M. G. Krein, *Topics in differential and integral equations and operator theory.* Edited by I. Gohberg. Operator Theory: Advances and Applications, Vol. 7, Birkhäuser Verlag, Basel, 1983.

[62] M. G. Krein. *Stability Theory of Differential Equations in Banach Spaces.* Kiev, 1964. (Russian). ([16] is an expanded English translation.)

[63] L. Kronecker. Algebraische Reduktion der Scharen bilinearer Formen. *S.-B. Akad. Berlin*, 763–77, 1890.

[64] P. Lancaster. *Lambda-matrices and Vibrating Systems.* Pergamon Press, 1966; Dover, 2002.

[65] P. Lancaster. Lectures on Linear Algebra, Control and Stability. *Dept. of Mathematics and Statistics, University of Calgary*, 1999.

[66] P. Lancaster and L. Rodman. Existence and uniqueness theorems for algebraic Riccati equations. *International Journal of Control*, 32:285-309, 1980.

[67] P. Lancaster and L. Rodman. *Algebraic Riccati Equations.* Clarendon press, Oxford, 1995.

[68] P. Lancaster and L. Rodman. Canonical forms for hermitian matrix pairs under strict equivalence and congruence. *SIAM Review*, to appear.

[69] P. Lancaster and L. Rodman. Canonical forms for symmetric/skew-symmetric real matrix pairs under strict equivalence and congruence. *Linear Algebra and Appl.*, to appear.

[70] P. Lancaster and M. Tismenetsky. *The Theory of Matrices.* Academic Press, Orlando, 1985.

[71] H. J. Landau. Polynomials orthogonal in an indefinite metric operator theory. *Operator theory: Advances and Applications*, 34:203-214, Birkhäuser Verlag, Basel, 1988.

[72] A. J. Laub. A Schur method for solving algebraic Riccati equations. *IEEE Trans. Automat. Control*, 24:913–921, 1979.

[73] A. J. Laub and K. Meyer. Canonical forms for symplectic and Hamiltonian matrices. *Celestial Mech.*, 9:213-238, 1974.

[74] P. D. Lax. *Linear Algebra.* John Wiley and Sons, New York, 1997.

[75] N. Levinson. The stability of linear, real, periodic self-adjoint systems of differential equations. *J. Math. Anal. Appl.*, 6:473-482, 1963.

[76] C.-K. Li. A simple proof of the elliptical range theorem. *Proc. Amer. Math. Soc.*, 124:1985-1986, 1996.

[77] V. B. Lidskii and M. G. Neĭgauz. On the boundedness of the solutions to linear systems of differential equations with periodic coefficients. *Doklady Akad. Nauk. SSSR*, 77:189-192, 1951. (Russian).

[78] W.-W. Lin, V. Mehrmann, and H. Xu. Canonical forms for Hamiltonian and symplectic matrices and pencils. *Linear Algebra and Appl.*, 302/303:469-533, 1999.

[79] B. Lins, P. Meade, C. Mehl, L. Rodman. Normal matrices and polar decompositions in indefinite inner products. *Linear and Multilinear Algebra*, 49:45-89, 2001.

[80] A. I. Mal'cev, *Foundations of Linear Algebra.* W. H. Freeman, San Francisco and London, 1963. (Translation from Russian.)

[81] A. S. Markus, *Introduction to the Spectral Theory of Polynomial Operator Pencils.* AMS, Providence, RI, 1988.

[82] J. E. Marsden and M. J. Hoffman. *Basic Complex Analysis.* 2nd Ed. W. H. Freeman, New York, 1987.

[83] C. Mehl, A. C. M. Ran, and L. Rodman. Semidefinite invariant subspaces: degenerate inner products. *Operator Theory: Advances and Appl.*, 149:467-486, Birkhäuser, Basel, 2004.

[84] C. Mehl, A. C. M. Ran, and L. Rodman. Hyponormal matrices and semidefinite invariant subspaces in indefinite inner products. *Electronic Journal of Linear Algebra*, 11:192-204, 2004.

[85] C. Mehl and L. Rodman. Classes of normal matrices in indefinite inner products, *Linear Algebra and Appl.*, 336:71-98, 2001.

[86] A. Ostrowski and H. Schneider. Some theorems on inertia of general matrices. *J. of Math. Anal. and Appl.*, 4:72-84, 1962.

[87] C. Paige and C. Van Loan. A Schur decomposition for Hamiltonian matrices. *Linear Algebra and Appl.*, 41:11–32, 1981.

[88] R. C. Penney. *Linear Algebra. Ideas and Applications.* Second edition, Wiley Interscience, 2004.

[89] H. A. Priestley. *Introduction to Complex Analysis.* 2nd edition, Oxford University Press, New York, 1990.

[90] A. C. M. Ran and L. Rodman. Stability of invariant maximal semidefinite subspaces. I. *Linear Algebra and Appl.*, 62:51-86, 1984).

[91] L. Rodman. Maximal invariant neutral subspaces and an application to the algebraic Riccati equation. *Manuscripta Math.*, 43:1-12, 1983.

[92] L. Rodman. *An Introduction to Operator Polynomials*. Operator theory: Advances and Applications, Vol. 38, Birkhäuser Verlag, Basel, 1989.

[93] G. E. Shilov. *An Introduction to the Theory of Linear Spaces*. Dover Publications, New York, 1974. (Translation from Russian.)

[94] L. Smith. *Linear Algebra*. Springer-Verlag, New York, 1984.

[95] Yu. L. Shmulyan. Finite dimensional operators depending analytically on a parameter. *Ukrainian Math. Journal*, IX:195-204, 1957. (Russian.)

[96] G. Szegő. *Orthogonal Polynomials*. Amer. Math. Soc. Colloquium Publications, v. 23. Amer. Math. Soc., New York, 1939.

[97] O. Taussky. A generalization of a theorem of Lyapunov. *J. of Soc. Ind. Appl. Math.*, 9:640-643, 1961.

[98] O. Taussky. Matrices C with $C^n \to 0$. *Journal of Algebra*, 1:5–10, 1964.

[99] R. C. Thompson. Pencils of complex and real symmetric and skew matrices. *Linear Algebra and Appl.*, 147:323-371, 1991.

[100] G. R. Trott. On the canonical form of a nonsingular pencil of hermitian matrices. *Amer. J. of Math.*, 56:359–371, 1934.

[101] F. Uhlig. Inertia and eigenvalue relations between symmetrized and symmetrizing matrices for the real and the general field case. *Linear Algebra and Appl.*, 35:203–226, 1981.

[102] R. Webster. *Convexity*. Oxford University Press, Oxford, New York, Tokyo, 1994.

[103] K. Weierstrass. Zur Theorie der quadratischen und bilinearen Formen. *Monatsber. Akad. Wiss.*, Berlin, 310–338, 1868.

[104] J. Williamson. The equivalence of nonsingular pencils of hermitian matrices in an arbitrary field. *Amer. J. of Math.*, 57:475–490, 1935.

[105] J. Williamson. On the algebraic problem concerning the normal forms of linear dynamical systems. *Amer. J. of Math.*, 58:141–163, 1936.

[106] H. Wimmer. On the Ostrowski-Schneider inertia theorem. *J. of Math. Anal. and Appl.*, 14:164–169, 1973.

[107] H. Wimmer and A. D. Ziebur. Remarks on inertia theorems for matrices. *Czechoslovak Math. J.*, 25:556-561, 1975.

[108] V. A. Yakubovich and V. M. Starzhinskii. *Linear Differential Equations with Periodic Coefficients*. Vols. 1 and 2. Halsted Press, Jerusalem-London, 1975. (Translation from Russian.)

Index